Advances in Solid State Circuit Technologies

Advances in Solid State Circuit Technologies

Editor:

Nelson Bolívar, Ph.D

ArclerPress

www.arclerpress.com

Advances in Solid State Circuit Technologies

Editor: Nelson Bolívar, Ph.D

© 2017 Arcler Press LLC
708 3rd Avenue, 6th Floor
New York
NY 10017
United States of America
www.arclerpress.com

ISBN: 978-1-68094-397-9

Library of Congress Control Number: 2016951059

About the Editor

Nelson Bolívar, Ph.D

Nelson Bolivar has a PhD. in physics from the University of Lorraine in France finished in 2014. His expertise is in quantum systems and condensed matter. His interest includes spintronic devices and correspondences between general relativity and condensed matter. He is currently an associate professor at the Central University of Venezuela

Contents

List of Contributors

Michał Tadeusiewicz
Department of Electrical, Electronic, Computer and Control Engineering, Ło´dz´ University of Technology, ul. Stefanowskiego 18/22, 90-924 Ło´dz´, Poland

Stanisław Hałgas
Department of Electrical, Electronic, Computer and Control Engineering, Ło´dz´ University of Technology, ul. Stefanowskiego 18/22, 90-924 Ło´dz´, Poland

Cosmin Popa
Faculty of Electronics, Telecommunications and Information Technology University Politehnica of Bucharest

Adisak Monpapassorn
Department of Electronic Engineering, South-East Asia University, Bangkok, Thailand

Fujihiko Matsumoto
Department of Applied Physics National Defense Academy of Japan

Syuzo Nishioka
Department of Applied Physics National Defense Academy of Japan

Takeshi Ohbuchi
Department of Applied Physics National Defense Academy of Japan

Tatsuya Fujii
Department of Applied Physics National Defense Academy of Japan

Vijaya Bhadauria
Electronics and Communication Engineering Department, Motilal Nehru National Institute of Technology, Allahabad, India

Krishna Kant
Department of Computer Engineering and Application, GLA University, Mathura, India

Swapna Banerjee
Electronics and Electrical Communication Engineering Department, Indian Institute of Technology, Kharagpur, India

Joshua Weber
Department of Electrical and Computer Engineering, Illinois Institute of Technology, Chicago, USA

Erdal Oruklu
Department of Electrical and Computer Engineering, Illinois Institute of Technology, Chicago, USA

Omid Mirmotahari
Nanoelectronics System Group, Department of Informatics, Universisty of Oslo, Oslo, Norway

Yngvar Berg
Department of Micro- and Nanosystems Technology, Buskerud & Vestfold University College, Horten, Norway

Jhin-Fang Huang
Department of Electronic Engineering, National Taiwan University of Science and Technology, Chinese Taipei

Yan-Cheng Lai
Department of Electronic Engineering, National Taiwan University of Science and Technology, Chinese Taipei

Wen-Cheng Lai
Department of Electronic Engineering, National Taiwan University of Science and Technology, Chinese Taipei

Ron-Yi Liu
Department of Electronic Engineering, National Taiwan University of Science and Technology, Chinese Taipei

Arash Ahmadpour
Department of Electronic Engineering, Lahijan Branch, Islamic Azad University, Lahijan, Iran

Sannian Song
State Key Laboratory of Functional Materials for Informatics, Shanghai Institute of Micro-system and Information Technology, Chinese Academy of Sciences, 865 Changning Road, Shanghai 200050, China

Zhitang Song
State Key Laboratory of Functional Materials for Informatics, Shanghai Institute of Micro-system and Information Technology, Chinese Academy of Sciences, 865 Changning Road, Shanghai 200050, China

Cheng Peng
State Key Laboratory of Functional Materials for Informatics, Shanghai Institute of Micro-system and Information Technology, Chinese Academy of Sciences, 865 Changning Road, Shanghai 200050, China

Lina Gao
Division of Nuclear Materials Science and Engineering, Shanghai Institute of Applied Physics, Chinese Academy of Sciences, 2019 Jialuo Road, Jiading District, Shanghai 201800, China

Yifeng Gu
State Key Laboratory of Functional Materials for Informatics, Shanghai Institute of Micro-system and Information Technology, Chinese Academy of Sciences, 865 Changning Road, Shanghai 200050, China

Zhonghua Zhang
State Key Laboratory of Functional Materials for Informatics, Shanghai Institute of Micro-system and Information Technology, Chinese Academy of Sciences, 865 Changning Road, Shanghai 200050, China
National Laboratory for Infrared Physics, Shanghai Institute of Technical Physics, Chinese Academy of Sciences, Shanghai 200083, China

Yegang Lv
State Key Laboratory of Functional Materials for Informatics, Shanghai Institute of Micro-system and Information Technology, Chinese Academy of Sciences, 865 Changning Road, Shanghai 200050, China

Dongning Yao
State Key Laboratory of Functional Materials for Informatics, Shanghai Institute of Micro-system and Information Technology, Chinese Academy of Sciences, 865 Changning Road, Shanghai 200050, China

Liangcai Wu
State Key Laboratory of Functional Materials for Informatics, Shanghai Institute of Micro-system and Information Technology, Chinese Academy of Sciences, 865 Changning Road, Shanghai 200050, China

Bo Liu
State Key Laboratory of Functional Materials for Informatics, Shanghai Institute of Micro-system and Information Technology, Chinese Academy of Sciences, 865 Changning Road, Shanghai 200050, China

Jagan Singh Meena
Department of Electronics Engineering and Institute of Electronics National Chiao Tung University

Simon Min Sze
Department of Electronics Engineering and Institute of Electronics National Chiao Tung University

Umesh Chand
Department of Electronics Engineering and Institute of Electronics National Chiao Tung University

Tseung-Yuen Tseng
Department of Electronics Engineering and Institute of Electronics National Chiao Tung University

Lei Wang
School of Information Engineering, Nanchang HangKong University, Nanchang 330063, Peoples' Republic China

Ci Hui Yang
School of Information Engineering, Nanchang HangKong University, Nanchang 330063, Peoples' Republic China

Jing Wen
School of Information Engineering, Nanchang HangKong University, Nanchang 330063, Peoples' Republic China

Si Di Gong
School of Information Engineering, Nanchang HangKong University, Nanchang 330063, Peoples' Republic China

Yuan Xiu Peng
School of Information Engineering, Nanchang HangKong University, Nanchang 330063, Peoples' Republic China

Jin Young Choi
Electronic & Electrical Engineering Department, Hongik University, Chungnam, South Korea

Choongkoo Park
Electronic & Electrical Engineering Department, Hongik University, Chungnam, South Korea

Amir Homaee
Staff in Iranian Offshore Oil Company (IOOC), Tehran, Iran

Debanjan Jana
Thin Film Nano Tech. Lab., Department of Electronic Engineering, Chang Gung University, 259 Wen-Hwa 1st Rd., Kwei-Shan, Tao-Yuan 333, Taiwan

Sourav Roy
Thin Film Nano Tech. Lab., Department of Electronic Engineering, Chang Gung University, 259 Wen-Hwa 1st Rd., Kwei-Shan, Tao-Yuan 333, Taiwan

Rajeswar Panja
Thin Film Nano Tech. Lab., Department of Electronic Engineering, Chang Gung University, 259 Wen-Hwa 1st Rd., Kwei-Shan, Tao-Yuan 333, Taiwan

Mrinmoy Dutta
Thin Film Nano Tech. Lab., Department of Electronic Engineering, Chang Gung University, 259 Wen-Hwa 1st Rd., Kwei-Shan, Tao-Yuan 333, Taiwan

Sheikh Ziaur Rahaman
Thin Film Nano Tech. Lab., Department of Electronic Engineering, Chang Gung University, 259 Wen-Hwa 1st Rd., Kwei-Shan, Tao-Yuan 333, Taiwan

Rajat Mahapatra
Thin Film Nano Tech. Lab., Department of Electronic Engineering, Chang Gung University, 259 Wen-Hwa 1st Rd., Kwei-Shan, Tao-Yuan 333, Taiwan
Department of Electronics and Communication Engineering, National Institute of Technology, Durgapur 713 209, India

Siddheswar Maikap
Thin Film Nano Tech. Lab., Department of Electronic Engineering, Chang Gung University, 259 Wen-Hwa 1st Rd., Kwei-Shan, Tao-Yuan 333, Taiwan

Yang Li
Key Laboratory of Microelectronics Devices and Integrated Technology, Institute of Microelectronics, Chinese Academy of Sciences, Beijing 100029, China
Lab of Nanofabrication and Novel Device Integration, Institute of Microelectronics, Chinese Academy of Sciences, Beijing 100029, China

Shibing Long
Key Laboratory of Microelectronics Devices and Integrated Technology, Institute of Microelectronics, Chinese Academy of Sciences, Beijing 100029, China
Lab of Nanofabrication and Novel Device Integration, Institute of Microelectronics, Chinese Academy of Sciences, Beijing 100029, China

Yang Liu
Department of Materials Physics and Chemistry, University of Science and Technology Beijing, Beijing 100083, China

Chen Hu
Department of Materials Physics and Chemistry, University of Science and Technology Beijing, Beijing 100083, China

Jiao Teng
Department of Materials Physics and Chemistry, University of Science and Technology Beijing, Beijing 100083, China

Qi Liu
Key Laboratory of Microelectronics Devices and Integrated Technology, Institute of Microelectronics, Chinese Academy of Sciences, Beijing 100029, China
Lab of Nanofabrication and Novel Device Integration, Institute of Microelectronics, Chinese Academy of Sciences, Beijing 100029, China

Hangbing Lv
Key Laboratory of Microelectronics Devices and Integrated Technology, Institute of Microelectronics, Chinese Academy of Sciences, Beijing 100029, China
Lab of Nanofabrication and Novel Device Integration, Institute of Microelectronics, Chinese Academy of Sciences, Beijing 100029, China

Jordi Suñé
Departament d'Enginyeria Electrònica, Universitat Autònoma de Barcelona, Bellaterra 08193, Spain.

Ming Liu
Key Laboratory of Microelectronics Devices and Integrated Technology, Institute of Microelectronics, Chinese Academy of Sciences, Beijing 100029, China
Lab of Nanofabrication and Novel Device Integration, Institute of Microelectronics, Chinese Academy of Sciences, Beijing 100029, China

Abdou Karim Diallo
Departement de Physique Appliquée, Université de Gaston Berger, Saint Louis, Sénégal

Abdoul Kadri Diallo
Laboratoire de Chimie et de Physique des Matériaux (LCPM), Université Assane Seck de Ziguinchor, Ziguinchor, Sénégal

Diouma Kobor
Laboratoire de Chimie et de Physique des Matériaux (LCPM), Université Assane Seck de Ziguinchor, Ziguinchor, Sénégal

Marcel Pasquinelli
Aix Marseille Université, Domaine Universitaire de Saint Jérôme, Marseille, France

Veerappan Saminathan
Department of ECE, Maharaja Engineering College, Avinashi, India

Kuppusamy Paramasivam
Department of ECE, Karpagam College of Engineering, Coimbatore, India

D. V. Kazantsev
University of Erlangen-Nuremberg, Erlangen, Germany
Institue for Theoretical and Experimental Physics, Moscow, Russia

H. Ryssel
University of Erlangen-Nuremberg, Erlangen, Germany

Sandy Tith
Department of Nuclear Engineering, Faculty of Engineering, Chulalongkorn University, Bangkok, Thailand

Nares Chankow
Department of Nuclear Engineering, Faculty of Engineering, Chulalongkorn University, Bangkok, Thailand

Raveendran Arun Prasath
Department of ECE, Anna University Regional Campus-Madurai, Madurai, India

Parasuraman Ganesh Kumar
K.L.N. College of Engineering, Pottapalayam, Sivagangai, India

Preface

Solid-state transistors and integrated circuits has spawned the information age in the last 50 years. Information is at people's hands and communications take seconds. Such rapid development stems from tremendous developments in both hardware and software such as solid-state circuits. Approaches such as parallel processing, new circuit design, and particularly novel materials are necessary.

Advances in Solid State Circuit Technologies provides concepts and current improvements of circuit technology. The applications of solid state in circuit's elements go from simple led technology to complex transistors and chips. The advances take the new technologies to smaller sizes and faster operations. First chapter presents fault diagnosis of DC analog CMOS circuits and second chapter explores on high-accuracy function synthesizer circuit with applications in signal processing. Third chapter provides a new current-controlled-power technique for small signal applications. In fourth chapter, a design of a symmetry-type floating impedance scaling circuit and the improvement method of its operation bandwidth are proposed. Fifth chapter elaborates on the design and analysis of a power efficient linearly tunable cross-coupled transconductor having separate bias control, and sixth chapter presents a novel and unique polymorphic processor design. In seventh chapter, the differential ULV inverter is presented. The improvement in term of stability and delay relative to the ULV is elaborated. Furthermore, the reliability, yield and the defect tolerance of the differential ULV inverter compared with both standards CMOS and with ULV floating-gate inverter are examined. Chip design of a low-voltage wideband continuous-time sigma-delta modulator with DWA technology for WIMAX applications are highlighted in eighth chapter. Ninth chapter introduces the system architecture of the wideband continuous-time sigma-delta modulator and ninth chapter presents the design of a novel two-stage bulk-input pseudo differential Operational Transconductance Amplifier. Tenth chapter explores on the performance improvement of phase-change memory cell using AlSb3Te and atomic layer deposition TiO2 buffer layer and eleventh chapter highlights emerging nonvolatile memory technologies. Probe-based storage technologies are illustrated in twelfth chapter and thirteenth chapter presents a comparative analysis result of the proposed scheme and the conventional thyristor protection scheme incorporating a clamp NMOS device. Fourteenth chapter presents a CMOS 3.1 - 10.6 GHZ UWB LNA employing modified derivative superposition method. In fifteenth chapter, the challenges and opportunities of CBRAM devices using different switching materials such as chalcogenides, oxides, and bilayers in different structures have been reviewed. In sixteenth chapter, we focus our attention on the recent development of the research on the QC effect in CF-based non-volatile RS devices including basic QC phenomenon in resistive random access memory (RRAM), RS mechanisms, device structures, materials, theory, and modeling of conductance quantization in RRAM. In seventeenth chapter, we report on electrical instabilities in pentacene-based transistors with Mylar and PMMA/Mylar gate dielectrics transferred by a lamination process in ambient environment. Special emphasis is given on comparing the two types of devices in ON state and subthreshold regions under gate bias stress. Eighteenth chapter presents an introduction of memristor and modelling of memristor with nonlinear dopant drift with new parameter. Ninteenth chapter focuses on canning head for the apertureless near field optical microscope and measurement of gamma-rays using smartphones is highlighted in twentieth chapter. Last chapter presents an approach on design of low power level shifter circuit with sleep transistor using multisupply voltage scheme. These chapters contain applications, written by renowned experts in the respective fields on to integrated circuits and materials science. It is intended for a broad readership which includes electrical engineers and material scientists. Readers will be able to familiarize themselves with the latest technologies in the various fields.

Editor
Nelson Bolívar, Ph.D

INTRODUCTION

Integrated circuit composed of a single solid (monolithic) piece of material (usually silicon crystal) on which electronic components and their connecting paths are etched or printed. Solid-state refers to electronic components, devices, and systems based entirely on the semiconductor. The expression was especially prevalent in the late 1950s and early 1960s, during the transition from vacuum tube technology to the semiconductor diode and transistor. More recently, the integrated circuit (IC), the light-emitting diode (LED), and the liquid-crystal display (LCD) have evolved as further examples of solid-state devices. In a solid-state component, the current is confined to solid elements and compounds engineered specifically to switch and amplify it. Current flows in two forms: as negatively charged electrons, and as positively charged electron deficiencies called holes. In some semiconductors, the current consists mostly of electrons; in other semiconductors, it consists mostly of holes. Both the electron and the hole are called charge carriers. An example of a non-solid-state component is a cathode-ray tube (CRT). In this device, electrons flow freely through a vacuum from an electron gun, through deflecting and focusing fields, and finally to a phosphorescent screen.

What does solid-state mean in relation to electronics?

Solid-state electronic devices are part of our everyday lives. The transistor, invented in 1947 by Bell Labs, was the first solid-state device to come into commercial use in the 1960s. Solid-state electronic devices have replaced vacuum tubes in just about all electronics devices. Vacuum tubes are still used in the transmitters of radio stations you listen to, many guitar amplifiers and some audiophile equipment. Vacuum tubes are the opposite of "solid-state" because tubes burnout, break, etc.

One of the first solid-state devices was a crystal radio. In a crystal radio, a piece of wire positioned on a crystal's surface is able to separate the lower-frequency audio from the higher-frequency transmitted radio carrier wave. This form of signal detection is due to the crystal's ability to pass a current in only one direction.

Three common solid state components are displayed. Clockwise from the top: inch-long circuit chip, red light-emitting diode and a transistor. See more pictures of what's inside electronics.

Solid-state gets its name from the path that electrical signals take through solid pieces of semi-conductor material. Prior to the use of solid-state devices, such as the common transistor, electricity passed through the various elements inside of a heated vacuum tube. Solid-state devices, such as a transistor, use conductors to control the flow of signals through a circuit.

- In a transistor amplifier, a small change on the input signal's amplitude is immediately reflected in larger amplitude in the output within a transistor.
- In a vacuum tube amplifier, after the tube warms up, a signal is applied to the "grid" of a tube and the resultant output of the same frequency is at a much higher amplitude.

In the sandwich-like construction of a transistor, the emitter, base, and collector do a similar task at much lower DC voltages, with no "warm-up" time! In digital circuits, an integrated circuit chip is nothing more than a collection of transistors and wires that hook them together.

Solid state devices called diodes have a replaced rectifier vacuum tubes, used to transform alternating current (AC) to direct current (DC). Cool-running light-emitting diodes (LEDs), another solid-state device used for indicators on the front panel of your computer and monitor, have replaced the earlier incandescent bulbs. Multiple bright LEDs are also used for the third stoplight on many U.S. vehicles and for traffic signals.

Solid-state miniature electronic components are in many places:

- Mounted on flexible thin film printed circuits in cameras and disk drives.
- The beeping sound made by a cell phone, page or auto dashboard alarm.
- The voice chip in an answering machine.
- TV remote control.
- Laser pointer.
- The inside of an MP3 player.
- A quartz watch.
- The image sensor in a digital camera and a camcorder.
- The computer monitor you are viewing.
- The mouse you will click to the next HSW screen!

Solid-state electronics

Solid-state electronics are those circuits or devices built entirely from solid materials and in which the electrons, or other charge carriers, are confined entirely within the solid material. The term is often used to contrast with the earlier technologies of vacuum and gas-discharge tube devices, and it is also conventional to exclude electro-mechanical devices (relays, switches, hard drives and other devices with moving parts) from the term solid state. While solid-state can include crystalline, polycrystalline and amorphous solids and refer to electrical conductors, insulators and semiconductors, the building material is most often a crystalline semiconductor. Common solid-state devices include transistors, microprocessor chips, and RAM. A specialized type of RAM called flash memory is used in flash drives and, more recently, solid-state drives to replace mechanically rotating magnetic disc hard drives. A considerable amount of electromagnetic and quantum-mechanical action takes place within the device. The expression became prevalent in the 1950s and the 1960s, during the transition from vacuum tube technology to semiconductor diodes and transistors. More recently, the integrated circuit (IC), the light-emitting diode (LED), and the liquid-crystal display (LCD) have evolved as further examples of solid-state devices.

What Does an Integrated Circuit Manufacturer Do?

An integrated circuit manufacturer produces semiconductor electronic circuits, which are small solid-state electronic components used in mobile phones, televisions and many other electronic devices. Solid-state devices have no moving parts, and are made from silicon-based chips that provide computing and electrical processing capability. Semiconductors replaced vacuum tubes and early mechanical computer switches in the mid-20th century, allowing devices to become very small, yet contain a great deal of computing power. Production of integrated circuits requires several steps and carefully controlled manufacturing conditions to prevent contamination. The integrated circuit manufacturer builds clean rooms that use very high levels of air filtration to remove dust and other contaminants that can ruin a circuit. Employees wear coverings that prevent dust, skin or hair from entering the process area, and will remove and replace these coverings if they must leave and re-enter the production area.

The first step in producing integrated circuits is creating silicon wafers used as the base of the circuit. An integrated circuit manufacturer or an outside contractor purifies silicon from natural sands, removing any impurities to create a cylinder-shaped crystal made from pure silicon. The cylinder is then cut into thin wafers, which may be several inches or centimeters across. These wafers are chemically cleaned, and then one side is polished to a mirror-like finish that will be the base of the integrated circuit.

Polished wafers are then exposed to pure oxygen in a process that creates a very thin layer of silicon dioxide. Oxygen reacts with the pure silicon on the wafer, which consumes a small amount of the silicon on the wafer surface. The resulting silicon dioxide is molecularly bound to the pure silicon beneath it, and will not be removed later except by chemical processing.

The next steps used by an integrated circuit manufacturer are a repeated process of masking, light exposure, etching and cleaning. An integrated circuit is an electronic circuit formed by thin layers of electrically conducting material separated by layers of non-conductors. Masking places a template or design of the first circuit layer onto the silicon wafer.

First, a material called photoresist is placed on the wafer surface, with the mask over it. The photoresist is exposed to light, which causes a chemical curing reaction of any exposed material. When the mask is removed, the uncured photoresist can be removed in a process called etching.

An integrated circuit manufacturer repeats these steps, adding layers of conductor or non-conductor onto the wafer to build electronic circuits. Many small integrated circuits are made on a wafer at one time, and are cut apart in later steps once the circuits have been completed. The final steps are adding electrical connections to the individual circuits so they can be placed on circuit boards used in computers and other electronic devices.

What Is a Silicon Integrated Circuit?

A silicon integrated circuit (IC) is a compact functional electronic circuit that contains silicon-based semiconductors. Semiconductors have conductivity characteristics that can be controlled in various ways. The integrated circuit, also known as chip or microchip, is seen as a small component with terminals ranging from eight pins or more. Very large-scale integration ICs are used in computing devices that contain more than a million silicon transistors. Transistors are three-terminal devices that act as current-controlled current valves and are like electromagnetic relays, but with continuous current control instead of two-state or on/off control.

The semiconductor used in a silicon integrated circuit is the main entity referred to as a solid-state device. Before the first solid-state device was put into commercial use, the way to control the flow of electrons was to use a vacuum tube with a heater, cathode, plate, and at least a control grid. The size and power requirement of vacuum tubes now make them popular only in specialized fields such as extra-high power amplifiers for audio and radio applications. The silicon integrated circuit is very popular also due to the availability of silicon, as a common compound of silicon is found everywhere as ordinary sand. By proper heating in the absence of oxygen, semiconductor manufacturers are able to produce pure silicon, which can be further processed as diodes, transistors, and ICs.

The silicon integrated circuit is not used for radio signal applications because of the limitations of silicon semiconductor devices that are covered by the better signal performance of other semiconductors such as germanium semiconductors. In cell phones, the audio and alternating current (AC)/direct current (DC) power circuits use silicon semiconductor ICs. Computers and other digital devices make use of many silicon ICs. From the convenience outlet, the power supply uses silicon diodes capable of rectifying directly the main AC voltage. These diodes are usually in an encased switching power supply that employs silicon-based power regulator chips that are usually of the pulse-width modulation (PWM) type, which feedbacks the output DC levels as changes in the duty cycle of the power pulses.

Majority of the electronic circuits for relatively low-frequency applications use silicon-based semiconductors. The silicon diode is the most popular low-frequency power diode available in the market. Other switching semiconductors, such as the silicon-controlled rectifier (SCR), the three-terminal AC (TRIAC) device, and most audio power transistors, are made of silicon semiconductors. In solar photovoltaic panels, it is observed that silicon junctions that are exposed to sunlight produce an electrical potential across the terminals. This leads to the wide use of silicon-based solar panel devices that promise to become cheaper and cheaper as more research and manufacturing breakthroughs are achieved.

CHAPTER 1

FAULT DIAGNOSIS OF DC ANALOG CMOS CIRCUITS

Michał Tadeusiewicz and Stanisław Hałgas

Keywords: Analog nonlinear circuits, Fault diagnosis, Multiple soft faults, Nanometer technology, Nelder–Mead method

INTRODUCTION

Fault diagnosis of analog circuits is an important problem in the design and testing of electronic devices [1–22]. Generally, fault diagnosis includes detecting faulty circuits, locating faulty parameters and evaluating their values. If a faulty parameter is drifted from its tolerance range but does not lead to some topological changes, the fault is said to be soft or parametric. If a fault is open circuit or short circuit, it is called hard or catastrophic. In integrated circuits physical imperfections, such as near–opens or near–shorts may occur as spot defects [7, 10, 21, 22]. The methods dedicated to soft fault diagnosis usually exploit the simulation after test approach, where circuit simulations take place after any testing. They are based on measurements of the voltages at accessible points of the circuit, leading to equations with the tested parameters as unknown variables.

In current CMOS technology the global variations of parameters are measured by dedicated test structures included in the wafer. However, the problem is how to identify the random local variations of the process parameters. The local variations are due to fabrication or due to aging phenomenon. They affect the components across the die independently. Examples of local variations in ICs include local geometrical deformations, such as variations in the channel length and width, the oxide thickness, etc.

Many concepts and methods focused on parametric fault diagnosis are presented in references [1–3, 5, 9, 11, 13, 16–20]. Most of the works, dealing with soft fault diagnosis of analog circuits, address only the case when just one parameter is faulty. Fewer papers are devoted to the multiple fault diagnosis, where several parameters can be faulty. In real circuits the test equations, that express the measured voltages in terms of the parameters are nonlinear and cannot be presented in explicit analytical form. These equations may actually have multiple solutions, which means that several sets of the parameter values meet the test. To find the multiple solutions the parametric homotopy [17], the simplicial homotopy [18], or the block relaxation method [19] were proposed. To determine the actual solution a new efficient approach was proposed in Ref. [20] as follows. Two tests of the circuit are arranged, one used to find the solutions and the other to check if the obtained solution is the actual one. To compute the solution the extended systematic search method was developed [20]. In this paper the Nelder–Mead optimization method is applied with the objective function properly modified during the computation process and similarly as in [20] the obtained result is checked using the validation test. If the obtained solution passes this test the algorithm terminates, otherwise another solution is calculated and verified. The procedure is carried out as long as the solution which meets the validation test is obtained.

DIAGNOSTIC TESTS

Let us consider a nonlinear DC circuit, with n parameters $x_1, \ldots, x_n$ considered as potentially faulty, having one or more input nodes accessible for excitation and one output node accessible for measurement. We connect to the output node a resistor R_o and apply DC voltage sources to the input nodes (see Fig. 1). We choose n sets of the input voltage values and read the corresponding values of the output voltage. They are labelled $v_o^{(1)}, \ldots, v_o^{(n)}$ and used to form vector $\mathbf{v}^{(o)} = \left[v_o^{(1)} \cdots v_o^{(n)}\right]^{\mathrm{T}}$, where T means transposition. Each of the voltages is a function of the circuit parameters $x_1, \ldots, x_n$, $v_o^{(j)} = \tilde{g}_j(\mathbf{x})$, where $\mathbf{x} = [x_1 \cdots x_n]^{\mathrm{T}}$ is the vector consisting of the parameters considered as potentially faulty. Thus, it holds

Citation: Michał Tadeusiewicz, Stanisław Hałgas, fault diagnosis of DC analog CMOS circuits, DOI: 10.1007/s10470-016-0752-y.

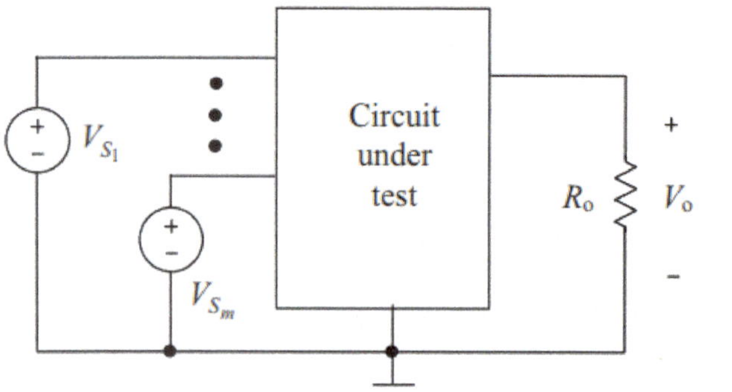

Figure 1: Diagnostic test

$$\mathbf{v}^{(o)} = \tilde{\mathbf{g}}(\mathbf{x}), \qquad (1)$$

where $\tilde{\mathbf{g}}(\mathbf{x}) = [\tilde{g}_1(\mathbf{x}) \cdots \tilde{g}_n(\mathbf{x})]^{\mathrm{T}}$. Equation (1) can be rewritten in the compact form

$$g(x)=0, \qquad (2)$$

where $\mathbf{g}(\mathbf{x}) = [g_1(\mathbf{x}) \cdots g_n(\mathbf{x})]^{\mathrm{T}} = \tilde{\mathbf{g}}(\mathbf{x}) - \mathbf{v}^{(o)}$, and named a test equation.

The diagnostic method developed in Sect. 3 requires two tests arranged using the above–described approach. As a result the voltages $\hat{v}_o^{(1)}, \ldots, \hat{v}_o^{(n)}$ are measured in the circuit as depicted in Fig. 1, but driven by different sets of the input voltage sources, forming vector $\hat{\mathbf{v}}^{(o)} = \left[\hat{v}_o^{(1)} \cdots \hat{v}_o^{(n)}\right]^{\mathrm{T}}$. The first test leading to Eq. (2) will be named a principal test (PT), whereas the second one will be named a validation test (VT).

Unfortunately, in real electronic circuits the function g(x) cannot be presented in explicit analytical form. However, the values of $\tilde{g}_i(\mathbf{x})$ $i = 1, \ldots, n$, for given xx, can be found by performing the analyses of the circuits driven by the sources as in the test, with the parameters being the elements of vector x.

FAULT DIAGNOSIS ALGORITHM

An algorithm that allows finding actual values of the parameters $x_1, \ldots, x_n$ is developed in this section. The algorithm solves the PT Eq. (2) and verifies the obtained solutions applying the VT. Its core is the Nelder–Mead optimization method [23–25]. The algorithm takes into consideration the possibility of existing several solutions of the nonlinear Eq. (2). Each of the solutions is a set of the parameters that meet the PT.

Since the algorithm exploits the Nelder–Mead method, a background of this method as well as the version that is used in this paper is described below. The Nelder–Mead method is designed to solve the unconstrained optimization problem of minimizing given nonlinear function f(x): $f(\mathbf{x}) : R^n \to R$ In this paper the method is adapted to solve Eq. (2). For this purpose the function

$$f(\mathbf{x}) = \sqrt{\alpha_1 g_1^2(\mathbf{x}) + \cdots + \alpha_n g_n^2(\mathbf{x})} \qquad (3)$$

is formed, where $\alpha_1,\ldots,\alpha_n$ are coefficients equal to zero or one. If all the coefficients are equal to one the function f(x) will be called a complete function, otherwise a reduced function. The complete function is identical to the Euclidean norm of $\mathbf{g}(\mathbf{x})$, i.e., $f(\mathbf{x}) = \|\mathbf{g}(\mathbf{x})\|_2$. The Nelder–Mead method uses only the function values at some points in R^n and does not require gradients at the points. This is why this method is very useful to solve Eq. (2), with g(x not given in explicit analytical form. The Nelder–Mead method is simplex–based [23–25]. An *m*-simplex is a figure formed by m+1 independent points, written as $S = \{\mathbf{x}^0, \ldots, \mathbf{x}^m\}$, where $\mathbf{x}^i = \left[x_1^i \cdots x_n^i\right]^{\mathrm{T}}$, $i = 0, 1, \ldots, m$, are called vertices. It is a convex hull of the m+1independent points $\mathbf{x}^i$ $(i = 0, 1, \ldots, m)$. For example, 2-simplex is a triangle (see Fig. 2).

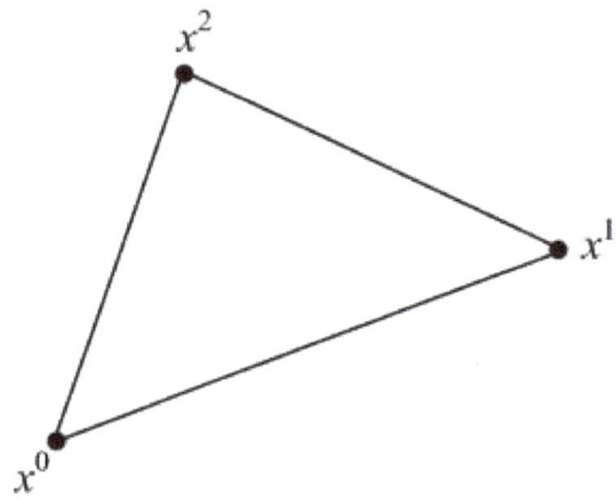

Figure 2: Exemplary 2-simplex.

The method exploits simplices having n+1 vertices $S=\{x^0,\ldots,x^n\}$. The corresponding function values at these vertices are labeled $f^0=f(x^0),\ldots,f^n=f(x^n)$. At any stage of the computation process the method generates a new simplex, aimed at decreasing the function values at its vertices. To construct this simplex one or more new points are searched comparing their function values with those at the vertices. Generally the method terminates when the simplex becomes sufficiently small, or the simplex is flat or degenerated [24]. The crucial point of the Nelder–Mead method is creating an adjusted simplex to the current simplex S. In this paper the approach described in [23] is adopted. Choose the indices h, s, l, of the worst, second worst, and the best vertex of S so that $f^h = \max_i f^i,\ f^s = \max_{i \neq h} f^i, f^l = \min_i f^i$. Calculate the centroid c of the side opposite to the worst vertex h, $\mathbf{c} = \frac{1}{n}\sum_{i \neq h} \mathbf{x}^i$ and find the reflection point $\mathbf{x}^r = \mathbf{c} + \left(\mathbf{c} - \mathbf{x}^h\right)$ as well as $f^r = f(\mathbf{x}^r)$ (see Fig. 3).

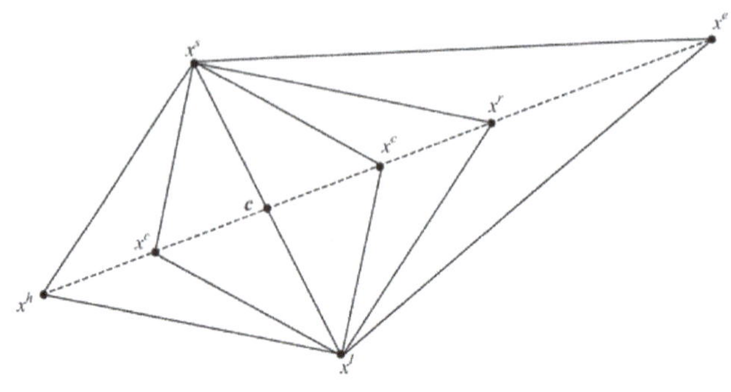

Figure 3: Basic simplex operations.

If $f^l \leq f^r < f^s$ choose x^r as the new vertex of the adjusted simplex. Otherwise, continue the procedure depending on whether $f^r < f^l$ or $f^r \geq f^s$.

- If $f^r < f^l$, compute the point $\mathbf{x}^e = \mathbf{c} + \gamma(\mathbf{x}^r - \mathbf{c})$ and $f^e = f(\mathbf{x}^e)$. If $f^e < f^r$ choose x^e as the new vertex, otherwise, choose x^r as a new vertex.
- If $f^r \geq f^s$, compute a point x^c using the following approach.
- If $f^r < f^h$, compute $x^c = c + \beta(x^r - c$ and $f^c = f(x^c)$. If $f^c \leq f^r$, choose x^c as the vertex, otherwise, perform a shrink operation as described below.
- If $f^r \geq f^h$, compute $x^c = c + \beta(x^h - c)$ and $f^c = f(x^c)$. If $f^c \leq f^h$, choose x^c as the vertex. Otherwise, perform a shrink operation.

Shrink operation: Compute n new vertices $\mathbf{x}^i = \mathbf{x}^l + \delta(\mathbf{x}^i - \mathbf{x}^l)$, for $i = 0, \ldots, n$, $i \neq l$, (see Fig. 4).

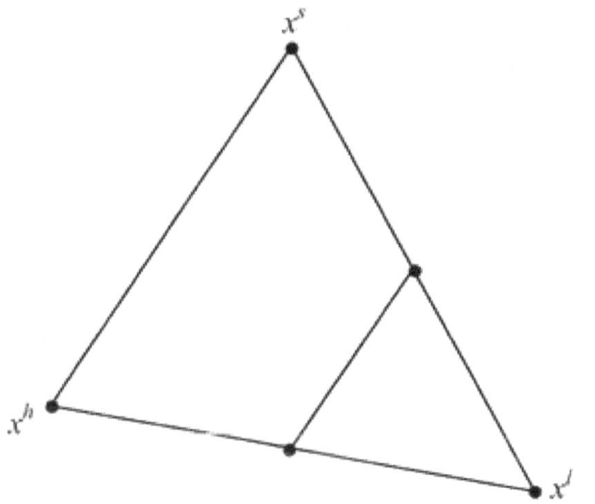

Figure 4: Shrink operation.

Note

The coefficients γ, β, δ are chosen as proposed in Ref. [25]: $\gamma = 1 + \frac{2}{n}$, $\beta = 0.75 - \frac{1}{2n}$, $\delta = 1 - \frac{1}{n}$.

As the initial simplex we choose the regular one using the procedure described in Ref. [24].

It should be emphasized that the function f(x) is not given in explicit analytical form. In consequence, to find the value of the complete function f(x) at given x, n analyses of the circuit must be performed applying the sources as in PT. This is time consuming process. The time is shrunk if f(x) is the reduced function. Since the Nelder–Mead method requires large number of the values of f(x) at various points x, the reduced function is exploited at some stages of the algorithm proposed in this paper. Moreover, for different reduced functions the method searches for the solution (a set of the parameters) in different directions. This observation is used to find a new solution, when the obtained one does not pass the VT.

SKETCH OF THE ALGORITHM

- Pick the required measurement accuracy of the voltages $v_o^{(1)}, \ldots, v_o^{(n)}$ and $\hat{v}_o^{(1)}, \ldots, \hat{v}_o^{(n)}$, the tolerance ε_1 such that the inequality $\|\mathbf{g}(\mathbf{x})\|_2 < \varepsilon_1$ is a good approximation of $\|g(x)\|_2 = 0$, the tolerance ε_2 used in Step 4, a maximum number of the generated simplices M, and the side a of the initial regular simplex.
- Arrange two diagnostic tests, PT and VT, and form the vectors $v^{(o)}$ and $\hat{\mathbf{v}}^{(o)}$ consisting of the measured voltages.
- Create the reduced function f(x) specified by Eq. (3) with $\alpha_1 = \cdots = \alpha_{n-1} = 1$, $\alpha_n = 0$ and apply the Nelder–Mead method. If during the process a simplex is obtained so that the value of f(x) at the best vertex is $< 100\varepsilon_2$, the computation process is modified as follows. The obtained xx is considered as an approximate solution. Next the complete function f(x) is created by setting $\alpha_1 = \cdots = \alpha_n = 1$, a new regular simplex is constructed around this best vertex and the procedure is continued. If during the process a vertex, at which the value of f(x) is less than ε_1 appears, the parameters $x_1, \ldots, x_n$ corresponding to this vertex meet the PT. Otherwise, go to Step 5.
- Check if the obtained parameters satisfy the VT. For this purpose analyse the circuit with these parameters, driven by the sources as in VT, find the voltages labelled $\tilde{v}_o^{(1)}, \ldots, \tilde{v}_o^{(n)}$ and form vector $\mathbf{r} = [r_1 \cdots r_n]^T$, where $r_j = \tilde{v}_o^{(j)} - \hat{v}_o^{(j)}$, $j = 1, \ldots, n$. If $\|\mathbf{r}\|_2 = \sqrt{r_1^2 + \cdots + r_n^2} \leq \varepsilon_2$, the parameters are the actual ones. In such a case the algorithm terminates. Otherwise, they are virtual and we proceed to Step 5.
- Modify the function f(x) by setting $\alpha_{n-1} = 0$, $\alpha_1 = \cdots = \alpha_{n-2} = \alpha_n = 1$ and repeat the steps 3–4 adapted to this case.

This procedure can be continued if the actual parameters have not been found, by setting in succession $\alpha_{n-2} = 0, \cdots, \alpha_1 = 0$. In any case the remaining coefficients are equal to one and the number of the generated simplices cannot exceed maximum value MM.

Note

If at any stage of the algorithm the small or the flat simplex [24] appears, the set of the coefficients $\alpha_1, \ldots, \alpha_n$ is changed as described in step 5. If the degenerated simplex [24] appears, a new regular simplex is created around the best vertex of this simplex and the process is continued.

NUMERICAL EXAMPLES

The proposed algorithm has been implemented in the joined environments: DELPHI and IsSPICE 4, and tested numerically using MOS circuits designed in nanometer technology. The calculations were executed on PC with the processor Intel Core (TM) i7-2600. To illustrate effectiveness of the algorithm we consider three exemplary circuits designed in nanometer technology. In all the examples the transistors are characterized by the BSIM 4.6 model implemented in IsSPICE 4, Level 14 [26]. The nominal values of the channel lengths of the transistors are indicated in Figs. 5, 6 and 7, whereas the nominal values of the oxide thicknesses are $(TOX)_p = (TOX)_n = 1.4$nm. The other quantities are as follows: XL=5nm, XW=15nm. The discussed in this section soft faults are variations in the channel length ΔL and in the oxide thickness ΔTOX, considered separately for NMOS and PMOS transistors.

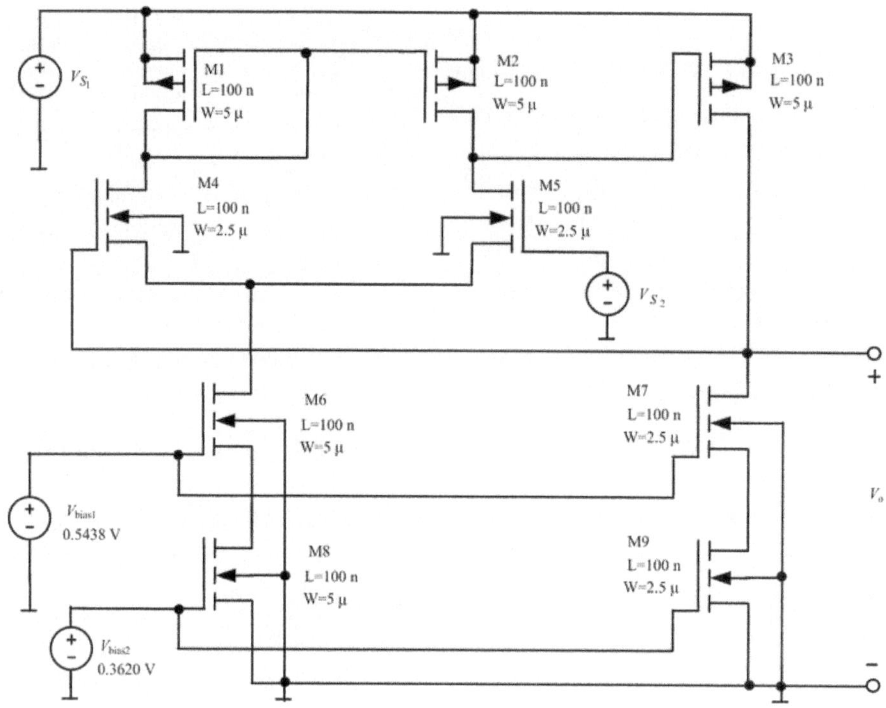

Figure 5: Bias two-stage operational amplifier.

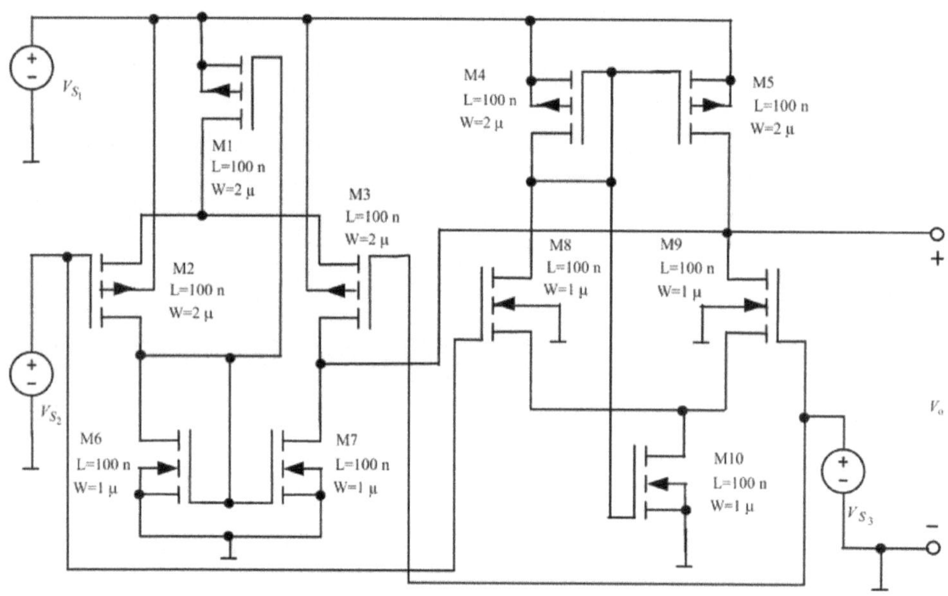

Figure 7: A rail–to–rail input buffer.

Example 1

Let us consider the bias two–stage op–amp shown in Fig. 5 [27]. To arrange the diagnostic tests (PT and VT) the input voltages V_{S_1} and V_{S_2} as well as the output voltage V_o are selected. It is assumed that the measurement accuracy of V_o is $1\ \mu V$, $M = 2000$, $\varepsilon_1 = 10^{-6}$, $\varepsilon_2 = 10^{-5}$, $a = 0.05$, the output resistance R_o=10kΩ. At the preliminary stage of the diagnosis process the sensitivities of the voltage V_o due to variations of the parameters L and TOX of all the transistors, for different values V_{S_1} and V_{S_2}, are calculated. The analysis shows that the sensitivities of V_o due to the parameters of the transistors M6 and M7 are approximately 100 times smaller than due to the parameters of the other transistors. Thus, deviations of the parameters of transistors M6 and M7 have slight influence on the tested voltage V_o and they cannot be diagnosed in real conditions. In the sequel we consider the soft faults of the following parameters:

- the channel lengths and the oxide thicknesses of PMOS transistors M1, M2, M3,
- the channel lengths and the oxide thicknesses of NMOS transistors M4, M5, M8, M9.

On the basis of sensitivity analyses and numerical experiments the following sets of the values of V_{S_1} and V_{S_2} have been chosen to perform of the PT and VT.

PT	VT
$V_{S_1} = 1.30\text{ V } V_{S_2} = 1.25\text{ V},$	$V_{S_1} = 1.20\text{ V } V_{S_2} = 0.90\text{ V},$
$V_{S_1} = 1.10\text{ V } V_{S_2} = 0.95\text{ V},$	$V_{S_1} = 0.90\text{ V } V_{S_2} = 0.70\text{ V},$
$V_{S_1} = 0.55\text{ V } V_{S_2} = 0.50\text{ V},$	$V_{S_1} = 0.70\text{ V } V_{S_2} = 0.50\text{ V},$
$V_{S_1} = 0.80\text{ V } V_{S_2} = 0.65\text{ V},$	$V_{S_1} = 0.60\text{ V } V_{S_2} = 0.40\text{ V}.$

All the four sets are exploited to test the parameters of the four NMOS transistors M4, M5, M8, M9 and the first three of them are used to test the parameters of the three PMOS transistors M1, M2, M3.

The results of different multiple soft fault diagnoses are summarized in Tables 1, 2, 3, 4. They comprise three sets of the simultaneous variations of all the channel lengths in NMOS transistors M4, M5, M8, M9 (Table 1), and in PMOS transistors M1, M2, M3 (Table 2), three sets of the simultaneous variations of all the oxide thicknesses in NMOS transistors M4, M5, M8, M9 (Table 3) and in PMOS transistors M1, M2, M3 (Table 4). In all the cases the obtained values of the parameters are very close to their actual values.

Table 1: Results of the fault diagnosis of the channel lengths in NMOS transistors M4, M5, M8, M9

Actual values of the parameters	Values of the parameters determined by the method	Number of generated simplices	Computation time in seconds
$\Delta L_{M4} = 6.00$ nm	$\Delta L_{M4} = 6.02$ nm	191	59.8
$\Delta L_{M5} = 4.00$ nm	$\Delta L_{M5} = 4.02$ nm		
$\Delta L_{M8} = 2.50$ nm	$\Delta L_{M8} = 2.49$ nm		
$\Delta L_{M9} = 6.00$ nm	$\Delta L_{M9} = 5.97$ nm		
$\Delta L_{M4} = 15.00$ nm	$\Delta L_{M4} = 15.02$ nm	196	61.7
$\Delta L_{M5} = 2.00$ nm	$\Delta L_{M5} = 2.03$ nm		
$\Delta L_{M8} = 15.00$ nm	$\Delta L_{M8} = 15.00$ nm		
$\Delta L_{M9} = 4.50$ nm	$\Delta L_{M9} = 4.47$ nm		
$\Delta L_{M4} = 2.50$ nm	$\Delta L_{M4} = 2.55$ nm	742	227.5
$\Delta L_{M5} = 2.00$ nm	$\Delta L_{M5} = 2.06$ nm		
$\Delta L_{M8} = 1.75$ nm	$\Delta L_{M8} = 1.73$ nm		
$\Delta L_{M9} = 20.00$ nm	$\Delta L_{M9} = 19.90$ nm		

Table 2: Results of the fault diagnosis of the channel lengths in PMOS transistors M1, M2, M3

Actual values of the parameters	Values of the parameters determined by the method	Number of generated simplices	Computation time in seconds
$\Delta L_{M1} = 4.50$ nm	$\Delta L_{M1} = 4.56$ nm	79	26.0
$\Delta L_{M2} = 3.00$ nm	$\Delta L_{M2} = 3.07$ nm		
$\Delta L_{M3} = 1.00$ nm	$\Delta L_{M3} = 1.00$ nm		
$\Delta L_{M1} = 15.00$ nm	$\Delta L_{M1} = 14.82$ nm	92	28.2
$\Delta L_{M2} = 10.00$ nm	$\Delta L_{M2} = 9.85$ nm		
$\Delta L_{M3} = 2.50$ nm	$\Delta L_{M3} = 2.50$ nm		
$\Delta L_{M1} = 1.50$ nm	$\Delta L_{M1} = 1.49$ nm	152	48.8
$\Delta L_{M2} = 25.00$ nm	$\Delta L_{M2} = 24.98$ nm		
$\Delta L_{M3} = 1.00$ nm	$\Delta L_{M3} = 1.00$ nm		

Table 3: Results of the fault diagnosis of the oxide thicknesses in NMOS transistors M4, M5, M8, M9

Actual values of the parameters	Values of the parameters determined by the method	Number of generated simplices	Computation time in seconds
$\Delta TOX_{M4} = 140.0$ pm	$\Delta TOX_{M4} = 140.1$ pm	207	67.8
$\Delta TOX_{M5} = -56.0$ pm	$\Delta TOX_{M5} = -55.9$ pm		
$\Delta TOX_{M8} = -28.0$ pm	$\Delta TOX_{M8} = -27.8$ pm		
$\Delta TOX_{M9} = 98.0$ pm	$\Delta TOX_{M9} = 97.9$ pm		
$\Delta TOX_{M4} = 280.0$ pm	$\Delta TOX_{M4} = 280.1$ pm	346	106.9
$\Delta TOX_{M5} = -56.0$ pm	$\Delta TOX_{M5} = -55.9$ pm		
$\Delta TOX_{M8} = 280.0$ pm	$\Delta TOX_{M8} = 280.1$ pm		
$\Delta TOX_{M9} = -42.0$ pm	$\Delta TOX_{M9} = -42.0$ pm		
$\Delta TOX_{M4} = 70.0$ pm	$\Delta TOX_{M4} = 70.0$ pm	407	120.1
$\Delta TOX_{M5} = -56.0$ pm	$\Delta TOX_{M5} = -56.0$ pm		
$\Delta TOX_{M8} = -49.0$ pm	$\Delta TOX_{M8} = -49.0$ pm		
$\Delta TOX_{M9} = 560.0$ pm	$\Delta TOX_{M9} = 559.6$ pm		

Table 4: Results of the fault diagnosis of the oxide thicknesses in PMOS transistors M1, M2, M3

Actual values of the parameters	Values of the parameters determined by the method	Number of generated simplices	Computation time in seconds
$\Delta TOX_{M1} = 112.0$ pm	$\Delta TOX_{M1} = 111.4$ pm	124	37.6
$\Delta TOX_{M2} = 56.0$ pm	$\Delta TOX_{M2} = 55.5$ pm		
$\Delta TOX_{M3} = -28.0$ pm	$\Delta TOX_{M3} = -28.0$ pm		
$\Delta TOX_{M1} = -42.0$ pm	$\Delta TOX_{M1} = -42.8$ pm	321	102.5
$\Delta TOX_{M2} = 420.0$ pm	$\Delta TOX_{M2} = 419.2$ pm		
$\Delta TOX_{M3} = 28.0$ pm	$\Delta TOX_{M3} = 28.0$ pm		
$\Delta TOX_{M1} = 280.0$ pm	$\Delta TOX_{M1} = 278.0$ pm	333	103.2
$\Delta TOX_{M2} = -70.0$ pm	$\Delta TOX_{M2} = -71.1$ pm		
$\Delta TOX_{M3} = 280.0$ pm	$\Delta TOX_{M3} = 280.0$ pm		

In this example the transistors M6 and M7 cannot be tested, because their parameters have very slight influence on the output voltage. Numerical experiments show that even in some idealized circumstances, under very high accuracy of the measurements of VoVo, equal to 0.01 µV, and M=5000M=5000 some determined parameters of these transistors can be erroneous and the CPU time long. Some exemplary results conforming this statement are included in Table 5.

Table 5: Results of the fault diagnosis of all NMOS transistors under the measurement accuracy 0.01µV

Actual values of the parameters	Values of the parameters determined by the method	Number of generated simplices	Computation time in seconds
$\Delta L_{M4} = 7.00$ nm	$\Delta L_{M4} = 7.00$ nm	2295	793.1
$\Delta L_{M5} = 7.00$ nm	$\Delta L_{M5} = 7.00$ nm		
$\Delta L_{M6} = 1.00$ nm	$\Delta L_{M6} = 0.92$ nm		
$\Delta L_{M7} = 2.80$ nm	$\Delta L_{M7} = 2.71$ nm		
$\Delta L_{M8} = 2.00$ nm	$\Delta L_{M8} = 2.00$ nm		
$\Delta L_{M9} = 4.00$ nm	$\Delta L_{M9} = 4.00$ nm		
$\Delta TOX_{M4} = 21.0$ pm	$\Delta TOX_{M4} = 21.0$ pm	1049	323.1
$\Delta TOX_{M5} = 280.0$ pm	$\Delta TOX_{M5} = 280.0$ pm		
$\Delta TOX_{M6} = -28.0$ pm	$\Delta TOX_{M6} = -28.2$ pm		
$\Delta TOX_{M7} = 70.0$ pm	$\Delta TOX_{M7} = 57.5$ pm		
$\Delta TOX_{M8} = -56.0$ pm	$\Delta TOX_{M8} = -56.0$ pm		
$\Delta TOX_{M9} = 42.0$ pm	$\Delta TOX_{M9} = 42.6$pm		

Example 2

Figure 6 shows the CMOS circuit, being the voltage reference, designed in nanometer technology [27]. The measurement accuracy and the constants of the computation process are the same as in Example 1. To arrange the diagnostic tests the input voltage V_{S_1} and the output voltage V_o are selected. At the preliminary stage of the diagnosis process the sensitivities of the output voltage due to variations of the parameters L and TOX of all the transistors, for different values V_{S_1}, are calculated. The analysis reveals that the sensitivities due to the parameters of the transistors M1, M2, M6, M7, and M8 are 100–100,000 times smaller than due to the parameters of the other transistors. Thus, they cannot be tested. This is why we perform the fault diagnosis of the parameters of transistors M3, M4, M5, M9, M10, and M11, separately for the PMOS (M3, M4, M5) and NMOS (M9, M10, M11) ones. On the basis of sensitivity analyses and numerical experiments the following values of VS1VS1 have been chosen to perform of the PT and VT.

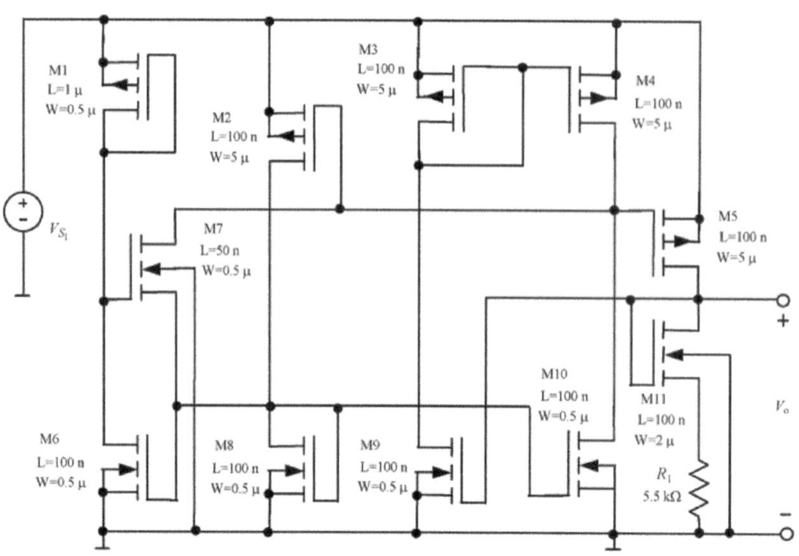

Figure 6: Voltage reference circuit.

PT	VT
$V_{S_1} = 0.45\ V,$	$V_{S_1} = 0.40\ V,$
$V_{S_1} = 0.60\ V,$	$V_{S_1} = 0.50\ V,$
$V_{S_1} = 0.70\ V,$	$V_{S_1} = 0.80\ V.$

The results of various multiple soft fault diagnoses are summarized in Tables 6, 7, 8, 9. They comprise three sets of the simultaneous variations of all the channel lengths in NMOS transistors (Table 6), in PMOS transistors (Table 7), and three sets of the simultaneous variations of all the oxide thicknesses in NMOS transistors (Table 8) and the PMOS transistors (Table 9). In all the cases the obtained values of the parameters are very close to their actual values.

Table 6: Results of the fault diagnosis of the channel lengths in NMOS transistors M9, M10, M11

Actual values of the parameters	Values of the parameters determined by the method	Number of generated simplices	Computation time in seconds
$\Delta L_{M9} = 15.00$ nm	$\Delta L_{M9} = 15.00$ nm	90	31.4
$\Delta L_{M10} = 3.00$ nm	$\Delta L_{M10} = 3.00$ nm		
$\Delta L_{M11} = 15.00$ nm	$\Delta L_{M11} = 15.00$ nm		
$\Delta L_{M9} = 5.00$ nm	$\Delta L_{M9} = 4.98$ nm	85	29.6
$\Delta L_{M10} = 2.00$ nm	$\Delta L_{M10} = 1.99$ nm		
$\Delta L_{M11} = 7.50$ nm	$\Delta L_{M11} = 7.53$ nm		
$\Delta L_{M9} = 1.75$ nm	$\Delta L_{M9} = 1.76$ nm	67	23.2
$\Delta L_{M10} = 10.00$ nm	$\Delta L_{M10} = 10.00$ nm		
$\Delta L_{M11} = -1.50$ nm	$\Delta L_{M11} = -1.51$ nm		

Table 7: Results of the fault diagnosis of the channel lengths in PMOS transistors M3, M4, M5

Actual values of the parameters	Values of the parameters determined by the method	Number of generated simplices	Computation time in seconds
$\Delta L_{M3} = 2.00$ nm	$\Delta L_{M3} = 2.00$ nm	94	32.2
$\Delta L_{M4} = 6.00$ nm	$\Delta L_{M4} = 6.00$ nm		
$\Delta L_{M5} = 6.00$ nm	$\Delta L_{M5} = 6.00$ nm		
$\Delta L_{M3} = 17.5$ nm	$\Delta L_{M3} = 17.5$ nm	92	30.8
$\Delta L_{M4} = 1.00$ nm	$\Delta L_{M4} = 0.98$ nm		
$\Delta L_{M5} = 1.50$ nm	$\Delta L_{M5} = 1.50$ nm		
$\Delta L_{M3} = 15.00$ nm	$\Delta L_{M3} = 14.98$ nm	79	26.6
$\Delta L_{M4} = 15.00$ nm	$\Delta L_{M4} = 14.99$ nm		
$\Delta L_{M5} = -3.00$ nm	$\Delta L_{M5} = -3.00$ nm		

Table 8: Results of the fault diagnosis of the oxide thicknesses in NMOS transistors M9, M10, M11

Actual values of the parameters	Values of the parameters determined by the method	Number of generated simplices	Computation time in seconds
$\Delta TOX_{M9} = -140.0$ pm	$\Delta TOX_{M9} = -140.0$ pm	113	38.9
$\Delta TOX_{M10} = -56.0$ pm	$\Delta TOX_{M10} = -55.9$ pm		
$\Delta TOX_{M11} = -140.0$ pm	$\Delta TOX_{M11} = -140.0$ pm		
$\Delta TOX_{M9} = 490.0$ pm	$\Delta TOX_{M9} = 489.7$ pm	136	44.6
$\Delta TOX_{M10} = -28.0$ pm	$\Delta TOX_{M10} = -31.0$ pm		
$\Delta TOX_{M11} = 42.0$ pm	$\Delta TOX_{M11} = 41.8$ pm		
$\Delta TOX_{M9} = 70.0$ pm	$\Delta TOX_{M9} = 69.8$ pm	91	32.5
$\Delta TOX_{M10} = 70.0$ pm	$\Delta TOX_{M10} = 69.6$ pm		
$\Delta TOX_{M11} = -42.0$ pm	$\Delta TOX_{M11} = -42.0$ pm		

Table 9: Results of the fault diagnosis of the oxide thicknesses in PMOS transistors M3, M4, M5

Actual values of the parameters	Values of the parameters determined by the method	Number of generated simplices	Computation time in seconds
$\Delta TOX_{M3} = -49.0\,\text{pm}$	$\Delta TOX_{M3} = -48.5\,\text{pm}$	155	51.8
$\Delta TOX_{M4} = 98.0\,\text{pm}$	$\Delta TOX_{M4} = 98.5\,\text{pm}$		
$\Delta TOX_{M5} = 112.0\,\text{pm}$	$\Delta TOX_{M5} = 112.0\,\text{pm}$		
$\Delta TOX_{M3} = 350.0\,\text{pm}$	$\Delta TOX_{M3} = 350.6\,\text{pm}$	191	60.6
$\Delta TOX_{M4} = 28.0\,\text{pm}$	$\Delta TOX_{M4} = 28.5\,\text{pm}$		
$\Delta TOX_{M5} = -42.0\,\text{pm}$	$\Delta TOX_{M5} = -42.0\,\text{pm}$		
$\Delta TOX_{M3} = -49.0\,\text{pm}$	$\Delta TOX_{M3} = -47.9\,\text{pm}$	149	48.6
$\Delta TOX_{M4} = 210.0\,\text{pm}$	$\Delta TOX_{M4} = 211.1\,\text{pm}$		
$\Delta TOX_{M5} = 210.0\,\text{pm}$	$\Delta TOX_{M5} = 210.0\,\text{pm}$		

Example 3

Let us consider the rail–to–rail input buffer [27] shown in Fig. 7. To arrange the diagnostic tests the input voltages V_{S_1}, V_{S_2} and V_{S_3} as well as the output voltage V_o are selected. The constants of the computation process are the same as in Example 1, R_o=100kΩ. Sensitivity analyses of the output voltage due to variations of the parameters LL and TOXTOX of the transistors reveal that all the transistors can be tested.

The following sets of the values of the input voltages have been chosen.

$$\begin{array}{ll}
\text{PT} & \text{VT} \\
V_{S_1} = 1.00\text{ V } V_{S_2} = 0.50\text{ V } V_{S_3} = 0.35\text{ V}, & V_{S_1} = 1.00\text{ V } V_{S_2} = 0.30\text{ V } V_{S_3} = 0.25\text{ V}, \\
V_{S_1} = 1.00\text{ V } V_{S_2} = 0.50\text{ V } V_{S_3} = 0.42\text{ V}, & V_{S_1} = 1.00\text{ V } V_{S_2} = 0.40\text{ V } V_{S_3} = 0.40\text{ V}, \\
V_{S_1} = 1.00\text{ V } V_{S_2} = 0.50\text{ V } V_{S_3} = 0.45\text{ V}, & V_{S_1} = 1.00\text{ V } V_{S_2} = 0.50\text{ V } V_{S_3} = 0.45\text{ V}, \\
V_{S_1} = 1.30\text{ V } V_{S_2} = 0.80\text{ V } V_{S_3} = 0.70\text{ V}, & V_{S_1} = 0.80\text{ V } V_{S_2} = 0.40\text{ V } V_{S_3} = 0.40\text{ V}, \\
V_{S_1} = 1.30\text{ V } V_{S_2} = 0.80\text{ V } V_{S_3} = 0.75\text{ V}, & V_{S_1} = 0.80\text{ V } V_{S_2} = 0.50\text{ V } V_{S_3} = 0.50\text{ V}.
\end{array}$$

Some results of multiple soft fault diagnosis are summarized in Tables 10, 11. The obtained values of the parameters are very close to their actual values.

Table 10: Results of the fault diagnosis of the channel lengths in NMOS and PMOS transistors

Actual values of the parameters	Values of the parameters determined by the method	Number of generated simplices	Computation time in seconds
$\Delta L_{M6} = 8.00\,\text{nm}$	$\Delta L_{M6} = 7.98\,\text{nm}$	544	170.6
$\Delta L_{M7} = -2.00\,\text{nm}$	$\Delta L_{M7} = -2.02\,\text{nm}$		
$\Delta L_{M8} = 1.00\,\text{nm}$	$\Delta L_{M8} = 0.99\,\text{nm}$		
$\Delta L_{M9} = 8.00\,\text{nm}$	$\Delta L_{M9} = 8.00\,\text{nm}$		
$\Delta L_{M10} = 6.00\,\text{nm}$	$\Delta L_{M10} = 5.99\,\text{nm}$		
$\Delta L_{M1} = 15.00\,\text{nm}$	$\Delta L_{M1} = 15.03\,\text{nm}$	482	146.8
$\Delta L_{M2} = 3.00\,\text{nm}$	$\Delta L_{M2} = 3.00\,\text{nm}$		
$\Delta L_{M3} = 15.00\,\text{nm}$	$\Delta L_{M3} = 15.00\,\text{nm}$		
$\Delta L_{M4} = 3.00\,\text{nm}$	$\Delta L_{M4} = 3.00\,\text{nm}$		
$\Delta L_{M5} = 2.00\,\text{nm}$	$\Delta L_{M5} = 2.00\,\text{nm}$		

Table 11: Results of the fault diagnosis of the oxide thicknesses in NMOS and PMOS transistors

Actual values of the parameters	Values of the parameters determined by the method	Number of generated simplices	Computation time in seconds
$\Delta TOX_{M6} = 42.0\,\text{pm}$	$\Delta TOX_{M6} = 42.1\,\text{pm}$	1316	408.1
$\Delta TOX_{M7} = 168.0\,\text{pm}$	$\Delta TOX_{M7} = 168.1\,\text{pm}$		
$\Delta TOX_{M8} = -42.0\,\text{pm}$	$\Delta TOX_{M8} = -42.0\,\text{pm}$		
$\Delta TOX_{M9} = 168.0\,\text{pm}$	$\Delta TOX_{M9} = 168.0\,\text{pm}$		
$\Delta TOX_{M10} = 42.0\,\text{pm}$	$\Delta TOX_{M10} = 42.0\,\text{pm}$		
$\Delta TOX_{M1} = 28.0\,\text{pm}$	$\Delta TOX_{M1} = 27.9\,\text{pm}$	530	181.3
$\Delta TOX_{M2} = -14.0\,\text{pm}$	$\Delta TOX_{M2} = -14.0\,\text{pm}$		
$\Delta TOX_{M3} = 98.0\,\text{pm}$	$\Delta TOX_{M3} = 98.0\,\text{pm}$		
$\Delta TOX_{M4} = 56.0\,\text{pm}$	$\Delta TOX_{M4} = 56.0\,\text{pm}$		
$\Delta TOX_{M5} = -28.0\,\text{pm}$	$\Delta TOX_{M5} = -28.0\,\text{pm}$		

CONCLUSION

The method developed in this paper allows effective diagnosing multiple soft faults of the process parameters in small and middle–size ICs designed in nanometer technology. The method does not require access to internal nodes of the circuit. The set of the parameter values is obtained by solving nonlinear equations, not given in explicit analytical form, that may have more than one solution. The proposed approach, whose core is the Nelder–Mead optimization method, is capable of finding the multiple solutions and select the actual one. Numerical examples reveal that the accuracy of the determined parameter values is very good. The method does not allow testing the transistors whose parameters have slight influence on the output voltage. This is why the fault diagnosis process should be preceded by the sensitivity analyses.

The proposed method has been compared with the methods recently published in references [18] and [20] devoted to similar problem. They will be named M2016, M2014 and M2015, respectively.

At first we concentrate on the methods M2016 and M2014. To perform the comparison, method M2014 has been implemented to allow diagnosing CMOS circuits designed in nanometer technology, comprising the transistors characterized by the model BSIM 4.6. All 30 fault diagnoses included in Tables 1, 2, 3, 4, 5, 6, 7, 8, 9, 10, 11 providing the results found by method M2016, have been performed using method M2014 with the same diagnostic tests and the assumed measurement accuracy and the parameters ϕϕ, ξξ, kmaxkmax as in [18] (Example 1). On the basis of these experiments the following conclusion can be drawn. Method M2014 is very sensitive to the tests and in numerous cases several tries must be taken to select the proper one. Moreover, sometimes this method requires more than one measurement node. In the great majority of cases method M2014 is more time consuming than method M2016. To be specific, let us consider in detail Example 1. For the diagnoses included in Tables 1, 2, 3 the results given by method M2014 are very similar to the ones provided by method M2016 (see Tables 12, 13,14). In all cases, except one, method M2014 consumes more CPU time which is 1.48–6.64 times longer. For the cases presented in Tables 4, 5 method M2014 fails. However, it is possible to arrange another tests so that the method works and gives correct results. Unfortunately, the tests relating to the diagnoses indicated in Table 5 requires access to two measurement nodes.

Table 12: Results of the fault diagnosis of the channel lengths in NMOS transistors M4, M5, M8, M9 given by method M2014

Actual values of the parameters	Values of the parameters determined by the method	Computation time in seconds
$\Delta L_{M4} = 6.00$ nm	$\Delta L_{M4} = 5.99$ nm	178.2
$\Delta L_{M5} = 4.00$ nm	$\Delta L_{M5} = 3.99$ nm	
$\Delta L_{M8} = 2.50$ nm	$\Delta L_{M8} = 2.50$ nm	
$\Delta L_{M9} = 6.00$ nm	$\Delta L_{M9} = 6.02$ nm	
$\Delta L_{M4} = 15.00$ nm	$\Delta L_{M4} = 15.02$ nm	180.3
$\Delta L_{M5} = 2.00$ nm	$\Delta L_{M5} = 1.98$ nm	
$\Delta L_{M8} = 15.00$ nm	$\Delta L_{M8} = 15.04$ nm	
$\Delta L_{M9} = 4.50$ nm	$\Delta L_{M9} = 4.53$ nm	
$\Delta L_{M4} = 2.50$ nm	$\Delta L_{M4} = 2.47$ nm	183.8
$\Delta L_{M5} = 2.00$ nm	$\Delta L_{M5} = 1.96$ nm	
$\Delta L_{M8} = 1.75$ nm	$\Delta L_{M8} = 1.76$ nm	
$\Delta L_{M9} = 20.00$ nm	$\Delta L_{M9} = 20.06$ nm	

Table 13: Results of the fault diagnosis of the channel lengths in PMOS transistors M1, M2, M3 given by method M2014

Actual values of the parameters	Values of the parameters determined by the method	Computation time in seconds
$\Delta L_{M1} = 4.50$ nm	$\Delta L_{M1} = 4.57$ nm	172.7
$\Delta L_{M2} = 3.00$ nm	$\Delta L_{M2} = 3.05$ nm	
$\Delta L_{M3} = 1.00$ nm	$\Delta L_{M3} = 1.01$ nm	
$\Delta L_{M1} = 15.00$ nm	$\Delta L_{M1} = 15.01$ nm	178.4
$\Delta L_{M2} = 10.00$ nm	$\Delta L_{M2} = 9.99$ nm	
$\Delta L_{M3} = 2.50$ nm	$\Delta L_{M3} = 2.51$ nm	
$\Delta L_{M1} = 1.50$ nm	$\Delta L_{M1} = 1.46$ nm	171.2
$\Delta L_{M2} = 25.00$ nm	$\Delta L_{M2} = 25.03$ nm	
$\Delta L_{M3} = 1.00$ nm	$\Delta L_{M3} = 1.00$ nm	

Table 14: Results of the fault diagnosis of the oxide thicknesses in NMOS transistors M4, M5, M8, M9 given by method M2014

Actual values of the parameters	Values of the parameters determined by the method	Computation time in seconds
$\Delta TOX_{M4} = 140.0$ pm	$\Delta TOX_{M4} = 141.3$ pm	190.2
$\Delta TOX_{M5} = -56.0$ pm	$\Delta TOX_{M5} = -55.6$ pm	
$\Delta TOX_{M8} = -28.0$ pm	$\Delta TOX_{M8} = -26.9$ pm	
$\Delta TOX_{M9} = 98.0$ pm	$\Delta TOX_{M9} = 97.9$ pm	
$\Delta TOX_{M4} = 280.0$ pm	$\Delta TOX_{M4} = 281.4$ pm	158.6
$\Delta TOX_{M5} = -56.0$ pm	$\Delta TOX_{M5} = -59.2$ pm	
$\Delta TOX_{M8} = 280.0$ pm	$\Delta TOX_{M8} = 282.7$ pm	
$\Delta TOX_{M9} = -42.0$ pm	$\Delta TOX_{M9} = -40.7$ pm	
$\Delta TOX_{M4} = 70.0$ pm	$\Delta TOX_{M4} = 70.8$ pm	179.5
$\Delta TOX_{M5} = -56.0$ pm	$\Delta TOX_{M5} = -56.0$ pm	
$\Delta TOX_{M8} = -49.0$ pm	$\Delta TOX_{M8} = -48.1$ pm	
$\Delta TOX_{M9} = 560.0$ pm	$\Delta TOX_{M9} = 559.1$ pm	

Method M2015 comprises very large class of circuits, including bipolar and CMOS ones, designed in micrometer and submicrometer technologies. In Ref. [20] the MOS transistors designed in nanometer technology are characterized by PSP103.1 model introduced into DELPHI environment. Thanks to this the circuit analyses are executed using a dedicated program written in DELPHI, that considerably improves the computation process and makes the method sound. To compare with M2016 the method M2015 was implemented including BSIM 4.6 modeled MOS transistors and circuit analyses were performed in IsSPICE 4. Method M2015 requires numerous sensitivity analyses that in this case must be performed using the brute–force incremental approach. In consequence this method needs large computing power and consumes much CPU time. Thus, method M2016 is less universal than M2015 but, in the case of CMOS circuits with BSIM 4.6 modeled transistors, faster and easier to implement. To be specific, all the cases presented in Tables 1, 2, 3, 4 were retaken using method M2015 with h=0.05h=0.05. The method gives the results very close to the ones provided by M2016, but the CPU time is 7–45 times longer.

REFERENCES

1. Aminian, M., & Aminian, F. (2007). A modular fault-diagnostic system for analog electronic circuits using neural networks with wavelet transform as a preprocessor. *IEEE Transactions on Instrumentation and Measurement, 56*, 1546–1554.
2. Bhunia, S., Raychowdhury, A., & Roy, K. (2005). Defect oriented testing of analog circuits using wavelet analysis of dynamic supply current. *Journal of Electronic Testing, 21*, 147–159.
3. Catelani, M., & Fort, A. (2002). Soft fault detection and isolation in analog circuits: some results and a comparison between a fuzzy approach and radial basis function networks.*IEEE Transactions on Instrumentation and Measurement, 51*, 196–202.
4. Economikos, L., Morrison, T., & Crnic, F. (1994). Electrical test of multichip substrates.*IEEE Transactions on Components, Packaging, and Manufacturing Technology, Part B: Advanced Packaging, 17*, 56–61.
5. Fedi, G., Giomi, R., Luchetta, A., Manetti, S., & Piccirilli, M. C. (1998). On the application of symbolic techniques to the multiple fault location in low testability analog circuits. *IEEE Transactions on Circuits and Systems II: Analog and Digital Signal Processing, 45*, 1383–1388.
6. Gizopoulos, D. (Ed.). (2006). *Advances in electronic testing: Challenges and methodologies*. Dordrecht: Springer.
7. Huang, K., Stratigopoulos, H. G., Mir, S., Hora, C., Xing, Y., & Kruseman, B. (2012). Diagnosis of local spot defect in analog circuits. *IEEE Transactions on Instrumentation and Measurement, 61*, 2701–2712.
8. Jahangiri, M., & Razaghian, F. (2014). Fault detection in analogue circuits using hybrid evolutionary algorithm and neural network. *Analog Integrated Circuits and Signal Processing, 80*(3), 551–556.
9. Kabisatpathy, P., Barua, A., & Sinha, S. (2005). *Fault diagnosis of analog integrated circuits*. Dordrecht: Springer.
10. Kim, B., Swaminathan, M., Chatterjee, A., & Schimmel, D. (1997). A novel test technique for MCM substrates. *IEEE Transactions on Components, Packaging, and Manufacturing Technology, Part B: Advanced Packaging, 20*, 2–12.
11. Papakostas, D. K., & Hatzopoulos, A. A. (2008). A unified procedure for fault detection of analog and mixed-mode circuits using magnitude and phase components of the power supply current spectrum. *IEEE Transactions on Instrumentation and Measurement, 57*, 2589–2995.
12. Provost, B., & Sanchez-Sinencio, E. (2003). On-chip ramp generators for mixed-signal BIST and ADC self-test. *IEEE Journal of Solid-State Circuits, 38*(2), 263–273.
13. Robotycki, A., & Zielonko, R. (2002). Fault diagnosis of analog piecewise linear circuits based on homotopy. *IEEE Transactions on Instrumentation and Measurement, 51*, 876–881.
14. Segura, J., Keshavarzi, A, Soden, J, & Hawkins, C. (2002). Parametric failures in CMOS ICs—a defect-based analysis. In *Proceedings of the International Test Conference* (pp.90–99).
15. Sunter, S., & Nagi, N. (1999). Test metrics for analog parametric faults. In *Proceedings ofthe 17th IEEE VLSI Test Symposium* (pp.226–234).

16. Tadeusiewicz, M., Hałgas, S., & Korzybski, M. (2002). An algorithm for soft-fault diagnosis of linear and nonlinear circuits. *IEEE Transactions on Circuits and Systems I: Fundamental Theory and Applications, 49*, 1648–1653.
17. Tadeusiewicz, M., & Hałgas, S. (2012). Multiple soft fault diagnosis of nonlinear circuits using the continuation method. *Journal of Electronic Testing: Theory and Applications,28*, 487–493.
18. Tadeusiewicz, M., & Hałgas, S. (2014). Global and local parametric diagnosis of analog short-channel CMOS circuit using homotopy-simplicial algorithm. *International Journal of Circuit Theory and Applications, 42*, 1051–1068.
19. Tadeusiewicz, M., & Hałgas, S. (2014). Multiple soft fault diagnosis of BJT circuit.*Metrology and Measurement Systems, 21*, 663–674.
20. Tadeusiewicz, M., & Hałgas, S. (2015). A new approach to multiple soft fault diagnosis of analog BJT and CMOS circuits. *IEEE Transactions on Instrumentation and Measurement,64*, 2688–2695.
21. Tadeusiewicz, M., Kuczyński, A., & Hałgas, S. (2015). Spot defect diagnosis in analog nonlinear circuits with possible multiple operating points. *Journal of Electronic Testing: Theory and Applications, 31*, 491–502.
22. Wang, Z., Gielen, G., & Sansen, W. (1998). Probabilistic fault detection and the selection of measurements for analog integrated circuits. *IEEE Transactions on Computer-Aided Design of Integrated Circuits and Systems, 17*(9), 862–872.
23. Singer, S., & Nelder, J. (2009). Nelder–Mead algorithm. *Scholarpedia, 4*(7), 2928.
24. Luersen, M.A., & Le Riche, R. (2004). Globalized Nelder–Mead method for engineering optimization. *Computers & Structures, 82*, 2251–2260.
25. Gao, F., & Han, L. (2012). Implementing the Nelder–Mead simplex algorithm with adaptive parameters. *Computational Optimization and Applications., 51*, 259–277.
26. IsSPICE4 User's Guide, vol.1,2, rev. 04/08 Build 3247, Intusoft 2008.
27. Baker, R. J. (2010). *CMOS circuit design, layout, and simulation*. New York: Wiley-IEEE Press.

CHAPTER 2

HIGH-ACCURACY FUNCTION SYNTHESIZER CIRCUIT WITH APPLICATIONS IN SIGNAL PROCESSING

Cosmin Popa

Keywords: Signal synthesizing, Approximation error, Exponential function generator, Current-mode operation, Continuous mathematical function, CMOS analog designs

INTRODUCTION

Analog signal processing represents an important area of analog integrated circuits analysis and design, with a multitude of applications in many domains. Multiplier and exponential circuits are useful in telecommunication circuits[1–4], medical equipments[5, 6], hearing devices[7,8], or disk drives[9, 10]. Squaring circuits represent the core for implementing any continuous function, using the limited Taylor series expansion. The Euclidean distance function is very important in instrumentation circuits[11, 12], communication[1, 2], neural networks[13, 14], display systems[15, 16], or classification algorithms[17], being also useful for vector quantization or nearest neighbor classification[18, 19].

The main problem in designing analog signal processing structures is how to implement with minimal effort a large number of nonlinear mathematical functions[3]. The requirements for an analog signal processing structure are mainly related to the circuit accuracy, to the possibility of achieving a multitude of circuit functions with reasonable design costs and to the controllability of the implemented function. In this context, analog processing performed in focal-plane Vision Systems-on-Chip[20] can represent an interesting choice. Important functions from the perspective of their applications are multiplying/dividing[21–30], exponential[31–34], squaring/square-rooting[30, 35–41], or Euclidean distance[42, 43] functions.

In bipolar technology, the multiplying/dividing function can easily be obtained from the logarithmical characteristic of the bipolar transistor. Important errors still remain because of the nonzero values of the base currents and of the temperature dependence of the bipolar transistor parameters (the thermal voltage is linearly increasing with temperature and the saturation current has an exponential dependence on temperature). In order to achieve a low-power operation of complementary metal oxide semiconductor (CMOS) designs, the subthreshold biasing of metal oxide semiconductor (MOS) transistors is an interesting choice. Based on the logarithmical law of a MOS transistor in weak inversion, the implementation of a CMOS Multiplier/Divider circuit becomes very simple (even with respect to the bipolar version). In consequence, the result will be smaller silicon area consumption, the circuit being also compatible with low-power very large-scale integration (VLSI) designs. For obtaining an important increasing of the circuit frequency response, the multiplying/dividing function can be achieved by employing the square law model of MOS transistors biased in saturation.

The exponential function is available in bipolar technology using the exponential characteristic of the bipolar transistor. In CMOS technology, the exponential law can directly be implemented exploiting the weak inversion operation of the MOS transistor. In order to obtain the exponential function using the squaring characteristic of the MOS transistor in saturation (for improving the circuit frequency response), the classical method is to approximate the exponential function with its n th-order expansion (the polynomial series). The approximation error will be proportional with the number of terms neglected in the expansion.

There are many possibilities of implementing squaring and square-rooting functions using the quadratic characteristic of the MOS transistor biased in saturation. The main goals of this class of circuits are the silicon-occupied area, the independence of the output current on the technological parameters and on temperature and the small sensitivity to the multitude of second-order effects.

The classical approach in analog signal processing circuits is to implement, for each circuit function, a class of

Citation: Cosmin Popa, High-accuracy function synthesizer circuit with applications in signal processing, doi:10.1186/1687-6180-2012-146.

computational structures. The proposed method for obtaining a multitude of continuous mathematical functions is to design a complex computational structure, named function synthesizer circuit that is able to generate these functions approximating them using an original superior-order approximation function. The accuracy of computation is correlated with the order of approximation. In conclusion, a tradeoff between the complexity of the structure and its intrinsic accuracy must be made. The advantages of the proposed method are mainly related to the possibility of strongly reduce the power consumption and circuit complexity per implemented function.

It exists in literature a relatively small number of function synthesizer circuits implemented in CMOS technology[4, 33, 44–46], these applications being dedicated to the realization of a limited number of mathematical functions. In[44], the approximation of the implemented function can be obtained by adding the weighted output currents of a number of basic building blocks, built around a basic current squarer, and a constant current, the circuit presenting the disadvantage of a relatively large complexity. The circuit proposed in[45] is based on approximating the required function using the first three terms of its Taylor series expansion. The approximations can be implemented by adding the output currents of a weighted current square, a weighted current amplifier (or attenuator), and a constant current. The errors are mainly caused by the small value (two) of the approximation order and, in consequence, they are relatively large. Additionally, from the same reason, the range of the input signal is strongly restricted. A structure for synthesis of analog exponential functions, based on approximating the exponential function using rational functions, is proposed in[46]. The circuit presents the important limitations of realizing only the exponential function and of an implementation in bipolar technology.

The article is multidisciplinary, starting from a rigorous mathematical analysis and continuing with original implementations of the proposed electronic computational circuits. A new implementation of a current-mode function synthesizer circuit with an extended capability of generating continuous mathematical functions will be presented. The proposed structure is based on a fourth-order original approximation function, having the important advantages of an improved accuracy and of reconfiguration capability.

MAIN TEXT

Fundamental Methodology

In order to implement an improved accuracy analog function synthesizer circuit, the proposed method is to use a general continuous approximation function, $g(x)$, having a Taylor series that can be made to fourth-order match a multitude of continuous mathematical functions, $f(x)$.

$$g(x) = \frac{1 + a_0 x}{1 + a_1 x} + \frac{a_2 x}{1 + x} + a_3 x + a_4 \quad (1)$$

$a_0 \dots a_4$ represent constant coefficients, having values imposed by the necessity that the Taylor series of $g(x)$ function to be identical (in a fourth-order approximation) with the Taylor series of $f(x)$ function, that can be expressed by the following polynomial relation:

$$f(x) = m + nx + px^2 + qx^3 + rx^4 + sx^5 + tx^6 + \dots \quad (2)$$

$m \div t$ represents constant coefficients of the expansion, depending on the expression of $f(x)$ function that must be implemented.

The motivation for choosing this particular $g(x)$ approximation function is correlated with the possibilities of its facile implementation in CMOS technology (the multiplying/dividing function has a relatively simple realization). In comparison with this original approximation function, the Pade approximation[24] (having also good accuracy) presents a much more complicated implementation in CMOS technology, requiring squaring and superior-order computational circuits that strongly increase the overall complexity of the function synthesizer. The requirements for an extremely accuracy of the function synthesizer circuit impose a relatively high order of approximation for $g(x)$ function. The increasing of the order of approximation strongly increases the complexity of the designed circuit, being necessary to realize a tradeoff between the circuit complexity and its overall accuracy. From this point of view, an optimal choice that allows to obtain a very good accuracy and a relative large dynamic range of the function synthesizer using a reasonable circuit complexity is based on a fourth-order approximation.

The fourth-order identity between the Taylor series of $f(x)$ and $g(x)$ functions is equivalent with the identity between the first five terms from the Taylor series of the previous functions, being evident the reason of choosing five coefficients $a_0 \dots a_4$ for defining $g(x)$ function. It results

$$a_0 = \frac{(p+q)^3}{(r+q)(r+p+2q)} - \frac{r+q}{p+q} \quad (3)$$

$$a_1 = -\frac{r+q}{p+q} \quad (4)$$

$$a_2 = \frac{(p+q)^2}{r+p+2q} - p \quad (5)$$

$$a_3 = n + p - \frac{(p+q)^2}{r+q} \quad (6)$$

$$a_4 = m - 1 \quad (7)$$

The approximation error will have the following general expression:

$$\epsilon_{f(x)}^{g(x)}(x) \cong \left[\frac{q^2 - 2rp - qr}{p+q} - s\right] \frac{x^5}{f(x)} \quad (8)$$

In order to further reduce the approximation error of the synthesizer circuit, it is possible to increase the order of approximation. For example, a possible fifth-order approximation function that requires a reasonable complexity of CMOS implementation can be generally expressed as follows:

$$g'(x) = \frac{b_0 x}{1 + b_1 x} + \frac{b_2 x}{1 + x} + b_3 x^2 + b_4 x + b_5 \quad (9)$$

$b_0 \dots b_4$ represent constant coefficients, having values imposed by the necessity that the Taylor series of $g'(x)$ function to be identical (in a fifth-order approximation) with the Taylor series (2) of $f(x)$ function. The additional complexity of the circuit that implements $g'(x)$ function comparing with the computation structure required by $g'(x)$ is represented by a current-mode squaring circuit (for obtaining the $b_3 x^2$ term).

Current-mode implementation of the function synthesizer circuit

Block diagram of the Function Synthesizer with Fourth-Order Approximation

The block diagram of the function synthesizer, based on the original proposed approximation function (1), is presented in Figure1. The "Multiplier/Divider" circuits are current-mode structures [30, 47, 48], having the new implementation and the description of their operation presented in the following paragraph. The expressions of I_{OUTa} and I_{OUTb} currents are

$$I_{OUTa} = \frac{I_O I_{1a}}{I_{2a}} \tag{10}$$

and

$$I_{OUTb} = \frac{I_O I_{1b}}{I_{2b}} \tag{11}$$

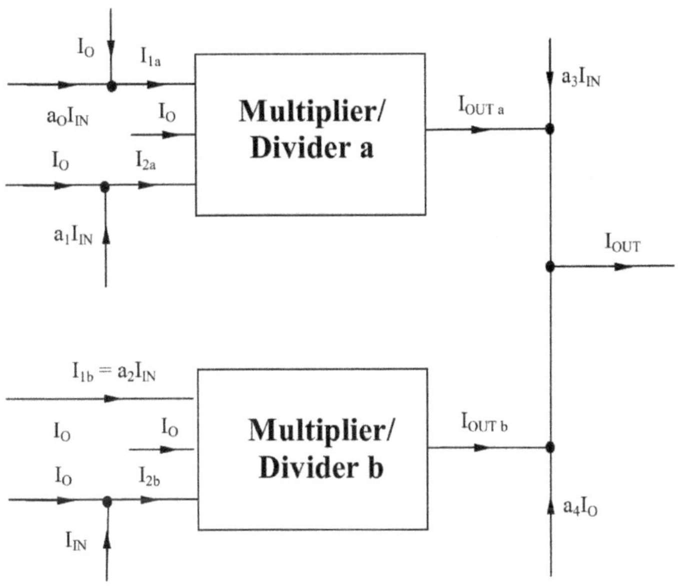

Figure 1: The block diagram of the function synthesizer with fourth-order approximation.

The output current of the circuit having the block diagram presented in Figure1 will have the following expression:

$$I_{OUT} = I_{OUTa} + I_{OUTb} + a_3 I_{IN} + a_4 I_O \tag{12}$$

Using the notation $x = I_{IN}/I_O$ and relation (1), it results that I_{OUT} current represents the fourth-order approximation of $f(x)$ function:

$$I_{OUT} = I_O\left(\frac{1 + a_O x}{1 + a_1 x} + \frac{a_2 x}{1 + x} + a_3 x + a_4\right) = I_O g_2(x) \cong I_O f(x) \tag{13}$$

The implementation of the "X^2" block

The functional core for designing the "Multiplier/Divider" circuit is represented by a current-mode squaring circuit, having the original implementation presented in Figure2. The aspect ratios of MOS transistors are $1\mu/0.54\mu$ for M1–M4 transistors and $0.8\mu/0.54\mu$ for M5–M7 transistors.

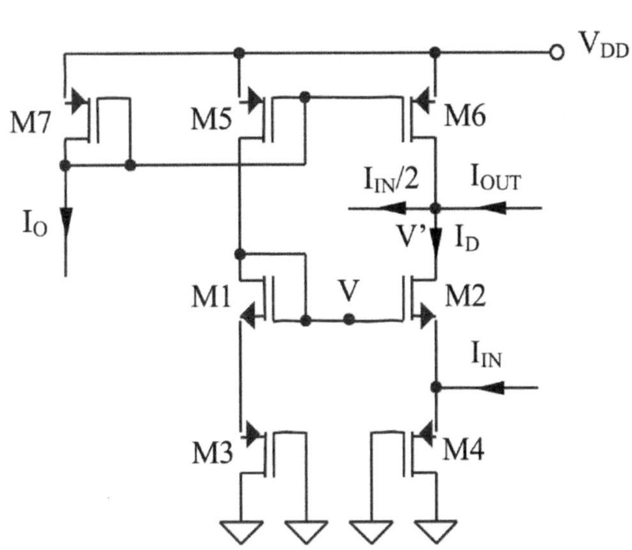

Figure 2: The implementation of the "X^2" block.

Noting with $V_{GS}(I)$ the absolute value of the gate–source voltage of a MOS transistor biased at a drain current equal with I the equation of the translinear loop can be expressed as follows:

$$2V_{GS}(I_O) = V_{GS}(I_D) + V_{GS}(I_D + I_{IN}) \tag{14}$$

resulting

$$2\sqrt{I_O} = \sqrt{I_D} + \sqrt{I_D + I_{IN}} \tag{15}$$

Reducing the radicals by squaring the terms and simplifying, it results

$$4I_O - 2I_D - I_{IN} = 2\sqrt{I_D(I_D + I_{IN})} \tag{16}$$

equivalent with:

$$16I_O^2 + I_{IN}^2 - 16I_D I_O - 8I_O I_{IN} = 0 \tag{17}$$

So

$$I_D = I_O - \frac{I_{IN}}{2} + \frac{I_{IN}^2}{16I_O} \tag{18}$$

The expression of the output current will be

$$I_{OUT} = I_D + \frac{I_{IN}}{2} - I_O = \frac{I_{IN}^2}{16I_O} \tag{19}$$

The previous relations can be used only for strong saturated devices and for not too small devices. In order to increase the accuracy of the computations, quasi-identical source–drain voltages for M1–M2 and M3–M4 transistors, respectively, must be imposed. The source–drain voltage of M3 transistor can be expressed as follows

$$V_{SD3} = V - V_{GS1} \cong V - V_{GS2} = V_{SD4} \tag{20}$$

because, considering $I_{IN} << I_O$, M1 and M2 transistors are identical and biased at drain currents that differ only with a small amount. Additionally, for biasing M1 and M2 transistors at approximately equal drain–source voltages, V' potential must be imposed, from the I_{OUT} external terminal, to be equal with V potential. This particular biasing of M1 and M2 transistors will impose quasi-identical source–drain voltages also for M5 and M6 transistors, this fact increasing the accuracy of M5–M6 current mirror.

The M3–M7 transistors are not affected by the substrate effect, as their bulks are connected to their sources terminals. The M1 and M2 transistors have the bulk connected to the ground, so their non-zero bulk-source voltage will be responsible for small errors introduced by the substrate effect. As a result, the expression of I_{OUT} current will slightly be affected by undesired dependencies on technological parameters. The impact of these errors on the overall accuracy of the proposed function synthesizer circuit is small and they can be compensated using specific design techniques. The biasing current of the squaring circuit from Figure2 is approximately $200\mu A$.

The Implementation of the "Multiplier/Divider" Block

The original proposed "Multiplier/Divider" circuit is presented in Figure3a, being designed using two current-mode squaring circuits, similar with the structures presented in Figure2 (M1–M4 and M3–M6, respectively). The full CMOS implementation of the "Multiplier/Divider" circuit is shown in Figure3b. The complexity of the "Multiplier/Divider" circuit is minimized by re-using M3–M4 transistors for both squaring circuits. The aspect ratios of MOS transistors are $1\mu/0.54\mu$ for M1–M6 transistors and $0.8\mu/0.54\mu$ for M7–M8 transistors. The output current has the following expression:

$$I_{OUT} = I_{D1} - I_{D2} + 2I_O \tag{21}$$

where two translinear loops similar with the loop analyzed for the "X^2" circuit implement the following expressions of I_{D1} and I_{D2} currents (derived from (19)):

$$I_{D1} = I_2 - (I_1 + I_O) + \frac{(I_1 + I_O)^2}{4I_2} \tag{22}$$

and

$$I_{D2} = I_2 - (I_1 - I_O) + \frac{(I_1 - I_O)^2}{4I_2} \tag{23}$$

resulting an output current of the circuit having the following expression:

$$I_{OUT} = I_O \frac{I_1}{I_2} \tag{24}$$

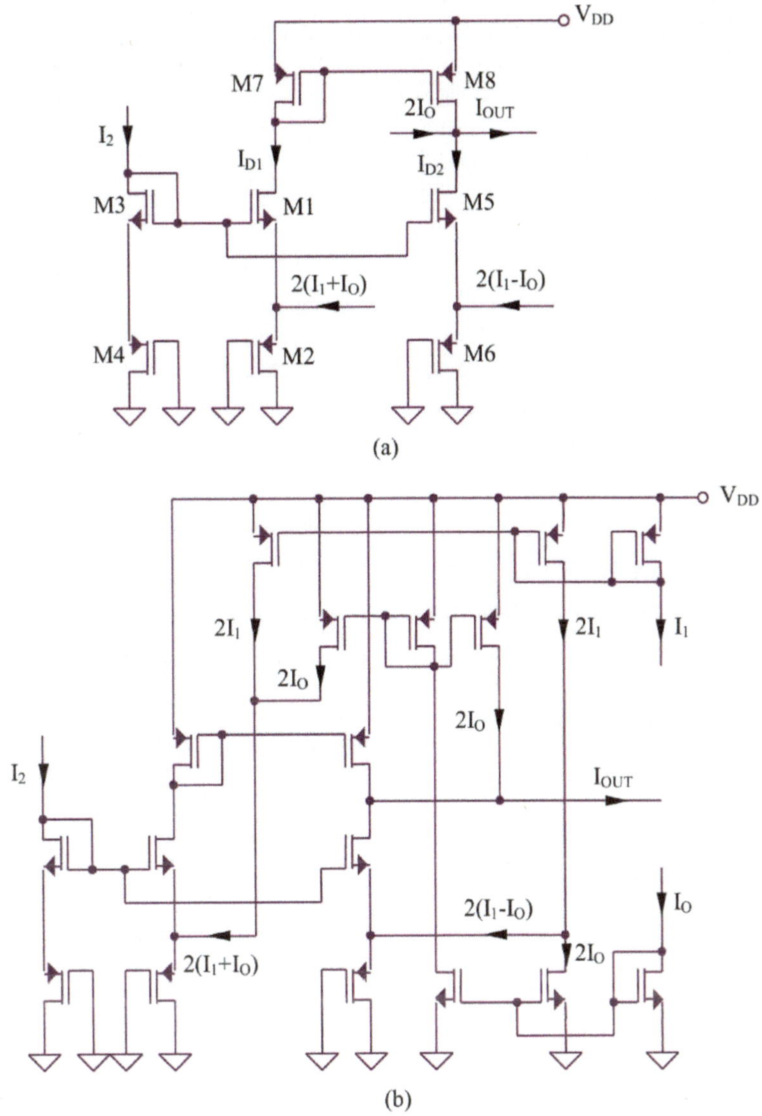

Figure 3: (a) The core of the Multiplier/Divider circuit. (b) The implementation of the Multiplier/Divider circuit.

The original proposed Multiplier/Divider circuit presents two important advantages. First, the current-mode operation increases the frequency response of the computational structure. Second, the independence of the output current expressed by (24) on technological parameters removes, in a first-order analysis, the temperature- and technological-caused errors, the overall accuracy of the proposed circuit being increased in this way. Additionally, the resulted concrete implementation of the function synthesizer circuit based on the block diagram proposed in Figure4 has a smaller complexity comparing with previously reported similar computational structures.

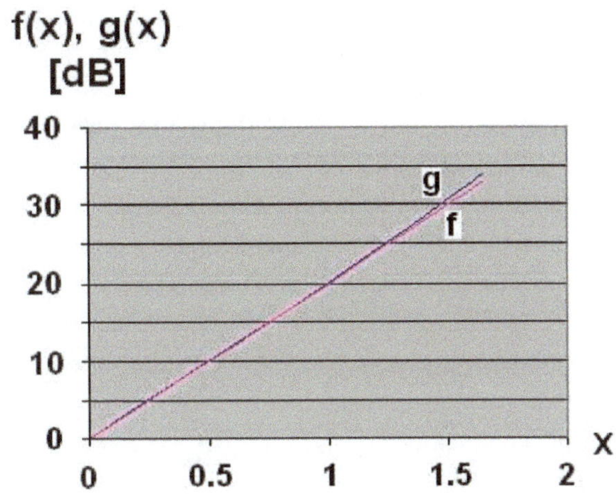

Figure 4: Graphical comparison betweeng(x)approximation function and$f$(x)=exp(x).

Using MOS transistors implemented in $0.18\mu m$ CMOS technology, the maximal frequency of operation of the proposed function synthesizer circuit is approximately hundreds *kHz* - *MHz*, depending on the particular model of MOS active devices. The biasing current of the "Multiplier/Divider" circuit is approximately $300\mu A$, while the maximal biasing current of the function synthesizer circuit is smaller than $800\mu A$.

ERROR MECHANISMS FOR FUNCTION SYNTHESIZER CIRCUIT

Real circuits are affected by a multitude of errors[3, 4] that slightly affects their overall accuracy. The most important errors that must be taken into account for evaluating the function synthesizer accuracy are presented in the following paragraphs.

The deviation of the MOS Transistor Characteristic from the Square-Law, Bulk Effect, Leakage

The saturated MOS transistor squaring characteristic is affected by the second-order effects: mobility degradation (25), channel-length modulation (26), and bulk effect (27):

$$K = \frac{K_O}{[1+\theta_G(V_{GS}-V_T)](1+\theta_D V_{DS})} \tag{25}$$

$$I_D = \frac{K}{2}(V_{GS}-V_T)^2(1+\lambda V_{DS}) \tag{26}$$

$$V_T = V_{TO} + \gamma\left(\sqrt{\Phi - V_{BS}} - \sqrt{\Phi}\right) \tag{27}$$

The errors introduced by the bulk effect[3, 4] can be reduced by proper designs that avoid the dependence of the circuit parameters on the threshold voltage. The design of circuits for obtaining a zero bulk-source voltage cancels out the errors introduced by the bulk effect. Additional errors produced by the second-order effects are given by the dependence of K transconductance parameter on V_{GS} voltage. Taking into account only the $K(V_{GS})$ dependence, a small changing of $K(V_{GS})$ voltage comparing with the analysis based on the first-order model of MOS transistors can be determined:

$$V_{GS} = V_T + \sqrt{\frac{2I_D}{K} + \frac{\theta_G}{K} I_D} \tag{28}$$

For latest CMOS VLSI nanometer designs, the leakage[3, 4] becomes more important, the leakage current depending on the properties of the layout and also on the device structure. The determination of the total leakage cannot be made by adding individual leakage currents, because of a correlation that exists between leakage currents.

The subthreshold leakage is the current produced by minority electrons flowing through p substrate from source to drain, being modeled by an exponential function:

$$I_{SL} = I_{DO}\frac{W}{L}\exp\left(\frac{V_{GS}-V_T}{nV_{TH}}\right)\left[1-\exp\left(-\frac{V_{DS}}{V_{TH}}\right)\right] \tag{29}$$

W/L is the aspect ratio of the MOS transistor, V_{TH} is the thermal voltage, n is a parameter, and I_{DO} current is an additional parameter. The subthreshold-off current (obtained for $V_{GS}=0$ and $V_{DS}>>V_T$) can be expressed as follows:

$$I_{SLoff} = I_{DO}\frac{W}{L}\exp\left(\frac{-V_T}{nV_{TH}}\right) \tag{30}$$

The gate leakage current can be carried by tunneling electrons or holes, the carriers leaking to source, drain, and channel. For junction leakage, low currents are carried by minority carriers drifting across the junction, by the electron–hole generation in junction or by the impact ionization at high reverse bias, the electrons being able to pass the barrier by tunneling through it—Band-To-Band Tunneling current[3, 4].

Mismatches in Current Mirrors

Current mirrors implemented using real circuits present some errors caused by the mismatches between the composing MOS transistors and also by the channel-length modulation:

$$\frac{I_O}{I_{REF}} = \frac{(W/L)_O}{(W/L)_{REF}}\frac{1+\lambda V_{DSO}}{1+\lambda V_{DS\,REF}} \tag{31}$$

The method for reducing the errors introduced by the channel-length modulation is to impose to the current mirror an output voltage that set $V_{DSO} \cong V_{DSREF}$.

Different threshold voltages and K_n/K_p constants for NMOS/PMOS devices

Real circuits present differences between the threshold voltages and K_n/K_p constants for NMOS and PMOS transistors. In order to avoid additional inaccuracies introduced by the practical situation, translinear loops that represent the functional basis of many computational circuits must contain only NMOS or PMOS active devices or must be implemented using quasi-symmetrical structures from the point of view of this complementarily (the same number of NMOS and PMOS transistors).

Variations over the Process, Temperature, and Supply Voltage

Most of MOS transistors parameters are affected by the process in which the computational circuits are implemented. On the other side, technological parameters present important temperature dependencies. From this perspective, in order to avoid important errors introduced by process and temperature, the practical realizations of analog computational circuits must be done in such a way that minimizes the number of technological parameters in the expression of the output signal. Additionally, the errors caused by the supply voltage variations can be minimized using self-biased cascode configurations. In this context, a tradeoff between power supply rejection ratio and minimal supply voltage has to be considered.

APPLICATIONS

An original exponential function generator circuit can be designed using the fourth-order approximation function (1), replacing in general relations (3)–(7) the particular values of constant coefficients of expansion (2): $m=1$, $n=1$, $p=1/2$, $q=1/6$ and $r=1/24$. The $g(x)$ function that fourth-order approximates the exponential function can be expressed as follows:

$$g(x) = \frac{16+21x}{16-5x} + \frac{1}{126}\frac{1}{1+x} - \frac{7x}{11} \tag{32}$$

Using (8), the approximation error of the exponential function generator will have the following general expression:

$$\epsilon_{\exp(x)}^{g(x)}(x) \cong \frac{x^5}{120\exp(x)} \qquad (33)$$

A comparison between $g(x)$ approximation function and $f(x)=\exp(x)$ is shown in Table1.

Table 1: Comparison betweeng(x) approximation function and$f(x)=\exp(x)$

x	g(x) (dB)	f(x) (dB)	? (dB)
0	0	0	0
0.2	4.075	4	0.075
0.4	8.029	8	0.029
0.6	12.011	12	0.011
0.8	16.028	16	0.028
1.0	20.095	20	0.095
1.2	24.239	24	0.239
1.4	28.495	28	0.495
1.6	32.914	32	0.914
1.65	34.001	33	1.001

As a result of using a fourth-order approximation function, the dynamic range of the proposed exponential function generator is approximately 33 dB, for an error smaller than 1 dB.

SIMULATIONS

The proposed function synthesizer circuit is designed for implementing in 0.18μm generic CMOS technology, the simulations being based on the square-law SPICE LEVEL 3 model. The parasitic and noise effects have been neglected in the performed analysis, but it could be estimated that their impact to the overall accuracy of the proposed function synthesizer circuit are relatively small comparing with other causes of errors (approximation error or second-order effects). The overall accuracy of the proposed fourth-order function synthesizer circuit is 0.3% for an extended range of the input signal.

SPICE simulation $I_{OUT}(I_2)$ for the Multiplier/Divider circuit proposed in Figure3 is presented in Figure5. The I_O and I_1 currents have the following values: $I_O=50\mu A$ and $I_1=20\mu A$, while the range of I_2 current was chosen to be between $30\mu A$ and $100\mu A$.

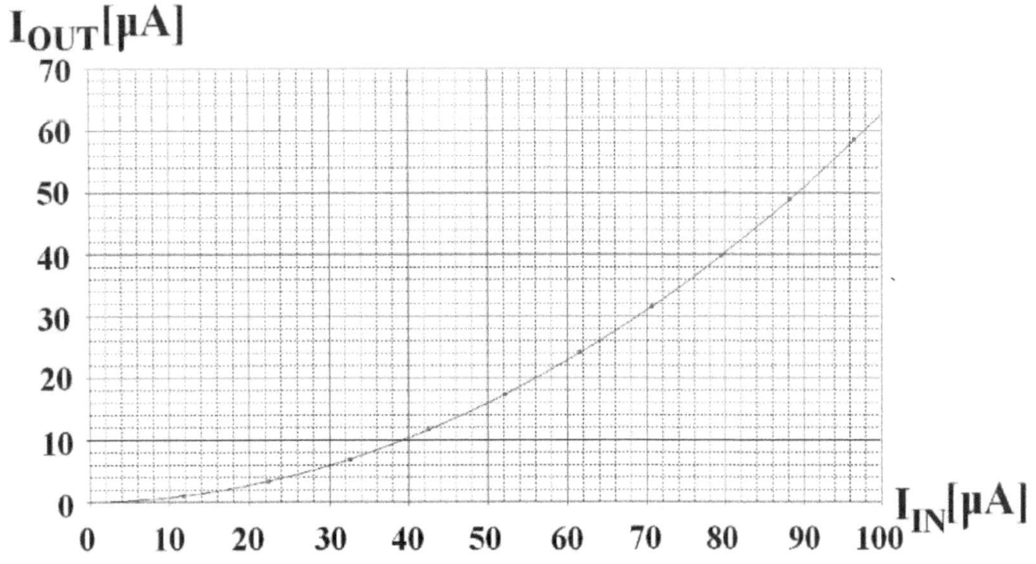

Figure 5: SPICE simulation$I_{OUT}(I_{IN})$ for the squaring circuit.

A comparison between the simulated and the theoretical estimated results for the proposed current squarer is shown in Figure6.

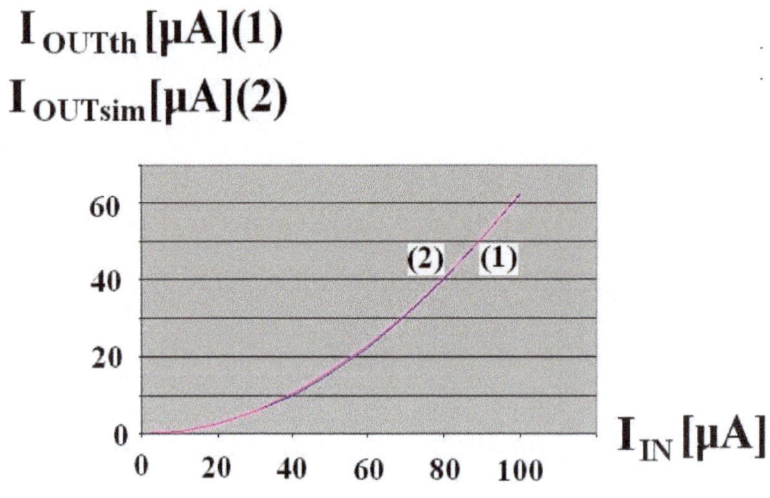

Figure 6: Comparison between the simulated and the theoretical estimated results for the current squarer.

SPICE simulation $I_{OUT}(I_1)$ for the Multiplier/Divider circuit proposed in Figure3 is presented in Figure7. The I_O and I_2 currents have the following values: $I_O=10\mu A$ and $I_2=100\mu A$, while the range of I_1 current was chosen to be between 0 and $100\mu A$.

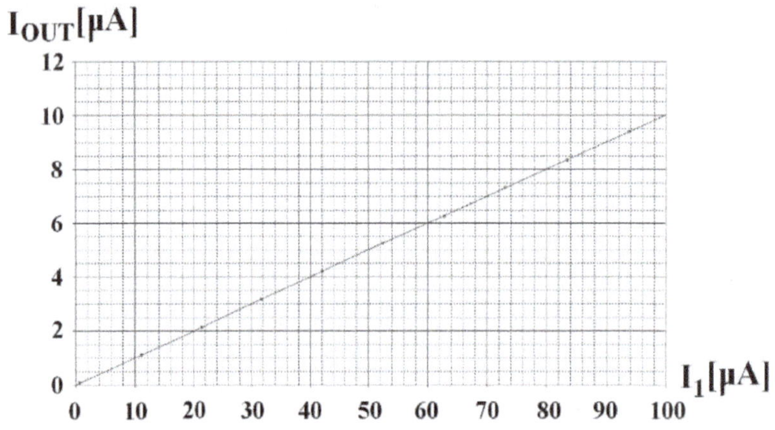

Figure 7: SPICE simulation $I_{OUT}(I_1)$ for the Multiplier/Divider circuit.

SPICE simulation $I_{OUT}(I_2)$ for the Multiplier/Divider circuit proposed in Figure3 is presented in Figure8. The I_O and I_1 currents have the following values: $I_O=50\mu A$ and $I_1=20\mu A$, while the range of I_2 current was chosen to be between $30\mu A$ and $100\mu A$.

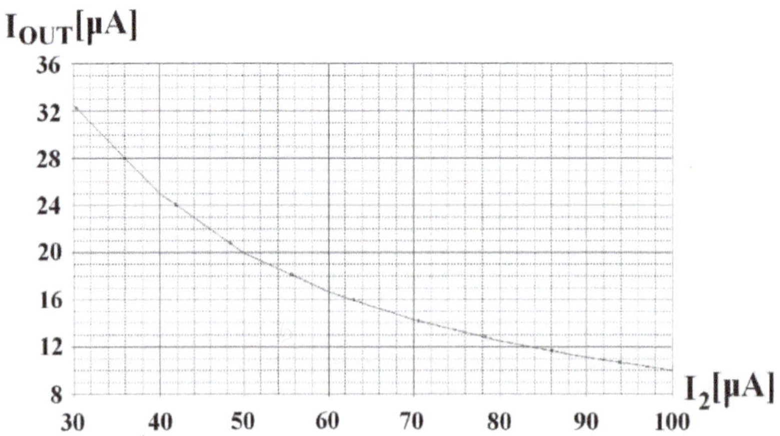

Figure 8: SPICE simulation $I_{OUT}(I_2)$ for the Multiplier/Divider circuit.

A comparison between the simulated and the theoretical estimated results for the proposed current Multiplier/Divider circuit is shown in Figure9. The values of I_O and I_1 currents are $I_O=50\mu A$ and $I_1=20\mu A$, while I_2 current has values between $30\mu A$ and $100\mu A$.

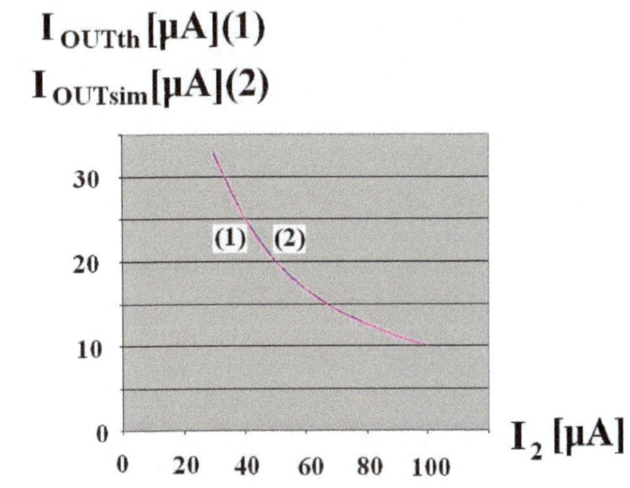

Figure 9: Comparison between the simulated and the theoretical estimated results for the proposed current Multiplier/Divider circuit.

As a result of the particular architecture proposed for the realization of the function synthesizer structure, it presents a low-voltage operation (a minimal supply voltage of *IV* for an implementation in the mentioned technology). This value of the minimal supply voltage was obtained using the theoretical analysis of the proposed function synthesizer circuit ($V_{DD\ \min}=2V_{SG}+V_{DSsat}$) and also by performing simulations based on the square-law SPICE LEVEL 3 model.

METHODS

The necessity of implementing a multitude of continuous mathematical functions in CMOS technology has been solved by the original proposed function synthesizer circuit. A very important advantage of this computational structure is mainly related to its large capability of generating continuous mathematical functions: exponential, multiplying/dividing, or squaring/square-rooting functions. The original operating method was based on a new fourth-order approximation function, having a superior-order polynomial series that match the polynomial series of the approximated function. The current-mode operation and the independence of the circuit performances on the technological errors are responsible for an additional improvement of structure accuracy. The proposed function synthesizer circuit allows a relatively simple implementation in CMOS technology using only two current-mode Multiplier/Divider circuits.

The designed function synthesizer circuit has the advantages of increased modularity and controllability and of minimal design costs per implemented mathematical function. The function synthesizer is designed for implementing in $0.18\mu m$ CMOS technology and it is supplied at *IV*. SPICE simulations confirm the theoretical estimated results, showing an accuracy of 0.3% for an extended range of the input signal. This accuracy is exclusively referring to the precision of generating the previous functions, not to the additional circuits that deserve the proposed function synthesizer structure.

The speed of the original proposed function generator circuit is correlated with its overall accuracy, expecting the necessity of making a tradeoff between the previous performance parameters. Additionally, especially for extreme small values of the approximation error, the noise of the circuit can slightly degrade the circuit performances.

As an immediate application of the function synthesizer circuit, it was proposed a new exponential function generator based on a particular fourth-order approximation function. The circuit has a dynamic range of approximately 33 dB, in the conditions of limiting the error to 1 dB.

ACKNOWLEDGMENTS

The author thanks Professor Maurits Ortmanns from Institute of Microelectronics, University of Ulm for reviewing the article and for helpful discussions, the collaboration with him bringing many improvements to the article.

Competing Interests

The author declares that they have no competing interests.

REFERENCES

1. Rittenhouse G, Goyal S, Neilson DT, Samuel S: *Sustainable telecommunications*. Technical Symposium at ITU Telecom World, Geneva, Switzerland; 2011:19-23.
2. Altiparmak F, Dengiz B, Smith AAE: General neural network model for estimating telecommunications network reliability. *IEEE Trans. Reliab.* 2009, 58(1):2-9.
3. Popa C: *Superior-Order Curvature-Correction Techniques for Voltage References*. Springer, New York; 2009.
4. Popa C: *Synthesis of Computational Structures for Analog Signal Processing*. Springer, New York; 2011
5. Du Preez CC, Sinha S, du Plessis M: CMOS ECG, EEG and EMG waveform bio-simulator.*International Semiconductor Conference* 2006, 1: 27-29. September 2006, Sinaia, Romania, pp. 29–38
6. Pini F, McCarthy K: *Capacitive instrumentation amplifier for low-power bio potential signal detection*. Paper presented at Signals and Systems Conference, Cork, Ireland; 2010:54-58. 23–24 June 2010
7. Azzolini C, Ricciardi A, Boi A: *Very low-cost CMOS audio amplifier for 1-V portable applications*. Paper presented at 3rd International Conference on Design and Technology of Integrated Systems in Nanoscale Era, Tozeur, Tunisia; 2008:25-27. pp. 1–4
8. C Azzolini, A Boni, A 1-V CMOS audio amplifier for low cost hearing aids, in Paper presented at 15th IEEE International Conference on Electronics, Circuits and Systems, 31 August – 3: *St.* (Julian's, Malta; September 2008):562-565.
9. Wei DC, Sun DQ, Abidi AA: A 300-MHz fixed-delay tree search-DFE analog CMOS disk-drive read channel. *IEEE J. Solid State Circuits* 2001, 36(11):1795-1807. 10.1109/4.962303
10. Ndjountche T, Fa-Long L, Bobda C: *A CMOS front-end architecture for hard-disk drive read-channel equalizer*. Paper presented at IEEE International Symposium on Circuits and Systems, Kobe, Japan; 2005:2184-2187. vol. 3, 23–26 May

11. Hwang-Cherng C, Jia-Yu W: *High CMRR instrumentation amplifier for biomedical applications*. Paper presented at 9th International Symposium on Signal Processing and Its Applications, 12–15 February 2007, Sharjah, United Arab Emirates; 2007:1-4.
12. Haider MR, Islam SK, Mostafa S, Mo Z, Taeho O: Low-power low-voltage current readout circuit for inductively powered implant system. *IEEE Trans. Biomed. Circuits Syst.* 2010, 4(4):205-213.
13. Dlugosz R, Talaska T, Pedrycz W, Wojtyna R: Realization of the conscience mechanism in CMOS implementation of winner-takes-all self-organizing neural networks. *IEEE Trans. Neural Netw.* 2010, 21(6):961-971.
14. Ebong IE, Mazumder P: CMOS and memristor-based neural network design for position detection. *Proc. IEEE* 2011, 99: 1-11.
15. Ming-Dou K: *ESD-aware circuit design in CMOS integrated circuits to meet system-level ESD specification in microelectronic systems*. Paper presented at International Conference of Electron Devices and Solid-State Circuits, Tianjin, China; 2011:1-2. 17–18 November 2011
16. Seung-Chul L, Young-Deuk J, Jong-Kee K, Jongdae K: A 10-bit 205-MS/s 1.0-mm2 90-nm CMOS pipeline ADC for flat panel display applications. *IEEE J. Solid State Circuits* 2007, 42(12):2688-2695.
17. Sudha N, Sridharan K: A high-speed VLSI design and ASIC implementation for constructing Euclidean distance-based discrete Voronoi diagram. *IEEE Trans. Robot. Autom.* 2004, 20(2):352-358. 10.1109/TRA.2004.824638
18. Villmann T, Schleif FM, Hammer B: *Fuzzy Labeled Soft Nearest Neighbor Classification with Relevance Learning*. Paper presented at Fourth International Conference on Machine Learning and Applications, Los Angeles, USA; 2005:11-15. 15–17 December 2005
19. Triguero I, Derrac J, Garcia S, Herrera F: A taxonomy and experimental study on prototype generation for nearest neighbor classification. *IEEE Trans. Syst. Man Cybern. C: Appl. Rev.* 2012, 42(1):86-100.
20. Rodríguez-Vázquez A, Domínguez-Castro R, Jiménez-Garrido F, Morillas S, García-Ortiz A, Utrera C, Pardo MD, Listan J, Romay R: *Cellular Nanoscale Sensory Wave Computing—A CMOS Vision System On-Chip with Multi-Core, Cellular Sensory-Processing Front-End*. Edited by: Baatar C, Porod W, Roska T. Springer, New York; 2010:129-146.
21. Sawigun C, Mahattanakul J: *A 1.5 V, wide-input range, high-bandwidth, CMOS four-quadrant analog multiplier*. IEEE International Symposium on Circuits and Systems, Washington, USA; 2008:2318-2321. 18–21 May 2008
22. Akshatha BC, Akshintala VK: *Low voltage, low power, high linearity, high speed CMOS voltage mode analog multiplier*. Paper presented at 2nd International Conference on Emerging Trends in Engineering and Technology, Nagpur, India; 2009:149-154. 16–18 December 2009
23. Mahmoud SA: *Low voltage low power wide range fully differential CMOS four-quadrant analog multiplier*. Paper presented at 52nd IEEE International Midwest Symposium on Circuits and Systems, Cancun, Mexico; 2009:130-133. 2–5 August 2009
24. Naderi A, Khoei A, Hadidi K: *High speed, low power four-quadrant CMOS current-mode multiplier*. Paper presented at 14th IEEE International Conference on Electronics, Circuits and Systems, Marrakech, Morocco; 2007:1311. 11–14 December 2007
25. Gravati M, Valle M, Ferri G, Guerrini N, Reyes N: *A novel current-mode very low power analog CMOS four quadrant multiplier*. Paper presented at 31st European Solid-State Circuits Conference, Grenoble, France; 2005:495-498. 12–16 September 2005
26. Sawigun C, Serdijn WA: Ultra-low-power, class-AB. *CMOS four-quadrant current multiplier. Electron. Lett.* 2009, 45(10):483-484.
27. Hidayat R, Dejhan K, Moungnoul P, Miyanaga Y: *OTA-based high frequency CMOS multiplier and squaring circuit*. Paper presented at International Symposium on Intelligent Signal Processing and Communications Systems, Bangkok, Thailand; 2009:1-4. 8–10 Ferbruary 2009
28. Machowski W, Kuta S, Jasielski J, Kolodziejski W: *Quarter-square analog four-quadrant multiplier based on CMOS invertes and using low voltage high speed control circuits*. Paper presented at 17th International Conference Mixed Design of Integrated Circuits and Systems, Wroclaw, Poland; 2010:333-336. 24–26 June 2010
29. Ehsanpour M, Moallem P, Vafaei A: *Design of a novel reversible multiplier circuit using modified full adde*. Paper presented at International Conference on Computer Design and Applications, Qinhuangdao, China; 2010:V3-230-V3-234. vol. 3, 25–27 June 2010
30. Naderi A, Mojarrad H, Ghasemzadeh H, Khoei A, Hadidi K: *Four-quadrant CMOS analog multiplier based on new current squarer circuit with high-speed*. IEEE International Conference on "Computer As A Tool, Saint Petersburg, Russia; 2009:282-287. 18–23 May 2009
31. Kao CH, Tseng CC, Hsieh CS: Low-voltage exponential function converter. *IEE Proc. Circuits Dev. Syst.* 2005, 152(5):485-487. 10.1049/ip-cds:20045110
32. Ethier S, Sawan M: Exponential Current pulse generation for efficient very high-impedance multisite stimulation. *IEEE Trans. Biomed. Circuits Syst* 2010, 99: 1-9.
33. Hedayati H, Bakkaloglu B: *A 3 GHz wideband* Σ *fractional-N synthesizer with voltage-mode exponential CP-PFD*. Paper presented at IEEE Radio Frequency Integrated Circuits Symposium, Boston, USA; 2009:325-328. 7–9 June 2009
34. Moro-Frias D, Sanz-Pascual MT, de la Cruz-Bias CA: *Linear-in-dB Variable Gain Amplifier with PWL exponential gain control*. Paper presented at IEEE International Symposium on Circuits and Systems, Paris, France; 2010:2824-2827. 2010

35. CA De La Cruz Blas, O Feely, Limit cycle behavior in a class-AB second-order square root domain filter, in Paper presented at 15th IEEE International Conference on Electronics, Circuits and Systems, 31 August – 3: *St.* (Julian's, Malta; September 2008):117-120.

36. Boonchu B, Surakampontom W: *A CMOS current-mode squarer/rectifier circuit*. Paper presented at International Symposium on Circuits and Systems, Bangkok, Thailand; 2003:I-405-I-408. vol. 1, 25–28 May 2003

37. CA De La Blas, A Lopez, A novel two quadrant MOS translinear Squarer-divider cell, in Paper presented at 15th IEEE International Conference on Electronics, Circuits and Systems, 31 August – 3 Sptember: *St.* (Julian's, Malta; 2008):5-8.

38. De La Cruz-Blas CA, Lopez-Martin AJ, Carlosena A: 1.5-V square-root domain second-order filter with on-chip tuning. *IEEE Trans. Circuits Syst. I: Regular Papers* 2005, 52(10):1996-2006.

39. Raikos G, Vlassis S: *Low-voltage CMOS voltage squarer*. Paper presented at 16th IEEE International Conference on Electronics, Circuits, and Systems, Yasmine Hammamet, Tunisia; 2009:159-162. 13–16 December 2009

40. Garofalo V, Coppola M, De Caro D, Napoli E, Petra N, Strollo AGM: *A novel truncated squarer with linear compensation function*. Paper presented at IEEE International Symposium on Circuits and Systems, Paris, France; 2010:4157-4160. 30 May – 2 June 2010

41. Kircay A, Keserlioglu MS: *Novel current-mode second-order square-root-domain highpass and allpass filter*. Paper presented at International Conference on Electrical and Electronics Engineering, Bursa, Turkey; 2009:242-246. 5–8 November 2009

42. Popa C: *A New FGMOST Euclidean Distance Computational Circuits Based on Algebraic Mean of the Input Potentials*. Springer, Berlin; 2009:459-466. Lecture Notes in Computer Science

43. Hyo-Jin A, Chang-Seok C, Hanho L: *High-speed low-complexity Folded Degree-Computationless Modified Euclidean algorithm architecture for RS decoders*. Paper presented at 12th International Symposium on Integrated Circuits, Singapore, Singapore; 2009:582-585. 14–16 December 2009

44. Abuelma'atti MT, Al-Yahia NM: *An improved universal CMOS current-mode analog function synthesizer*. Paper presented at International Symposium on Integrated Circuits, Singapore; 2007:580-583. 26–30 September 2007

45. Abuelma'atti MT: Universal CMOS current-mode analog function synthesizer. *IEEE Trans. Circuits Syst. I: Fund. Theory Appl* 2002, 49(10):1468-1474. 10.1109/TCSI.2002.803356

46. Abuelma'atti MT: *A translinear current-mode programmable analog exponential function synthesizer*. Paper presented at Eleventh International Conference on Microelectronics, Kuwait; 1999:209-212. 22–24 November 1999

47. Psychalinos C, Vlassis S: A systematic design procedure for square-root-domain circuits based on the signal flow graph approach. *IEEE Trans. Circuits Syst. I: Fund. Theory Appl.*2002, 49(12):1702-1712. 10.1109/TCSI.2002.805695

48. Baker GA, Graves-Morris P: *Padé Approximants (Cambridge University Press*. Cambridge, MA; 1996.

CHAPTER 3 A NEW CURRENT-CONTROLLED-POWER TECHNIQUE FOR SMALL SIGNAL APPLICATIONS

Adisak Monpapassorn

Keywords: Current-Controlled-Power; Full-Wave Rectifier; Analog Signal Processing

INTRODUCTION

As is well known, it is hard to find the device that operates as a silicon-controlled rectifier (SCR) [1] for controlling the power in small signal applications. In this paper, the author presents a new current-controlledtechnique for small signal applications. The proposed technique yields the advantages:

- The proposed technique needs no a capacitor and a resistor but the well-known SCR technique needs; therefore, without passive devices, the proposed technique uses a much smaller chip area in the IC process [2].
- The proposed technique works with a small signal but the SCR technique cannot work because of the threshold voltage of the SCR [1].
- The proposed technique can also work with a large signal by driving a power bipolar transistor.
- The proposed technique with current-controlled is very suitable for an automatic control system by the current feedback. Whereas the SCR technique uses the RC time constant for clipping time [1], it is hard for the automatic control system.
- The example application as a new current-controlledpower CMOS full-wave rectifier is also presented. It provides the advantages as follows.
- The new full-wave rectifier works both full-wave rectification and current-controlled-power operation. But the previous proposed full-wave rectifiers cannot control the output power.
- The new rectifier uses all MOSs with typical structure thus it is very suitable for IC processes [3].
- The new rectifier, its core operates in balanced mode resulting in a low noise output [4].
- Using the new rectifier, one can independently control the positive part and the negative part of the input signal, this yields more precision control.

PROPOSED CURRENT-CONTROLLED-POWER TECHNIQUE

The proposed technique is shown in **Figure 1**(a) and its signals are shown in **Figure 1**(b). The operation is as follows. I_{CL1} is fed to shift down the input differential current signal (I_{in+}). It will decrease the amplitude and area of the positive signal and will increase those of the negative signal. The bias source (V_b) is the voltage or current, for turning-on the current mirror (MN1 and MN2) all the time for very small signal operation. The V_b must be the minimum voltage or current to make the minimum offset at the output (I_{out}). The I_{out} is the positive signal rectified by the current mirror. Because the input signal is shifted down, one must compensate this for (I_{in-}) by feeding I_{CL2} for balancing the input signal.

Note the proposed technique that the output is a horizontal clipped signal but the well-known SCR technique the output is a vertical clipped signal. The proposed technique with all active devices is better than the SCR technique with the resistor and capacitor devices in view of IC fabrication. Additionally, the proposed technique with current-controlled is very suitable for automatic control system by a current feedback. Whereas the SCR technique uses the RC time constant for clipping time, it is hard for automatic control system. Moreover, the proposed technique can be used in high power applications; it can drive the high current bipolar transistor.

Citation: A. Monpapassorn, "A New Current-Controlled-Power Technique for Small Signal Applications," Circuits and Systems, Vol. 3 No. 2, 2012, pp. 161-165. doi: 10.4236/cs.2012.32021.

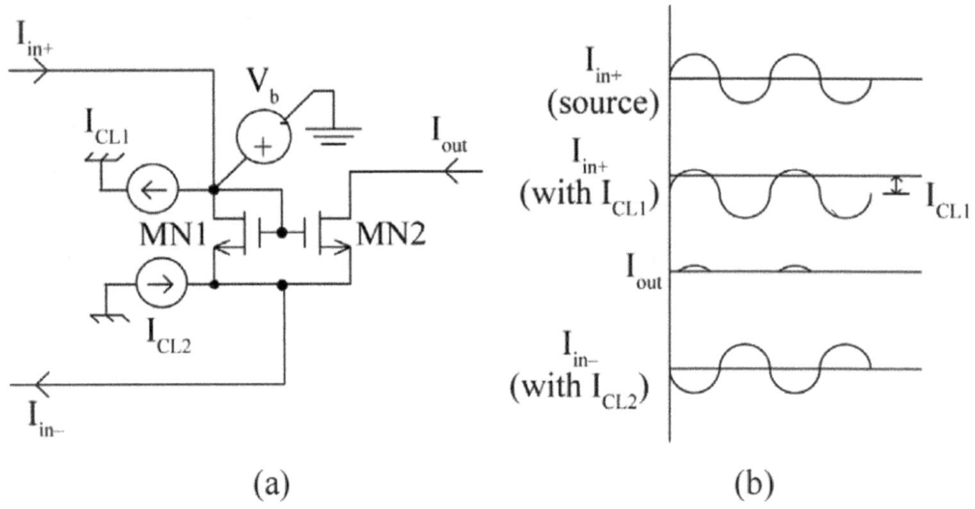

Figure 1. Proposed technique: (a) circuit and (b) operation.

APPLICATION EXAMPLE AS A NEW CURRENT-CONTROLLED-POWER FULL-WAVE RECTIFIER

The circuit in **Figure 2** operates as a full-wave rectifier where $I_{CL+1} = I_{CL+2} = I_{CL-1} = I_{CL-2} = 0$. Its operation is as follows. The constant current source I_C is mirrored by MI1 and MI3 to the drain of M3 (I_{D3}); and is mirrored by MI1, MI2, MI4 and MI5 to the drain of M1 (I_{D1}). The W/L ratios of MI1 to MI5 are equal, making $I_{D1} = I_{D3}$. Moreover, using matched MOS transistors M1 to M4 [5], the input voltage is thus followed to a MOS resistor (R_A) [6], the resistance of which is given by

$$R = \frac{1}{2KV_{DT}} \tag{1}$$

where MA1 and MA2 have the same characteristics; $K = mC_{ox}W/L$, m is the carrier mobility, C_{ox} is the gate capacitance per unit area, W and L are the channel width and length respectively; $V_{DT} = V_{DD} - V_T = -(V_{SS} + V_T)$, $V_{DD} = -V_{SS}$, V_T is the threshold voltage.

Equation (1) is valid when both MA1 and MA2 maintain in the saturation region, which is true if

$$|V_A| \leq V_{DT} \tag{2}$$

This input voltage creates the current I_{in} flowing through R_A.

$$I_{in} = V_{in} 2KV_{DT} \tag{3}$$

In **Figure 2**, I_{in} is mirrored by M5, M6, M9, and M10 as I_1; and is mirrored by M5, M7, M9, and M11 as $I_4 = -I_1$. Furthermore, I_{in} is mirrored by M5, M8, M9, M12, M13, M14, M16, and M17 as I_2; and is mirrored by M5, M8, M9, M12, M13, M15, M16, and M18 as I_3. One assumes that the gain of all current mirrors is unity, thus

$$\left.\begin{array}{l} I_1 = I_2 \\ I_3 = I_4 \end{array}\right\} \tag{4}$$

When the input voltage is positive, the current mirror (M19 and M20) turns on and the current mirror (M21 and M22) turns off. Inversely, when the input voltage is negative, the current mirror (M19 and M20) turns off and the current mirror (M21 and M22) turns on, resulting in

$$\left.\begin{array}{l} V_{in} > 0;\ I_1 = I_2 = I_{in};\ I_3 = I_4 = 0 \\ V_{in} < 0;\ I_1 = I_2 = 0;\ I_3 = I_4 = I_{in} \end{array}\right\} \tag{5}$$

The drain currents of M20 and M22 are finally mirrored by M23 and M24 and then are converted to the voltage by R_B. Using $R_A = R_B$ by setting MA1 = MA2 = MB1 = MB2, the relation between input and output voltages (full-wave rectification) can be written as

$$\left.\begin{array}{l} V_{in} > 0;\ V_{out} = V_{in} \\ V_{in} < 0;\ V_{out} = -V_{in} \end{array}\right\} \tag{6}$$

The bias sources (V_{b1} and V_{b2}) in **Figure 1** may be the voltage or current source. This bias source makes the MOSs in core 1 and core 2 (M19, M20, M21, and M22) turning-on all the time, to reduce the error at zero-crossing of the full-wave output signal.

Three main errors of the operation of the proposed rectifier can be considered. Assume that the error of W/L ratios of MOS transistors is ignored since it is very small in the present technology.

The first error is for transferring an input voltage to node A [5] because the currents I_{D1} and $I_{D3}$are slid resulting from the error of simple current mirrors (MI1 to MI5). This error can be minimized by using the better current mirrors (cascode or Wilson type); however, it requires higher supply voltage [7].

The second error is for mirroring the input current from R_A to core 1 and core 2. And the last error is the error of core 1 and core 2. The second and the last errors can be reduced by using the better current mirrors as mentioned above.

The best way for setting three above errors to minimize, is compensation. The compensation can be done by adjusting the gain of core 1 and core 2 through the W/L ratios (of M20 for positive input and of M22 for negative input). In addition, adjusting the resistant of R_B through W/L ratios of MB1 and MB2 is also compensation.

M2 and M4 must be in the saturation mode as M1 and M3; this yields the input operation range,

$$\left.\begin{aligned} V_{in(min)} &= \frac{V_{SS}+|V_{eff4}|+|V_{eff9}|+|V_{TP}|+|V_{TN}|}{\alpha} \\ V_{in(max)} &= \frac{V_{DD}-|V_{eff2}|-|V_{eff5}|-|V_{TN}|-|V_{TP}|}{\alpha} \end{aligned}\right\} \quad (7)$$

where $V_{eff} = V_{GS} - V_T = \sqrt{\frac{2I_D}{\mu C_{OX}(W/L)}}$ [7], and α is the voltage gain between input and node A, ideally it must be unity.

In the same way, M6, M7, M8, M10, M11, M12, M14, M15, M17, and M18 have to be in the saturation mode. If these MOSs have the same parameters, one can write the operation range of signals at nodes B, C, D, and E as

$$\left.\begin{aligned} V_{B,C,D,E(min)} &= V_{SS}+|V_{effN}|+|V_{TN}| \\ V_{B,C,D,E(max)} &= V_{DD}-|V_{effP}|-|V_{TP}| \end{aligned}\right\} \quad (8)$$

Using the same characteristics M5 to M18, considering Equations (2), (7), and (8), the operation range of the proposed full-wave rectifier is considered as (7).

Note that two signals of core 1 at nodes B and D, two signals are 180 degrees out-of-phase; hence, some inphase outside noise around the core will be cancelled, as well-known. It is also for core 2. This indicates that the proposed rectifier operates in a balanced mode, canceling some outside noise around the core.

As mentioned above, if we feed the control current ($I_{CL+1} = I_{CL+2}$), we can control the peak area of the positive input signal (see **Figure 1**(b)). Moreover, by feeding $I_{CL-1} = I_{CL-2}$, we also can control the peak area of the negative input signal. It is evident that we can independently control the positive part and the negative part of the input signal, this yields more precision control.

At this point, we can see that one application example of the new current-controlled-power technique is the new full-wave rectifier. The new rectifier can be realized by using all MOS transistors with typical structure. It is very suitable for the process of IC fabrication. Furthermore, not only full-wave rectification but also current-controlledpower operation, the proposed rectifier can work. The previous proposed rectifiers cannot operate this currentcontrolled-power function. Additionally, the balanced mode operation of the core gains a lower noise output.

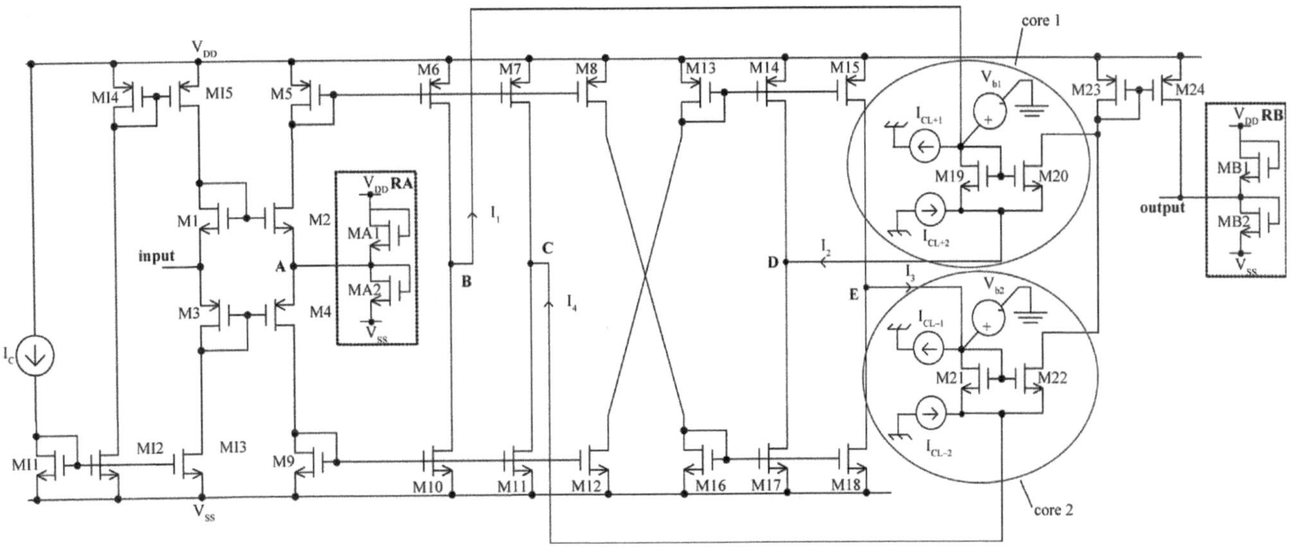

Figure 2. Application example as a new current-controlled-power full-wave rectifier.

RESULTS

To verify the theoretical design, the application example as a new current-controlled-power full-wave rectifier was simulated by using parameters extracted from its layout (including parasitic capacitance) in a 0.5 μm AMI MOS transistor technology, through a level-49 model.

The supply voltage is ±1.2 V. For full-wave rectifier simulation, I_C is 20 mA, $V_{b1} = V_{b2}$ are the current sources of 5 μA, $I_{CL+1} = I_{CL+2} = I_{CL-1} = I_{CL-2}$ are 0 μA. The W/L ratios of MA1 = MA2 = 1.5 μm/1.5 μm, MB1 = MB2 = 1.5 μm/3.9 μm, and other MOSs = 20 μm/0.6 μm, were chosen.

Figure 3 shows the output of the proposed full-wave rectifier with a sine wave input signal (100 mV_{peak}, 100 kHz). The amplitude error of the output signal is almost zero because of compensation by using $R_B > R_A$, by adjusting their W/L ratios as mentioned in Section 3. Also, the gain compensation can be done through the W/L ratios of M20, M22, and M24. The author tried to decrease the amplitude of the input signal; one found that the proposed rectifier can rectify a minimum voltage of 350 μV_{peak}. Changing frequency was done, the maximum operation frequency of the proposed rectifier was 30 MHz (–3 dB point).

For current-controlled-power operation, the author set $I_{CL+1} = I_{CL+2} = I_{CL-1} = I_{CL-2}$ following to the left column of **Table 1**, the right column shows the rectified output peak area amplitude. According to the theory, it was (namely, more $I_{CL+1} = I_{CL+2} = I_{CL-1} = I_{CL-2}$, more shifted-down signal, less output peak area amplitude).

As mentioned in Section 3 that we can independently control the positive part and the negative part of the input signal, this yields more precision control. We will show it here. The rectifier output can be divided to two parts for controlling (A. the positive input part and B. the negative input part shifted the phase by 180 degrees, by rectifier). A is controlled by $I_{CL+1} = I_{CL+2}$ and B is controlled by $I_{CL-1} = I_{CL-2}$. **Figure 4** shows the full-wave rectified output with $I_{CL+1} = I_{CL+2} = 1$ μA and $I_{CL-1} = I_{CL-2} = 3.5$ μA (changing the frequency to 1 MHz). It also shows that we can independently control the positive part and the negative part of the input signal, giving more precision control. Moreover, the proposed current-controlledpower full-wave rectifier is very suitable for an automatic control system by the current feedback. Whereas the SCR technique uses the RC time constant for clipping time, it is hard for the automatic control system.

Absolutely, not only full-wave rectification but also current-controlled-power operation, the proposed rectifier can work. The previous proposed rectifiers cannot operate this current-controlled-power function.

CONCLUSIONS

In this paper, the author has reported a new current-controlled-power technique that can be useful in analog small signal processing. The proposed technique needs no passive devices; it is very suitable for the IC process. The proposed technique can also work with a large signal by driving a power bipolar transistor. The proposed technique is very suitable for an automatic control system by the current feedback.

An application example as a new current-controlled power full-wave rectifier has also been proposed. The new full-wave rectifier works both full-wave rectification

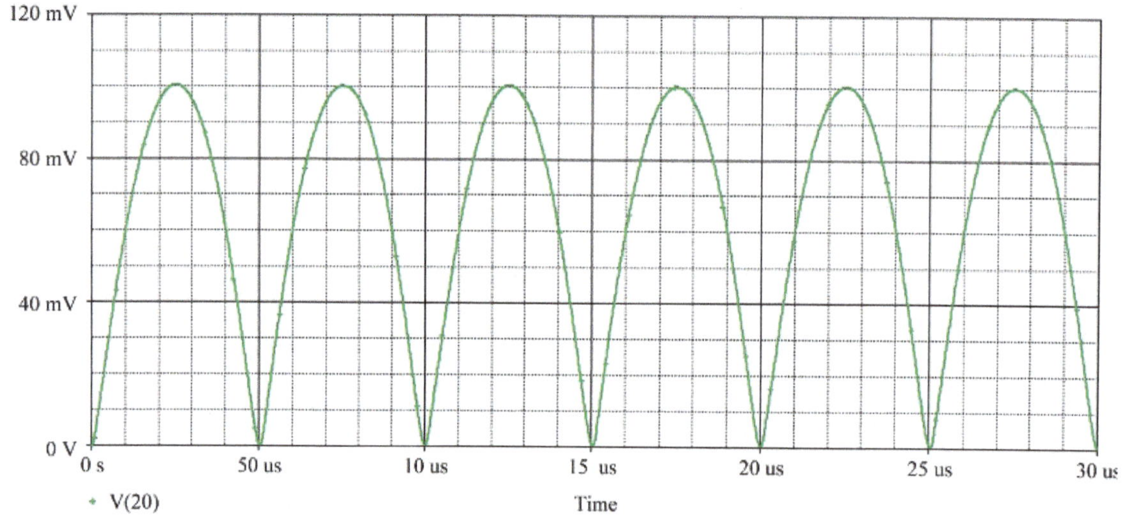

Figure 3. Output of the proposed current-controlled-power full-wave rectifier ($I_{CL+1} = I_{CL+2} = I_{CL-1} = I_{CL-2} = 0$) with an input sine wave of 100 mV_{peak}, 100 kHz.

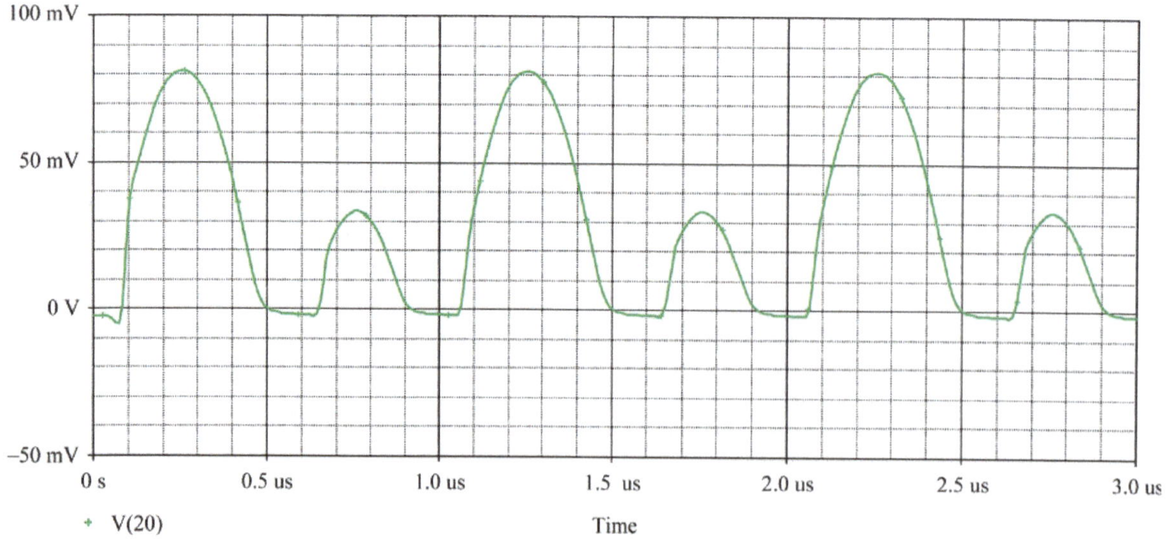

Figure 4. Output of the proposed current-controlled-power full-wave rectifier with an input sine wave, 100 mV_{peak}, 1 MHz, ($I_{CL+1} = I_{CL+2} = 1$ μA and $I_{CL-1} = I_{CL-2} = 3.5$ μA).

Table 1. Relation between the controlled current and the output.

$I_{CL+1} = I_{CL+2} = I_{CL-1} = I_{CL-2}$ (μA)	Output peak area amplitude (mV)
0	100
0.5	91
1	81.7
1.5	72.3
2	62.8
2.5	53.1
3	43.3
3.5	33.4
4	23.4
4.5	13.5
5	1.9
5.5	0

and current-controlled-power operation. The new rectifier uses all MOSs with typical structure thus it is very suitable for the IC process. The new rectifier, its core operates in balanced mode resulting in a low noise output. Using the new rectifier, one can independently control the positive part and the negative part of the input signal, this yields more precision control.

It should be noted on the use of the new rectifier that it is suitable for a high impedance load. If the low impedance load is applied, it needs a voltage buffer at the output.

REFERENCES

1. G. B. Rutkowski and J. E. Oleksy, "Solid-State Electronics," McGraw-Hill, Singapore, 1992.
2. A. Hastings, "The Art of Analog Layout," Prentice-Hall, Upper Saddle River, 2006.
3. B. Razavi, "Design of Analog CMOS Integrated Circuits," McGraw-Hill, Singapore City, 2001.
4. F. W. Hughes, "Op Amp Handbook," Prentice-Hall, Upper Saddle River, 1993.
5. E. Bruun, "CMOS High Speed, High Precision Current Conveyor and Current Feedback Amplifier Structures," International Journal of Electronics, Vol. 74, No. 1, 1993, pp. 93-100. doi:10.1080/00207219308925816
6. Z. Wang, "2-MOSFET Transresistor with Extremely Low Distortion for Output Reaching Supply Voltages," Electronics Letters, Vol. 26, No. 13, 1990, pp. 951-952. doi:10.1049/el:19900620 [Paper reference:1]
7. D. Johns and K. Martin, "Analog Integrated Circuit Design," John Wiley & Sons, New York, 1997

CHAPTER 4

DESIGN OF A SYMMETRY-TYPE FLOATING IMPEDANCE SCALING CIRCUITS FOR A FULLY DIFFERENTIAL FILTER

Fujihiko Matsumoto, Syuzo Nishioka, Takeshi Ohbuch and Tatsuya Fujii

Keywords: Impedance scaling circuit, Analog filters, Analog integrated circuits, Common-mode rejection circuit

INTRODUCTION

Filters are used to process various signals such as visuals, audios, and communications. Low frequency filter processing biomedical signals [1, 2] from a few Hz to hundreds Hz requires a large time constant. In order to implement a large time constant, a large capacitance and a large resistance are needed. However, large capacitances and large resistances are impracticable in integrated circuits, which are required to be low-powered and downsized.

The large resistance is implemented by a transconductor of low transconductance [3]. The area occupied by a transconductor is much smaller than that of a resistor with the same resistance. In contrast, it is difficult to shrink the area occupied by large capacitances. As a method to realize a large capacitance, an impedance scaling circuit has been proposed [4]. The impedance scaling is a technique to reduce apparent impedance using current feedback.

Several gmgm-*C* filters using impedance scaling circuits have been proposed [5–7]. In those filters, a grounded impedance scaling circuit or a pair of the grounded circuits are employed.

The authors have proposed a floating impedance scaling circuit [8]. Floating impedance scaling circuits are useful for not only low-frequency band-pass and high-pass filters but also fully differential low-pass filters. If the floating impedance scaling circuit is employed in a fully differential low-pass filter, the required capacitance is no more than 1/4 comparing with using a pair of grounded capacitances.

However, the circuit in [8] has a problem that the different level between the DC operating points of two terminals prevents realization of fully differential filters. The problem is due to the asymmetric structure of the current amplification stage.

In this paper, a design of a symmetry-type floating impedance scaling circuit and the improvement method of its operation bandwidth are proposed. The proposed circuit is applied to a fully differential 3rd-order low-pass filter (LPF). The fully differential filter requires introducing of a common-mode rejection circuit. The interaction of the common-mode rejection circuit and the proposed floating impedance scaling circuit is confirmed by simulation. The common-mode impedance characteristics of the proposed impedance scaling circuit are verified by small-signal analysis. The effectiveness of the application to the LPF of the proposed circuit is confirmed by simulation.

CONVENTIONAL FLOATING IMPEDANCE SCALING CIRCUIT

Figure 1 shows the block diagram of the conventional floating impedance scaling circuit [8].

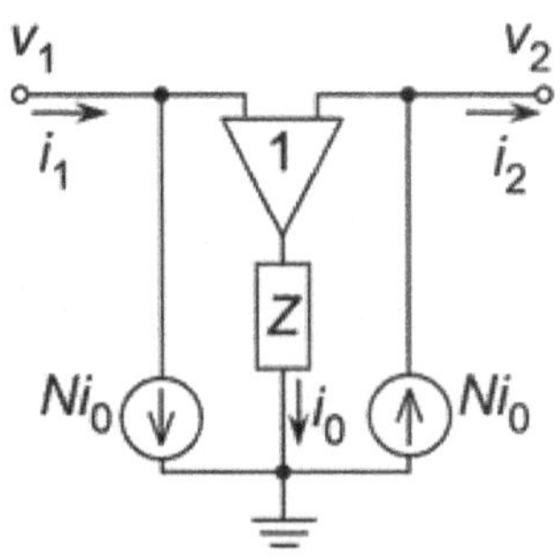

Figure 1: Block diagram of the conventional floating impedance scaling circuit.

Citation: Fujihiko Matsumoto, Syuzo Nishioka, Takeshi Ohbuchi, Tatsuya Fujii, Design of a symmetry-type floating impedance scaling circuits for a fully differential filter, doi:10.1007/s10470-015-0569-0.

This circuit is composed of differential unity gain amplifier (DUA) and a differential output current amplifier. The output voltage of the voltage follower is a voltage difference between v_1 and v_2, and applied to the impedance component Z. The output current of the voltage follower is expressed as

$$i_0 = \frac{v_1 - v_2}{Z}. \quad (1)$$

The differential output current amplifier multiplies a signal current i_0 flowing through an impedance Z by N times for current feedback. The relationships among signal current i_0, i_1, and i_2 are given by

$$i_1 = i_2 = Ni_0. \quad (2)$$

The apparent impedance of the conventional circuit is expressed as

$$Z_{sc} = \frac{v_1 - v_2}{Ni_0} = \frac{Z}{N}. \quad (3)$$

The impedance is reduced by current feedback. If the impedance component Z is a capacitor which has capacitance C, Z_{sc} is expressed as

$$Z_{sc} = \frac{1}{sNC}. \quad (4)$$

This means that the circuit performs as floating capacitor having capacitance NC.

Figure 2 shows the circuit diagram of the conventional circuit. This circuit has asymmetric structure at the current amplification stage. The terminal 'i_2/v_2' is set up by adding M25–M31 to the output circuit for 'i_1/v_1'. Even though cascode stages are introduced, the current amplification factors of the current mirrors deviate from the ideal one. Furthermore, the currents flowing M28–M31 are 50 times as much as the currents of M25–M27. Consequently, the degrees of the current mismatches at the 'i_1/v_1' and 'i_2/v_2' terminals tend to be much different. This makes the DC operating points at the terminals different each other. This problem causes serious trouble in operation of fully differential filters, in which a certain node bias voltage is different from one at the counterpart of the opposite node.

In previous authors' study, application to fully differential filters employing the conventional circuit has been examined. However, the filers never worked because of the unbalanced DC bias condition. If both terminals are of the same structure, namely symmetrical topology like an ordinary differential input/output OTA, DC bias deviation from the proper operating point can be about the same level, and the problem can be alleviated.

In the next section, a design of a symmetry-type floating impedance scaling circuit is proposed.

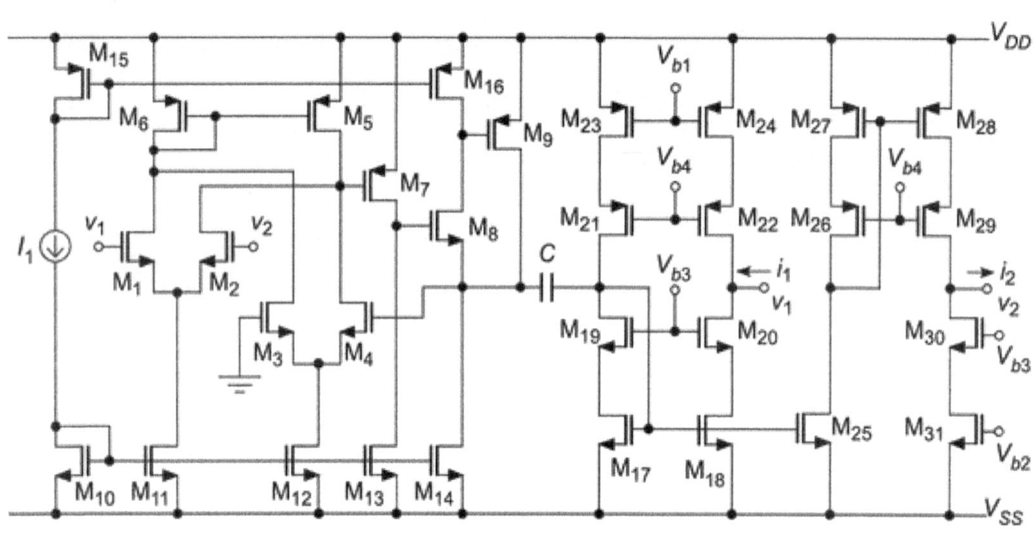

Figure 2: Circuit diagram of the conventional floating impedance scaling circuit.

PROPOSED SYMMETRY-TYPE FLOATING IMPEDANCE SCALING CIRCUIT

Figure 3 shows the block diagram of the proposed symmetry-type floating impedance scaling circuit.

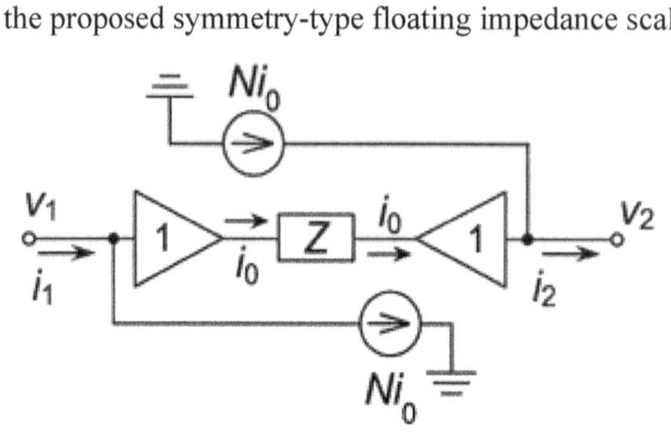

Figure 3: Block diagram of the proposed symmetry-type floating impedance scaling circuit.

This circuit is composed of two voltage followers and two current amplifiers. The output voltages of the voltage followers are applied to the impedance element *Z*. The signal current flowing through the impedance element is amplified with the current amplifier for current feedback.

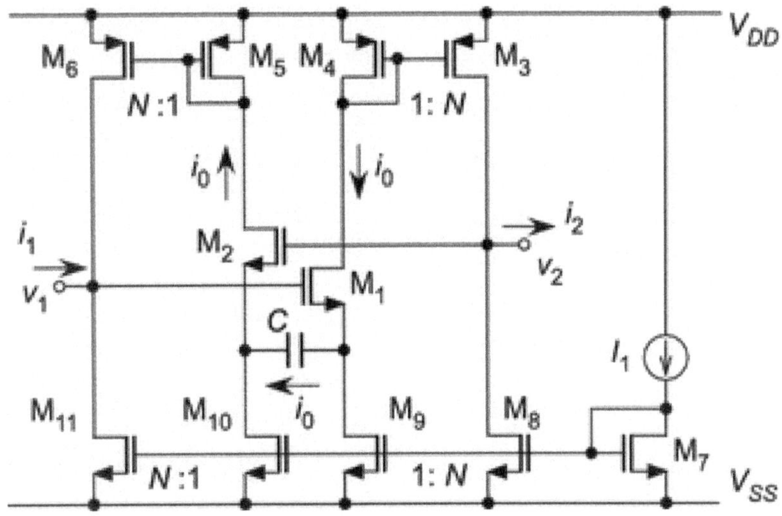

Figure 4: Circuit diagram of the symmetry-type floating impedance scaling circuit ("Proposed 1").

Basic Symmetry-type Floating Impedance Scaling Circuit

The circuit diagram of the symmetry-type floating impedance scaling circuit based on the block diagram of Fig. 3 is shown in Fig. 4. Hereinafter, this circuit is called as "Proposed 1". The transistors M_1 and M_2 are source followers. The input voltage signal is applied to the capacitor *C* through the source followers. The current mirrors of M_4–M_3 and M_5–M_6 perform as the current amplifier of *N* times. The small signal current flowing through the capacitor is amplified by the current amplifier. Assuming that the drain resistance of the bias current sources M_9 and M_{10} are enough large, the signal current flowing through the capacitor, i_0, is expressed as

$$i_0 = (v_1 - v_2)sC. \tag{5}$$

Assuming that the drain resistance of the bias current sources M_8 and M_{11} are enough large, the small signal currents i1i1 and i2i2 are equal to Ni0Ni0, and expressed as

$$i_1 = i_2 = N(v_1 - v_2)sC. \tag{6}$$

The impedance of "Proposed 1" circuit is given by

$$Z_{sc} = \frac{v_1 - v_2}{N(v_1 - v_2)sC} = \frac{1}{sNC}. \tag{7}$$

Therefore, the apparent capacitance is increased by *N* times.

To confirm the frequency characteristics of "Proposed 1" shown in Fig. 4, the small-signal characteristics are analyzed. Assuming that the fully differential voltages are applied to the both terminals, the input voltages are expressed as

$$v_1 = \frac{1}{2}v_{in} \tag{8}$$

$$v_2 = -\frac{1}{2}v_{in}. \tag{9}$$

The differential-mode half circuit of "Proposed 1" is shown in Fig. 5. The resistance r_{d10} indicates the drain resistance of the bias current source M_{10}. The resistance R_d indicates the resistance component observed from terminal v_1. The combined resistance R_d is the drain resistances connected in parallel and expressed as $R_d = r_d/(2N)$. Assuming that $g_m r_d \gg 1$, the impedance Z_1 of the half circuit is given by

$$Z_1 = \frac{v_1}{i_1} \simeq \frac{g_{m5}\left(s + \frac{g_{m2}}{2C}\right)}{g_{m2}g_{m6}\left[s + \frac{1}{2C}\left(\frac{1}{r_d} + \frac{g_{m5}}{g_{m6}R_d}\right)\right]}, \tag{10}$$

The pole frequency ω_p and the zero frequency ω_z are given by

$$\omega_p \simeq \frac{1}{2C}\left(\frac{1}{r_d} + \frac{g_{m5}}{g_{m6}R_d}\right) \simeq \frac{1}{2NC}\left(\frac{N}{r_d} + \frac{1}{R_d}\right) \tag{11}$$

$$\omega_z \simeq \frac{g_{m2}}{2C} \simeq \frac{g_m}{2C}, \tag{12}$$

where $g_{m5} = g_{m2} = g_m$, and $g_{m6} = Ng_m$. Because of $r_d \gg 1/g_m$, the zero frequency ω_z is much higher than the pole frequency ω_p. The pole ω_p depends on the resistance component R_d in the input terminal, and the zero ω_z depends on the conductance g_m of the input MOS transistors M_1 and M_2 associated with capacitor employed in "Proposed 1".

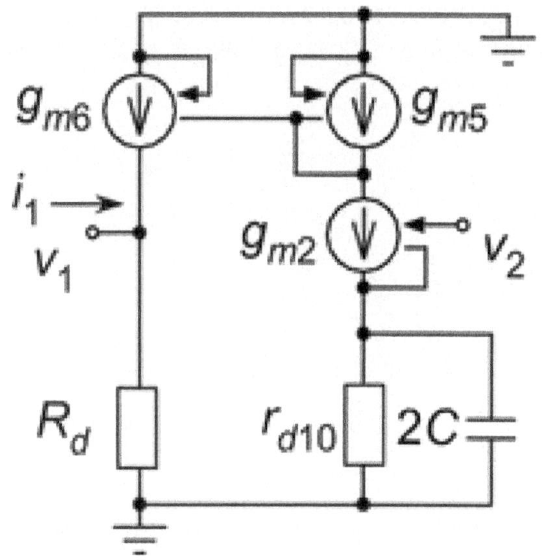

Figure 5: Small signal equivalent circuit of "Proposed 1".

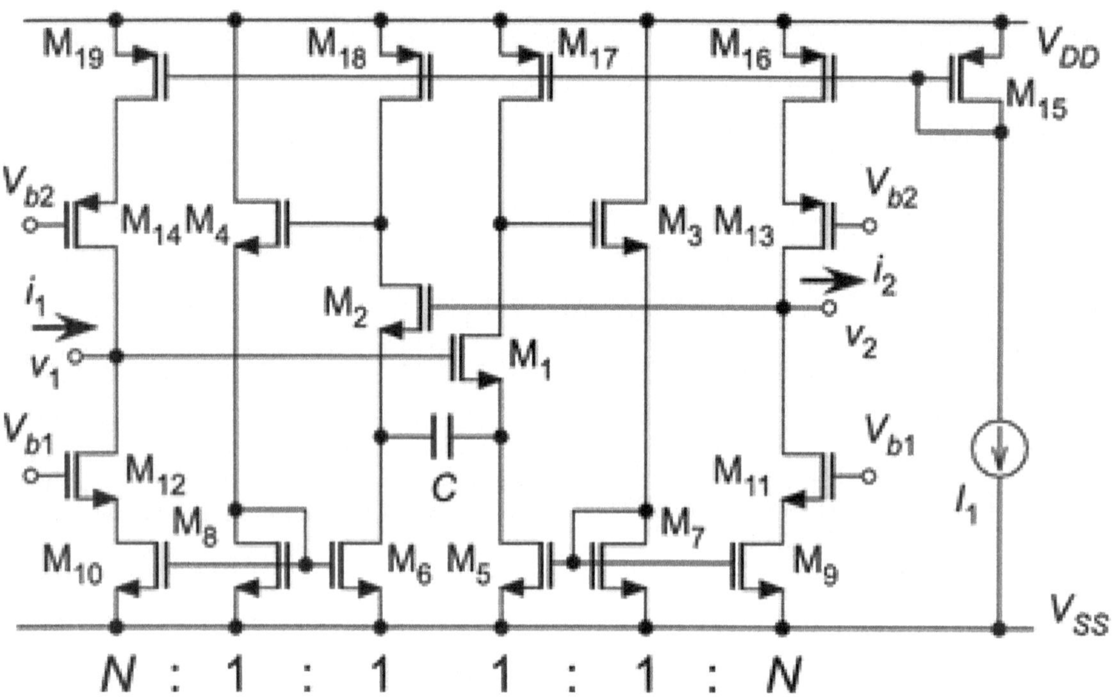

Figure 6: Circuit diagram of the symmetry-type floating impedance scaling circuit ("Proposed 2").

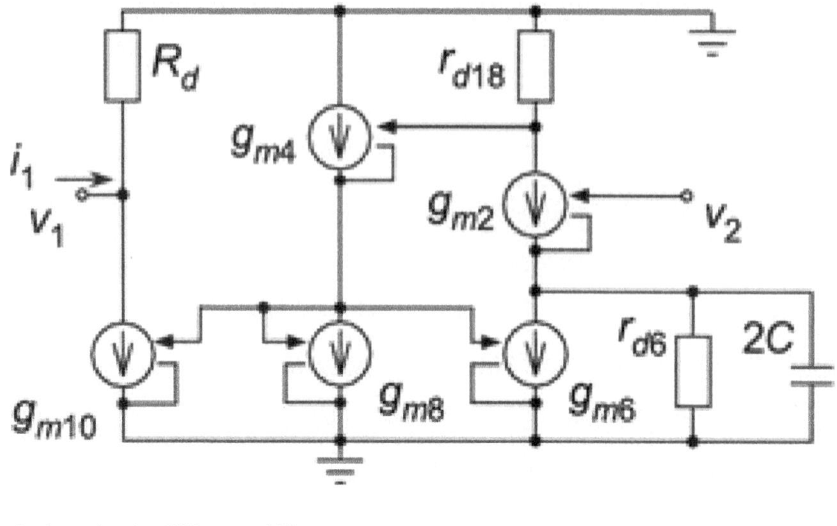

Figure 7: Small signal equivalent circuit of "Proposed 2".

Improved Symmetry-Type Floating Impedance Scaling Circuit

From the small signal analysis related to the pole and the zero of "Proposed 1" given by (11) and (12), an improvement technique of the operation bandwidth is provided. The higher the resistance R_d at the terminal v_1 and v_2 are, the lower the pole frequency is, and the higher the conductance components g_{m1} and g_{m2} associated with the capacitor are, the higher the zero frequency is. The circuit diagram of the symmetry-type floating impedance circuit with improved operation bandwidth is shown in Fig. 6. Hereinafter, this is called as "Proposed 2". In order to make the resistance component at the terminals v_1 and v_2 high, the cascode stages configured with the common-gate M_{11}–M_{14} are introduced. In order to make the conductance component gmgm high, the negative feedback circuits are composed of M_3–M_8. The small-signal half circuit are analyzed to confirm the frequency characteristics of "Proposed 2". The small-signal half circuit of "Proposed 2" is shown in Fig. 7. Because the drain resistances of M_4 and M_8 are much higher than 1/gm81/gm8, they can be ignored. The drain resistance of M_{10} is included in R_d. The drain resistances of M_2, M_6 and M_{18} are indicated by r_{d2}, r_{d6} and r_{d18}. Assuming that $g_{mx}r_{dy} \gg 1\ (x, y : 2, 8, 16)$, the impedance Z_1 is approximately expressed as

$$Z_1 \simeq \frac{4}{Ng_m^2 r_{d18}} \cdot \frac{s + \frac{g_m^2 r_{d2} r_{d18}}{4C(r_{d2}+r_{d18})}}{s + \frac{1}{2NC}\left(\frac{1}{R_d} + \frac{N}{r_{d6}}\right)}, \tag{13}$$

where $g_{m2} = g_{m4} = g_{m6} = g_{m8} = g_m$ and $g_{m10} = Ng_m$. The pole frequency ω_p and the zero frequency ω_z are given by

$$\omega_p \simeq \frac{1}{2NC}\left(\frac{N}{r_d} + \frac{1}{R_d}\right) \tag{14}$$

$$\omega_z \simeq \frac{g_m^2 r_{d2} r_{d18}}{4C(r_{d2} + r_{d18})} \simeq \frac{g_m^2 r_d}{8C}. \tag{15}$$

where r_{d2} and r_{d18} are roughly approximated to be equal, and are r_d. Comparing "Proposed 2" with "Proposed 1", R_d becomes $g_m r_d$ times with the cascode stages, the pole moves to lower frequency. Comparing (15) with (12), the zero frequency is multiplied by $g_m r_d/4$ times and moves to higher frequency. The impedance Z_1 in the range between the pole and the zero frequencies, $\omega_z \gg \omega \gg \omega_p$, is given by

$$Z_{sc} = 2Z_1 \simeq 2 \cdot \frac{4}{Ng_m^2 r_{d18}} \cdot \frac{\omega_z\left(\frac{s}{\omega_z} + 1\right)}{s\left(1 + \frac{\omega_p}{s}\right)} \tag{16}$$

$$\simeq \frac{8}{Ng_m^2 r_d} \cdot \frac{\frac{g_m^2 r_d}{8C}}{s} \tag{17}$$

$$\simeq \frac{1}{sNC}. \tag{18}$$

The "Proposed 2" performs as an N-times capacitance in wider range than "Proposed 1".

Simulation Results

The validity of the proposed circuits is confirmed using simulation software LTspice (Linear Technology). The transistor model used in the simulation is BISIM3 0.18 μm process model. The supply voltage is 1.8 V. The reference current I_1 is 30 nA. In order to realize the scaled capacitance of 500 pF, N=50 and the base capacitance is set to 10 pF. The tables of transistor size are shown in Tables 1 and 2.

Table 1: Transistor size of proposed circuit (Proposed 1)

Transistors	*W* (μμm)/*L* (μμm)
M33, M66, M88, M1111	90.0/1.8
The others	1.8/1.8

Table 2: Transistor size of proposed circuit (Proposed 2)

Transistors	*W* (μμm)/*L* (μμm)
M99–M1414, M1616, M1919	90.0/1.8
the others	1.8/1.8

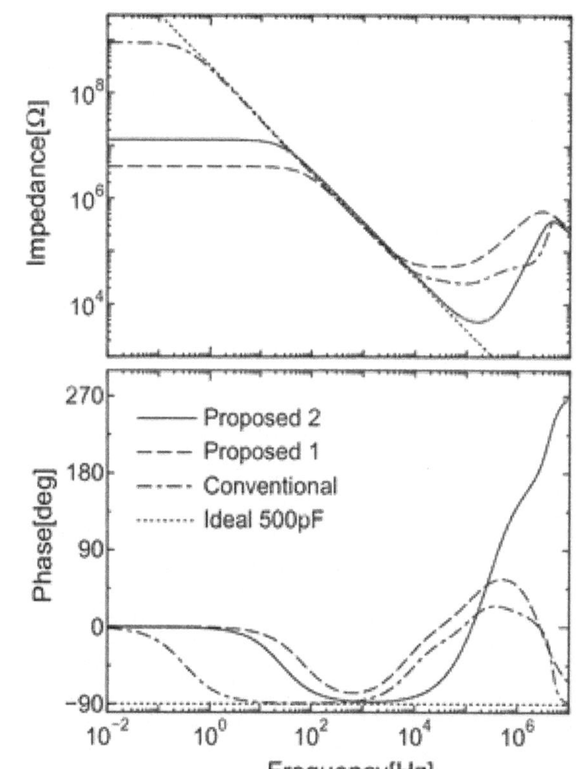

Figure 8: Frequency characteristics.

The simulation results of impedance and phase frequency characteristics are shown in Fig. 8. The solid line indicates the proposed circuit of "Proposed 2", the broken line indicates the proposed circuit of "Proposed 1", the dash-dotted line indicates the conventional impedance scaling circuit [8], and the dotted line indicates the ideal capacitor of 500 pF. From the impedance characteristic of "Proposed 1", it is seen that the pole and the zero frequencies are near each other.

Table 3: Power consumption and area

	Power (μμW)	Area (μμm22)	Range (Hz)
Conventional	6.05	2310	1.86–2.09 k
Proposed 1	5.94	1120	N/A
Proposed 2	5.72	2220	158–11.5 k

Table 3 shows the comparison of the power consumption, the occupied area, and the frequency range. The definition of the frequency range is one in which the phase is between −80° and −90°.

The power consumption and area of "Proposed 2" are reduced by 5.5 and 3.9% compared to "Conventional", respectively. Although the operation frequency range of the proposed circuit is narrower than the conventional one, the proposed circuit has an advantage that the problem of DC operating points can be improved. In the next section, application of the proposed circuit to a 3rd-order fully differential filter is shown. While the conventional circuit can be applied to a 2nd-order bandpass filter [8], even a 1st-order fully differential filter cannot work. This is because a node bias voltage is different from one at the counterpart of the opposite node.

The phase characteristic of "Proposed 1" do not reach −90° in the range of 100–10 kHz, in which the minimum value of phase is −77.2°. The impedance characteristic of "Proposed 2" shows the capacitor characteristics in the range from 158 Hz to 11.5 kHz. In summary, comparing "Proposed 2" with "Proposed 1", the pole frequency is lower and the zero frequency is higher.

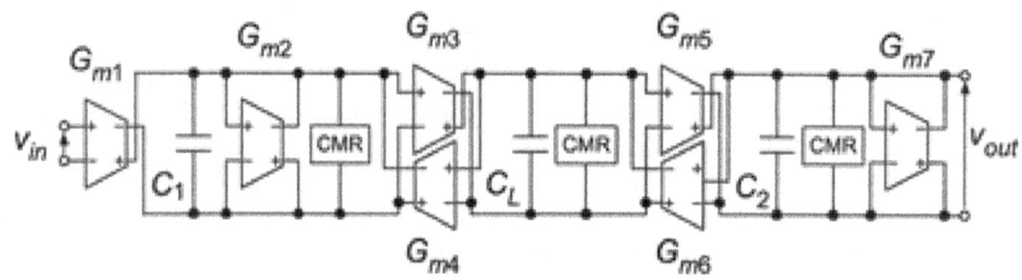

Figure 9: 3rd-order LPF employing symmetry-type floating impedance scaling circuit ("Proposed LPF").

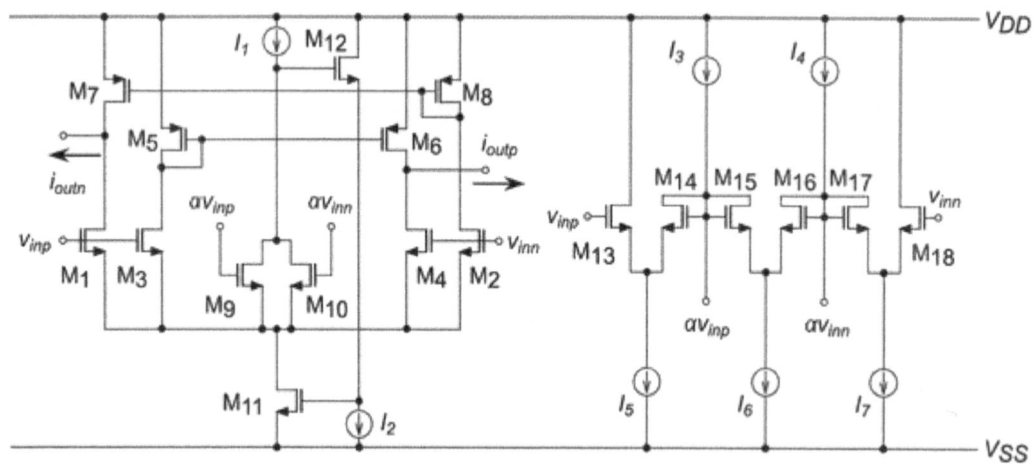

Figure 10: Low-gmgm linear OTA.

APPLICATION TO FULLY DIFFERENTIAL FILTER

Configuration of Fully Differential LPF

Employing the proposed circuits, a fully differential 3rd-order butterworth LPF shown in Fig. 9designed in this study. The capacitors C_1, C_2, and C_L are implemented employing the floating impedance scaling circuit. The low-g_m linear OTA [9] shown in Fig. 10 is used as G_{m1}–G_{m7}. Table 4 shows the bias currents in the OTA. All transconductances of the OTAs are 606.7 nS by simulation. The fully differential filter requires introducing of a common-mode rejection (CMR) circuit. The CMR circuit employed in the LPF is shown in Fig. 11. The reference current I1I1 in the CMR circuit is 300 nA. In order to realize the 3rd-order butterworth LPF of f_c = 100Hz, $C_1 = C_2$ = 965 pF, and C_L = 1.93 nF. By setting $N = 50$, the capacitances in the floating impedance scaling circuit are $C_1/50 = C_2/50$ = 19.3 pF, and $C_L/50$ = 38.6 pF.

Table 4: The bias current source

Current source	Current (nA)
I1=I8I1=I8	50
I2=I5=I7=I9I2=I5=I7=I9	100
I3=I4I3=I4	68.3
I6I6	36.6

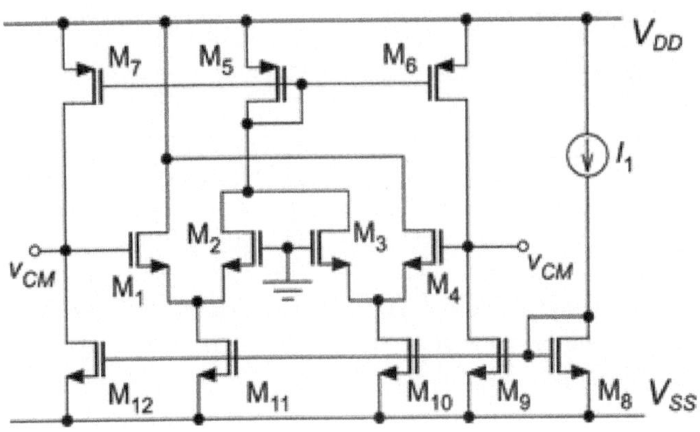

Figure 11: Common-mode rejection circuit.

As is mentioned previously, the conventional circuit is not suitable for fully differential filters. For comparison, the conventional floating circuit is used as a grounded circuit. The terminal 'i_2/v_2' shown in Fig. 2, which is difficult to be proper operating condition, is grounded. A scaled capacitance C_x(x:1,2,L) realized by the proposed circuit is replaced with two grounded conventional circuits with 4 times capacitance as shown in Fig. 12.

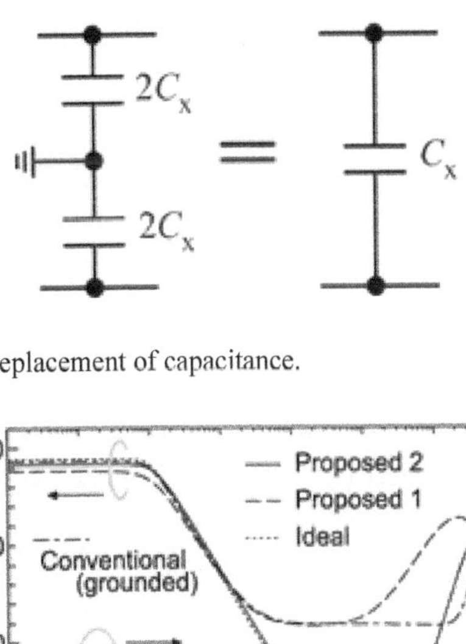

Figure 12: Replacement of capacitance.

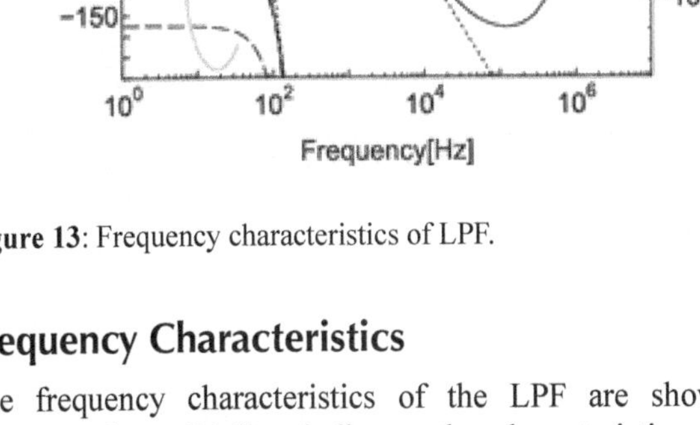

Figure 13: Frequency characteristics of LPF.

Frequency Characteristics

The frequency characteristics of the LPF are shown in Fig. 13. The solid line indicates the characteristics of the LPF employing "Proposed 2", the dashed line indicates the characteristics of the LPF employing "Proposed 1", the dash-dotted line indicates the characteristics of the LPF employing the conventional circuit as a grounded circuit, and the dotted line of "Ideal" indicates the ideal LPF. The passband gain and the cutoff frequency are listed in Table 5.

From the simulation results in Fig. 13, the proposed circuits show the characteristics of floating capacitors in stop band. The reduction of passband gain from ideal response are 0.66 dB (Conventional), 5.70 dB (Proposed 1) and 2.27 dB (Proposed 2). It is seen that the stopband characteristic of "Proposed 2" is superior to the conventional one and "Proposed 1".

The parasitic resistances of the conventional one, "Proposed 1", "Proposed 2", and CMR are 4.07, 13.1, 908, and 42.6 MΩ, respectively. Thus, it can be considered that the passband gain are reduced by the parasitic resistance component of the CMR circuit and the proposed impedance scaling circuits.

Table 5: Passband gain and cutoff frequency of LPF

	Gain (dB)	Frequency (Hz)
Ideal	--6.02	100
Conventional	--6.68	92.0
Proposed 1	--11.72	79.4
Proposed 2	--8.29	105.0

Common-Mode Impedance Characteristics

The current never flows when the same voltages are applied to the both terminals of passive elements, and the impedance is obviously infinite. However, because the impedance scaling circuit is synthesized with transistors, the impedance to the same voltage at the both terminals is finite, practically. Although it may not be correct expression, such an impedance is called "common-mode impedance" in this paper. Appling the common-mode voltage v_{CM} to the both terminals of "Proposed 2", the output voltages of the source followers M_1 and M_2 are equal, currents do not flow in the capacitor C associated with the source followers. Thus, the capacitor C can be considered as an open element. The common-mode half circuit is shown in Fig. 14. The resistances r_{d6}, r_{d18}, and R_d indicate the drain resistances of the transistor M_6, the bias current source M_{18}, and the resistance component at the terminal v_1, respectively. Assuming that $g_m r_d \gg_1$, the impedance Z_{CM} of the half circuit is given by

$$Z_{CM} \simeq \frac{R_d r_{d6} g_{m6}}{g_{m6} r_{d6} - g_{m10} R_d} \tag{19}$$

$$\simeq \frac{R_d \cdot \left(-\frac{r_{d6}}{N}\right)}{R_d - \frac{r_{d6}}{N}} \tag{20}$$

$$= R_d \parallel \left(-\frac{r_{d6}}{N}\right), \tag{21}$$

where $g_{m6} = g_m$, $g_{m10} = N g_m$. Moreover, we have Rd= $(g_m r_d^2)/N$ and $R_d \gg r_{d6}/N$, (22)

For the common-mode signal, the "Proposed 2" symmetry-type impedance scaling circuit performs as a negative resistance −rd6/N−rd6/N due to positive current feedback.

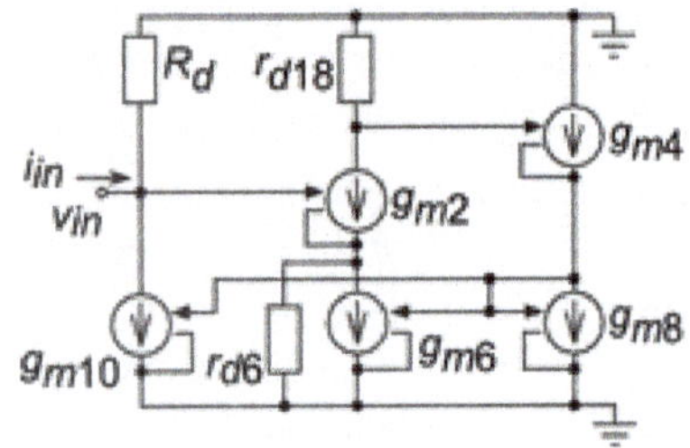

Figure 14: Small signal equivalent circuit of the floating type scaling capacitor circuit for common-mode signal ("Proposed 2").

The common-mode impedance of the CMR circuit Z_{cmr} is expressed as

$$Z_{cmr} = \frac{1}{g_m} \tag{23}$$

where $g_{m1} = g_{m2} = g_{m3} = g_{m4} = g_{m6} = g_{5m}/2 = g_m$.
The synthetic admittance Y is expressed as

$$Y = \frac{1}{Z_{cmr}} + \frac{1}{Z_{CM}} = g_m - \frac{N}{r_{d6}} \tag{24}$$

The reason why I_1 in the CMR circuit is set to be 300 nA, which is 10 times as much as I_1 in "Proposed 2", is not only to obtain high common-mode rejection but also to hold Y>0.

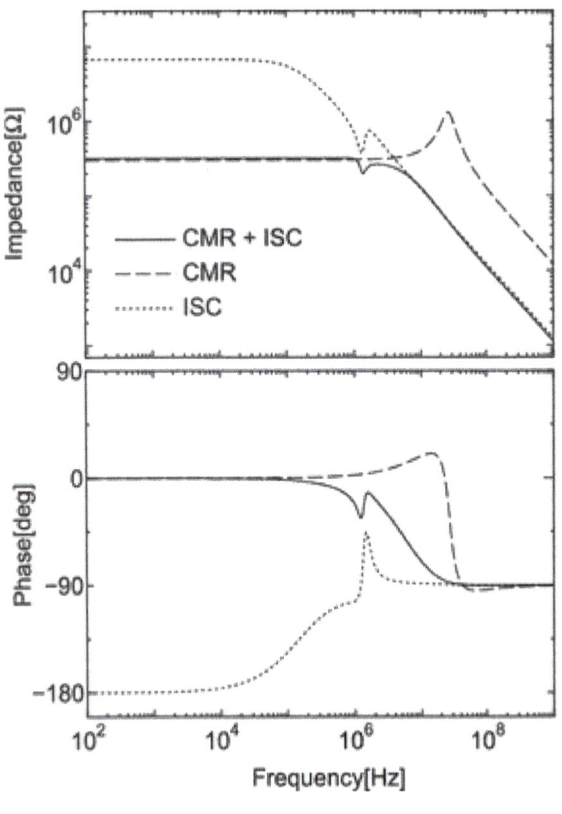

Fig.ure15: Common-mode impedance.

The simulation results of the common-mode impedance and the phase characteristics are shown in Fig. 15. The solid line of "CMR + ISC" indicates the result of the parallel of the common-mode rejection circuit and "Proposed 2" circuit, the dashed line of "CMR" indicates the common-mode rejection circuit, the dotted line of "ISC" indicates "Proposed 2" circuit. From the phase characteristic of "ISC", it is shown that the symmetry-type floating impedance scaling circuit performs as the negative resistance in lower frequency range. The phase characteristic of "ISC + CMR" performs as the positive resistance. Consequently, because of the combined impedance of "CMR" and "ISC" is positive from the characteristics of "CMR + ISC", the operation of the LPF is stable.

CONCLUSION

In this paper, to implement large capacitance for the application of the biomedical signal processing such a LPF, a design of a symmetry-type floating impedance scaling circuit and the improvement methods of those operation bandwidth had been proposed. Floating impedance scaling circuits are useful for not only low-frequency band-pass and high-pass filters but also fully differential low-pass filters. The conventional floating impedance scaling circuit has a problem that the different level between the DC operating points of two terminals prevents realization of fully differential filters.

The proposed circuit is composed of two source followers and two current amplifiers. From the small signal analysis, the improvement methods of the operating bandwidth are found. In order to lower the pole frequency, the resistance component in the parts of input terminals has been increased by introducing the common gate, and to enhance the zero frequency, the conductance component associated to the capacitor has been enhanced by negative feedback. To realize the scaled capacitance of 500 pF, *N* and the capacitance are set to 50 and 10 pF, respectively. The impedance characteristic of the synthesized impedance scaling circuit had shown the capacitor characteristics in the range from 158 Hz to 11.5 kHz. The power consumption and area are reduced comparing the conventional circuit. The fully differential 3rd-order butterworth LPF with the fc= 100 Hz has been designed employing the proposed circuit. The appropriate operation of the scaling capacitor composed of the proposed circuit and the LPF employing the proposed circuit has been confirmed by the simulation results. By the small signal analysis, it has been shown that the improved proposed circuit becomes the negative conductance when the same voltages are applied to both terminals. This negative resistance can be ignored due to the small common-mode resistance of the common-mode rejection circuit. The future work is the analysis of the characteristics under PVT variation, the improvement of the pass band gain of the filter, the application for the higher-order filters, and the prototype of the proposed circuit.

ACKNOWLEDGMENTS

This study was supported by Japan Society for the Promotion of Science (JSPS) Grant-in-Aid for Scientific Research (C 24560436).

Conflict of Interest

The authors declare that they have no conflict of interest.

REFERENCES

1. Hsu, C.-L., Ho, M.-H., Wu, Y.-K. & Chen, T.-H. (2006). Design of low-frequency low-pass filters for biomedical

applications. In *IEEE Asia Pacific Conference on Circuits and Systems, 2006 (APCCAS* (pp. 690–695).

2. Silva-Martinez, J., & Solis-Bustos, S. (1999). Design considerations for high performance very low frequency filters. In *Proceedings of the 1999 IEEE International Symposium on Circuits and Systems, 1999 (ISCAS '99)* (vol. 2, pp. 648–651).

3. Veeravalli, A., Sanchez-Sinencio, E., & Silva-Martinez, J. (2002). A CMOS transconductance amplifier architecture with wide tuning range for very low frequency applications. *IEEE Journal of Solid-State Circuits, 37*(6), 776–781.

4. Silva-Martinez, J., & Vazquez-Gonzalez, A. (1998) . Impedance scalers for ic active filters. In *Proceedings of the 1998 IEEE International Symposium on Circuits and Systems, 1998 (ISCAS '98)* (vol. 1, pp. 151–154).

5. Domenech-Asensi, G., Martinez-Viviente, F., Illade-Quinteiro, J., Zapata-Perez, J., Ruiz-Merino, R., Lopez-Alcantud, J. A., Martinez-Alajarin, J., Fernandez-Luque, F., Carrillo, J. M., & Dominguez, M. A. (2012). A fourth order cmos band pass filter for pir sensors. In*19th IEEE International Conference on Electronics, Circuits and Systems (ICECS), 2012*(pp. 268–271).

6. Solis-Bustos, S., & Silva-Martinez, J. (1998). A 4 hz low-pass continuous-time filter. In*IEEE International Conference on Electronics, Circuits and Systems, 1998* (vol. 1, pp. 169–172).

7. Padilla-Cantoya, I. (2013). Low-power high parallel load resistance current-mode grounded and floating capacitor multiplier. *IEEE Transactions on Circuits and Systems II: Express Briefs*, *60*(1), 16–20.

8. Matsumoto, F., Fujii, T., Nishioka, S., Abe, T., & Ohbuchi, T. (2013). Design of a floating-type impedance scaling circuit for large capacitances. In *International Symposium on Intelligent Signal Processing and Communications Systems (ISPACS), 2013* (pp. 391–396).

9. Matsumoto, F., Sugimoto, R., Ohbuchi, T. & Abe, T. (2011). A synthesis of linear transconductors using MOSFETs operating in weak-inversion region based on SINH circuit. In *Proceedings of the IEEJ International Analog VLSI Workshop* (pp. 15–20). Bali, Indonesia

CHAPTER 5

DESIGN AND ANALYSIS OF A POWER EFFICIENT LINEARLY TUNABLE CROSS-COUPLED TRANSCONDUCTOR HAVING SEPARATE BIAS CONTROL

Vijaya Bhadauria, Krishna Kant, Swapna Banerjee

Keywords: Analog Electronics, Low Power Analog CMOS Circuit, Operational Transconductance Amplifier (OTA),; Multiple-Output OTA (MOTA), MOS Transconductors, Linearly Tunable G_m, Current Efficiency, Linearization Techniques, Harmonic Distortion Analysis

INTRODUCTION

Transconductors interface a current mode circuit to a voltage signal, making it a fundamental element of analog circuits [1-4]. Performance of a transconductor deteriorates due to higher total harmonic distortion (THD), more so at high signal level. Several techniques are devoted to improve THD of transconductors [5-20]. Sánchez-Sinencio and J. Silva-Martínez [4] have classified these as input signal attenuation [5-7], cancellation of nonlinear terms [7-12], and source degeneration [7, 11-19]. Besides these, adaptive biasing [16,19] and mobility compensation [20] techniques are also reported to improve THD.

One of the important building blocks in many analog circuits is a tunable linear transconductor [9,13,15,17, 21-25], which may be used for tuning center frequency and Q-factor of filters. Moreover it helps to compensate fabrication tolerances and environment parameters, especially temperature.

A method of tuning transconductance, G_m is to control the bias current (I_{Bias}) applied to the differential amplifier. But G_m is proportional to the square root of I_{Bias} due to which the allowable input signal swing is limited to small value [1]. In [17], the allowable input swing for a constant value of I_{Bias} is made independent of variation in G_m by using current mirrors with source degeneration.

Another tuning method uses source degeneration resistance (R). In this method due to degeneration, THD improves but at the cost of transconductance. Further, the condition of improvement, ($1/G_m < R$) [1] occupies large area and add noise if passive resistance is used. In [13], tunability has been achieved using active resistance. But G_m adjustment in this tuning method affects the bandwidth, which has been compensated in [15] by employing a separate source follower biased with constant current source.

Linearly adjustable G_m is realized using control voltage at inputs of one of the cross-coupled differential amplifier [8,9]. In both the papers authors have used identical transistors in the cross-coupled amplifiers to obtain linear tuning of G_m at supply voltage 5 V or higher. Different bias strategies are given in [25] to achieve tunability for ultra low range transconductance.

A low voltage transconductor on 0.18 μm technology is given by us in [23] in which bias currents of crosscoupled differential amplifiers and aspect ratio of their transistors are adjusted to cancel third harmonic distortion. Shift level biasing is used in the cross-coupled differential amplifier to obtain tunability. The circuit is designed for resistive load, not suitable for current mode signal processing.

The topology cited in [23] is analyzed in detail in this paper. In addition diode mode MOSFETs are used as load for current mode signal processing in place of active resistance. A triode mode tail transistor is used in MA for better linear tuning instead of constant current source. Design steps in detail with algorithm are given for low HD_3, high linearly tunable G_m and high frequency of operation at low voltage. Efforts have also been made to maintain low power operation and high current efficiency. The circuit is developed for 1 $V_{p\text{-}p}$ input signal for frequency range upto 50 MHz.

The paper is organized as follows. In Section 2, block diagram, circuit design alongwith its algorithm and analysis of

Citation: V. Bhadauria, K. Kant and S. Banerjee, "Design and Analysis of a Power Efficient Linearly Tunable Cross-Coupled Transconductor Having Separate Bias Control," Circuits and Systems, Vol. 3 No. 1, 2012, pp. 99-106. doi: 10.4236/cs.2012.31013.

the proposed circuit are given. In Section 3, simulation results are discussed and finally conclusions are drawn in Section 4.

DESIGN AND ANALYSIS OF TUNABLE TRANSCONDUCTOR

Block Diagram of Tunable Transconductor

The proposed transconductor is designed using main differential amplifier (MA) biased using tail transistors in triode mode, compensatory differential amplifier (CA) biased using separate constant current source for minimizing HD_3 and level shifters (LS) are used for tunability. Separate controls of the two bias currents of MA and CA provide flexibility, improvement in HD_3 and better tuning in comparison to others [8,9]. Block diagram of the proposed transconductor is shown in **Figure 1** and the complete circuit diagram is furnished in **Figure 2**.

The square law model of MOSFETs is used in this work for design and analysis. Design steps are given below:

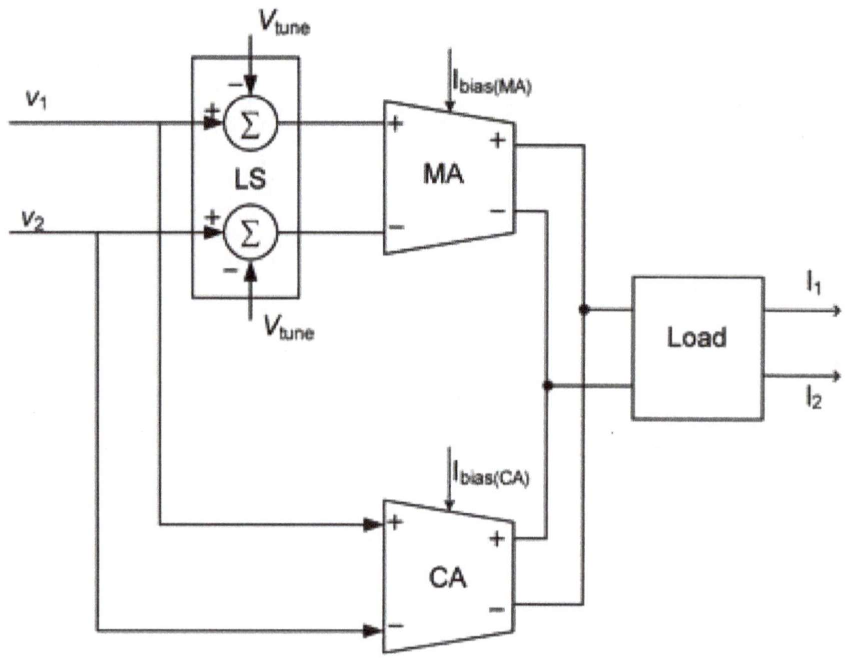

Figure 1. Functional block diagram of the trnsconductor.

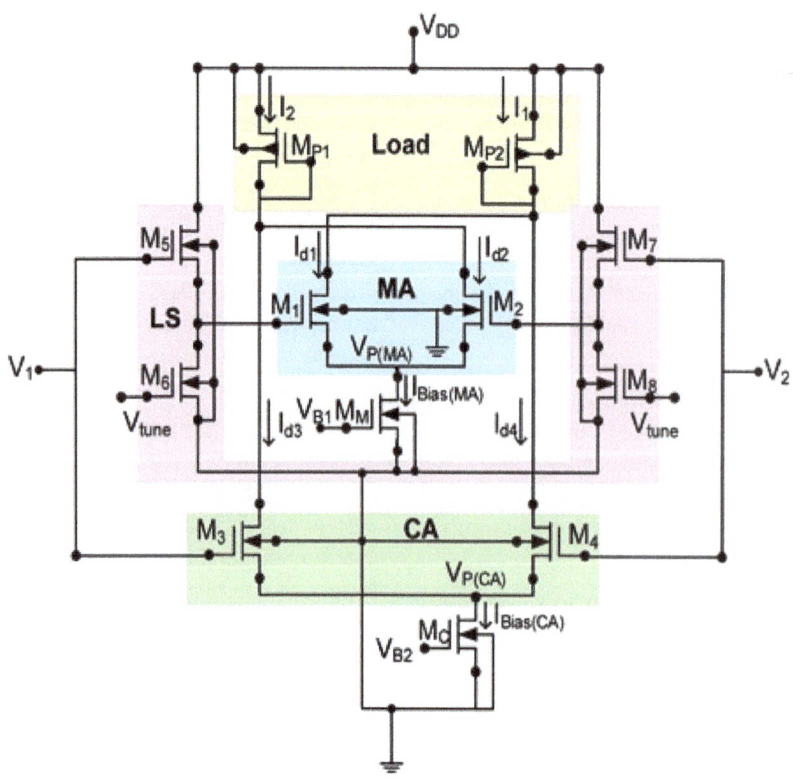

Figure 2. Circuit diagram of the transconductor.

Design of MA

Upper limit and lower limit of biasing current ($I_{Bias(MA)}$) for MA are given as follows [3]:

$$I_{Bias(MA)_{(max)}} \ll P_{Diss}/V_{DD} \quad (1)$$

$$I_{Bias(MA)_{(min)}} \geq (2 * \omega_{3dB} * C_l)/(\lambda_n + \lambda_p) \quad (2)$$

where P_{Diss} is the specified total power consumption, C_l is the load capacitor, ω_{3dB} is the 3 dB frequency, λ_n and λ_p are the channel length coefficients of NMOS and PMOS transistors respectively. MA is designed using NMOS transistors and load using PMOS transistors.

Transconductance, $G_{m(MA)}$ of MA and input swing, V_{swing} are given in [1, 3] and are reproduced in Equation (3) and Equation (4).

$$G_{m(MA)} = \sqrt{\mu_n C_{ox} (W/L)_{MA} I_{Bias(MA)}} \quad (3)$$

$$V_{swing} = \sqrt{(2I_{Bias(MA)})/\{\mu_n C_{ox} (W/L)_{MA}\}} \quad (4)$$

From Equation (3) $G_{m(MA)}$ is the function of $I_{Bias(MA)}$ and $(W/L)_{MA}$. To keep high $G_{m(MA)}$, higher values of $I_{Bias(MA)}$ are not used as it would lead to loss of power. The second option is to increase the aspect ratio but that may deteriorate HD_3 for specified input swing {Equation (4)}.

The second option is preferred wherein, for better HD_3, the method of non-linear terms cancellation with the help of compensatory amplifier is used but at the cost of area occupied by it. However, the use of CA (**Figure 2**) reduces overall transconductance. To maintain transconductance G_m of the OTA at the desired level, $G_{m(MA)}$ is adjusted {Equation (7)}. For high current efficiency (*i.e.* G_m/I_{DD}), $I_{Bias(MA)}$ is selected slightly greater than $I_{Bias(MA)_{min}}$ and aspect ratio $(W/L)_{MA}$ is determined from Equation (3) for the modified value of $G_{m(MA)}$. In addition, $I_{Bias(MA)}$ is allowed to vary independently of CA as a function of tuning voltage by operating tail transistor of MA in triode mode.

Design of CA

The ratio "p" of bias currents of MA and CA are adjusted [11,12] in accordance with Equation (5):

$$p = q^3 \quad (5)$$

where $p = I_{Bias(MA)}/I_{Bias(CA)}$ and $q = (W_{MA}/L_{MA})/(W_{CA}/L_{CA})$. A lower value of "q" not only increases power dissipation (due to high bias current of CA), but also lowers the overall transconductance, thereby, decreasing the current efficiency. On the other hand, due to higher values of "q" the harmonics will not be suppressed effectively. Moreover, it is difficult to bias the CA transistor in saturation region in the specified low voltage operation due to skewed overdrive voltages of the MA and CA. In the proposed circuit choosing a moderate "q" as 2 and equal channel length of MA and CA gives:

$$W_{MA}/W_{CA} = 2 \quad (6)$$

and $I_{Bias(MA)}/I_{Bias(CA)} = 8$. This ratio of bias currents is adjusted for the centre value of the tuning voltage range at which specified G_m is expected to occur.

Overall transconductance G_m due to cross-coupling of MA and CA is the difference of $G_{m(MA)}$ and $G_{m(CA)}$, hence $G_{m(MA)}$ is selected to get the specified G_m in accordance of Equation (7).

$$G_{m(MA)} = \{q^2/(q^2-1)\}G_m \quad (7)$$

Design of Load

The PMOS transistors, M_{p1} and M_{p2} are selected in diode mode as load. It is possible to easily convert the proposed transconductor to Multiple-output OTA suitable for current mode signal processing by incorporating three current mirrors. The aspect ratio of load transistors is computed using the Equation (8), where in Equation (8) $V_{CM_{max}}$ is the maximum common mode input voltage at MA, V_{thn} and V_{thp} are the threshold voltages of NMOS and PMOS transistors respectively.

$$(W/L)_{load} = \frac{\left(I_{Bias(MA)} + I_{Bias(CA)}\right)}{\left\{\mu_p C_{ox} * \left(V_{DD} - V_{CM_{max}} - V_{thn} - V_{thp}\right)^2\right\}} \quad (8)$$

Design of LS

Linearly tunable G_m is obtained by modulating common mode voltage with V_{tune} in two level shifters. The transistors (M_5 and M_6) of a level shifter of LS are selected to have same aspect ratio to maintain same V_{tune} across gate to source in both the transistors. They are biased in saturation region by following the conditions given in Equation (9) and Equation (10).

$$V_1 (\text{or } V_2) < V_{DD} + V_{thn} \quad (9)$$

$$V_{thn} < V_{tune} < \{V_1 (\text{or } V_2) + V_{thn}\}/2 \quad (10)$$

where $V_1 = \left(V_{DC} + \frac{V_{id}}{2}\right)$ and $V_2 = \left(V_{DC} - \frac{V_{id}}{2}\right)$ are the signal inputs applied at the gates of M_5 and M_7 of level shifters, respectively. Same input V_1 and V_2 are applied to the gate of M_3 and M_4 of CA. As per the Equation (10) higher values of V_1 (or V_2) give better tuning range, accordingly, V_{DC} is selected slightly less than to V_{DD}. Further, MA remains in saturation if and only if condition given in Equation (11) is satisfied.

$$V_{tune} \leq V_1 (or V_2) - V_{thn} - V_{P(MA)} \quad (11)$$

where $V_{P(MA)}$ is the drain to source voltage of tail transistor of MA Thus, the limits on V_{tune} are obtained as in Equation (12). Transistors of other level shifter of LS are also made identical on the same ground.

$$V_{thn} < V_{tune} \leq \min\left[\{V_1 (\text{or } V_2) + V_{thn}\}/2, \; V_1 (\text{or } V_2) - V_{thn} - V_{P(MA)}\right] \quad (12)$$

Design of Tail Transistor of MA

As the bias current supplied to MA by its tail transistor is a function of tuning voltage, it is operated in triode mode. It needs $V_B > V_{P(MA)} + V_{thn}$, where V_{B1} gate voltage of tail transistor of MA is. The aspect ratio of tail transistor of MA is derived for the mentioned operating mode and is given in Equation (13):

$$(W/L)_{T(\text{MA})} = I_{Bias(\text{MA})} \Big/ \{(\mu_n C_{ox})(V_{B1} - V_{thn}) V_{P(\text{MA})}\} \quad (13)$$

Design of Tail Transistor of CA

Tail transistor of CA is chosen as constant current source independent of $I_{Bias(MA)}$ by biasing the transistor in saturation. The aspect ratio of it is given in Equation (14):

$$(W/L)_{T(\text{CA})} = \frac{2 I_{Bias(\text{CA})}}{(\mu_n C_{ox})(V_{B2} - V_{thn})^2} \quad (14)$$

where V_{B2} is gate voltage of tail transistor of CA.

Algorithm for Designing the Proposed Transconductor

The algorithm to design proposed transconductor is summarized as below:

Step 1	Compute $I_{Bias(\text{MA})_{(max)}}$	/* from specified P_{Diss}: refer Equation (1)*/
Step 2	Compute $I_{Bias(\text{MA})_{(min)}}$	/* from specified ω_{3dB}: refer Equation (2)*/
Step 3	Select $I_{Bias(\text{MA})} \geq I_{Bias(\text{MA})_{(min)}}$	/* to increase current efficiency*/
Step 4	Compute $G_{m(\text{MA})}$	/* for specified G_m : refer Equation (7)*/
Step 5	Compute $(W/L)_{\text{MA}}$	/* using Equation (3)*/
Step 6	Calculate $(W/L)_{\text{CA}}$	/* using Equation (6)*/
Step 7	Select diode mode MOSFET load	/* for current mode signal processing*/
Step 8	Compute $(W/L)_{load}$	/* using Equation (8)*/
Step 9	Compute $(W/L)_{LS}$	/*minimum size transistors*/
Step 10	Design tail transistor of MA	/*refer (13)*/
Step 11	Design tail transistor of CA	/*refer (14)*/
	END	

Analysis of the Circuit

The analysis of the circuit is given below:

In **Figure 2**, applying the square law to the transistors of MA and CA, which are biased in saturation region, the currents, I_1 and I_2 through load transistors of the cross-coupled amplifiers (MA-CA) are given by:

$$I_1 = I_{d1} + I_{d4} = \beta_{\text{MA}} \left(V_1 - V_{tune} - V_{P(\text{MA})} - V_{thn}\right)^2 + \beta_{\text{CA}} \left(V_2 - V_{P(CA)} - V_{thn}\right)^2 \quad (15)$$

$$I_1 = I_{d2} + I_{d3} = \beta_{\text{MA}} \left(V_2 - V_{tune} - V_{P(\text{MA})} - V_{thn}\right)^2 + \beta_{\text{CA}} \left(V_1 - V_{P(CA)} - V_{thn}\right)^2 \quad (16)$$

where $\beta = (1/2)\mu C_{ox}(W/L)$ is the transconductance parameter. $V_{P(\text{MA})}$ and $V_{P(\text{CA})}$ are the drain to source voltage across tail transistor of MA and CA respectively.

The differential output current, i_o is given by

$$i_0 = I_1 - I_2 = 2(v_{id})\left[\{(\beta_{\text{MA}} - \beta_{\text{CA}})(V_{DC} - V_P - V_{thn})\} - \beta_{\text{MA}} V_{tune}\right] \quad (17)$$

where $V_{P(\text{MA})} \cong V_{P(\text{CA})} = V_P$ is assumed.

Since "q" is taken as 2 to minimize HD_3:

$$\beta_{\text{MA}} = 2\beta_{\text{CA}} \quad (18)$$

Combining Equation (17) and Equation (18), one gets:

$$i_0 = (v_{id})\{\beta_{\text{MA}}(V_{DC} - V_P - V_{thn}) - 2\beta_{\text{MA}} V_{tune}\} \quad (19)$$

From Equation (19) the transconductance G_m is calculated as,

$$G_m = \frac{\partial i_0}{\partial v_{id}} = \beta_{\text{MA}}(V_{DC} - V_P - V_{thn}) - 2\beta_{\text{MA}} V_{tune} \quad (20)$$

Equation (20) shows linear behaviour of G_m with V_{tune}, as first part, $\beta_{\text{MA}}(V_{DC} - V_P - V_{thn})$ on the right hand side of the equation is almost constant. Thus from Equation (20) the tuning range is obtained and is given in Equation (21).

$$G_{m_{range}} = |2\beta_{MA} V_{tune}| \quad (21)$$

SIMULATION RESULTS AND DISCUSSIONS

The proposed transconductor is simulated in Cadence VIRTUOSO environment using UMC 0.18 μm CMOS process technology. The transconductor is operated at 1.8 V and 27° Celsius. Bias currents $I_{Bias(\text{MA})}$ and $I_{Bias(\text{CA})}$ are kept at 200 μA and 25 μA, respectively to maintain $I_{Bias(\text{MA})} / I_{Bias(\text{CA})} = 8$. Channel length is taken as 0.9 μm which is five times the minimum as specified by the technology for all the MOSFETs of the transconductor to minimize channel length modulation effect [26]. The differential signal input voltage of 0.25 V is generated using voltage controlled voltage source (VCVS) as test bench setup.

The objective of the simulation is to demonstrate feasibility of power efficiency, linear tuning of G_m offering low harmonics with the help of separate bias control of cross-coupled amplifiers at 50 MHz and above. Different plots have been obtained to verify the designed aspects.

Plot in **Figure 3** shows the tunability of the proposed transconductor. The tuning voltage, V_{tune} is varied from 0.5 V to 1.0 V in steps of 0.1 V. The lower value (0.5 V) is kept slightly higher than the threshold voltage of input transistors M_5 and M_7 of LS and upper value is selected as per Equation (12) to keep them in saturation region. G_m is tuned between 390 μS to 195 μS. G_m is falling with increase in V_{tune} between ±0.25 V of v_{id} i.e. 1 $V_{p\text{-}p}$, However for higher values of v_{id}, (around ±0.5 V i.e. 2 $V_{p\text{-}p}$) the change in G_m w.r.t. V_{tune} is nonlinear and plots for different values of V_{tune} are almost merged at a point, indicates transconductance is independent of V_{tune}. The reason being that as per Equation (12), $V_1(or V_2) < V_{DD} + V_{thn}$, but at higher values, this condition is violated. At $V_1(or V_2) = V_{DD} + V_{thn}$ transistor M_5 and M_7 of **Figure 2** are at the verge of saturation, and variation in drain to source voltage across it is small enough, making output of level shifter is almost constant, i.e. independent of tuning voltage.

As shown in **Figure 4** the variation of transconductance with respect to tuning voltage is linear validating Equation (20). The small deviation may be attributed to body effect in LS transistors ($M_5$and M_7) and voltage drop in tail transistors.

Figure 5 shows HD_3 variation with respect to differential input signal amplitude for different values of tuning voltage. Distortion increases rapidly for the small values i.e. upto approximately 150 mV (0.6 $V_{p\text{-}p}$) of the differential input signal amplitude but after that it in creases

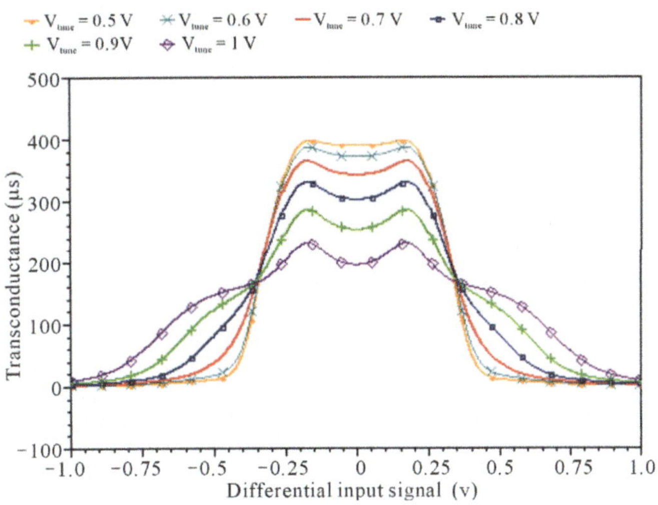

Figure 3. Transconductance (G_m) vs differential signal input (v_{id}) at different values of tuning voltage (V_{tune}).

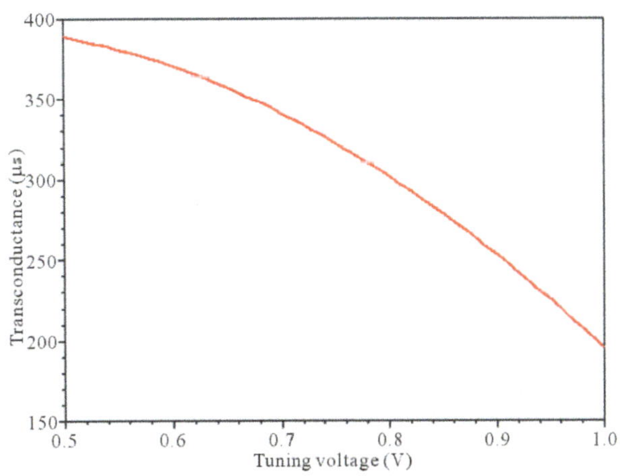

Figure 4. Transconductance (G_m) vs tuning voltage (V_{tune}) at differential signal input (v_{id}) = 0 V.

slowly due to cross-coupling effect. Harmonic distortion has also increased with the increase in tuning voltage due to decrease in DC current through MA. **Figure 6** shows variation of HD_3 with respect to V_{tune}.

Frequency response of the proposed transconductor for different values of tuning voltage is given in**Figure** 7. The 3 dB frequency of the transconductor is above 53 MHz for the complete range of variation of transconductance.

G_m, HD_3, P_{Diss} and G_m/I_{DD} for the three different values of G_m obtained from the simulation are given in **Table 1**. It may be noted that at lower values of transconductance current efficiency reduces. The reasons are twofold: firstly, transconductance is low and secondly lower transconductance obtained at higher value of tuning voltage, gives high overdrive voltage to LS transistors which in turn draws higher current through LS.

The performance is compared on eight metric points with other reported circuits as given in **Table 2**. The proposed transconductor offers better current efficiency

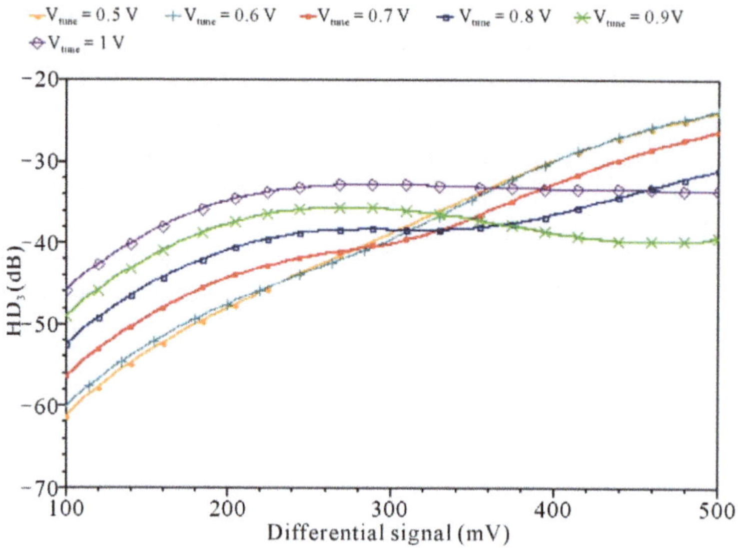

Figure 5. HD_3 vs differential signal input (v_{id}) at different tuning voltage (V_{tune}).

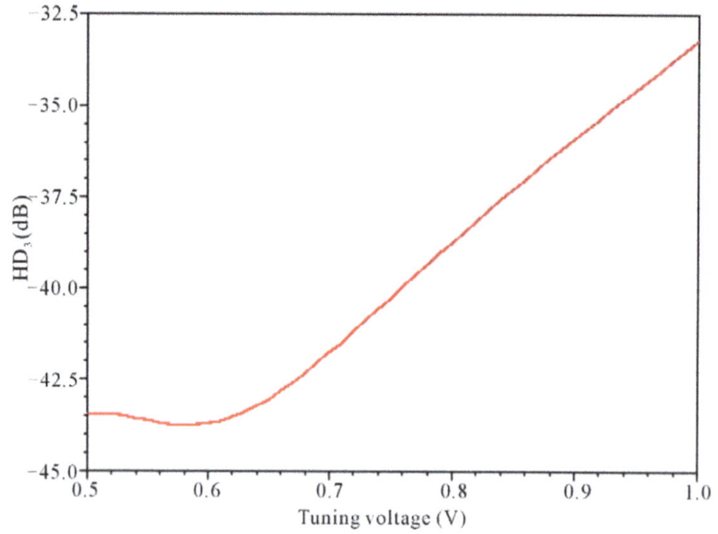

Figure 6. HD_3 vs Tuning voltage at differential signal input (v_{id}) = 0.25 V.

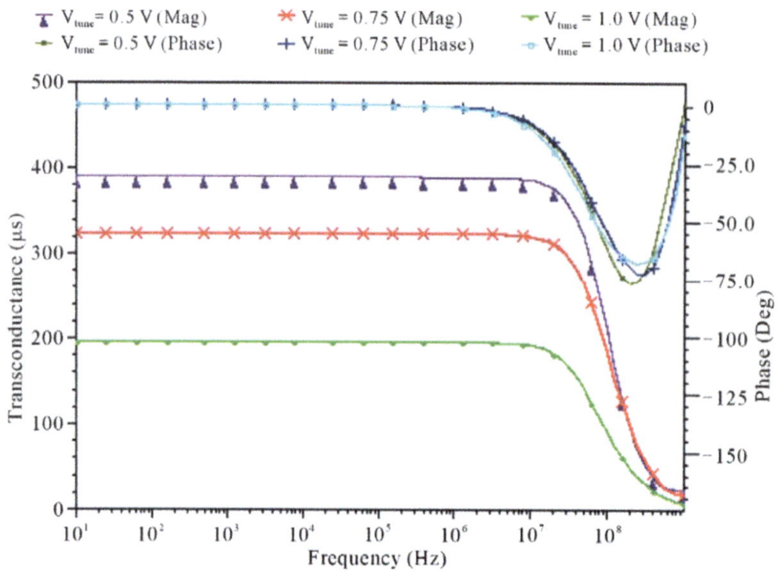

Figure 7. Transconductance vs frequency at different tuning voltage (V_{tune}).

(1.18 V^{-1}) and transconductance (390 μS) with lower harmonics (–43.7 dB), at input swing of 1 V_{p-p} @50 MHz in comparison to the others. However, the transconductance reported in [9] is higher than the proposed at the cost of high supply voltage (5 V) and very low current efficiency (0.1 $V^{-1}$to 0.07 V^{-1}).

CONCLUSIONS

A power efficient linearly tunable high G_m, cross-coupled transconductor with separate bias currents (for low third harmonic distortion) is designed and analyzed in this paper. Results have been obtained with diode mode transistors as load extendable for MOTA, which is necessary for current mode signal processing. Transconductance is varied in linear manner. G_m is tuned between 195 to 390 mS when the tuning voltage is varied from 1.0 V to 0.5 V. Maximum current efficiency of 1.18 V^{-1} and minimum HD_3 less than –43.7 dB is obtained at 1 V_{p-p} @ 50 MHz at the tuning voltage of 0.5 V. Comparative study reveals that the proposed circuit consumes low power and gives high current efficiency, with low HD_3 and high tunable transconductance for high frequency of operation compared to others.

ACKNOWLEDGEMENTS

This work has been performed using the resources of VLSI Laboratory developed under Special Manpower Development Programme for VLSI Design and related software (SMDP-II) project funded by Department of Information Technology, Ministry of Communication and Information Technology Government of India.

Table 1. Performance of proposed transconductor.

Performance Parameter	**Simulation results**		
Transconductance (G_m)	G_m = 390 μS (V_{tune}= 0.5 V)	G_m = 323 μS (V_{tune}= 0.75 V)	G_m = 195 μS (V_{tune} = 1.0 V)
(HD_3) at 1 V_{p-p}	– 43.7dB	– 40.5dB	– 33.2 dB
Power consumption (P_{diss})	0.60 mW	0.79 mW	1.20 mW
Current efficiency (G_m/I_{DD})	1.18 V^{-1}	0.74 V^{-1}	0.30 V^{-1}
3 dB Frequency	70 MHz	71 MHz	53 MHz

Table 2. Performance comparison of proposed transconductors with others.

Reference No.	[6]	[9]	[16]	[17]	[22]	[23]	Proposed
Supply voltage in V	1.8	5.0	3.3	2.5	1.8	1.8	1.8
Technology in μm	0.18	2.00	0.35	0.35	0.18	0.18	0.18
G_m in μS	8 to 131	166 to 575	160 to 340	3 to 30	110	310 to 100	390 to 195
Input Swing in V_{p-p}	1.0	1.0	1.4	1.0	0.1 0.9	1.0	1.0
HD_3/THD in dB	– 40	– 40	– 70	– 42 to –50	– 27 to – 43	– 30	– 43.7
Power consumption in mW	0.58[$]	8.24 to 41.50	6.60	0.38 to 0.24	0.42[$]	0.64[$$]	0.60 to 1.20
#Current efficiency in V^{-1}	0.02 to 0.41	0.10 to 0.07	0.08 to 0.17	0.02 to 0.31	0.47	0.60	1.18 to 0.30
Frequency in MHz	10	10	26	1	50	50	50

Legend: # Calculated from the given data/average value in the references. [$]: Authors did not mention G_m at which the Power consumption values are reported. [$$]: Power at a specific G_m.

REFERENCES

1. B. Razavi, "Design of Analog CMOS Integrated Circuits," Tata McGraw-Hill Publishing Company Limited, 2002.
2. D. A. Johns and K. Martin, "Analog Integrated Circuit Design," John Wiley and Sons, New York, 1997.
3. P. E. Allen and D. R. Holberg, "CMOS Analog Circuit Design," Oxford University Press, New York, 2004.
4. E. Sánchez-Sinencio and J. Silva-Martínez, "CMOS Transconductance Amplifiers, Architectures and Active Filters: A Tutorial," IEE Proceedings Circuits, Devices & Systems, Vol. 147, No. 1, 2000, pp. 3-12. doi:10.1049/ip-cds:20000055
5. J. Y. Kim and R. L. Geiger, "Characterisation of Linear MOS Active Attenuator and Amplifier," Electronic Letters, Vol. 31, 1995, pp. 511-513. doi:10.1049/el:19950353
6. X. Zhang and E. I. El-Masry, "A Novel CMOS OTA Based on Body-Driven MOSFETs and Its Applications in OTA-C Filters," IEEE Transactions on Circuits and System-I Regular Papers, Vol. 54, No. 6, 2007, pp. 1204-1211. doi:10.1109/TCSI.2007.897765
7. J. Chen, E. Sanchez-Sinencio and J. Silva-Martinez, "Frequency Dependent Harmonic Distortion Analysis of a Linearized cross Coupled CMOS OTA and Its Application to OTA C-Filter," IEEE Transactions on Circuits and System-I Regular Papers, Vol. 53, No. 3, 2006, pp. 499-510. doi:10.1109/TCSI.2005.859575

8. Z. Wang and W. Guggenbuhl, "A Voltage-Controlled Linear MOS Transconductor Using Bias Offset Technique," IEEE Journal of Solid-State Circuits, Vol. 25, 1990, pp. 315-317. doi:10.1109/4.50321
9. Y. Sun, C. Hill and S. Szczepanski, "Large Dynamic Range High Frequency Fully Differential CMOS Transconductance Amplifier," Analog Integrated Circuits and Signal Processing, Vol. 34, 2003, pp. 247-255.
10. D. V. Morozov and A. S. Kuroki, "Transconductance Amplifier with Low-Power Consumption," IEEE Transactions on Circuits and System-II Express Briefs, Vol. 52, No. 11, 2005, pp. 776-779. doi:10.1109/TCSII.2005.852190
11. A. Lewinski and J. Silva-Martinez, "OTA Linearity Enhancement Technique for High Frequency Application With IM3 Below-65 dB," IEEE Transactions on Circuits and System-II Express Briefs, Vol. 51, No 10, 2004, pp. 542-548. doi:10.1109/TCSII.2004.834531
12. S. Ouzounov, E. Roza, H. Hegt, G. V. D. Weide and A. V. Roermund, "Design of MOS Transconductors with Low Noise and Low Harmonic Distortion for Minimum Current Consumption," Integration, the VLSI Journal, Vol. 40, 2007, pp. 365-379.
13. F. Krummenacher and N. Joehl, "A 4-MHz CMOS Continuous-Time Filter with On-Chip Automatic Tuning," IEEE Journal of Solid-State Circuits, Vol. 23, No. 3, 1988, pp. 750-758. doi:10.1109/4.315
14. J. Silva-Martinez, M. S. J. Steyaert and W. M. C. Sansen, "A Large-Signal Very Low-Distortion Transconductor for High-Frequency Continuous-Time Filters," IEEE Journal of Solid-State Circuits, Vol. 26, No. 7, 1991, pp. 946-954. doi:10.1109/4.92014
15. M. Kachare, A. J. Lopez-Martin, J. Ramirez-Angulo, and R. G. Carvajal, "A Compact Tunable CMOS Transconductor with Linearity," IEEE Transactions on Circuits and System-II Express Briefs, Vol. 52, No. 2, 2005, pp. 82-84. doi:10.1109/TCSII.2004.842065
16. W. Huang and E. Sanchez-Sinencio, "Robust HighlyLinear High-Frequency CMOS OTA with IM3 Below-70 dB at 26 MHz," IEEE Transactions on Circuits and System-I Regular Papers, Vol. 53, No. 7, 2006, pp. 1433- 1447. doi:10.1109/TCSI.2006.875187
17. F. A. P. Baruqui and A. Petraglia, "Linearly Tunable CMOS OTA With Constant Dynamic Range Using Source-Degenerated Current Mirrors," IEEE Transactions on Circuits and System-II Express Briefs, Vol. 53, No. 9, 2006, pp. 797-801. doi:10.1109/TCSII.2006.881162
18. P. Monsurrò, S. Pennisi, G. Scotti and A. Trifiletti, "Linearization Technique for Source-Degenerated CMOS Differential Transconductors," IEEE Transactions on Circuits and System-II, Vol. 54, No. 10, 2007, pp. 848-852. doi:10.1109/TCSII.2007.906203
19. K. Kuo and A. Leuciuc, "A Linear MOS Transconductor Using Source Degeneration and Adaptive Biasing," IEEE Transactions on Circuits and System-II Express Briefs, Vol. 48, No. 10, 2001, pp. 937-943. doi:10.1109/82.974782
20. S. H. Yang, K. H. Kim, Y. You and K. R. Cho, "A Novel CMOS Operational Transconductance Amplifier Based on Mobility Compensation Technique," IEEE Transactions on Circuits and System-II Express Briefs, Vol. 52, No. 1, 2005, pp. 37- 42. doi:10.1109/TCSII.2004.839539
21. S. Koziel and S. Szczepanski, "Design of Highly Linear Tunable CMOS OTA for Continuous-Time Filters," IEEE Transactions on Circuits and System-II Analog and Digital Signal Processing, Vol. 49, No. 2, 2002, pp. 110-122. doi:10.1109/TCSII.2002.1002513
22. A. A. Fayed and M. Ismail, "A Low Voltage, Highly Linear Voltage-Controlled Transconductor," IEEE Transactions on Circuits and System-II Express Briefs, Vol. 52 No. 12, 2005, pp. 831-835. doi:10.1109/TCSII.2005.853511
23. V. Bhadauria and K. Kant, "A Novel Technique for Tuning Low Voltage Linear Transconductor," 2010 International Conference on Electronic Devices, System and Application (ICEDSA2010), 12-13 April 2010, Kuala Lumpur, Malaysia, pp. 22-25. doi:10.1109/ICEDSA.2010.5503108
24. V. Bhadauria K. Kant and S. Banerjee, "A Tunable Transconductor with High Linearity," Proceedings of Asia Pacific Conference on Circuits and Systems (APCCAS 2010), Kuala Lumpur, 6-9 December 2010, pp. 5-8. doi:10.1109/APCCAS.2010.5774749
25. P. Bruschi, F. Sebastiano and N. Nizza, "CMOS Transconductors with Nearly Constant Input Ranges over wide Tuning Intervals," IEEE Transactions on Circuits and System-II Express Briefs, Vol. 53, No. 10, 2006, pp. 1002- 1006. doi:10.1109/TCSII.2006.882126
26. R. J. Baker, "CMOS Circuit Design, Layout, and Simulation," IEEE Press, Wiley-Interscience, A John Wiley & sons, Inc., Publication, New York, 2005, Chapter 9, p. 291.

CHAPTER 6

A RECONFIGURABLE NETWORK-ON-CHIP DATAPATH FOR APPLICATION SPECIFIC COMPUTING

Joshua Weber and Erdal Oruklu

Keywords: Reconfigurable Computing; Network-on-Chip; Network Simulators; Polymorphic Computing

INTRODUCTION

Technological advances in Field Programmable Gate Arrays (FPGA) and performance improvement of reconfigurable systems are making a large impact on signal processing and computer processing. In addition, System-on-Chip (SoC) methodology facilitates tightly placement of reconfigurable arrays with embedded generalpurpose processors. In the past reconfigurable hardware has been used as a prototyping platform to aid in time to market development of application specific integrated circuits (ASIC). Increasingly, it is being shown that utilizing reconfigurable features in the field is producing designs with higher performance, lower cost, lower power, decreased design time, and increased flexibility.

Application-specific instruction set processors (ASIP) use extensions to standard processor instruction sets to achieve significant performance gains within various application domains. It has been shown that this approach will be a primary driver for future processor improvements [1]. Polymorphic processors [2,3] attempt to retain the flexibility advantages of general-purpose processors while providing the same advantages inherent in ASIPs and custom hardware logic. Polymorphism is a term borrowed from computer science, which is the ability for an object to utilize a common interface and yet execute custom type-specific actions based on the input type. Polymorphic computing provides the same functionality to a computer processor; creating a common programming interface, but allowing the execution of the algorithm to be application specific. Hence, the goal of a polymorphic processor is a general-purpose processor that can be reconfigured easily to accommodate application specific optimizations as needed.

This work presents a novel and unique polymorphic processor design. Integration of reconfigurable elements into a traditional general-purpose processor is achieved through replacement of the processor datapath with a network-on-chip (NoC) design. This NoC enables a higher level of fine-grained flexibility in the operation of the processor. The flexibility of the polymorphic NoC (PolyNoC) processor can be leveraged in many ways to achieve noticeable performance gains. A cycle-accurate simulator is produced to demonstrate the performance gains achievable by the PolyNoC architecture.

RELATED WORK

Reconfigurable computing has been the subject of much research [4] and it has been shown that reconfigurable computing can provide a significant improvement in performance over standard general-purpose processors [5,6]. Reconfigurable architectures offer a reduction in size and cost, improved time to market, and increased flexibility. All of which are especially important for embedded systems. The majority of systems focus on the integration of a general purpose processor with a reconfigurable resource array (most often an FPGA). The goal of polymorphic processor is to integrate these two units into a common design methodology and run-time control system.

An overview of the current trends and types of architectures has been studied and is shown in **Figure 1** [7]. In **Figure 1**(a), the reconfigurable array is communicating with the general purpose processor through the I/O data bus. In Figures 1(b) and (c), reconfigurable array is moved closer to the processor to decrease the communication cost. In **Figure 1**(d), the reconfigurable unit is coupled within the processor as a configurable functional unit. With the increasing size and complexity of programmable fabrics and the use of soft core

Citation: J. Weber and E. Oruklu, "A Reconfigurable Network-on-Chip Datapath for Application Specific Computing," Circuits and Systems, Vol. 4 No. 2, 2013, pp. 181-192. doi: 10.4236/cs.2013.42025.

processor, a processor core can also be embedded inside the reconfigurable fabric as shown in **Figure 1**(e).

These approaches have all been applied in very different application domains. Research has been performed to accelerate cryptographic systems [8], Euclidean distance transformation [9], matrix multiplication [10], and multimedia applications for imaging [11] and video processing [12]. In addition research has been performed into creation of polymorphic processors, which can be reconfigured and applied to multiple application domains [2,3, 13-15].

A common problem for all polymorphic processors is the integration of design flow for both software and reconfigurable hardware. Some work has been performed in auto generation of hardware design targeted at reconfigurable polymorphic processors [16-18]. Most of these approaches use the standard C language to capture program functionality and then through compiler optimization extract sections to be targeted at reconfigurable resources.

RECONFIGURABLE DATAPATH DESIGN

The proposed PolyNoC processor attempts to closely integrate a general-purpose processor with reconfigurable elements. This tight coupling produces a polymorphic reconfigurable processor, which can be reconfigured and extended in real time to include application specific optimizations. Enabling the ability to tailor the processor architecture to the specific application at run time is an important feature of this design.

Prior approaches at integration of reconfigurable elements to general-purpose processor have had various types of implementation ranging from simple attachment as a peripheral to tighter coupling to the memory system or co-processor usage. The PolyNoC processor approaches the design goals by integrating the reconfigurable elements into the foundation of the processor architecture. Unlike previous designs, the datapath of the processor is replaced with a network-on-chip (NoC). Instead of traditional direct wire communication with pipeline registers, the PolyNoC processor utilizes a NoC for all element-to-element communication. The use of a NoC provides many advantages to the design and enables the PolyNoC processor to achieve its main goal of real time reconfigurability.

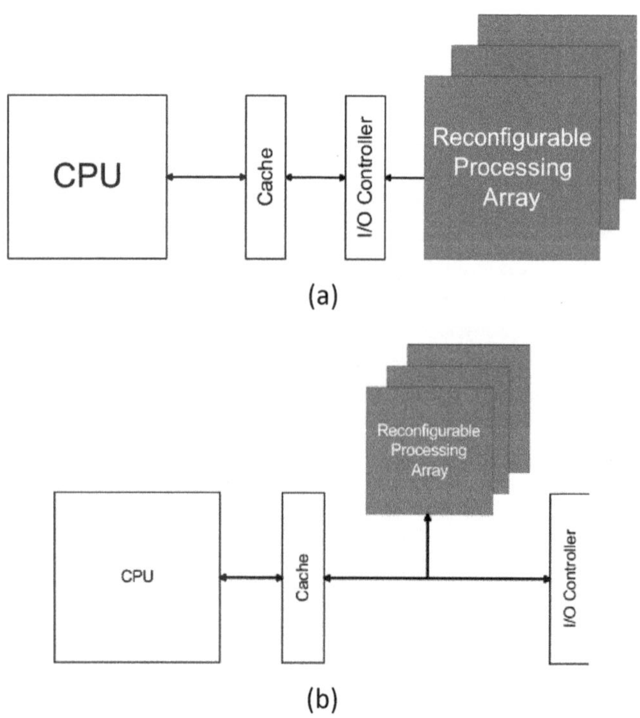

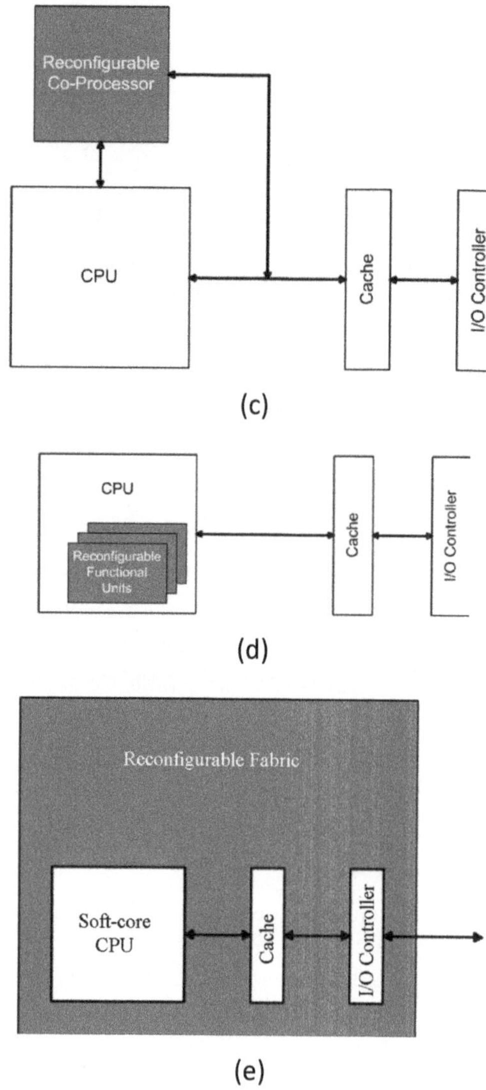

Figure 1. Integration of processors and reconfigurable logic.

The NoC provides a flexible interconnect and allows new functional computation elements to be added and subtracted in real time. Furthermore, the NoC relieves the designer from strict constraints on placement and ordering of functional elements, as the NoC will provide element-to-element communication regardless of each elements placement. However, the NoC also place many constrains on the overall processor operation. In particular, the latency penalty from the NoC needs to be overcome by the additional advantages provided by the PolyNoC processor, namely the ability to reconfigure the processor to have application specific optimizations.

POLYNOC PROCESSOR ARCHITECTURE

The PolyNoC processor architecture can be seen in Figure 2. It consists of a standard instruction and data cache connected to a instruction fetch and decode unit. Next, the instruction is passed to a packetizer. The packetizer wraps the instruction into a NoC packet and then transmits it through the NoC. The packetizer also keeps track of all instruction executions and issues new instructions. The NoC delivers instruction packets to functional units. Each instruction slowly progresses from functional unit to functional unit, stopping to get a subset of execution done.

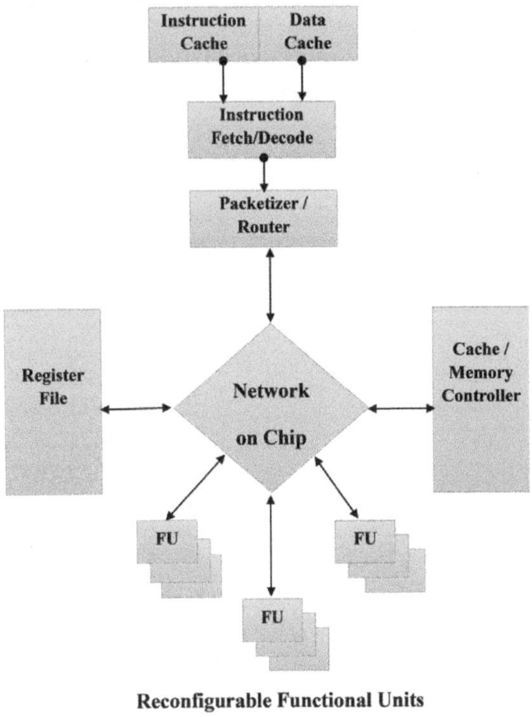

Figure 2. Network-on-chip datapath for polymorphic processors.

Rather than starting from scratch when designing the processor, an existing processor architecture has been adopted and extended. This enables the PolyNoC processor to focus on architectural issues, without dealing with the many contributing factors from compiler and instruction set changes. In addition, the use of a common target processor allows the PolyNoC processor to utilize all existing software and compiler tools available for the target processor. For these reasons, the PolyNoC processor is designed to execute SPARC code and emulate a SPARC processor [19]. Specifically the PolyNoC processor is based upon the LEON3 SPARC processor.

The base level of functional units represents roughly the stages of execution of the SPARC processor the design is modeled on. These base functional units consist of the register file, memory controller, ALU, and FPU. Each functional unit performs computations that are roughly equivalent to the pipeline stages of a LEON3 SPARC processor, the target processor being extended into the PolyNoC. As such, much of the implementation of the basic functional units can remain unchanged from a target general-purpose processor. Only modifications to the input and output communications are necessary.

Besides base functional units, additional functional units can be added to the system. These are represented by blocks of reconfigurable functional units, which can be configured depending on the current application. For example, they can be used to instantiate additional core units such as additional FPU functional elements. By instantiating additional FPU elements, the processor can be tailored to execute a floating-point heavy application. The reconfigurable functional blocks can also be configured to implement custom user logic. By enabling custom logic, the processor can execute new custom instructions. These instructions and logic can be heavily optimized and designed to support a specific application, enabling a great deal of performance increase.

FUNCTIONAL UNITS

Packetizer

The packetizer, as shown in **Figure 3**, functions as the primary control of instruction execution. It feeds a program counter (PC) value to the instruction cache unit and accepts the incoming instructions for execution. In order to overcome the increase in execution time caused by the NoC latency and to leverage the flexibility advantages of the NoC datapath, the packetizer allows for multiple instruction issues and out-of-order completion. To accomplish this, the packetizer must track the issuing and completion of all instructions and control the update of the PC.

The packetizer allows for an unlimited number of instructions in flight. This is constrained by the amount of instruction level parallelism present in the program. Before issuing any instruction, it is checked for data dependency. This dependency check is performed by the built in dependency tracker detailed in the Section 5.1.1. The dependency tracker analyses 1) data dependency among in-flight instructions and 2) control oriented dependencies from branch instructions. Once the dependency tracker declares the instruction as dependency free, it will be issued for execution and dependency tracker is updated with dependencies of the newly issued instruction.

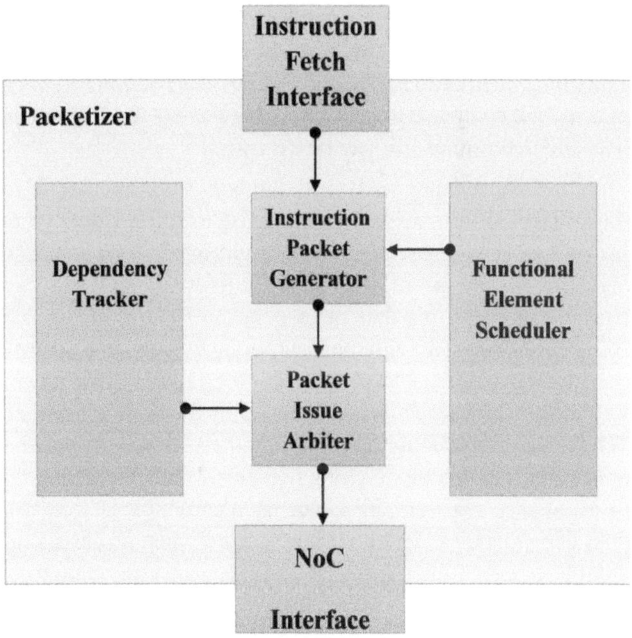

Figure 3. Packetizer block diagram.

In order to issue an instruction for execution, it needs to be packetized and transmitted to the NoC. The packetizer decodes the instruction to determine the list of necessary functional elements. For example, a basic ADD instruction will need to visit the register file to obtain input register operands, then the ALU unit for execution of the operation, then return to the register file to write the results into the output register and finally return to the packetizer to update execution completion status. Once an instruction is decoded, the functional element destinations are fed into the functional element scheduler, as detailed in Section 5.1.2. The functional element scheduler will return a network id for each functional element required.

A NoC packet is created to hold the necessary instruction contents. The critical data is the instruction itself and the destination list of functional elements. All of this data are wrapped into a NoC data packet, encoded with the destination of the first functional element and transmitted into the NoC. As instructions are completed they return to the packetizer. The packetizer checks the status of the packet to ensure it was correctly executed. Any instruction that failed execution will return as a trap condition (the packetizer will update the PC to a trap handler). If the instruction was a branch instruction, the new PC value will be included. As a final step, the packetizer will pass the completed instruction to the dependency tracker to remove any dependencies caused by this packet.

Dependency Tracker

The dependency tracker is responsible for determining if a new instruction is dependency free from all currently executing instructions. The tracker accomplishes this task by maintaining an internal list of all data and control dependencies of currently executing instructions. When a new instruction is to be executed, it compares the necessary registers to the dependency lists and determines if it can be executed.

The dependency lists are implemented as simple bit flags. Two 32 bit registers are created to track status of the instruction registers, eight global, eight local, eight input and eight output integer registers and 32 floating point registers. Each bit indicates if a data dependency exists for that register. As each instruction begins execution, it is registered with the dependency tracker. At this time all destination registers being written by the instruction are flagged in the tracking register. When a new instruction is to be executed, all source registers are compared with the tracking registers. If bits corresponding to all source registers are clear, the instruction is free from data dependencies. The final step occurs when an instruction completes execution. In this case the bits corresponding to the destination register are cleared from the tracking list to indicate that there is no longer a data dependency for those registers.

In addition to tracking data dependencies, the dependency tracker also tracks control dependencies. Unlike a traditional datapath, it is difficult to signal flush operations for instructions in flight. As such, no branch prediction is used and all branching cases cause a control dependency. When a control instruction is in-flight all future instructions hold off on execution until resolution of the branch instruction. The dependency tracker provides this tracking for the packetizer to control instruction issuing. In addition, as the register windows are affected by a SAVE or RESTORE instruction, they too exhibit a control dependency. All instruction must await completion before execution.

The dependency tracker is critical to the overall operation of the PolyNoC processor. Together with the packetizer it enables a multiple issue instruction datapath. Since, the NoC datapath imparts latency penalties to execution compared to a traditional processor, a large mitigating factor provided by the PolyNoC processor is the ability to extract instruction level parallelism from the instruction stream. The dependency tracker is crucial for this ability. It checks each subsequent instruction for dependencies. If the instruction is found to be free of dependencies then it is executed at the same time as the prior instruction. This behavior helps to keep the execution elements fully utilized and provides significant benefits to overall execution time.

Functional Element Scheduler

The functional element scheduler provides addressing and scheduling of functional elements to the packetizer. It maintains a list of all functional elements, their NoC addresses, and the functional element type. For simple cases with only one instance of each functional element, it simply returns the address of the element when a request is made for an element of that type. For cases when more than one element of a type has been created, it performs scheduling. The current PolyNoC processor performs scheduling through a round robin scheduling algorithm. This schedules all functional elements equally and distributes instruction load evenly. It also benefits from a very simple implementation and critically requires no signaling from any functional element. For these reasons, it was chosen as the scheduling algorithm for the PolyNoC processor. Further performance gains can be made through more advanced and intelligent scheduling algorithms since the round robin algorithm does not take into account current status of functional elements.

Register File

The register file implements and provides access to all processor registers. It implements the register definitions according to the SPARC architecture [19]. It defines a set of overlapping windows general-purpose instruction registers. These overlapping windows provide functionality to pass parameters between subroutine calls. The execution of SAVE and RESTORE instructions allow the adjustment to a new register window, which contains shared output and input registers for different window values.

The register file further implements 32 floating-point general-purpose registers. These registers are usable by all floating-point instructions. In addition to generalpurpose registers, the register file also implements the SPARC control and status registers. These control and status registers provide access to configuration and status of a SPARC processor and are necessary for compliance with SPARC execution model. Some status registers are critical in execution, for example the processor state register (PSR) holds the contents of the condition codes. The register file also supports the execution of the SAVE and RESTORE instructions, shifting the register window up or down. All of these options combine to make a fully compliant SPARC register set.

Memory Controller

The memory controller enables access to external addressed memory. This supports both reading and writing of external memory. The basic functionality of the memory controller is simple and its implementation is straightforward. The SPARC standard defines the primary external memory bus to use the AMBA AHB bus protocol. Therefore, the PolyNoC processor and the memory controller implements an AMBA AHB master controller for accessing the bus.

In addition to providing access to the external memory space, small amount of configuration memory space must also be implemented within the memory controller. This memory space provides plug-n-play capabilities for the AHB bus, enabling AHB bus slaves to be added and configuration options and memory address locations stored. This configuration space is crucial to the memory controller master controller and used to generate chip selects and initiate access to AHB slaves.

Arithmetic Logic Unit

The ALU performs all integer arithmetic and logic computations for instructions. In addition to execution of basic instructions, the ALU also computes memory addresses and branch and jump targets. The ALU must be robust enough to support arithmetic and logic operations for all data types as required by the SPARC instruction set. ALU performance can have a significant impact on over all processor performance but from the standpoint of the operation of the PolyNoC processor, it does not present any new challenges.

Floating-Point Unit

Like the ALU, the FPU performs all arithmetic and logic computations for floating point operations. This implements all basic computations on standard IEEE-754 floating-point numbers. The implementation can vary widely and have performance impacts on the PolyNoC processor. A potential advantage of the PolyNoC processor is the ability to incorporate multiple or even different FPU implementations allowing for design choices that are not feasible with a traditional general purpose processor.

NoC INTERCONNECT

The network-on-chip (NoC) interconnect links all functional elements together and enables communication among them. Any instruction will traverse the NoC multiple times during execution. For the basic SPARC instruction with just the core functional elements, it will require 3 to 5 trips through the NoC to complete execution of each instruction (similar to pipeline stages). Therefore, performance of the NoC links is very important to the overall performance of the PolyNoC processor.

The NoC should also be able to provide a scalable architecture with a dynamic layout. The PolyNoC processor's main advantages are in flexibility and reconfiguration. The greatest performance gains are obtained by real time reconfiguration of the number and type of functional elements. The NoC must be able to support to addition and subtraction of functional elements with no impact to overall NoC performance. In addition, due to the heavy dependency on latency, the NoC needs to scale well. As additional functional elements are added and the total number of nodes within the NoC increases, the average latency for travel over the NoC should not grow rapidly. If the NoC latency increases too quickly, any performance gain from additional functional elements can be lost due to the penalties from increased execution time.

The NoC also needs to be able to support a flexible packet type and size. Each NoC packet will need to encapsulate an instruction and all necessary information for execution of that instruction. Data carried by the packets will vary between each instruction. To save on overall packet size, the packet should allow for a flexible packet size to shrink or grow according to the demands of the individual instruction. Packet structure should have a minimum amount of overhead when encapsulating an instruction and data. Minimizing this overhead reduces the overall packet size and lower the overall bandwidth requirements for interconnect links.

The traffic generation for the PolyNoC processor is dependent on instruction execution. As a result, the traffic load is reduced compared to other NoC systems. The execution of instructions tends to be limited by the computational time of the functional elements. This stark limit on the number of instructions currently executing makes for a low overall bandwidth requirement. The low bandwidth requirement provides one of the main NoC design techniques to improve performance. The PolyNoC processor can run the NoC communications on a separate and higher frequency clock then the functional elements. Since latency is a critical NoC parameter, the higher NoC clock allows compensating for the NoC delay. The latency figure is computed based on the NoC clock. When running the NoC at a higher clock frequency, the overall NoC delay is drastically reduced. In many cases, it seems reasonable to run the NoC clock at 4 or 8 times the functional element clock. This provides the ability to have up to 4 or 8 clock cycles of latency for NoC transmission within a single functional element execution clock cycle, potentially hiding all latency penalties during execution.

The PolyNoC processor presents a very unique set of requirements for a NoC. In addition, the traffic generation is the result of instruction execution. This type of traffic load is uncommon and has not been studied in depth. For these reasons, it is difficult to approach the design of the NoC from theoretical statistical approach, as is often applied during NoC research. Instead, this work has focused on an experimental approach. A simulator for NoC design exploration has been created and various NoC design parameters have been modified and tested to determine the impact on NoC performance.

NoC ARCHITECTURE EXPLORATION

This section presents an exploration of the architectural and topology issues impacting the performance of the PolyNoC processor [3]. The Network Simulator 2 (ns-2) [20] is utilized as a network simulation platform. Ns-2 is extended to simulate the execution of the PolyNoC processor. With this simulation platform, instruction trace data from real world benchmark programs can be run through the simulator and it provides an excellent research tool for the study of the PolyNoC processor.

The ns-2 simulator is a discrete event simulator targeting networking research. It has been used in a wide variety of networking research including simulation of NoC architecture and protocols. It provides support for a wide range of networking techniques, in addition to robust access to the underlying framework to allow for custom modification and extension of the simulator. Ns-2 is implemented in C++ with an interface layer based upon OTcl. All major simulation objects are created as discrete objects in C++. It provides very fine grained access to all objects within a network, packet structure, link parameters, routing agents, network node modeling, and support for modeling of all layers of the OSI stack. Beyond the extensive list of built in networking objects, ns-2 provides the ability to extend the platform to incorporate new designs and techniques. This extensibility is heavily utilized for modeling the PolyNoC network execution.

Ns-2 was modified and extended to support simulation of a processor executing with a NoC datapath. To facilitate this, a new custom set of applications, agents, nodes, and packets were added to the ns-2 simulator. All functional units within the processor are represented in the ns-2 simulator as an application. These applications then communicate with each other using the ns-2 node network simulation.

Using the ns-2 simulator, four main network topologies were studied and compared as shown in **Figure 4**. These topologies consist of Ring, Modified Ring, 2D Mesh, and Crossbar architectures. The ring architecture provides a simple baseline implementation. Due to the fact that a processor datapath is highly sequential, it is very similar to the flow of data through a traditional datapath. The modified ring structure attempts to provide a more robust scalability than the standard ring. In this architecture similar functional units are placed in parallel on the ring structure. This incurs a minor increase in link cost, but provides the same total delay over the ring regardless of number of functional units instantiated. The next architecture studied is the standard 2D mesh architecture. The architecture is a rectangular layout with bidirectional links between neighboring nodes. The final architecture explored is a crossbar. The crossbar architecture is an extreme architecture in which every node has a link to every other node. This shrinks all communications to only a single hop away, but comes with a very sharp increase in link cost as the network scales in nodes. Figure 5 shows an example ns-2 script for evaluating the performance of a simple PolyNoC processor architecture (i.e., ring topology with only basic functional elements).

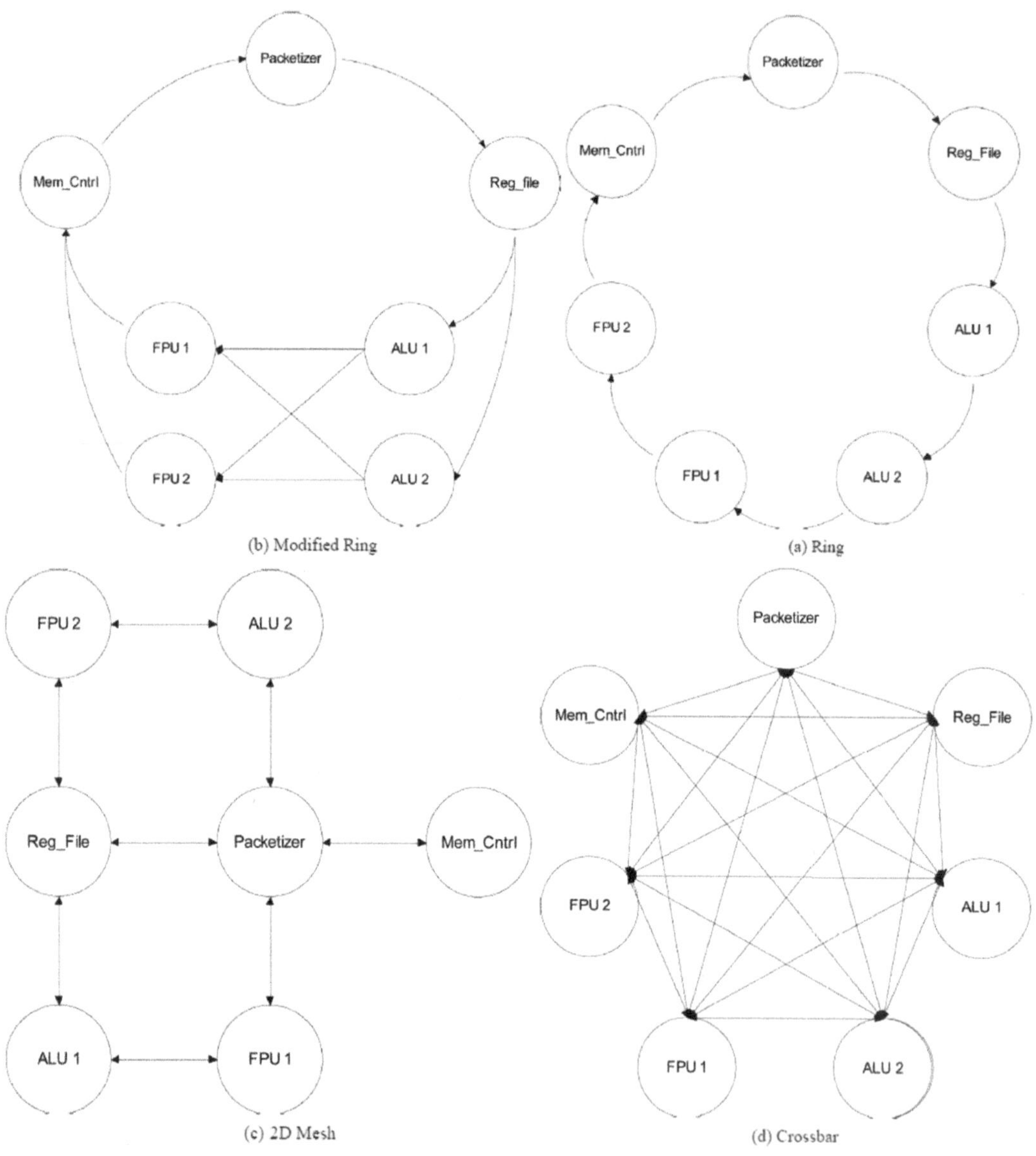

Figure 4. NoC topologies investigated.

In addition to exploration of the various topologies, the impact of increasing the number of functional units was also explored. The increase in the number of functional units and the ease of integration to the processor design is fundamental to a polymorphic processor. Increasing the number of computational functional units (both ALU and FPU) directly impacts the overall performance.

```
# This will simulate the CRC32 Algorithm
# with 1 ALU, 1 FPU, 1 RegFile, 1 MemCntrl
set FileName "CRC32_Ring_1_1_1_1"
# Select the benchmark to run
set Benchmark_Select_Num 4
# Link speed
set link_speed 6.4Gb
# Instruction Window Size
set inst_win_size 16
puts "Now starting execution of $FileName"

# Use only the needed packet headers.
remove-all-packet-headers
add-packet-header IP RTP NoC

# Create a new simulator object
set ns [new Simulator]
#Define a 'finish' procedure
proc finish {} {
        global ns nf FileName Benchmark_Select_Num
                link_speed inst_win_size
        $ns flush-trace
    # Print details of the execution to the screen
    # for reference
    puts "Executed file $FileName"
    puts "Benchmark selected was $Benchmark_Select_Num"
    puts "Link Speed was $link_speed"
    puts "Instruction Window Size was $inst_win_size"
        exit 0
}
#Create five nodes
set n1 [$ns node]
set n2 [$ns node]
set n3 [$ns node]
set n4 [$ns node]
set n5 [$ns node]

#Create links between the nodes
$ns duplex-link $n1 $n2 $link_speed 0ms DropTail
$ns duplex-link $n2 $n3 $link_speed 0ms DropTail
$ns duplex-link $n3 $n4 $link_speed 0ms DropTail
$ns duplex-link $n1 $n5 $link_speed 0ms DropTail

# Attach NoC Agents to all nodes
set noc1 [new Agent/NoC]
$ns attach-agent $n1 $noc1
set noc2 [new Agent/NoC]
$ns attach-agent $n2 $noc2
set noc3 [new Agent/NoC]
$ns attach-agent $n3 $noc3
set noc4 [new Agent/NoC]
$ns attach-agent $n4 $noc4
set noc5 [new Agent/NoC]
$ns attach-agent $n5 $noc5

#Attach a NoC Packetizer to node 1.
set packetizer [new Application/NoC/Packetizer]
$packetizer attach-agent $noc1

#Attach a PE application to nodes 2-5.
set reg_file [new Application/NoC/PE]
$reg_file set process_time_ 0.00000001
$reg_file attach-agent $noc2
set alu [new Application/NoC/PE]
$alu attach-agent $noc3
set fpu [new Application/NoC/PE]
$fpu attach-agent $noc4
set mem_cntrl [new Application/NoC/PE]
$mem_cntrl set process_time_ 0.00000004
$mem_cntrl attach-agent $noc5

$packetizer set benchmark_select_ $Benchmark_Select_Num
$packetizer set inst_window_ $inst_win_size
$packetizer loadtrace
puts "We are done Loading the Trace"

$packetizer scheduler add-packetizer [$packetizer get-addr]
        [$packetizer get-port]
$packetizer scheduler add-reg_file [$reg_file get-addr]
        [$reg_file get-port]
$packetizer scheduler add-alu [$alu get-addr] [$alu get-port]
$packetizer scheduler add-fpu [$fpu get-addr] [$fpu get-port]
$packetizer scheduler add-mem_cntrl [$mem_cntrl get-addr]
        [$mem_cntrl get-port]
puts "We are done setting the processing element address."

$ns compute-routes

$packetizer sim-trace

#Call the finish procedure after 5 seconds of simulation time
$ns at 1.0 "finish_stats"
$ns at 1.0 "finish"

proc finish_stats {} {
    global reg_file alu fpu mem_cntrl
    set reg_util [$reg_file get-util]
    puts "Reg File utilization is $reg_util"
    set alu_util [$alu get-util]
    puts "ALU File utilization is $alu_util"
    set fpu_util [$fpu get-util]
    puts "FPU utilization is $fpu_util"
    set mem_util [$mem_cntrl get-util]
    puts "Mem control utilization is $mem_util"
}

#Run the simulation
$ns run
```

Figure 5. Sample ns-2 script for instantiating a PolyNoC processor architecture (ring topology).

SIMULATION RESULTS

For performance analysis, benchmark programs have been compiled for execution on the PolyNoC processor simulator. The benchmarks come from the well-known MiBench benchmark suite [21]. MiBench provides a set of commercially representative embedded systems programs. Extracted from this set and used are the CRC32, FFT, IFFT, ADPCM.encode, ADPCM.decode, and BasicMath benchmark. This set of benchmarks provides a varied application load to get an initial impression on the performance of the proposed processor design.

- CRC32-Performs a 32-bit Cyclical Redundancy Check on input data. CRC checks are often used as error checking during data transmission.
- FFT/IFFT-Performs a fast Fourier transform or an inverse fast Fourier transform on an input data array. The FFT is used for frequency analysis during signal processing in a very wide range of application domains.
- ADPCM encode/decode-Adaptive Differential Pulse Code Modulation is a variation of the more common Pulse Code Modulation (PCM). This variation takes in 16-bit linear PCM samples and converts them to 4-bit samples providing significant compression. This algorithm is executed over samples of speech.
- BasicMath-Performs common mathematical computations that very frequently do not have hardware implementations. In this benchmark cubic function solving, integer square roots and angular conversion are computed.

Benchmark programs are first run on the TSIM LEON3 SPARC simulator. As the benchmarks are executed, a trace of all executed instructions is captured. This provides a cycle accurate, in-order instruction trace of the benchmark when it is executed on a commercially available SPARC processor. The instruction trace is then used as input to the ns-2 based simulator. The use of this trace provides an accurate traffic model of the execution of the PolyNoC processor. Trace based traffic model accurately represents the transmission of instruction packets into the NoC by the packetizer unit. This allows for very basic modeling of the instruction execution time, without being concerned with the full simulation of SPARC processor, significantly easing the implementation and execution time of the simulator.

The total execution time for all benchmarks can be seen in Tables 1-4. Table results show the total number of cycles to execute each benchmark with four different topologies. The total cycle count is an accurate measure of the performance of the processor. A higher performing processor will be able to execute more instructions in less

Table 1. Ring execution time (clock cycles).

Benchmark	1ALU 1FPU	2ALU 2FPU	ALU1 FPU2	ALU2 FPU2
FFT	6.126.842	4.852.776	6.085.547	5.012.044
IFFT	6.268.858	4.997.153	6.170.402	5.229.513
CRC32	55.454.286	56.015.776	37.911.220	39.543.279
BasicMath	55.457.834	55.999.913	48.527.041	48.667.159
ADPCM. decode	131.658.428	137.093.837	108.790.756	114.671.543
ADPCM. encode	137.409.561	141.663.042	108.037.207	112.377.209

Table 2. Modified ring execution time (clock cycles).

Benchmark	1ALU 1FPU	2ALU 2FPU	ALU1 FPU2	ALU2 FPU2
FFT	6.126.842	4.881.443	5.916.996	4.683.730
IFFT	6.268.858	5.024.278	6.001.722	4.737.821
CRC32	55.454.286	55.454.286	38.905.161	38.905.161
BasicMath	55.457.834	54.029.474	45.804.379	44.455.720
ADPCM. decode	131.658.428	131.658.428	100.332.329	100.332.329
ADPCM. encode	137.409.561	137.409.561	100.499.675	100.499.675

Table 3. 2D Mesh execution time (clock cycles).

Benchmark	1ALU 1FPU	2ALU 2FPU	ALU1 FPU2	ALU2 FPU2
FFT	5.767.886	4.351.394	5.495.290	4.082.665
IFFT	5.925.990	4.508.450	5.605.831	4.250.474
CRC32	51.971.730	51.971.730	33.941.356	33.941.356
BasicMath	51.937.483	50.494.185	41.226.881	39.269.764
ADPCM. decode	118.765.944	118.765.944	84.968.305	84.968.305
ADPCM. encode	129.321.184	129.321.184	88.790.061	88.790.061

Table 4. Crossbar execution time (clock cycles).

Benchmark	1ALU 1FPU	2ALU 2FPU	ALU1 FPU2	ALU2 FPU2
FFT	5.718.123	4.341.824	5.385.672	3.984.087
IFFT	5.876.096	4.498.707	5.500.106	4.099.194
CRC32	52.295.414	52.295.414	31.817.863	31.817.863
BasicMath	50.968.786	49.225.920	39.486.512	37.229.961
ADPCM. decode	118.781.718	118.781.718	85.392.11	82.539.211
ADPCM. encode	129.750.238	129.750.238	85.654.074	85.654.074

time. All the values presented were obtained using an instruction window size of 16, with an assumed NoC clock rate 8 times the datapath clock. It is reasonable assume the NoC clock can run at a higher rate then the functional elements datapath clock. This prevents the PolyNoC processor from becoming NoC bound wherein instruction execution time is limited by the NoC latency and the increase from additional functional units provides no performance gains.

Results show that all architectures exhibit scalability, and increased performance as the number of functional units increase. It can be seen that they all perform similar when only a single functional unit of each type is used. The advantages only become significant as the number of functional units increase. As the number of functional units is increased modestly from one to two ALUs/FPUs, significant performance improvements are achieved. It is important to note that a PolyNoC processor would allow this performance boost to be selectively applied during run time as reconfigurable elements are configured to act as additional functional units. The applications that are heavily FPU operation oriented benefit most from additional FPU elements, similarly ALU oriented benchmarks get more performance from an increase in number of ALU elements.

The ring architecture is simple and has very low resource cost however, its performance is lacking. The performance increase from additional function units is cannibalized by increased packet delay through the network. When more functional units are added, the latency of the overall NoC transmission increases due to the need to traverse the additional intermediate nodes. The additional penalty from NoC latency counteracts any performance gains from more functional elements, and in some cases actually lowers the overall performance of the PolyNoC processor. The modified ring architecture helps to mitigate the additional penalty from extra functional elements by eliminating the growth of delay when adding more functional units. In this case, the additional units do not increase the latency from NoC transmission. However, it still exhibits a similar weakness in overall structure; the ring based architecture delay is too high and prevents optimal processor execution either for the original ring or modified ring. On the other hand, the 2D Mesh architecture looks very promising as a candidate for NoC architecture. The max number of nodes compared to delay grows at a polynomial rate, producing a slow increase in delay relative to increase in total number of functional elements. Furthermore, for small mesh sizes, a large portion of communication occurs exclusively between neighboring nodes, allowing for an optimal 1 cycle delay time. This enables minimal latency. Near neighbor effect can also be maximized by careful layout of the 2D mesh. The mesh structure should be laid out such that node that frequently communicate with each other. For example, majority of instructions transits to the ALU after visiting the register file. Finally, the Crossbar architecture represents a reference point for comparison. It provides a connection from every node to every other node and it allows for a constant delay time of 1 clock for all packet transmissions. This can be seen as an upper limit to performance gains independent of architectural choices. This reference point is a valid comparison for the other topologies. With all packet delays at an optimal 1 cycle delay, it achieves the highest PolyNoC performance. The overall performance is now limited by the performance of the functional elements, and the overall parallelism that can be extracted from the instruction stream.

Both the performance and the hardware resource consumption of the architectures can be seen in **Figure 6**. This figure presents the cost of each architecture based on the total number of individual point to point unidirectional links. Results indicate the superior performance of the 2D Mesh architecture. With much lower total cost, it is able to achieve a performance that is only slightly reduced from the upper limit set by the Crossbar implementation.

In conclusion, the 2D mesh structure is able to outperform all other topologies studied. It is able to scale well with additional functional units. As the more units are added, the additional latency in NoC transmission is controlled and overall system performance goes up. In general, the performance of the 2D mesh is fairly close to the theoretical maximum demonstrated through the crossbar architecture. 2D mesh structure also performs best at cost scaling. The link cost for point to point links grows very slowly compared to other implementation, while still providing large gains in total PolyNoC performance.

The final PolyNoC processor design with incorporation of optimal parameters can be seen in **Figure 7**. It shows the detailed layout of the NoC architecture and the placement of reconfigurable arrays and basic functional units. From the topology results, it is clear that the 2D mesh architecture provides the optimal balance of performance and cost. For these reasons it is chosen as the basis of the NoC architecture. This design provides a basic 4 × 4 mesh layout. This is a relatively small layout which could easily be scaled to larger size. 4 × 4 layout provides space for all basic units and a total of 11 reconfigurable arrays for application specific functional units. The design makes use of 2-input, 3-input, and 4-input switches as the backbone of the communication infrastructure. Each switch is interfaced to a single functional unit. Functional units can be a basic unit supporting the execution of the standard SPARC ISA, or a reconfigurable unit supporting the creation of new application specific instruction.

CONCLUSION

In this work, a new reconfigurable processor architecture has been introduced. Traditional fixed pipeline based datapath is replaced with a reconfigurable NoC based communications channel. Functional elements can communicate with others elements and they can be added and reconfigured dynamically enabling polymorphic operations. Several NoC topologies were explored to find the

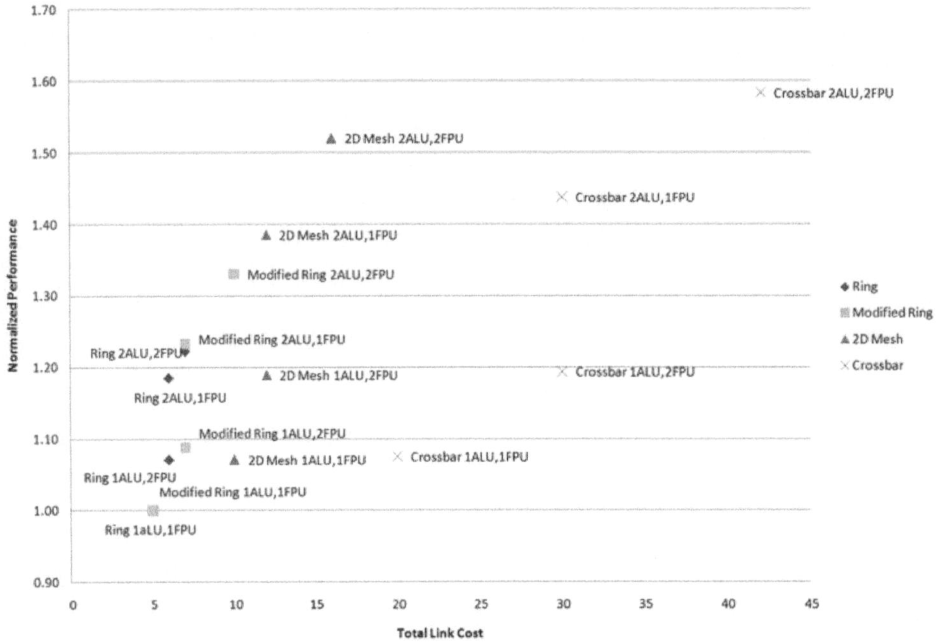

Figure 6. Cost vs performance analysis of different NoC topologies and number of functional units.

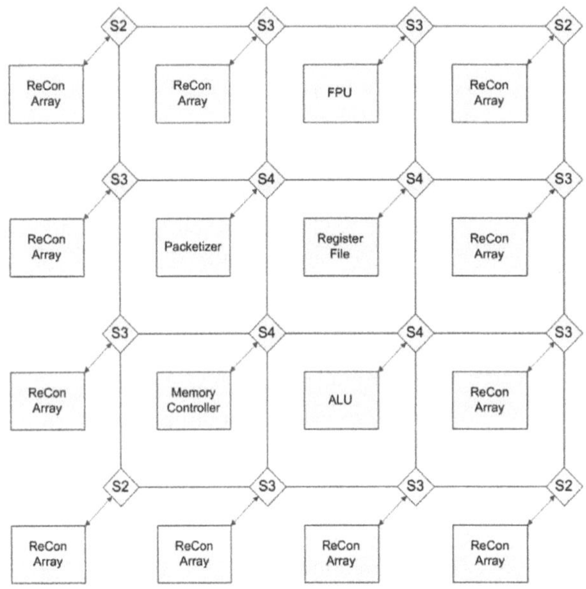

Figure 7. 4 × 4 2D mesh architecture.

optimal organization for a NoC based datapath. Industry standard MiBench benchmarks were used to study the execution of the PolyNoC processor. Results demonstrate that the PolyNoC processor provides significant advantages in flexibility of design. It is able to support both very large stream based optimizations (i.e., computation of ADPCM block based encoding), while also performing very fine grained, highly coupled custom instruction group optimizations, (i.e., optimization of the CRC32 hash update loop). Therefore, PolyNoC processor allows easy incorporation of reconfigurable arrays, both coarse and fine grained, for application specific optimizations.

REFERENCES

1. "International Technology Roadmap for Semiconductors," 2009. http://www.itrs.net/Links/2009ITRS/Home2009.htm
2. D. Hentrich, E. Oruklu and J. Saniie, "Polymorphic Computing: Definition, Trends, and a New Agent-Based Architecture," Circuits and Systems, Vol. 2, No. 4, 2011, pp. 358-364.doi:10.4236/cs.2011.24049
3. J. Weber, E. Oruklu and J. Saniie, "Architectural Topologies for NoC Datapath Polymorphic Processors," IEEE International Conference on Electro/Information Technology, Mankato, 15-17 May 2011, pp. 1-6.
4. K. Compton and S. Hauck, "Reconfigurable Computing: A Survey of Systems and Software," ACM Computing Surveys, Vol. 34, No. 2, 2002, pp. 171-210. doi:10.1145/508352.508353
5. Z. Guo, W. Najjar, F. Vahid and K. Vissers, "A Quantitative Analysis of the Speedup Factors of FPGAs over Processors," FPGA Proceedings of the ACM/SIGDA 12th International Symposium on Field

Programmable Gate Arrays, New York, 2004, pp. 162-170.

6. K. Underwood, "FPGAs vs CPUs: Trends in Peak Floating-Point Performance," FPGA Proceedings of the ACM/ SIGDA 12th International Symposium on Field Programmable Gate Arrays, New York, 2004, pp. 171-180.
7. T. Todman, G. Constantinides, S. Wilton, O. Mencer, W. Luk and P. Cheung, "Reconfigurable Computing: Architectures and Design Methods," IEEE Proceedings of Computers and Digital Techniques, Vol. 152, No. 2, 2005, pp. 193-207.
8. R. Cheung, N. Telle, W. Luk and P. Cheung, "Customizable Elliptic Curve Cryptosystems," IEEE Transactions on Very Large Scale Integration Systems, Vol. 13, No. 9, 2005, pp. 1048-1059.
9. P. Baglietto, M. Maresca and M. Migliardi, "Euclidean Distance Transform on Polymorphic Processor Array," CAMP Proceedings of Computer Architectures for Machine Perception, Como, 18-20 September 1995, pp. 288- 293.
10. G. Kuzmanov and W. van Oijen, "Floating-Point Matrix Multiplication in a Polymorphic Processor," ICFPT International Conference on Field-Programmable Technology, Kitakyushu, 12-15 December 2007, pp. 249-252.
11. S. Chai, S. Chiricescu, R. Essick, B. Lucas, P. May, K. Moat, J. Norris, M. Schuette and A. Lopez-Lagunas, "Streaming Processors for Next Generation Mobile Imaging Applications," Communications Magazine, Vol. 43, No. 12, 2005, pp. 81-89.doi:10.1109/MCOM.2005.1561924
12. H. Hubert and B. Stabernack, "Profiling-Based Hardware/ Software Coexploration for the Design of Video Coding Architectures," IEEE Transactions on Circuits and Systems for Video Technology, Vol. 19, No. 11, 2009, pp. 1680-1691.doi:10.1109/TCSVT.2009.2031522
13. C. Rupp, M. Landguth, T. Garverick, E. Gomersall, H. Holt, J. Arnold and M. Gokhale, "The Napa Adaptive Processing Architecture," IEEE Symposium on Proceedings of FPGAs for Custom Computing Machines, 17 April 1998, pp. 28-37.
14. S. Vassiliadis, S. Wong, G. Gaydadjiev, K. Bertels, G. Kuzmanov and E. Panainte, "The Molen Polymorphic Processor," IEEE Transactions on Computers, Vol. 53, No. 11, 2004, pp. 1363-1375. doi:10.1109/TC.2004.104
15. N. Vassiliadis, G. Theodoridis and S. Nikolaidis, "The Arise Approach for Extending Embedded Processors with Arbitrary Hardware Accelerators," IEEE Transactions on Very Large Scale Integration Systems, Vol. 17, No. 2, 2009, pp. 221-233.
16. M. Gokhale and J. Stone, "NAPA C: Compiling for a Hybrid Risc/Fpga Architecture," IEEE Symposium on Proceedings of FPGAs for Custom Computing Machines, 17 April 1998, pp. 126-135.
17. J. Cong, G. Han and Z. Zhang, "Architecture and Compiler Optimizations for Data Bandwidth Improvement in Configurable Processors," IEEE Transactions on Very Large Scale Integration Systems, Vol. 14, No. 9, 2006, pp. 986-997.
18. S. Gupta, N. Dutt, R. Gupta and A. Nicolau, "SPARK: A High-Level Synthesis Framework for Applying Parallelizing Compiler transformations," Proceedings of the 16th International Conference on VLSI Design, New Delhi, 4-8 January 2003, pp. 461-466.
19. The SPARC Architecture Manual, Version 8, Sun Microsystems, 1992. http://www.sparc.org/standards/V8.pdf [Paper reference:2]
20. Ns-2, The Network Simulator. http://nsnam.isi.edu/nsnam/index.php/Main_Page
21. M. Guthaus, J. Ringenberg, D. Ernst, T. Austin, T. Mudge and R. Brown, "Mibench: A Free, Commercially Representative Embedded Benchmark Suite," IEEE International Workshop on Workload Characterization, December 2001, pp. 3-14.

CHAPTER 7

RELIABILITY OF HIGH SPEED ULTRA LOW VOLTAGE DIFFERENTIAL CMOS LOGIC

Omid Mirmotahari and Yngvar Berg

Keywords: CMOS, Differential, Floating-Gate, Semi-Floating-Gate, Keeper, Recharge, Ultra Low Voltage, High Speed, Monte-Carlo, Cadence, STM, 90 nm

INTRODUCTION

In the last decade, sophisticated portable electronics has dramatically increased in the consumer market. The new trends are toward energy harvesting electronics and drive the need for low power and low voltage due to a limited budget of processed energy, while on the other hand demand for high-performance electronics regarding speed. Thus, in the last decade, major developments have made low power designs a key objective in addition to speed and silicon area, in order to lower the energy consumption several approaches exists. One of the most fundamental and effective approach is to lower the supply-voltage [1] [2]. Furthermore, it is called ultra low voltage (ULV) when the supply voltage is reduced to hundreds of milli-volts [3] [4]. However, the scaling of the supply-voltage has the adverse effect on the performance of the design concerning speed. The main challenge is to obtain high speed at as low as possible supply-voltages. To maintain good evaluation response time at ultra low supply-voltages, the threshold voltages of the transistors must also be reduced. Unfortunately, this requires a change in the CMOS fabrication process. Multiple-technique has been proposed for low voltage high performance circuits designs [5] without changing the fabrication process. Floating-gates (FG) have also been proposed for ultra low voltage and low power (LP) logic [6] . Unfortunately, modern processes face significant gate leakage due to the thin oxide. A ULV floating-gate inverter employing a frequent recharge technique has shown good properties in achieving high speed at ultra low voltages [7]. Even though the ULV gate has shown good performance, it also has limitations due to the leakage at the semi-floating-gates (SFG). A differential ULV gate has been proposed by including a keeper function [8] which is argued to have the speed of an ULV but the stability as a standard CMOS.

In this paper, the differential ULV inverter is presented. The improvement in term of stability and delay relative to the ULV is elaborated. Furthermore, the reliability, yield and the defect tolerance of the differential ULV inverter compared with both standards CMOS and with ULV floating-gate inverter are examined. The main goal is to find the designs behaviour and pinpoint the adjustment parameter giving the best yield and fault tolerance.

The outline of this paper is as follows. In Section 1, key aspects of the ultra low voltage (ULV) gate are presented. Section 3 elaborates on the leakage problem of the ULV gates and presents a solution with keeper functionality. In Section 4, a differential ULV gate including the keeper function is thoroughly presented and the key advantages are pointed out. Section 5 presents the aspects of reliability such as speed, stability and noise-margin of the differential ULV both compared with CMOS and with ULV floating-gate inverter. And in Section 5, a discussion on the achieved results is given. Finally, the paper concludes by pointing out the optimal design parameters. The simulation results demonstrated throughout this paper were obtained by simulation produced in a STM 90 nm process environment provided by Cadence.

ULTRA LOW VOLTAGE FLOATING-GATE INVERTER

A ultra low voltage (ULV) inverter were first introduced in [7] and is shown in Figure 1. The inverter has been demonstrated with measurements for supply voltage of 0.4 V. Moreover, its potential has been simulated to be more than ten times the operating frequency than compared footed domino logic gates

Citation: Mirmotahari, O. and Berg, Y. (2015) Reliability of High Speed Ultra Low Voltage Differential CMOS Logic. Circuits and Systems, 6, 121-135. doi: 10.4236/cs.2015.65013.

resulting in a improved EDP of 20 times better than standard inverter. Further comparisons to CMOS have been elaborated in [9] , which also strengthen the ULV inverters potential. Furthermore, this gate has been used in the field of power analysis

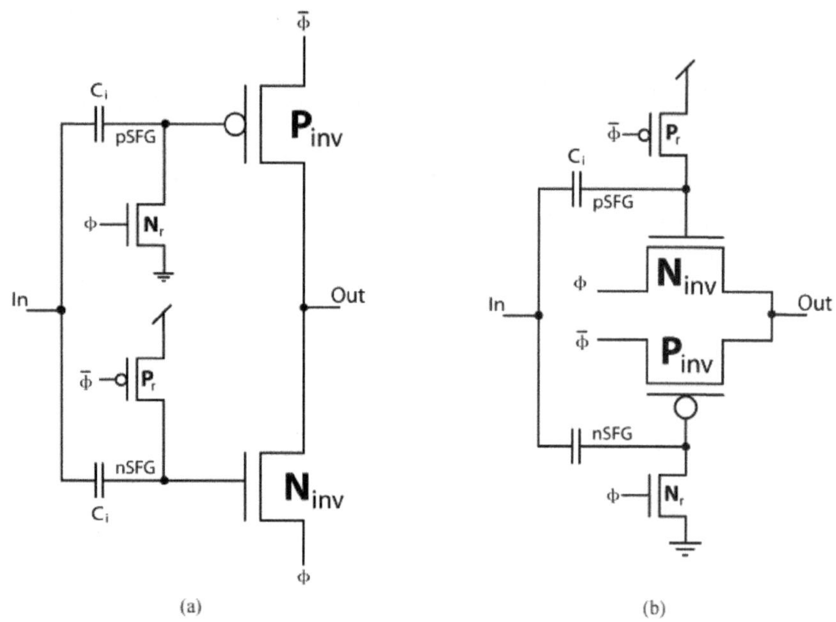

Figure 1. The figure illustrates the ultra low voltage gate. Both designs are logically and electrically equivalent. In (a) the design clearly shows the inverter and the biasing/recharging of the floating-gate, while (b) is designed to emphasise that the output is not directly connected to the supply voltage. (a) ULV design resembles a UV-floating-gate inverter; (b) ULV design with focus on the output regarding to supply-voltage.

and it has been found to be able to camouflage its instantaneous current dissipation due to the clock drivers [10] . Extracting the joint advantages of this particular ULV gate, we find the combination of high speed and low voltage giving quite good EDP numbers than other known similar logic, such as CMOS and footed domino logic. Although, the ULV inverter is presented here, there are published work on other logical gates, such as NAND and NOR [9] .

In general, all floating-gates suffer from leakage, some to more extent than other. It is a well-known phenomenon in the field of floating-gate that a frequent recharging strategy includes elements which would increase the leakage. The presented ULV gates are classified as semi-floating-gate (SFG) because of the direct connected recharge transistor. The leakage through the drain contact of the recharge transistor will set restrictions for the operating frequency both in terms of high cut-off and low cut-off. The leakage of the semi-floating-gate would also affect the inverters ability to reach the rails. Simulation results demonstrating the leakage as a function of the supply-voltage is given in Figure 2 and a corresponding deviation from the rails in Figure 3. With these two simulation result the lowest operating frequency can be calculated by the formula:

$$f_{low} = \frac{leakage}{NM - deviation} \tag{1}$$

where deviation represents the amount of voltage deviation from the rails, NM represents the noise margin (typically 25% of V_{dd}) and leakage is the amount given in μV/ns.

KEEPER FUNCTION

In order to face the problem with stability and leakage at the semi-floating-gate, that is to be able to reach the rails, a design solution is given in Figure 4. The inspiration may be claimed from the keeper function in domino logic. The additional transistors, labelled N_k and P_k, will at given time keep the charge on the semi floating-gate to a fixed potential. The behaviour during evaluation for a falling transition is given below:

- PER The semi floating-gates, nSFG and pSFG, are set to V_{dd} and Gnd, respectively. The recharge clock, ϕ, is turning off. The input signal, In, is at $V_{dd}/2$ and Out is $V_{dd}/2$.

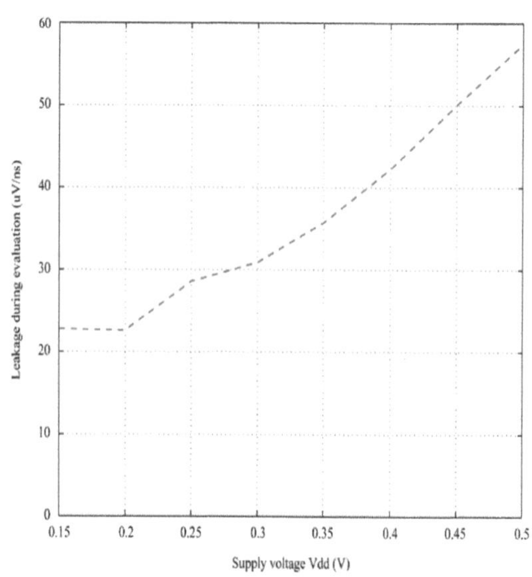

Figure 2. Simulation results show the leakage of the ULV gate during an evaluation period as a function of low supply-voltages. The leakage is given by μV/ns.

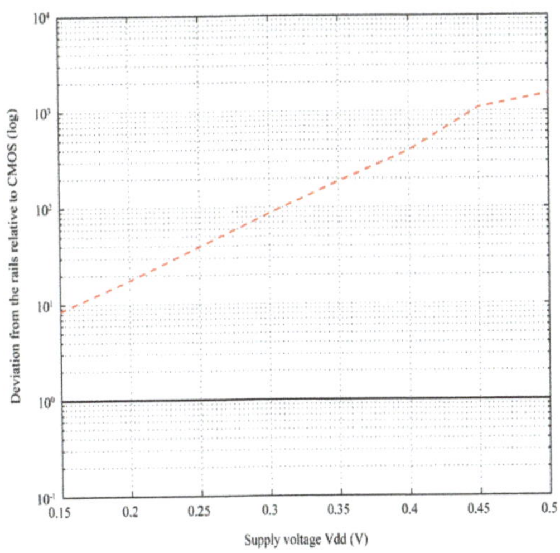

Figure 3. Simulation results demonstrating the amount of deviation relative to CMOS under given supply-voltages. This behaviour is expected due to the leakage through the diffusion contact of the recharge transistor.

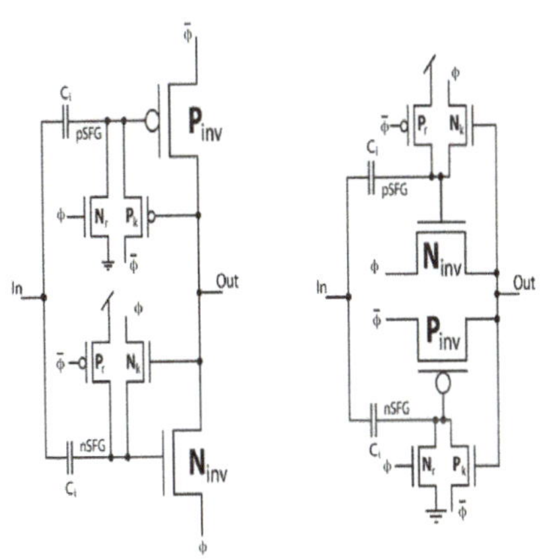

Figure 4. The ultra low voltage gate is modified by including a transistor which acts as an keeper. The keeper transistors are label N_k and P_k and would contribute to hold the semi- floating-gates potential, thus eliminating the leakage.

- DUR In starts on a falling transition. nSFG and pSFG see this change with an attenuation factor of C_i/C_{tot}, where C_{tot} is the total capacitance seen from the semi floating-gate. The output drivers, P_{inv} and N_{inv}, starts pulling the output node up. While evaluating, the semi floating-gates are leaking, for nSFG the leakage is from ($V_{dd} - \Delta V$) to Gnd, while for pMOS it is (Gnd $- \Delta V$) to Gnd. The voltage change ΔV is as result of the transition at the input.
- POST When In has reached its rail (Gnd), the output drivers are still evaluating the change and try to pull up the output to the rail (V_{dd}). It is here the keeper steps in and holds the nSFG to Gnd and makes sure that the output is as close to the rail as possible and hold the value until next recharge period. Note that the N_k is active and not P_k, due to the falling transition.

Simulation results also verify the theoretical approach and further states the improved stability and lower the leakage due to the included keeper function. Nevertheless, experience with semi floating-gates show that designs with a logical depth of 2 or more, some problems concerning signal delay and clock synchronisation occurs. In that particular case it can actually in worst-case scenario lead to a lock down for the gate. The worst- case in a design, with a logical depth of 2 or more, is where one of the gates output, due to some unexpected reason, is shifted away from $V_{dd} = 2$. Even a small voltage change from $V_{dd} = 2$ at the output would lead to a lock down to either rails. The lock-down can be avoided if the keeper transistors are not connected to the same reference as the recharge. One very interesting gate to examine in more details are a differential design of the ULV gate with keeper. A differential ULV gate has not, to our knowledge, been published.

DIFFERENTIAL ULV INVERTER WITH KEEPER

In a differential ULV gate, the keeper really gets to be used for more than its potential. The benefit of having the keeper is gained both through keeping a fixed potential for the semi floating-gate and to actually turn "more" off those transistors which should be off. The main benefit for having a differential design is that the opposite output signal is accessible and thus can be connected to the keeper transistors instead of a reference which could lead to a lock down. In Figure 5, a differential design of the ULV gate including keeper function is illustrated. From the simulation results presented in Figure 6, the keepers contribution in holding the nSFG down to Gnd is clearly visual. This differential design, called ULV-DIFF, would consume approximately the same amount of dynamic power as the ULV, but have the same static dissipation as CMOS. Furthermore, motivation for differential design can also be found in the context of power analysis, were one of the main countermeasure is to use differential design.

In the following the ULV-DIFF and the ULV are compared to standard CMOS. The simulation conditions has been the same for each logic style, though minimum matched output transistors. The recharge and keeper transistors are kept minimum and the input capacitance is 0.7fF and 1fF for nSFG and pSFG, respectively. Figure 7 and Figure 8 reinforce the real benefit of the stability and the leakage properties of the ULV-DIFF related to ULV. In Figure 9 and Figure 10, the current dissipation is shown. We like to stress that the dissipation behaviour of the ULV-DIFF is gaining the best from both logic styles, ULV and CMOS. The improvement and as a

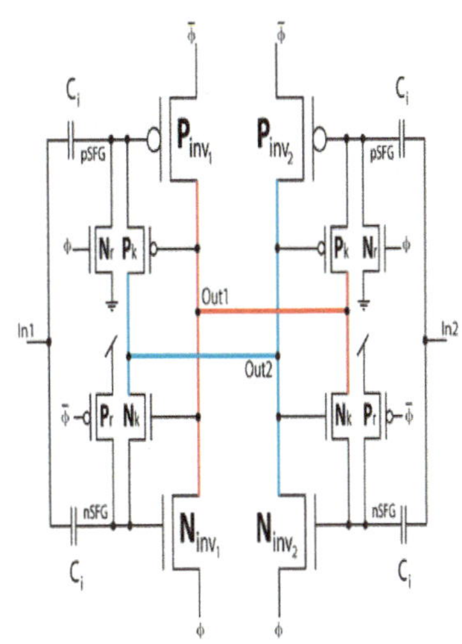

Figure 5. The ultra low voltage differential semi floating-gate inverter including a keeper function, P_k and N_k, is illustrated. Transistor sizes are kept minimum and matched, the input capacitance are scaled relative to obtain the same voltage attenuation for both nSFG and pSFG.

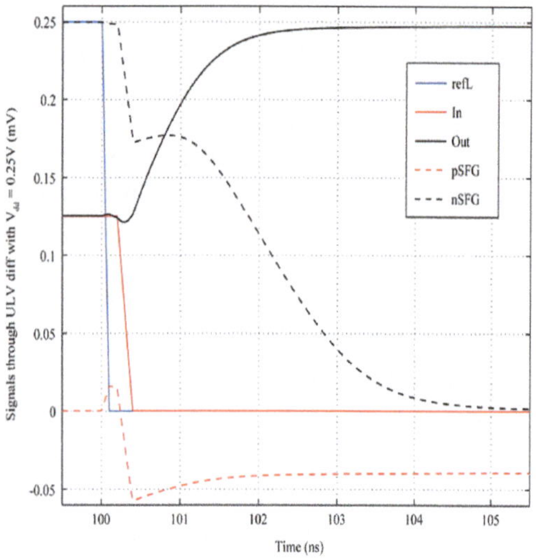

Figure 6. Simulation results showing all signals in the differential ULV gate during evaluation. As seen from the results the keeper functions steps in at approximately 101 ns and pulling the nSFG down to Gnd.

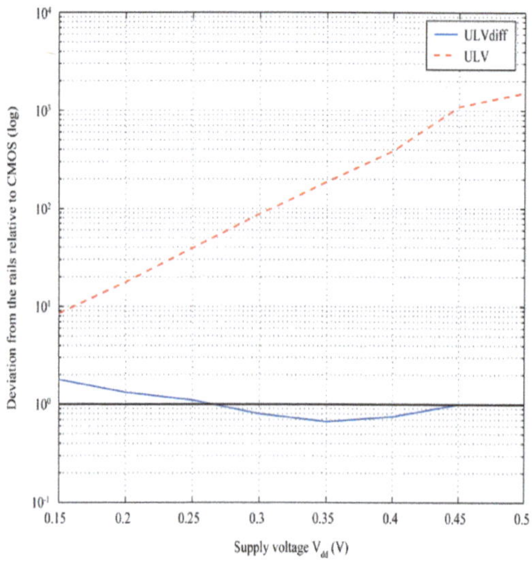

Figure 7. Simulation results showing the deviation for the ULV-DIFF and the ULV relative to CMOS.

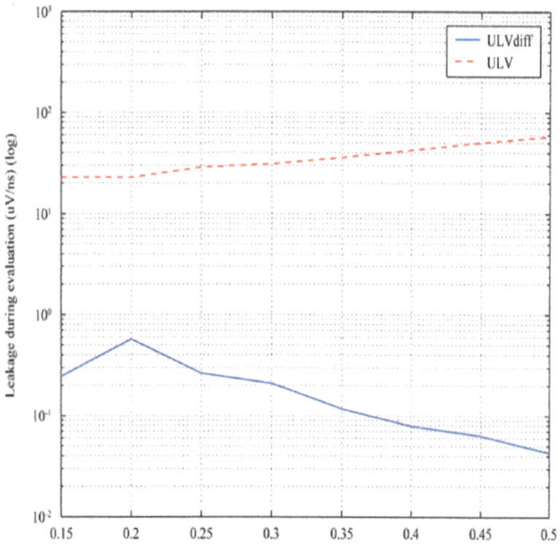

Figure 8. The leakage of the semi floating-gate affecting the output during the evaluation period.

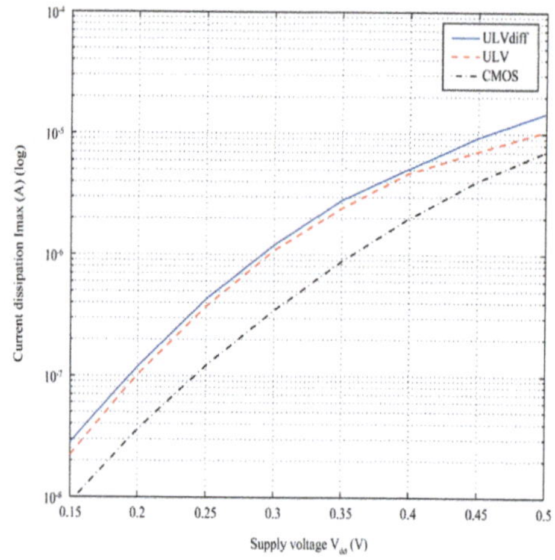

Figure 9. The maximum current dissipation (dynamic dissipation) for all three styles.

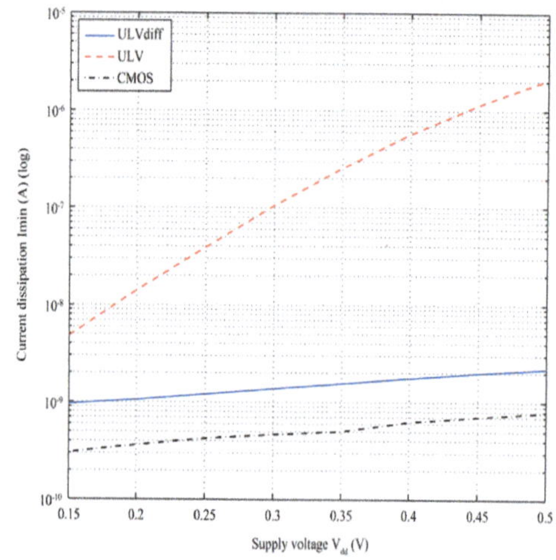

Figure 10. The minimum current dissipation (static dissipation) for all three logic styles.

figure of merit is the EDP for the ULV-DIFF and ULV relative to CMOS given in Figure 11. In order to include the stability factor and to have a more fair comparison, the deviation factor relative to CMOS is multiplied to the EDP. Figure 12 shows the real improvement in ULV-DIFF compared to ULV. The interesting point to extract from all these data is that the optimal point is to have V_{dd} = 0.35 V and that gives a improvement of approximately 5 times better for ULV-DIFF than CMOS and approximately 100 times better for ULV-DIFF than ULV.

RELIABILITY AND DEFECT TOLERANCE

The basis of the results presented in the following are obtained through Monte-Carlo simulations including both process and mismatch variations. The data is collected through 100 runs and a parametric simulation of the supply-voltage V_{dd}. The supply-voltage is swept from 0.15 V - 0.5 V with common conditions and has shown a 100% yield for all three gates. There are three aspects which are of interest that comply after the 100% yield, namely, 1) delay variation, 2) stability and 3) noise margin. In Figure 13, the simulation results of a 100 run Monte-Carlo is presented and the fields of interest are commented. Contrary to the standard CMOS both ULV gates have a recharge period and the recharge level represents the gates equilibrium state.

With the results in Figure 13, the mismatch variations of the transistors are evident and imply a direct consequence, which is altering the recharge level. Accordingly, the offset at the recharge level from $V_{dd}/2$ plays an important role for the succeeding gates. The fundamental behaviour of the floating-gate structures are based upon a input voltage transition from a specified level, in this case $V_{dd}/2$. An offset at the recharge level would decrease the voltage transition seen at the succeeding gates, and may at worst-case not be enough to represent an correct output. In other words, floating-gate structures which employ frequent recharge mechanism with a equilibrium state, $V_{dd}/2$, would complicate and may weaken the fault tolerance. In the light of Monte-Carlo simulation it is interesting to find out what the lowest voltage transition a differential ULV gate needs in order to be reliable also under such corner conditions. Data collected during the 100 runs have showed a maximum voltage deviation from the $V_{dd}/2$ at a $V_{dd} = 0.2$ V with a variation of $b \approx 30$ mV, which impose a 30% attenuation. Figure 14 shows how the recharge level changes, using corner simulation, at worst case for different supply-voltages. The higher supply- voltage the more, in terms of mV, the equilibrium state is shifted, but the lower percentage relative to V_{dd}. The

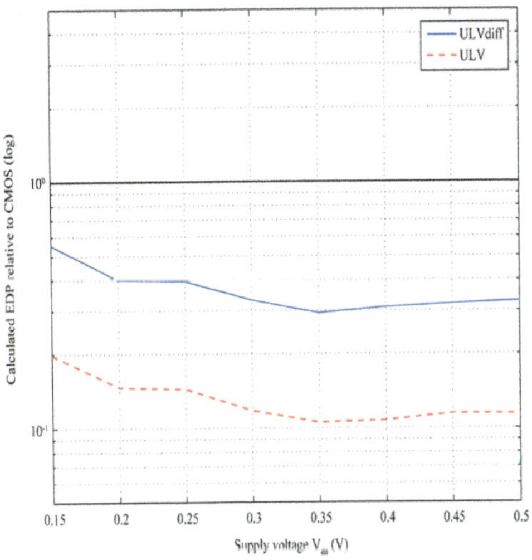

Figure 11. Simulation results for the EDP for the ULV-DIFF and the ULV relative to CMOS.

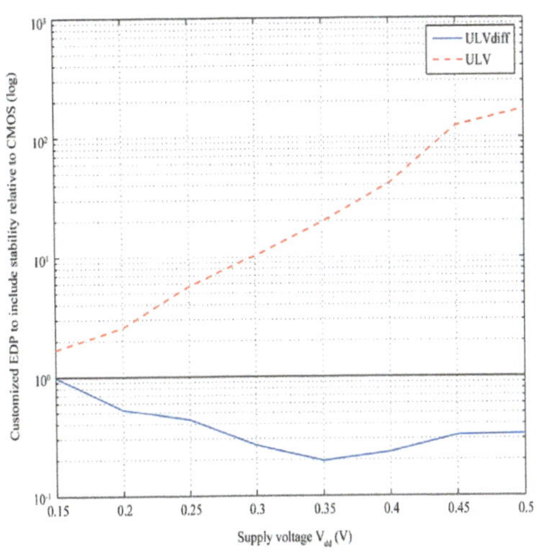

Figure 12. Simulation results for the EDP* stability for ULV diff and ULV relative to CMOS. The results show a optimum point at V_{dd} = 350 mV.

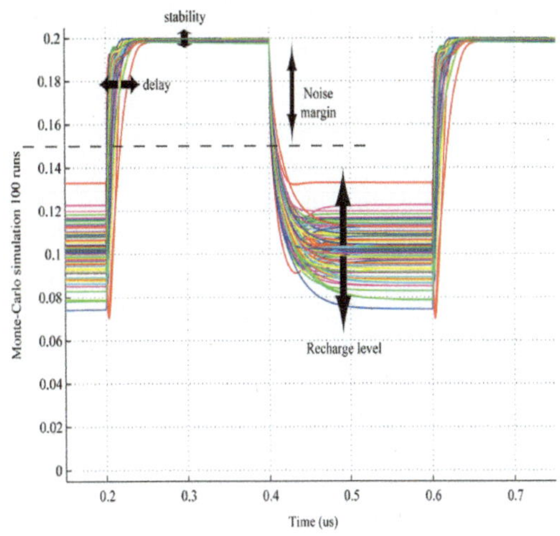

Figure 13. Simulation results for a Monte-Carlo corner simulation for 100 runs of the differential ULV gate shown in Figure 5. The environment is set to both process and mismatch variation. On the plot there are four commented areas to emphasise the discussion topics.

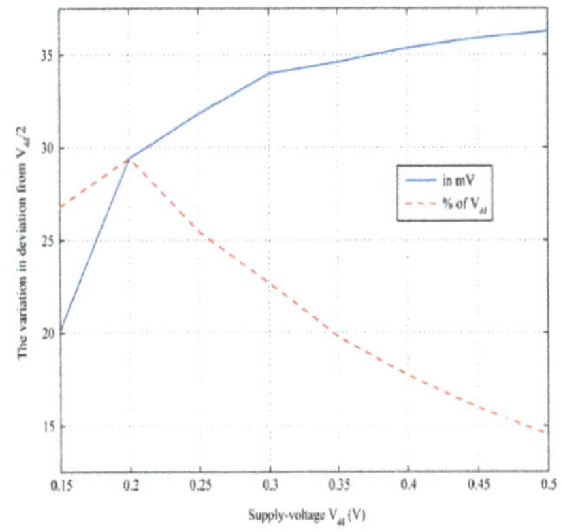

Figure 14. Simulation results have shown variation due to mismatch at the recharge level. The plot represents both the deviation from the optimal equilibrium state, $V_{dd}/2$, and the percentage within the supply-voltage V_{dd}.

plot in Figure 14 shows an intersect at V_{dd} = 200 mV and an attenuation of 30%. Although an attenuation of 30% is quite much, the differential ULV gate has not shown any signs of malfunction. The differential ULV gate has, as Figure 15 illustrates, been simulated with an input voltage transition which is attenuated 80% and still are able to output correctly and also to a very good stability. The main affect an attenuation would imply is the increase in delay, i.e. the more an input signal is attenuated the higher the evaluation delay would become. Furthermore, the attenuation is also a function of the relation to the capacitive ratio seen at the semi-floating-gates. The robustness to a large attenuation for the differential ULV gate should be credited to the keeper function. The keeper transistor would help to turn off the right transistor as long as there is a small change in the output. Actually, as long as the two differentiated output are shifting in the opposite direction from each other from the equilibrium state, the representative keeper transistors would help the output drivers to symmetrically pull their representative outputs to the rails. Therefore, the ULV gate which does not have the keeper function is not expected to sustain a corner situation where the recharge level is shifted and thus generating an attenuation of more than 30%. Furthermore, it is important to stress that the keeper function also provides high stability under any circumstances in terms of deviation from the rails.

The situation where worst case recharge level offset and the worst case mismatch of the transistors regarding delay arise on the same chip and for two succeeding gates is highly unlikely. In a simulation environment, such as Monte-Carlo, only one given aspects is examined. Referring back to Figure 13, the worst case of mismatch and process variation gives a change in the delay. The delay are measured based on a 50% to 50% input and output transition. The same conditions apply for all three designs. The maximum delay variation for a Monte-Carlo simulation with 100 runs is displayed in Figure 16. The delay variation would increase if the simulation environment has included a sweeping parameter to represent the offset at the recharge level. The need to easier evaluate and visualise the collected data and to lay a foundation for discussing the values relative to CMOS, the data is plotted in Figure 17. There is a minimum delay variation found at V_{dd} = 300 mV both for the ULV and the differential ULV.

Once the delay is considered, the next topic is the stability of the gates. The stability represents the deviation from the rails and can also be used to indicate the yield during corner simulation. Bear in mind that the data set

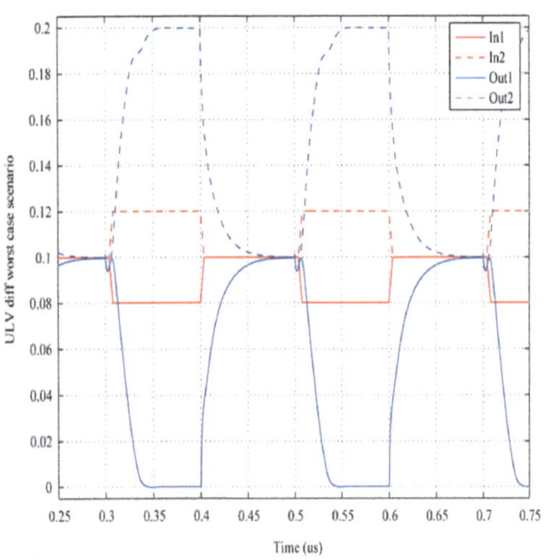

Figure 15. Simulation results for the differential ULV gate verifying the gates ability to operate on very high attenuation for the input signal. The simulated environment is reflecting an input transition which is attenuated 80% and the outputs are for a succeeding gate.

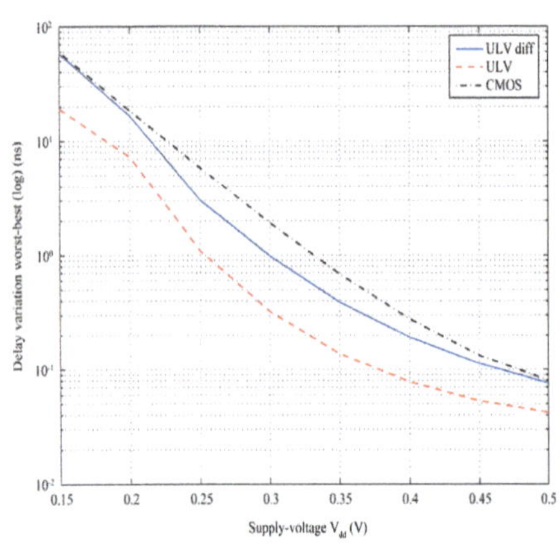

Figure 16. Corner simulations show a delay variation as a mismatch at the input capacitors and transistors. The delay variation is calculated from a 50% to 50% input and output transition.

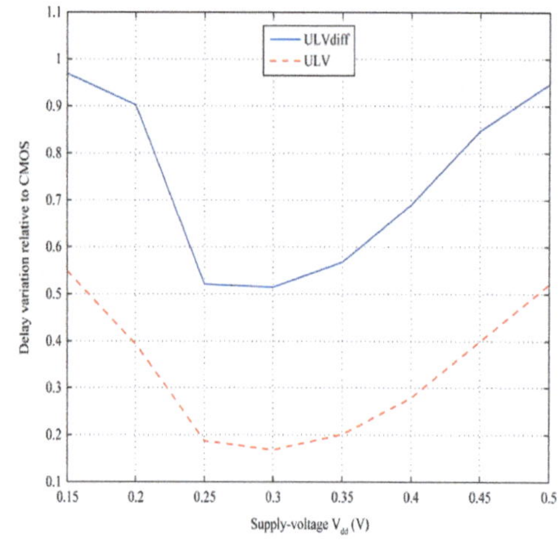

Figure 17. The delay variation obtained during Monte-Carlo simulation is plotted relative to CMOS. The main goal is to bring out the difference and the potential of the ULV gates. Both gates show a minima at V_{dd} = 300 mV.

used are for a V_{dd} ranging from 150 mV to 500 mV and that all gates have showed 100% yield. Previously, the ULV were implied to have a lower yield due to the leakage at the semi-floating-gates. Although this statement has not been proved, the indications showed by Figure 18 further strengthen the statement. The figure represents the stability variation relative to CMOS for different supply voltages. The stability variation is calculated as the difference between the best and worst stability during 100 runs.

DISCUSSION

All the simulation results speak in the favour of the differential ULV gate. Firstly and the most beneficial aspect is the reliability the keeper function adds, even though the recharge levels are shifted. Secondly, the delay variations show an optimal point at V_{dd} = 300 mV. Considering the threshold voltage, V_{th}, which for a 90 nm process is approximately 270 mV, a general optimal point is expected to be around the V_{th}. Previous work on differential ULV gate has showed an optimal point for EDP to be at V_{dd} = 350 mV [11] . Referring to Figure 17, there is not much of a difference between 300 mV and 350 mV. Thirdly, the stability of the gates has demonstrated quite a large difference. The leakage at the semi floating-gate for the ULV gate really rules out its candidate. Fortunately, the keeper function at the differential ULV gate ensures that its candidate is present alongside standard CMOS. It can actually be found that the stability for the differential ULV gate is actually better than CMOS for V_{dd} from 230 mV to 430 mV, with a local minimum at 300 mV. Referring to Figure 18, the plot clearly demonstrates the difference of the stability relative to CMOS. Ones again the results from corner simulations show an optimal point at 300 mV. Last but not least, the discussion about the yield of the gates is presented in Figure 19. The plot shows that the ULV gate is barely surviving the lowest supply-voltage, while the differential ULV gate at worst case is within 80% of the noise margin. An interesting discussion of the results are that the differential ULV gate becomes more and more stable and closer to obtaining 100% within the noise-margin, while the same cannot be seen for the ULV gate. The data show that the ULV gate has a top peak at V_{dd} = 350 mV. From all these results, there is a reason to believe that the optimal V_{dd} should be around 325 mV, in order to achieve the best EDP, reliability and fault tolerance. Furthermore, the main credit should be given to the keeper function. The kee-

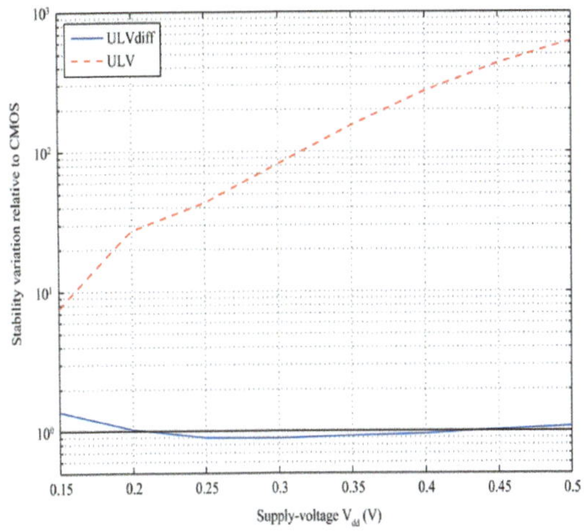

Figure 18. The stability variations during a Monte-Carlo simulation relative to CMOS are plotted. The ULV gates stability decreases as the stability variation increases as a function of V_d. The differential ULV gate shows an improvement of stability relative to CMOS for V_{dd} at the range 150 mV to 500 mV.

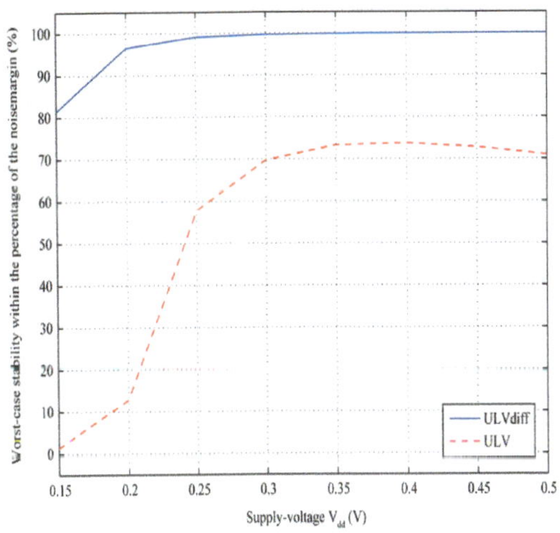

Figure 19. Simulation results for the lowest evaluation stability as percentage within the noise margin is presented. This plot can also be directly used to elaborate on the yield of the gates. In the V_{dd} range given, all gates have 100% yield, but the ULV is barely surviving corners as the V_{dd} decreases.

per function, as obtained from the results, can be regarded to protect the benefit, i.e. speed, of the ULV gate and to strengthen its weakness, i.e. stability, hence giving the differential ULV gates the best from both designs, ULV and standard CMOS.

Finally, the last aspect is the noise margin for the gates. The noise margin is actually closely linked to the yield. The data collected through corner simulation representing noise margin follow the formula below:

$$\frac{stability_{WC} - (V_{dd} \cdot 0.75)}{V_{dd} \cdot 0.25} \qquad (2)$$

Omid Mirmotahari, Yngvar Berg where is the lowest evaluation value through the 100 runs. The result will give a value which represents that the worst case stability is within the percentage of the noise-margin. The lower the result is, the closer the output is to the limits of the noise margin. In the context of yield, it can be used to denote that a result above 0 has a 100% yield. The result for corner simulation is given in Figure 19. As stated earlier for the ULV gate, the prediction for a 100% yield is diminishing. The result for a standard CMOS is very close to the differential ULV, as can be implied by Figure 18, and thus not been included in the plot of Figure 19.

REFERENCES

1. Verma, N., Kwong, J. and Chandrakasan, A. (1995) Nanometer MOSFET Variation in Minimum Energy Subthreshold Circuits. IEEE Transactions on Electron Devices, 55, 847-854.
2. Chandrakasan, A.P., Sheng, S. and Brodersen, R.W. (1992) Low-Power CMOS Digital Design. IEEE Journal of Solid-State Circuits, 27, 473-484.
3. Burr, J.B. and Peterson, A.M. (1991) Ultra Low Power CMOS Technology. NASA VLSI Design Symposium, Idaho, 30-31 October 1991, 4.2.1-4.2.13.
4. Burr, J.B. and Shott, J. (1994) A 200 mV Self-Testing Encoder/Decoder Using Stanford Ultra Low-Power CMOS. IEEE Solid-State Circuits Conference (ISSCC), San Francisco, 16-18 February 1994, 84-85. http://dx.doi.org/10.1109/ISSCC.1994.344717
5. Usami, K. and Horowitz, M. (1995) Clustered Voltage Scaling Technique for Low-Power Design. Proceedings of the 1995 International Symposium on Low Power, Dana Point, 23-26 April 1995, 3-8. http://dx.doi.org/10.1145/224081.224083
6. Berg, Y., Wisland, D.T. and Lande, T.S. (1999) Ultra Low-Voltage/Low-Power Digital Oating-Gate Circuits. IEEE Transactions on Circuits and Systems II: Analog and Digital Signal Processing, 46, 930-936.
7. Berg, Y., Mirmotahari, O., Norseng, P.A. and Aunet, S. (2006) Ultra Low Voltage CMOS Gates. IEEE International Conference on Electronics, Circuits and Systems (ICECS), Nice, 10-13 December 2006, 818-821.
8. Mirmotahari, O. and Berg, Y. (2009) Digital Ultra Low Voltage High Speed Logic. Proceedings of the International MultiConference of Engineers and Computer Scientists, Hong Kong, 18-20 March 2009, 1-1.
9. Berg, Y., Mirmotahari, O., Lomsdalen, J. and Aunet, S. (2008) High Speed Ultra Low Voltage CMOS Inverter. IEEE Computer Society Annual Symposium on VLSI, Montpellier, 7-9April 2008, 122-127.
10. Mirmotahari, O. and Berg, Y. (2008) Low Voltage Design against Power Analysis Attacks. IEEE International Symposium on Electronic Design, Test and Applications (DELTA), Hong Kong, 23-25 January 2008, 545-549.
11. Mirmotahari, O. and Berg, Y. (2008) Ultra Low Voltage High Speed Differential CMOS Inverter. IEEE PATMOS LNCS, 10-12 September 2008, Lisbon, Portugal, 1-1.

CHAPTER 8 CHIP DESIGN OF A LOW-VOLTAGE WIDEBAND CONTINUOUS-TIME SIGMA-DELTA MODULATOR WITH DWA TECHNOLOGY FOR WIMAX APPLICATIONS

Jhin-Fang Huang, Yan-Cheng Lai, Wen-Cheng Lai and Ron-Yi Liu

Keywords: ADC, Analog-to-Digital Conversion, Sigma-Delta Modulator, ΣΔ, DWA

INTRODUCTION

Sigma-delta modulation techniques have been extended in moderate and high accuracy analog/mixed-signal IC applications, such as analog-to-digital data converters (ADCs), digital-to-analog data converters (DACs), frequency synthesizers, and power amplifiers [1]. Moreover, ΣΔ modulators are widely used in receivers because of their ability to provide high-resolution with relatively low precision components and low power consumption [2,3]. Oversampling ΣΔ ADCs trade digital signal processing complexity for relaxed requirements on the analog components compared to Nyquist-rate ADCs [4]. Due to the over-sampling characteristics, ΣΔ modulators are limited on the application of voice band or lower frequency signals. As the ICs process is improved, recently it makes many researches transfer to wider bandwidth applications gradually, such as GSM, WCDMA, Bluetooth, WiFi, and WiMAX [5]. With the progress of wireless communication, ADCs need higher OSR in order to achieve higher speed and resolution in the system. When OSR is programmable, increasing OSR leads to higher power consumption due to the increased speed requirement for the integrators and comparators in ΣΔ modulators. Due to the requirements of low supply voltage and low power dissipation in the mobile communications, the low order ΣΔ modulators of lower SNR are not suitable for wide bandwidth applications. Therefore, the high order multi-bit ΣΔ modulator circuit is design to increase the SNR.

While most of current commercial ΣΔ ADCs for wireless applications were implemented by using switched capacitor (SC) techniques which are also known as discrete-time (DT) ΣΔ ADCs [6-8], mainly due to mature design methodologies and robustness, more and more continuous-time (CT) ΣΔ ADCs were reported and showed impressive performance. Compared with DT counterparts, the CT ΣΔ ADCs have two main advantages. First, the inherent anti-aliasing characteristics of the CT ΣΔ ADCs reduce the performance requirement of the anti-aliasing filter further and hence reduce the power consumption of the transceiver. Second, the bandwidth requirement of the operational amplifiers (op amps) in CT ΣΔ ADCs is much lower than that of the op amps in DT ones for a given sampling rate, so the CT ΣΔ ADCs are more suitable for broadband applications. Hence we propose a low-voltage, lower power consumption and high resolution CT ΣΔ modulator. Our target is to design a 10 MHz input signal bandwidth and 160MHz sample-rate ΣΔ modulator implemented in TSMC 0.18 μm CMOS process.

This paper begins with a brief summary of the innovative CT ΣΔ circuit design. Section 2 introduces the system architecture of the wideband CT ΣΔ modulator. Section 3 describes the design of building blocks of the modulator, while Section 4 presents the measured results of the prototype. Section 5 summarizes the paper.

SYSTEM CIRCUIT ARCHITECTURE

As high sampling frequency will restrict our design techniques, low OSR is more suitable for the bandwidth of 10 MHz structure. In order to achieve better resolution and

Citation: J. Huang, Y. Lai, W. Lai and R. Liu, "Chip Design of a Low-Voltage Wideband Continuous-Time Sigma-Delta Modulator with DWA Technology for WiMAX Applications," Circuits and Systems, Vol. 2 No. 3, 2011, pp. 201-209. doi: 10.4236/cs.2011.23029.

reduce quantization noise, at least a third-order noisetransfer function is indispensable. Figure 1 shows the proposed CT ΔΣ modulator architecture which consists of a 4-bit internal quantizer, operating at 160 MHz with an OSR of 8, and a third-order single-loop filter. In order to decrease power consumption and maintain a good alias filter characteristic, a combination of feed forward and feedback stabilized loop filters [9] is adopted. The 4-bit quantizer, including the NRZ feedback DAC is connected to the output of the loop filter. The quantizer delay is set to half of the sampling period. This large delay is compensated exactly by an additional feedback path K_{3fb}.

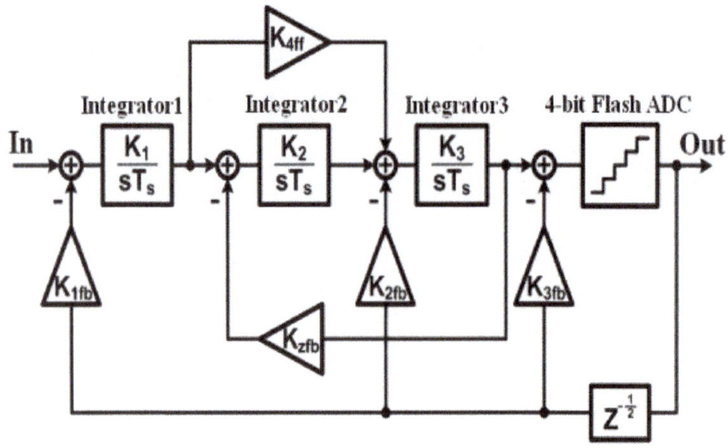

Figure 1. Continuous-time ΣΔ modulator architecture.

A possible design technique for CT modulators is described [10,11]. Specifying a DT modulator and trying to find the equivalent CT modulator between s-plane and z-plane can use the impulse-invariance transform expressed as:

$$Z^{-1}\{H(z)\} = L^{-1}\{R_D(s)H(s)\}\Big|_{t=n\cdot T_s} \tag{1}$$

where $R_D(s)$ is the Laplace transform of impulse response of the DAC and H(z) is the DT loop filter. Equation (1) is adopted to compensate the impairments of the circuit such that the resulting CT domain modulator still matches with the specified DT modulator. The followings outline the procedure used to determine the direct feedback coefficients such that quantizer delay is canceled exactly. First, a noise-transfer function (NTF) in the z-domain is chosen, then the loop transfer function $H_{loopz}(z)$ is derived as follows:

$$H_{Loopz}(z) = \frac{1-NTF(z)}{NTF(z)}. \tag{2}$$

Using the discrete-to-discrete (d2d) function in the MATLAB control system toolbox can easily transform z into $z^{1/2}$ shown in (3):

$$H'_{LoopZ}\left(z^{\frac{1}{2}}\right) = \frac{b'_{n-1}z^{\frac{n-1}{2}} + \cdots + b'_1 z^{\frac{1}{2}} + b'_0}{z^{\frac{n}{2}} + a'_{n-1}z^{\frac{n-1}{2}} + \cdots + a'_1 z^{\frac{1}{2}} + a'_0}. \tag{3}$$

After multiplying $z^{1/2}$ in the formula, a constant term b'_{n-1} can be easily separated from the transfer function $L_{filterD}(z^{1/2})$ of the loop filter as follows:

$$L_{filterD}\left(z^{\frac{1}{2}}\right) = \frac{b'_{n-1}z^{\frac{n-1}{2}} + \cdots + b'_1 z^{\frac{1}{2}} + b'_0}{z^{\frac{n}{2}} + a'_{n-1}z^{\frac{n-1}{2}} + \cdots + a'_1 z^{\frac{1}{2}} + a'_0} \cdot z^{\frac{1}{2}} \tag{4}$$

Using the discrete-to-continuous (d2c) function in the MATLAB tool box converts this transfer function $L_{filterD}(z^{1/2})$ to continuous time. A possible loop filter is then defined in the CT domain is RC loop filter.

CONTINUS-TIME ΣΔ MODULATOR IMPLEMENTATION

Continuous-Time ΣΔ Modulator Circuit

The 4-bit CT ΣΔ modulator circuit including the excess loop delay compensation is shown in Figure 2 [12]. The modulator consists of a 4-bit internal quantizer, operating at 160 MHz with an OSR of 8, and a third-order singleloop filter. The loop filter is realized as an active RC filter. Due to the low supply voltage and the high-linearity requirement two-stage op amps with CT common mode feedback (CMFB) are used. The 4-bit quantizer is connected to the DWA circuit followed by the feedback DAC. When multi-bit quantizer is used for better quan- 3.2. Loop Filter tization resolution, in-band tones are often observed due to the element mismatch in the feedback DAC. To solve the mismatch problem, dynamic element matching is used in the circuit design. The RC time constant in the circuit dominating the entire NTF pole function, will keep stable and therefore the circuit phase margin will also be stable. Due to the process variation, the capacitor tuning circuit is used in this modulator. The DAC1 and DAC2 circuits provide the first and second feedback paths K_{1fb} and K_{2fb}, respectively. The third K_{3fb} is the feedback path around the 4-bit quantizer and its output is connected to the DAC2 output. In order to reduce the loop filter capacitive loading effects, all the comparators inside flash ADC input transistors must be the minimumsize.

General "zero-order" feedback path requires additional summing amplifier and return-to-zero (RZ) DAC contains additional logic control circuits. However, this will cause additional loop delay, increase power consumption and complicate the circuit. Therefore, to improve these drawbacks, in this work, a feedback path is directly connected to the last integrator input, and then the additional summing amplifier is eliminated. Obviously this way reduces power consumption and excess loop delay.

Loop Filter

There are three types of commonly used CT integrators: active-RC integrators, Gm-C integrators and MOSFET-C integrators. In this design, an active-RC integrator is chosen for the three stages of the third-order loop filter because it has high linearity and easy interface with DACs compared to Gm-C integrators. If active-RC integrators were used, resistive loading increases the power requirements due to the need for buffer stages. A higher frequency range is additionally demanded in connection with a high linearity, the active-RC filters are the preferred structure. The third-order noise shaping loop filter is realized by an active-RC op amp circuit as shown in Figure 2. The advantages of this implementation are high linearity and high output signal swing, and it also provides a good virtual ground for the modulator feedback DACs. This eases design, especially with low supply voltages. Figure 3 show the architectures of the 1.2-V fully differential op amp [13] and the corresponding CMFB circuit is shown in Figure 4. The op amp shown in Figure 3 is a two-stage that consists of a folded cascode input stage, a common source output stage and a CT CMFB which is similar to a transimpedence amplifier.

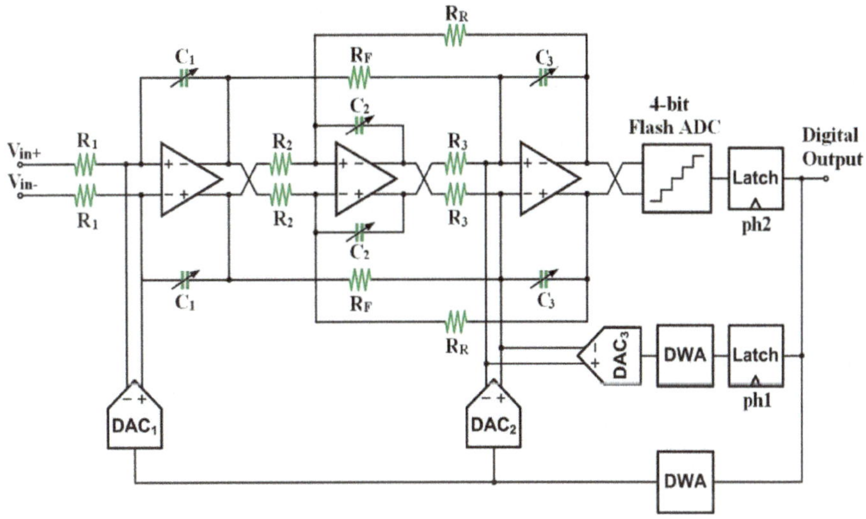

Figure 2. Four-bit CT ΣΔ modulator architecture.

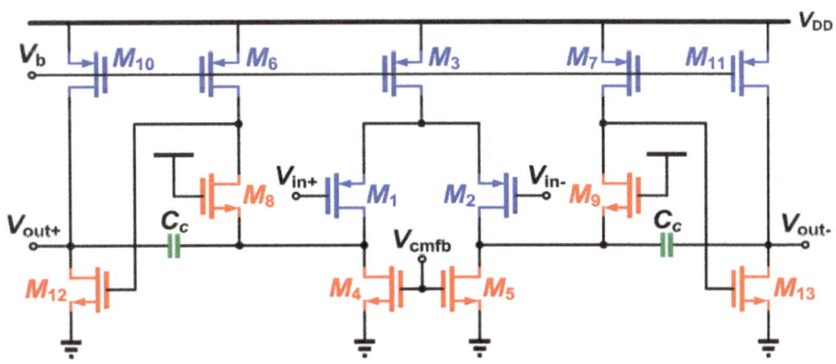

Figure 3. Fully differential 1.2-V op amp circuit.

As shown in Figure 4, two resistors with values equal to $2R_1$ are used to sense the output common mode voltage and produce a current I_1. This current is compared with I_2, which is set by the desired common-mode voltage ($V_{DD}/2$) and the resistor R_1. The difference between I_1 and I_2 is then converted into a control voltage labeled as V_{cmfb} by transistors M_3, M_6 and M_9. The control voltage will then be used to adjust the $V_{GS's}$ of M_4 and M_5, such that the output common-mode voltage is stabilized to about $V_{DD}/2$. The CMFB circuit has advantages of allowing rail-to-rail output swing. Furthermore, it does not need any level shift or attenuation on the common mode signal, unlike other CT CMFB circuits that use differential pairs.

Four-Bit Flash ADC

In ΣΔ modulators, the main specifications for the quantizer are offset, speed, area and power consumption requirements. Moreover the quantizer has to operate at the speed required by the oversampling process. Therefore it must be implemented as a flash ADC [14]. The block diagram of the 4-bit flash ADC used in the quantizer is shown in Figure 5. It consists of 15 differential comparators, a resistor ladder, and a thermal to binary encoder. These comparators compare the input signal with reference voltages by a resistor ladder biased by the full scale reference. Consequently, the comparator outputs constitute a thermometer code, which is converted to binary by the encoder. Since flash architectures employ comparators, they are susceptible to metastability errors. In order to lower the probability of metastable states, the thermometer-binary decoding can be pipelined so that potentially indeterminate outputs are allowed more regeneration time [15].

The clocked comparator is composed of a preamplifier and a regenerative latch. The schematic is shown in Figure 6. The comparator utilizes the advantages of the low kickback noise in static comparators and the high regeneration speed in dynamic comparators. On one hand, keeping the preamplifier continuously biased throughout the conversion period significantly reduces the kickback disturbance; on the other hand, the dynamic flip-flop in the latch circuit will shorten the regeneration and reset time [16].

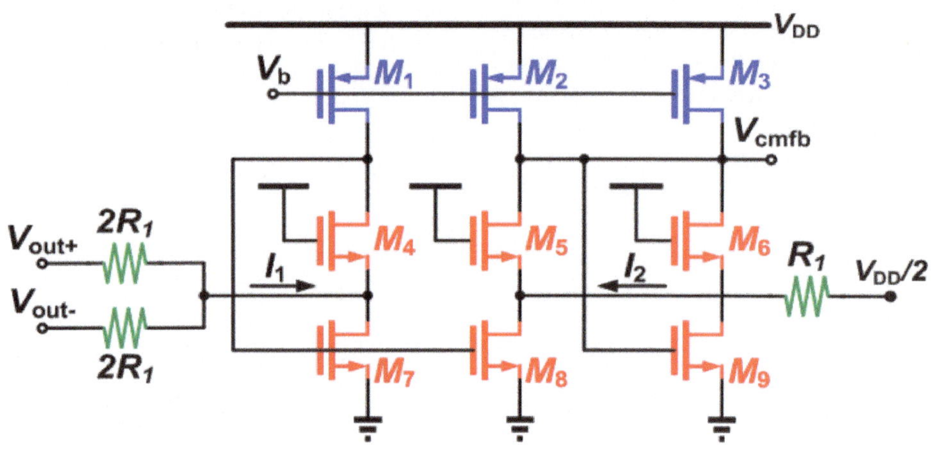

Figure 4. Continuous-time CMFB circuit.

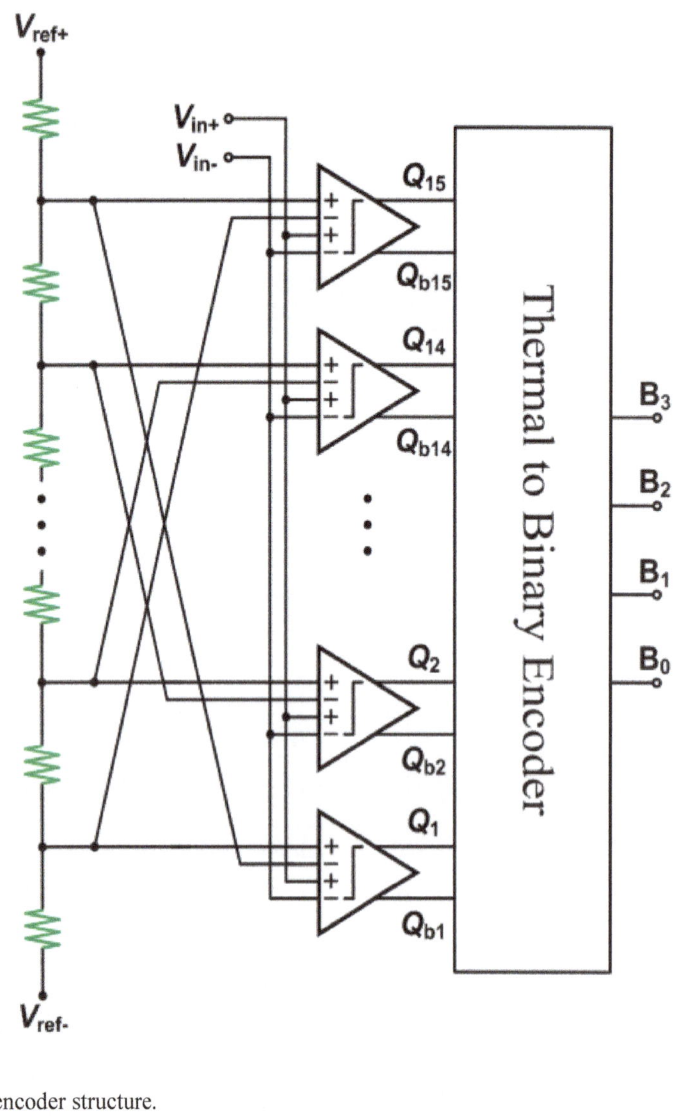

Figure 5. Four-bit quantizer and encoder structure.

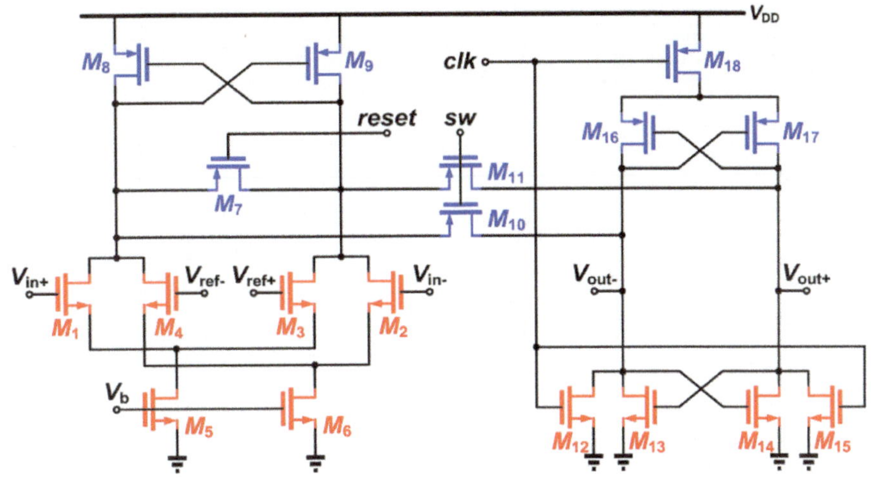

Figure 6. Schematic of the comparator circuit.

Feedback DACs

As shown in Figure 7, a multi-bit current-steering DAC, feeds current to the virtual grounds of the active-RC integrators, therefore good DAC linearity can be achieved. The improved DAC linearity is another advantage of active-RC integrators when compared to Gm-C integrators [17].

The CMFB circuit for the DAC along with the differential pairs connects to the op amp input. This DAC feedback circuit requires a 0.6-V voltage as a reference voltage, and this may operate in the virtual ground voltage. The reference voltage is connected by an external power supply. Two current sources inject common-mode currents to prevent a common-mode offset from appearing at the amplifier virtual grounds. The dynamic performance of current-steering DAC's is limited by the feedthrough of the control signals to the output lines. The coupling of the switching control signals to the output lines through the parasitic gate drain capacitance of the switching transistors is also a source of glitches. The lower part of Figure 7 is the simplified representation of the current-steering DAC. In this work, to minimize the feedthrough to the output lines, the drain of the switching transistors is isolated from the output lines by adding two cascaded transistors [18].

DWA Circuit

Combining ΣΔ modulators with multi-bit quantization is an effective means to achieve a high dynamic range and a wide bandwidth. The major obstacle in designing multi-bit ΣΔ modulators is that good component matching is required for internal DAC linearity. Good attenuation of DAC noise due to component mismatches can be provided by the DWA algorithm, which ideally can achieve a first-order DAC noise shaping. For DWA to be more useful in multi-bit SDM's, the DAC baseband tones must be removed. Conventionally, the problem is circumvented by adding dither. However, adding dither contributes additional noise to the base-band, degrades SNR and possibly destabilizes the modulator. A low- complexity high-speed circuit is proposed for the implementation of a DWA technique without adding dither, used for reducing DAC noise due to component mismatches [19].

The block diagram of the DWA logic is shown in Figure 8. The input of the DWA logic is connected to the four-bit quantizer output. The DWA logic converts the quantizer output code to the control signals, Si, i = 0, 1,..., 15, for the element selection of 4-bit DAC. A 4-bit adder and a 4-bit register produce two indexes which are converted to two sets of 16-bit thermometer codes by two binary to thermometer decoders. When the carry signal of the adder is low, the output control signals are the mutual XOR of the two 16-bit thermometer codes. When the carry signal is high, the control signals are the mutual XNOR of the two 16-bit codes. The DWA algorithm selects DAC components cyclically one by one. No unit is reselected before all the others are selected.

Time Constant Tuning Circuit

CMOS technologies usually do not have tight control over absolute values of R and C, so an automatic RC time constant tuning circuit is needed to ensure the ΣΔ modulator stability and SNR performance over large RC time constant variations. Therefore, a discrete capacitor tuning scheme is employed to calibrate the time constant of the active-RC integrators. The adjustable capacitor array is shown in Figure 9.

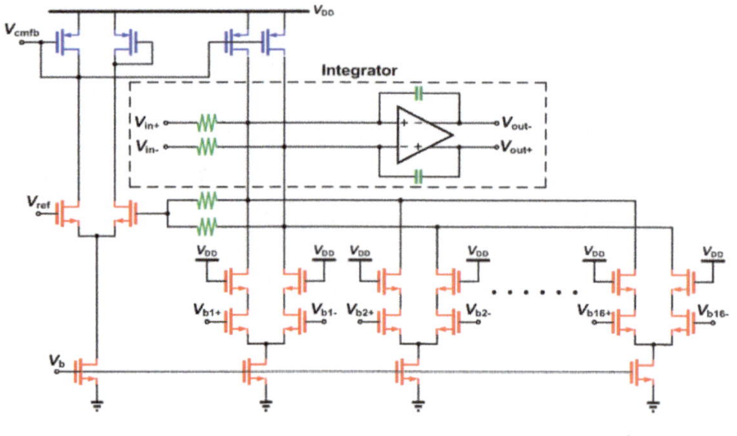

Figure 7. A multi-bit current-steering DAC schematic.

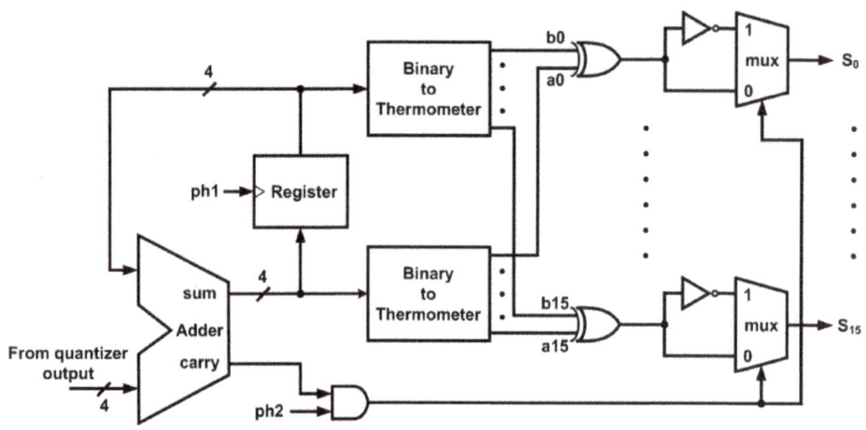

Figure 8. The block diagram for the DWA realization.

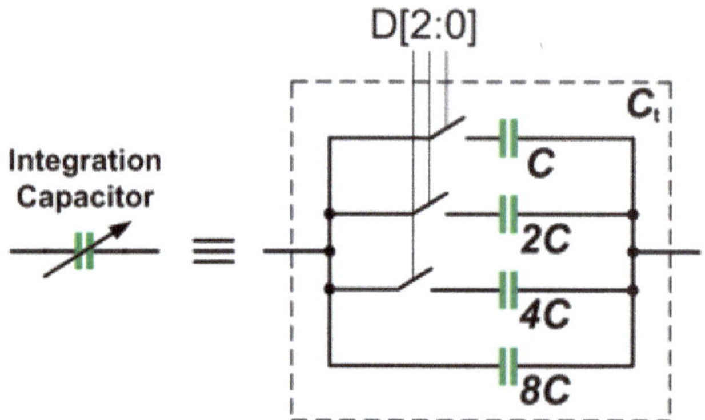

Figure 9. Tunable capacitor array.

The capacitors in the arrays are binary-weighted except the "always-in-use" capacitor which is equal to the most significant bit (MSB) capacitor, 8C. This sizing method provides constant tuning steps with the least number of capacitors. The 3-bit digital control codes are fed externally to choose which capacitors to use.

MEASUREMENT RESULTS

The proposed third-order multi-bit CT ΣΔ modulator in this paper is implemented in TSMC 0.18-μm CMOS process. Post-processing was performed using MATLAB before tapout. The modulator samples signals at 160 MHz with 10 MHz signal bandwidth and oversampling ratio of 8 and operates with a 1.2 V supply voltage. The total power consumption is 19.8 mW. Figure 10 shows the modulator die microphotograph including the wire bounding pads. The CT ΣΔ modulator is essentially a mixed-signal system which includes integrator, quantizer, and digital circuits. To achieve high resolution and linearity, caution should be taken in the layout design to reduce the effects of mismatch, parasitic and digital noise coupling to analog blocks. The total chip area including bonding pads is 0.9×1.284 mm^2.

The DWA circuit is located on the left of the chip. The noisy clock generator is placed in the bottom right-hand side to prevent interference with the weakly sensitive analog blocks. A single-to-differential circuit converts the single-ended input signal to a balanced differential signal input to the ADC. The output data stream of the modulator was captured using a logic analyzer. Figure 11 shows the digital outputs of the CT ΣΔ modulator measured by the logic analyzer for an input sinusoid at 3 MHz. The output spectrum density of the CT ΣΔ modulator analyzed by logic analyzer for an input sinusoidal signal of 3 MHz is shown in Figure 12. A peak SNDR of 48 dB which corresponds to a 7.7-bit within a bandwidth of 10 MHz is measured. The measured SNR and SNDR versus input signal level of the CT ΣΔ modulator for an input sinusoid at 3 MHz are plotted in Figure 13. The measured input peak dynamic range is 54 dB. Figure 14 summarizes the measured SNR and SNDR for varying input frequencies. The SNR and SNDR fall to 46 dB and 43 dB respectively for a 9 MHz input signal. Because the integrator is basically a low pass filter, the modulator acts as low-pass filtering characteristic. When input frequency is increased, the SNR/SNDR values will be decreased.

Figure 11. Measured digital output of the CT ΣΔ modulator at f = 3 MHz.

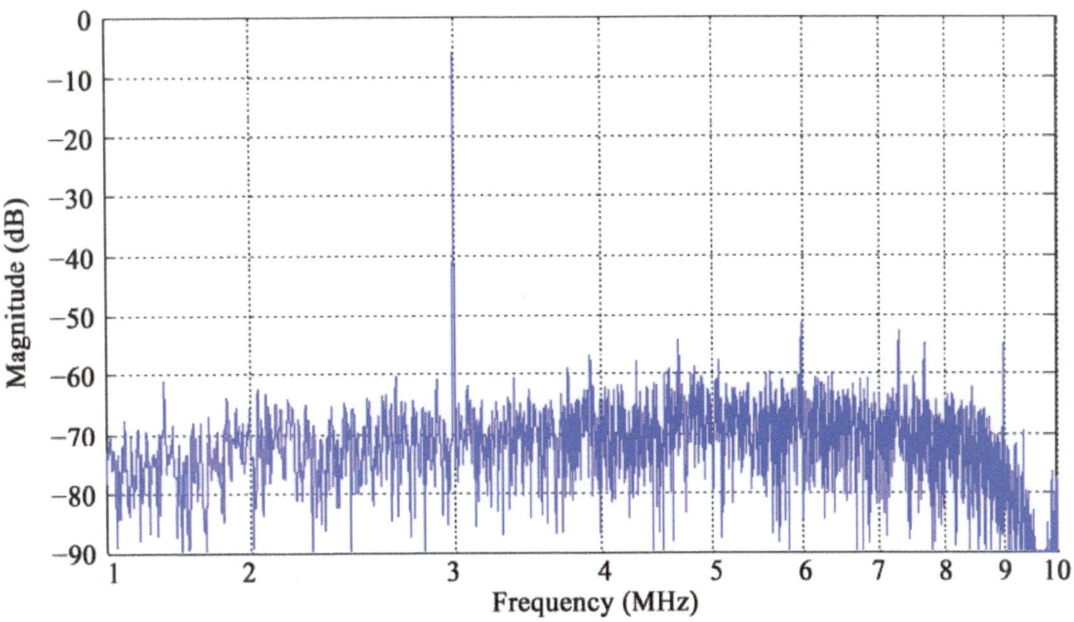

Figure 12. Measured output spectrum density.

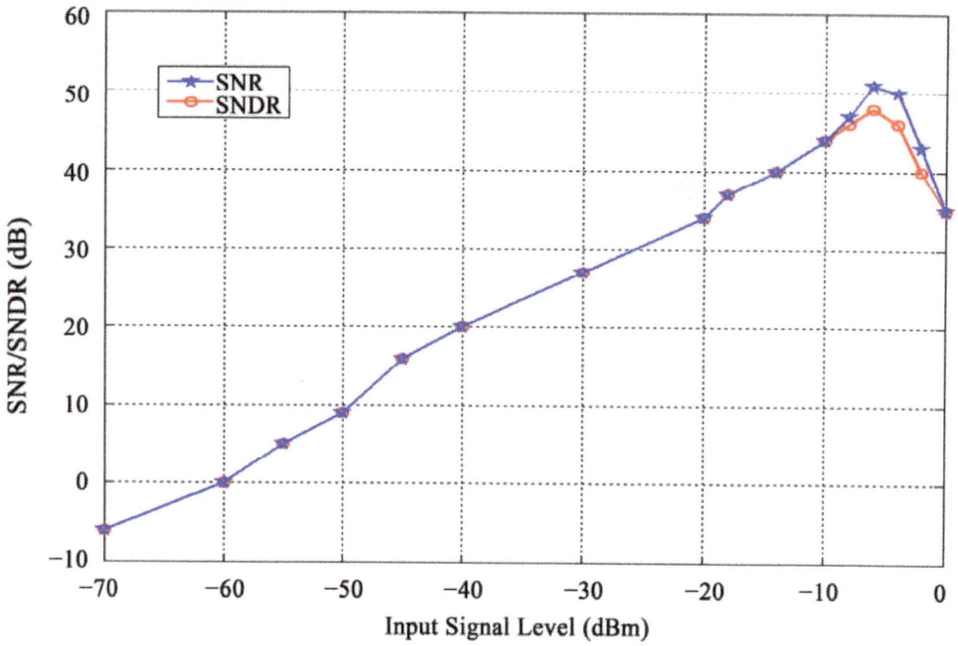

Figure 13. Measured SNR and SNDR vs. input signal level.

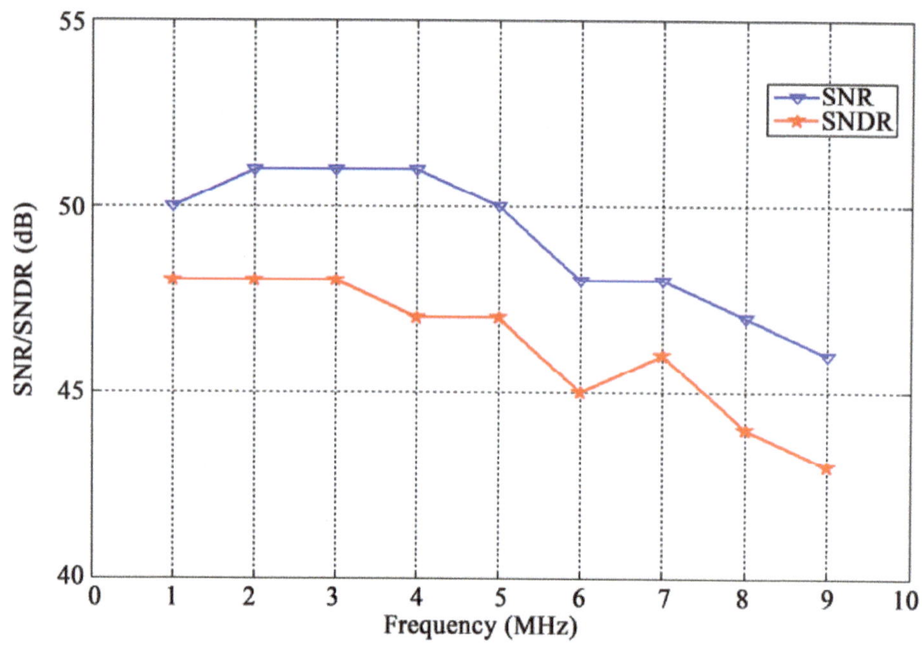

Figure 14. Measured SNR and SNDR vs. input frequency.

The performance parameters of this chip are summariz Table 2. Performance comparisons with other literatures. ed in Table 1. In this work, we use the design strategy for low-power CT ΣΔ modulator proposed by [20]. This concept is based on the figure of merit (FOM) which takes the overall power consumption, the dynamic range, and the signal bandwidth into account to find the most power-efficient ΣΔ modulator implementation with respect to these design parameters.

The FOM used is defined as

$$FOM = \frac{P}{2^B \cdot 2f_B}, \tag{5}$$

where P(mW) represents the power consumption, B is number of bits and f_B(MHz) is the bandwidth. The performance comparisons with other literatures are shown in Table 2. The smaller the FOM value is, the better the overall performance is. From this comparison table, it is confirmed that the proposed modulator with low voltage operations can achieve a wide bandwidth and lower power consumption.

Table 1. The performance summary of the CT ΣΔ modulator.

Parameters	Measured results
Sampling Frequency	160 MHz
Signal Bandwidth	10 MHz
SNR	51 dB
SNDR	48 dB
ENOB	7.7 bits
Dynamic Range	54 dB
Power Supply	1.2 V
Power Dissipation	19.8 mW
Chip Area	1.156 mm^2
Process	TSMC 0.18 um CMOS

Table 2. Performance comparisons with other literatures.

Parameter	This work	[21]	[22]	[23]	[24]
Technology	0.18 um	0.18 um	0.18 um	90 um	0.13 um
Voltage (V)	1.2	1.8	1.2	1.2	2.5
BW (MHz)	10	10	7.5	1.92	100
SNR (dB)	51	63	71	66.4	58.9
SNDR (dB)	48	56	67	62.4	53.1
ENOB (Bits)	7.7	9	10.8	10.1	8.5
Power (mW)	19.8	122.4	89	12.5	350
FOM (pJ/Conv.)	4.76	11.95	3.33	2.97	4.83

CONCLUSIONS

A low-voltage, low-power and wide bandwidth CT ΣΔ modulator has been implemented in a TSMC 0.18-um technology. The low-complexity high-speed implementation of the DWA technique for the reduction of baseband tones is used in this modulator. The excess loop delay set to half the sampling period of the quantizer has been used to avoid degradation of modulator stability in this architecture. All integration capacitors are tunable to overcome time constant variation. In addition, CT ΣΔ modulator provides a significant amount of inherent anti-aliasing, which is especially important when OSR is minimized in order to maximize the input bandwidth. The CT ΣΔ modulator itself occupies just 1.16 mm2 and consumes 19.8 mW. The modulator achieves a SNR of 51 dB and SNDR of 48 dB over 10 MHz signal bandwidth.

ACKNOWLEDGEMENTS

The authors would like to thank the staff of the CIC for the chip fabrication and technical supports with the number of T18-98D-158.

REFERENCES

1. J. Yu and B. Zhao, "Continuous-Time Sigma-Delta Modulator Design for Low Power Communication Applications," Proceedings of IEEE ASICON, October 2007, pp. 715-720.
2. R. H. M. van Veldhoven, B. J. Minnis, H. A. Hegt and A. H. M. van Roermund, "A 3.3-mW ΣΔ Modulator for UMTS in 0.18-um CMOS with 70-dB Dynamic Range in 2-MHz Bandwidth," IEEE Journal of Solid-State Circuits, Vol. 37, No. 12, 2002, pp. 1645-1652. doi:10.1109/JSSC.2002.804329
3. L. Dorrer, F. Kuttner, P. Greco, P. Torta and T. Hartig, "A 3-mW 74-dB SNR 2-MHz Continuous-Time Deltasigma ADC with a Tracking ADC Quantizer in 0.13-um CMOS," IEEE Journal of Solid-State Circuits, Vol. 40, No. 12, 2005, pp. 2416-2427. doi:10.1109/JSSC.2005.856282
4. S. R. Norsworthy, R. Schreier and G. C. Temes, "Delta-Sigma Data Converters: Theory, Design and Simulation," IEEE Press, New York, 1996.
5. B. J. Farahani and M. Ismail, "Adaptive Sigma Delta ADC for WiMAX Fixed Point Wireless Applications," IEEE Midwest Symposium on Circuits and Systems, Vol. 1, Covington, 7-10 August 2005, pp. 692-695.
6. T. L. Brooks, D. H. Robertson, D. F. Kelly, A. D. Muro and S. W. Harston, "A Cascaded Sigma-Delta Pipeline A/D Converter with 1.25 MHz Signal Bandwidth and 89

dB SNR," IEEE Journal of Solid-State Circuits, Vol. 32, No. 12, 1997, pp. 1896-1906. doi:10.1109/4.643648

7. R. Jiang and T. Fiez, "A 14-bit Delta-Sigma ADC with 8X OSR and 4-MHz Conversion Bandwidth in A 0.18 um CMOS Process," IEEE Journal of Solid-State Circuits, Vol. 39, No. 1, 2004, pp. 63-74. doi:10.1109/JSSC.2003.820877

8. A. Bosi, A. Panigada, G. Cesura and R. Castello, "An 80 MHz 4x Oversampled Cascaded ΔΣ-Pipelined ADC with 75 dB DR and 87 dB SFDR," IEEE International Solid-State Circuits Conference, 2005, pp. 174-175.

9. F. Munoz, K. Philips and A. Torralba, "A 4.7 mW 89.5 dB DR CT Complex DS ADC with Built-In LPF," IEEE International Solid-State Circuits Conference, February 2005, pp. 500-501.

10. J. A. Cherry and W. M. Snelgrove, "Continuous-Time Delta-Sigma Modulators for High-Speed A/D Conversion," Kluwer, Boston, 2000.

11. S. Paton, A. D. Giandomenico, L. Hernandez, A. Wiesbauer, T. Potscher and M. Clara, "A 70-mW 300-MHz CMOS Continuous-Time ΣΔ ADC with 15-MHz Bandwidth and 11 Bits of Resolution," IEEE Journal of Solid-State Circuits, Vol. 39, No. 7, July 2004, pp. 1056-1063. doi:10.1109/JSSC.2004.829925

12. G. Mitteregger, C. Ebner. S. Mechnig, T. Blon, C. Holu- igue and E. Romani, "A 20-mW 640-MHz CMOS Continuous-Time ΣΔ ADC with 20-MHz Signal Bandwidth, 80-dB Dynamic Range and 12-Bit ENOB," IEEE Journal of Solid-State Circuits, Vol. 41, No. 12, 2006, pp. 2641-2649. doi:10.1109/JSSC.2006.884332

13. S. Karthikeyan, S. Mortezapour, A. Tammineedi and E. Lee, "Low-Voltage Analog Circuit Design Based on Biased Inverting Opamp Configuration," IEEE Transactions on Circuits and Systems II: Analog and Digital Signal Processing, Vol. 47, No. 3, 2000, pp. 176-184. doi:10.1109/82.826743

14. L. Lamarre, M. Louerat and A. Kaiser, "A Simple 3.8 mW, 300 MHz, 4-Bit Flash Analog-to-Digital Converter," Proceedings of the SPIE, Vol. 5837, No. 51, 2005, pp. 825- 832. doi:10.1117/12.608343

15. B. Razavi, "Principles of Data Conversion System Design," IEEE Press, New York, 1995.

16. J. Chen, S. Kurachi, S. Shen, H. Liu, T. Yoshimasu and Y. Suh, "A Low-Kickback-Noise Latched Comparator for High-Speed Flash Analog-to-Digital Converters," Proceedings of IEEE ISCIT, Vol. 1, 12-14 October 2005, pp. 250-253.

17. Z. Li and T. S. Fiez, "A 14-Bit Continuous-Time Deltasigma A/D Modulator with 2.5 MHz Signal Bandwidth," IEEE Journal of Solid-State Circuits, Vol. 42, No. 9, 2007, pp. 1873-1883. doi:10.1109/JSSC.2007.903086

18. J. Bastos, A. M. Marques, M. S. J. Steyaert and W. Sansen, "A 12-Bit Intrinsic Accuracy High-Speed CMOS DAC," IEEE Journal of Solid-State Circuits, Vol. 33, No. 12, 1998, pp. 1959-1969. doi:10.1109/4.735536

19. T. H. Kuo, K. D. Chen and H. R. Yeng, "A Wideband CMOS Sigma-Delta Modulator with Incremental Data Weighted Averaging," IEEE Journal of Solid-State Circuits, Vol. 37, No. 1, 2002, pp. 11-17. doi:10.1109/4.974541

20. F. Gerfers, M. Ortmanns and Y. Manoli, "A 1.5-V 12-Bit Power-Efficient Continuous-Time Third-Order ΣΔ Modulator," IEEE Journal of Solid-State Circuits, Vol. 38, No. 8, 2003, pp. 1343-1352. doi:10.1109/JSSC.2003.814432

21. L. J. Breems, R. Rutten and G. Wetzker, "A Cascaded Continuous-Time ΣΔ Modulator with 67-dB Dynamic Range in 10-MHz Bandwidth," IEEE Journal of Solid-State Circuits, Vol. 39, No. 12, 2004, pp. 2152-2160. doi:10.1109/JSSC.2004.836245

22. S. D. Kulchycki, R. Trofin, K. Vleugels and B. A. Wooley, "A 77-dB Dynamic Range, 7.5-MHz Hybrid Continuous-Time/Discrete-Time Cascaded ΣΔ Modulator," IEEE Journal of Solid-State Circuits, Vol. 43, No. 4, 2008, pp. 796-804. doi:10.1109/JSSC.2008.917499

23. M. Anderson and L. Sundstrom, "Design and Measurement of a CT ΔΣ ADC with Switched-Capacitor Switched-Resistor Feedback," IEEE Journal of Solid-State Circuits, Vol. 44, No. 2, 2009, pp. 473-483. doi:10.1109/JSSC.2008.2010978

24. A. Hart and S. P. Voinigescu, "A 1 GHz Bandwidth Low-Pass ΔΣ ADC with 20-50 GHz Adjustable Sampling Rate," IEEE Journal of Solid- State Circuits, Vol. 44, No. 5, 2009, pp. 1401-1414. doi:10.1109/JSSC.2009.2015852

CHAPTER 9 AN ULTRA LOW-VOLTAGE AND LOW-POWER OTA USING BULK-INPUT TECHNIQUE AND ITS APPLICATION IN ACTIVE-RC FILTERS

Arash Ahmadpour

Keywords: Operational Transconductance Amplifier (OTA), Common-Mode Feedback (CMFB), Bulk-Input, Switched-Capacitor (SC) Integrator, Miller-Feed-Forward (MFF)

INTRODUCTION

There are growing strong demands for low-voltage supply and low-power consumption circuits and systems. This is especially true for very high integration level and very large scale integrated (VLSI) mixed-signal chips and system-on-a-chip. In mixed-signal systems, the analog circuits are combined with digital circuits in order to get the best performance with low-voltage supply and low power consumption. This combination should be in an optimal way and the optimization process is application dependent. Also modern portable applications such as medical devices and remote wireless sensors require extending the battery life as well. This trend has forced designers to develop new approaches more amenable to low-voltage and low-power integrated circuits and it poses lots of challenges for all involved such as processes, devices, circuits, and system architectures [1]. Operational transconductance amplifiers (OTAs) are the key active building blocks of continuous-time filters. They can be generally classified into single ended, fully differential and pseudo-differential OTAs. Fully differential OTAs are preferred because they provide larger signal swing, better distortion performance, better CM noise and supply noise rejection. The main drawback of using fully differential OTAs is that a common-mode feedback (CMFB) circuit must be added. The principle of the CM control circuit can be easily applied to the design of differential structures, and it is well suited for low-voltage pseudo-differential architectures. This extra circuit is used to establish the common-mode output voltage and suppress the common mode signal components [2]. The speed of the common-mode path should be comparable to that of the differential path; otherwise the common-mode noise may be significantly amplified such that the output signal becomes distorted. Also, the CMFB circuit is often a source of noise injection and increases the load capacitance that needs to be driven. Regardless of the limitations described above, fully differential OTAs work very well and can substantially improve the system's quality, especially in very unfriendly environments such as mixed-mode applications. However, at lower supply voltages, pseudo differential OTAs could be used to avoid the voltage drop across the tail current source used in the fully differential structures. Taking the tail current source achieves a larger signal swing, but it also results in larger CM gain. So it requires to carefully controlling the CM response for pseudo differential OTAs. The reduced supply voltage will cause many problems when designing analog circuits due to the reduction in available signal swing. This problem is magnified due to the fact that the threshold voltage (V_T) of the transistors has been reduced at a slower rate than the supply voltage. In other words, the threshold-voltage with decreasing the Power supply never reduces linearly. If standard analog design techniques continue to be used, the dynamic range and SNR of the circuits will degrade. The gain of devices such as amplifiers will also decrease because cascode transistors can no longer be used since they limit the output swing. The design of rail-to-rail input circuits is also made more difficult due to the large V_T which must be overcome. One solution for designing low-voltage analog circuits is to operate the

Citation: A. Ahmadpour, "An Ultra Low-Voltage and Low-Power OTA Using Bulk-Input Technique and Its Application in Active-RC Filters," Circuits and Systems, Vol. 2 No. 3, 2011, pp. 183-189. doi: 10.4236/cs.2011.23026.

transistors in the weak inversion region. The drawbacks to this technique include a limited input signal swing, an increase in the mismatch between transistors, a low slew-rate due to the low bias current levels, and large transistor sizes [3-6]. In order to avoid these drawbacks a bulk-input technique can be used, which allows for operation in the moderate inversion region at supply voltages equal to the V_T of the technology. The voltage applied to the bulk actually reduces the threshold voltage of the transistor, which increases the inversion level. One drawback of the bulk input technique is that the input transistor must sit in an n-well or p-well, so that its bulk is separate from the bulk of the rest of the transistors. This is not a problem in advanced technologies, which make use of triple well structures. In triple-well processes both the NMOS and PMOS can have isolated bulk terminals [5]. This paper presents the design of a novel two-stage bulk-input pseudo differential OTA, which operates at a supply voltage of 0.5 V. The circuit was designed using the 90 nm process in digital CMOS technology. Design procedure for this new merged OTA structure such as main OTA, CMFB and bias circuits are introduced in Section 2. In Section 3, SC integrator and active RC filter design are described. Section 4 presents simulation results. Finally, conclusions and acknowledge are given in Section 5 and 6, respectively.

BULK-INPUT OTA CIRCUIT DESIGN

Main Amplifier

A basic body-input stage, capable of operating with very low supply voltage is depicted in Figure 1. This structure is a two-path OTA built by the combination of the two different amplifiers. In this circuit a PMOS bodydriven OTA is implemented due to the action of M_{1+}, M_{1-}, M_{4+} and M_{4-}, and also a current-mirror bulk-input OTA is composed by transistors M_{2+}, M_{2-}, M_{3-} and M_{3+}. The OTA input consists of two bulk-input split differential pairs; some of the ac current generated by the input transistors is injected to the output transistors, providing a fast path for the current. The other part of the current is delivered to the current mirrors of M_{3+} and M_{3-}. These are the main transistors which the main portion of the currents by the factor n transfer to output transistors of M_{4+} and M_{4-}. If we suppose M_{4+} and M_4- have equal effective voltage to M_{1+} and M_{1-}, their transconductances are very much greater than bulk- transconductance of M_{1+} and M_{1-} ($g_{m4} >> g_{mb1}$). It is possible to utilize common-source amplifier which is parallel and in same phase with M1 when the input signal is implied to the gate of M4 via feed-forward (FF) compensate capacitor C_{ff} (not shown in Figure 1). Moreover, the transistor of the active load current source operates like a common source amplifier. This structure increases the gain and speed of the OTA. M_{3+} and M_{3-} with diode connections utilize for M_{4+} and M_{4-} biasing and use in M_{2+} and M_{2-} as a differential loads.

To further improve the differential gain PMOS devices (M_{5+}, M_{6+} and M_{5-}, M_{6-}) are added. The body-inputs of them are a new merged cross-coupled self-cascode pairs that adds a negative resistance to the output in the form of g'_{mb6} and boosts the differential DC gain. The g'_{mb6} is the g_{mb} of the self-cascode configuration of M_5, M_6. Also, the gate inputs biased at 100 mV, which biases them in moderate inversion. Moreover, these transistors are a self-cascode active load for transistors M_{4+} and M_{4-} which are common source amplifiers.

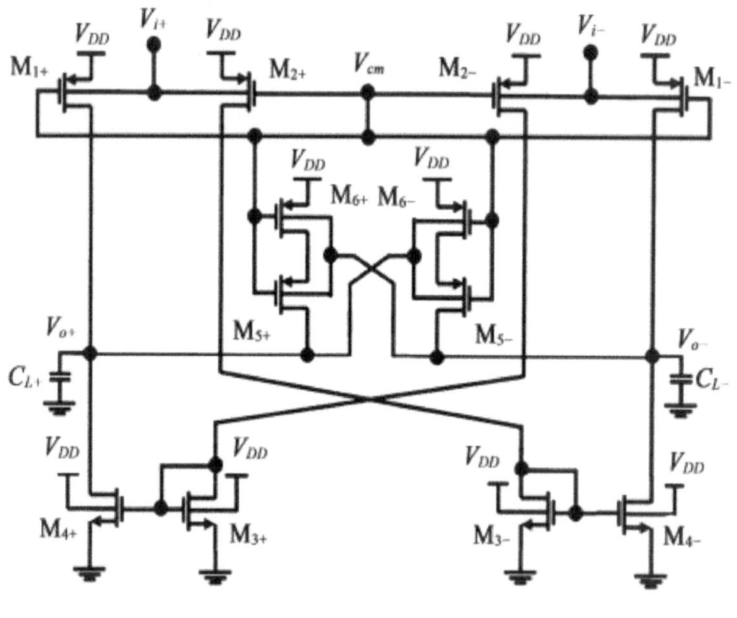

Figure 1. One stage of proposed bulk-input differential pair.

Bulk-Mode Common-Mode Feedback Circuit

The Fully differential OTAs require a common-mode feedback circuit. A CMFB circuit should behave linearly and only response to common-mode voltage. Lacking these features causes increase the THD of the circuit. Nowadays, designing a CMFB circuit which is able to operate under an ultra low-voltage supply is very difficult, mainly because of the difficulty of detecting of CM voltage. In Reference [5] a CMFB circuit which operated at 0.5 V by using two resistors to sense the output CM levels was designed. But this structure increases the die area and reduces the gain due to longer loads the OTA. To overcome some of these problems, a CMFB circuit has been reported [6]. This structure uses of four PMOS and two NMOS transistors. The NMOS devices is a bulk-input current mirror, which compares the current of the PMOS devices and then difference of these current is fed to the gate of the input transistors for controlling of the output CM voltages. This paper is given a novel continuous-time CMFB circuit that is able to operate with an ultra low-voltage supply. Figure 2 shows the proposed CMFB circuit for each stage.

In this circuit, a combination of two complementary NMOS and PMOS transistors is used in a bulk-driven configuration. The output voltage of the CMFB circuit can be found using the KCL at the output node (V_{cm}). The difference between currents of PMOS and NMOS transistors is fed to the gate of the input transistors of OTA. The bulks of the PMOS devices M_{cp+} (M_{cp-}) connected to the NMOS devices M_{cn+} (M_{cn-}) and are utilized to maintain the output common voltage at the required level (250 mV), while maximizing the output swing of the OTA. In other words, the inputs to the CMFB circuit are the outputs of the OTA, V_{o+} and V_{o-}. The CMFB circuit amplifies the difference between the average of V_{o+} and V_{o-} and the desired common level V_{cm} (250 mV), and sends a feedback signal V_{cm} to set the bias voltage at the gates of input transistors of the OTA. This structure is able to operate with an ultra-low supply voltage (as The Fully differential OTAs require a common-mode low as 0.3 V supply voltage).

In addition, this configuration with minimizing of the area cost and power consumption is a very cost-effective CMFB which is much more applicable regarding modern ultra-low-voltage and low-power mixed-signal SoCs. Simplicity and ability of using this new merged CMFB circuit to set the output CM voltage of OTA is superior compared with state-of-the-art CMFB. To obtain adequate gain, identical gain blocks can be cascaded so that a two-stage OTA is obtained. Figure 3 show the proposed OTA structure in first and second stages without bias and CMFB circuits. In order to improve the phase margin and speed of amplifier, we were used Feed-forward compensation. Finally, the OTA was stabilized by adding Miller compensation capacitors Cc with series resistors Rc for right half-plane zero cancelation. The OTA was designed and classified into two kinds of compensations, such as only Miller compensation and Miller-Feed-forward (MFF) compensation. The comparison of two these methods of compensations are given in Table 1. As can be seen, the MFF compensation method increases the phase margin and enhances the open-loop GBW about two times as much.

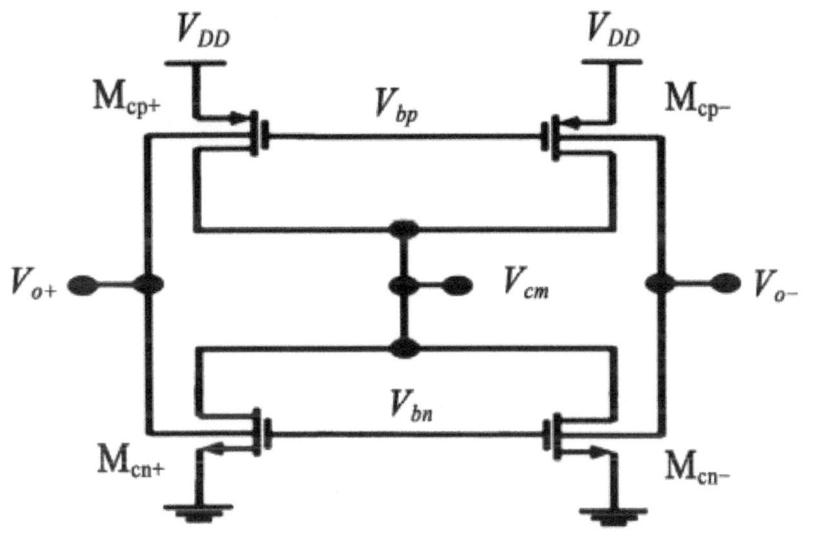

Figure 2. Ultra low-voltage bulk-mode CMFB circuit.

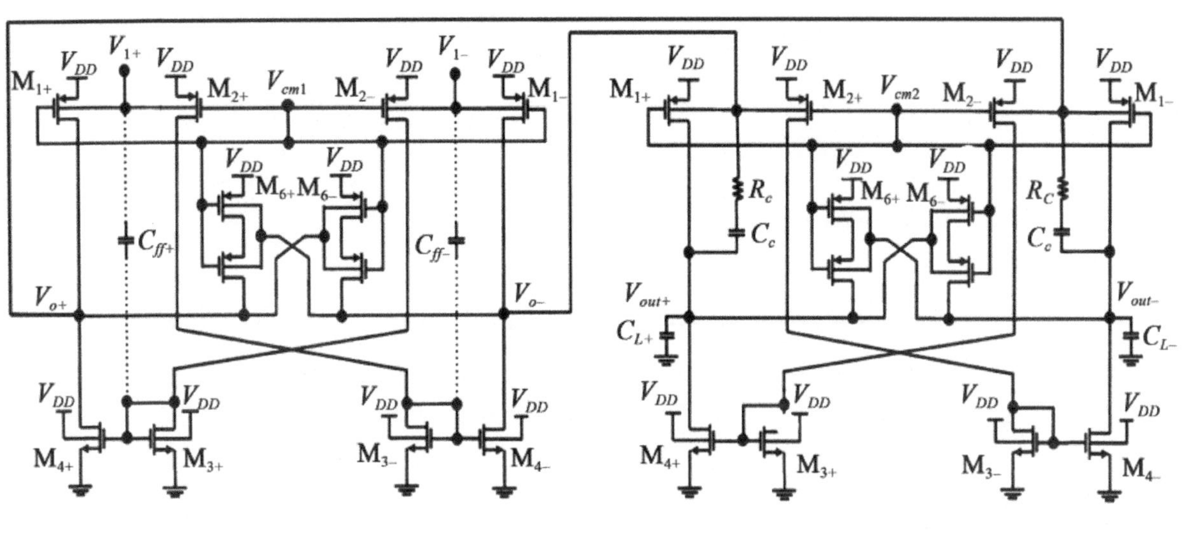

Figure 3. Two-stage bulk-input pseudo-differential OTA without bias and CMFB circuits.

Table 1. Comparison of two compensation methods.

Resistors and Capacitors	Miller-comp.	MFF-comp.
R_c	60 k	68 k
C_c	3 pf	1 pf
C_{ff}	-	3 pf
Open-loop GBW	962 KHz	2.5 MHz
Phase-Margin	65°	70.8°

Reference Current Generator and Bias Circuit

A low-sensitivity reference current generator and bias circuit are illustrated in Figure 4. Because the gate and source of M_{B3} and M_{B4} are common for both transistors, and the aspect ratios are equal, $I_{D\,MB3} = I_{D\,MB4}$. Also, note that $V_{GS\,MB2} = V_{GS\,MB1} + R_B \cdot I_{D\,MB1}$. Thus,

$$\sqrt{\frac{2I_{DM_{B3}}}{\mu_n \cdot C_{ox} \cdot (W/L)_{M_{B2}}}} = \sqrt{\frac{2I_{DM_{B3}}}{\mu_n \cdot C_{ox} \cdot K \cdot (W/L)_{M_{B2}}}} + R_B \cdot I_{DM_{B3}} \quad (1)$$

In the above mentioned equation, K is the ratio between the aspect ratios of M_{B1} and M_{B2}. Rearranging this expression,

$$I_{DM_{B3}} = \frac{1}{R_B^2} \frac{2}{\mu_n \cdot C_{ox} \cdot K \cdot (W/L)_{M_{B2}}} \left(1 - \frac{1}{K}\right) \quad (2)$$

In the target circuit, k = 1.13 and R_B = 1 k, thus $I_{D\,MB3}$ = 1.25 uA. As expected, the circuit is independent of the supply voltage. Transistor M_{B5} mirrors this current to generate a stable 2 uA reference current, which is used in the bias of PMOS devices. In order to ensure that the all transistors operate in the saturation region, fixed bias voltages are applied either to the gate of the PMOS and NMOS transistors in the main OTA and CMFB circuits. The gate voltage all of the NMOS and PMOS transistors biased in about V_{bn} = 350 mV and V_{bp} = 100 mV, respectively.

SC INTEGRATOR AND ACTIVE FILTER CIRCUIT

In order to show that the designed OTA will be useful in practical analog circuits, this structure was checked by a typical switched-capacitor (SC) integrator [7,8] with suitable input frequency 1 KHz and sampling frequency 1 MHz. Figure 5 is shown schematic of this structure. In switched-capacitor integrator design, we utilized two clock-frequencies which have no overlap and used Bootstrapped switches in Reference [9].

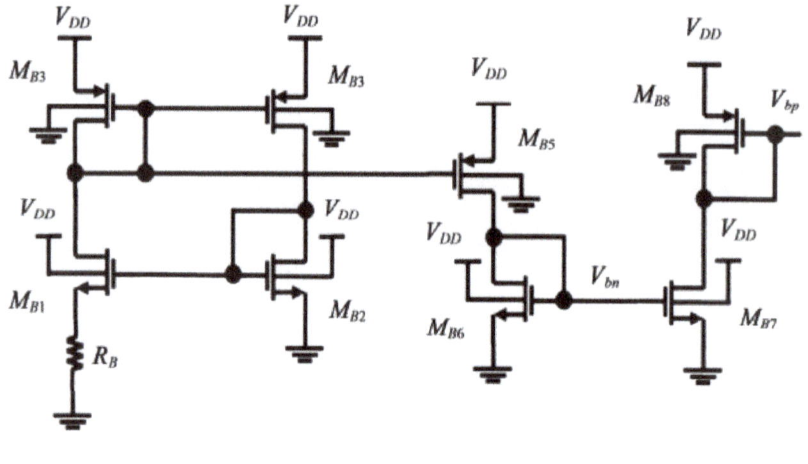

Figure 4. Reference current generator and bias circuit.

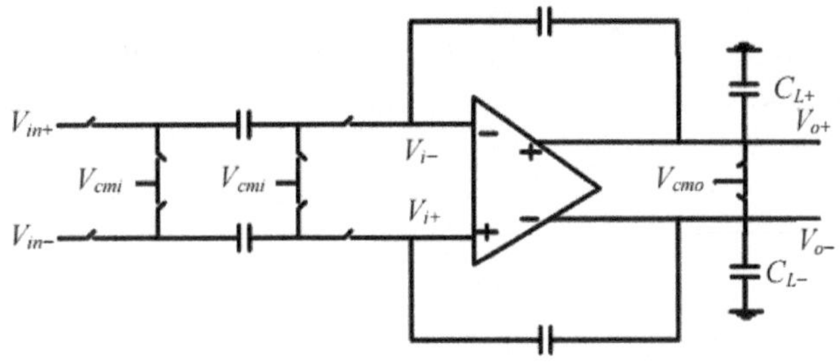

Figure 5. Fully differential switched-capacitor integrato.

Simulation result of this structure is depicted in Figure 11.

Filters are systems that can be used to manipulate the frequency spectrum of signals and they are essential in many different applications. The low-pass continuoustime filter is used to limit signal and noise bandwidth and provide anti-aliasing prior sampling. The analog-to- digital converter (ADC) digitizes the filtered output to take the advantages of the Digital Signal Processing (DSP) unit. The equalizer provides the equalization and the equalized signal goes to the decoder. The gain/timing control module is used to adjust the gain of the VGA. Although we are living in a digital age, many digital systems interfacing with the real analog world might use continuous-time filters. There are mainly two types of filters: digital filters and analog filters. While the data samples are discrete for digital filters, analog filters process continuous signals. Analog filters can be further divided into passive filters and active filters. While passive filters comprise passive components only such as resistors, capacitors, and inductors, active filters use active devices such as operational amplifiers and/or operational transconductance amplifiers (OTAs). Active filters can also be classified into Active-RC, Switched-Capacitor (sampled-data filters), OTA-C/Gm-C, and LC filters. Passive filters do not employ amplifiers and usually they are off-chip filters and are not suitable for integrated circuits. But active-RC and SC filters are suitable for low to medium frequency applications.

This OTA was tested in an 8th order active Biquadrate RC filter for maximize the attainable swing. The filter was built by connecting four 2nd orders Thomas BiQuad stage to implement the Butterworth function, shown in Figure 6. The equations of the Two-Thomas biquad filter as follow:

$$\omega_o^2 = \frac{1}{R_{s2} \cdot R_{f2} \cdot C_{f1} \cdot C_{f2}}, Q = \frac{R_{f1}}{\sqrt{R_{s2} \cdot R_{f2}}}\sqrt{\frac{C_{f1}}{C_{f2}}}, A_o = \frac{R_{f2}}{R_{s1}} \tag{3}$$

where A_o is the gain of integrator and ω_o and Q are the cut-off frequency and quality factor of integrator, respectively. Assume $R_{s2} = R_{f2}$, $C_{f1} = C_{f2}$, Equation (3) can be written as:

$$\omega_o = \frac{1}{R_{f2} \cdot C_{f1}}, Q = \frac{R_{f1}}{R_{f2}}, A_o = \frac{R_{f2}}{R_{s1}} \tag{4}$$

In the above mentioned equation, Q and A_o are determined by the ratio of resistors and both Q and ω_o can be tuned independently [10]. This 8th order Butterworth filter consists of four cascaded Thomas biquads; and the cut-off frequency is about 400 KHz. The filter was simulated with components values given in Table 2. Also, a plot of the frequency response for input CM voltage from rail-to-rail of this filter is shown in Figure 12.

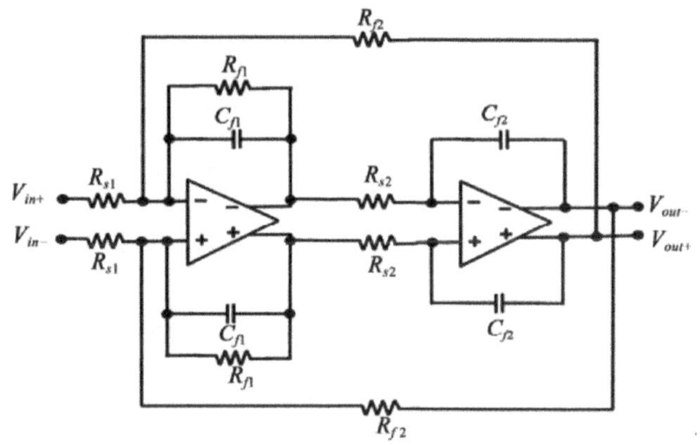

Figure 6. 2nd order Thomas Biquad filter stage.

Table 2. Filter component values.

Components	Stage (1)	Stage (2)	Stage (3)	Stage (4)
R_{s1}	56 k	39 k	47 k	56 k
R_{s2}	56 k	39 k	47 k	56 k
R_{f1}	30 k	87 k	87 k	30 k
R_{f2}	56 k	39 k	47 k	56 k
C_{f1}	10 pf	10 pf	10 pf	12 pf
C_{f2}	10 pf	8 pf	5.4 pf	10 pf

SIMULATION RESULTS

To test op-amp's performance, various configurations were implemented to simulate application condition. Based on the analysis procedure described in the previous sections, an ultra low-voltage OTA at a single 0.5 V supply voltage was designed. The OTA has been designed in 90 nm CMOS process, and then simulated by Hspice. Then this structure was checked by a typical SC integrator with a 1 MHz clock frequency. Also a 8th order active RC-filter with a 400 KHz cut-off frequency was designed. The simulation results are shown in Figures 7 to 12. Furthermore, Table 3 summarizes the simulated performance of this OTA and comparisons of characteristics of proposed OTA with state-of-the-art OTAs. The open-loop frequency response of the OTA for the common-mode input V_{cm} varying from rail-to-rail by a step 0.01 V was tested. The amplitudes and the phases of the proposed OTA were almost independent of the applied V_{cm}. Figure 5 shows the frequency response of the OTA for V_{cm} = 0.25 V. The result showed a 70.2 dB DC gain, a 2.5 MHz bandwidth and a of 70.8° phase margin. For the closed-loop simulation, the OTA was configured in unity-gain with 1 pf capacitors.

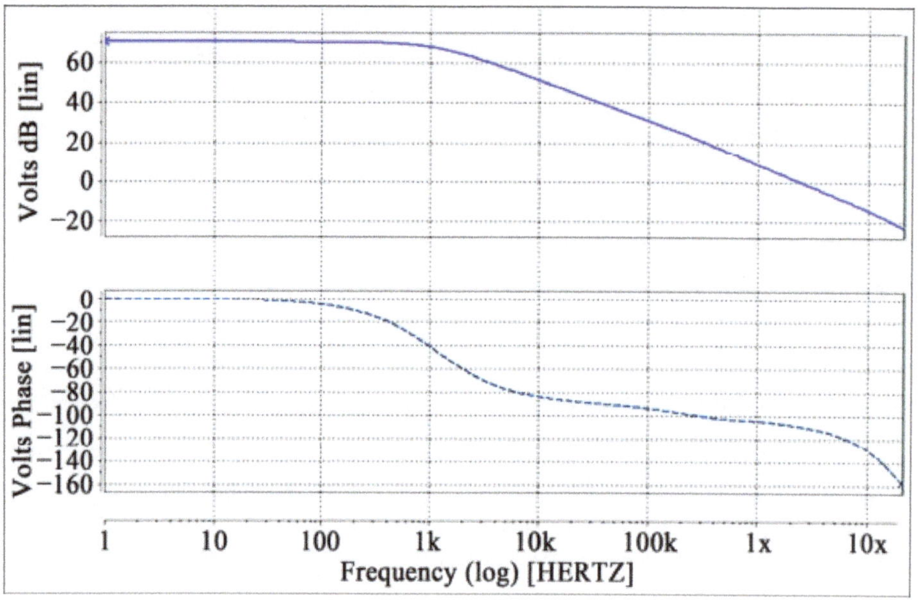

Figure 7. Open-loop frequency response of OTA for V_{cm} = 0.25 V.

Table 3. Comparison of characteristics of proposed OTA with state-of-the-art OTAs.

Parameters	This work	Ref. [5]	Ref. [6]
Technology	90 nm	0.18 μm	0.18 μm
Supply Voltage	0.5 V	0.5 V	0.5 V
Open-loop DC Gain	70.2 dB	62 dB	65 dB
Open-loop GBW	2.5 MHz	10 MHz	550 KHz
Closed-loop GBW	645 KHz	-	475 KHz
Phase-Margin	70.8°	60°	50°
Input ref. noise@1 KHz	139 nV/Hz$^{1/2}$	-	432 nV/Hz$^{1/2}$
FOM	100	133.3	19.6
THD@ 500 mV	0.119%	1%	0.13%
Load Capacitors	20 pf	20 pf	20 pf
Power Consumption	25 μw	110 μw	28 μw

The closed-loop frequency response of the OTA is shown in Figure 8 for rail-to-rail input common-mod ranges by a step of 50 mV. It is obvious that the frequency responses of OTA have no peak and result a bandwidth of 645 KHz. This shows that the OTA function correctly for rail-to-rail input common-mode voltage values without any considerable decrease in gain. The main optimization steps were done in transient tests because it is most important behavior. Two inputs were applied to see both small and large signal behavior of OTA in transient mode, when in closed-loop test; the OTA is configured in unity-gain with 1 pf capacitors. For an input common-mode voltage of 250 mV, when 0.1 V step was applied to inputs of OTA, the output voltage was settled to its final value in less than 735 ns (@ 0.01%). Identically, for large signal mode with a 0.3 V step and input common-mode voltage of 250 mV, output voltage was settled in less than 1500 ns for rising and less than 1170 ns for falling (@ 0.02%).

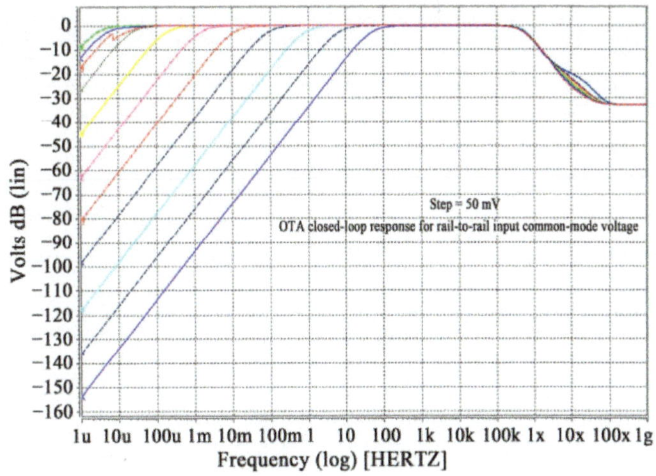

Figure 8. OTA closed-loop response for rail-to-rail common mode input value (Step = 50 mV).

In this case, positive and negative slew-rates are 0.67 V/μs and 0.8 V/μs, respectively. Figure 9 show transient response of OTA for 0.1 V and 0.3 V inputs, when input common-mode voltage varying from rail-to-rail by a step 50 mV. It is obvious that response has no ring or overshoot because of suitable bandwidth and phase margin.

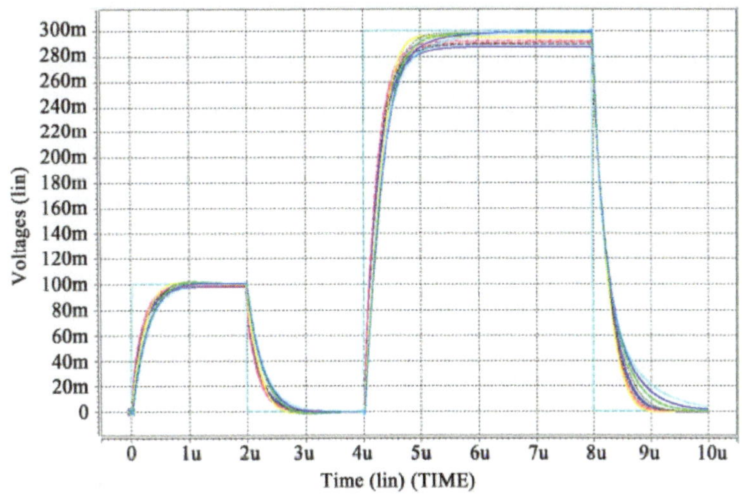

Figure 9. Settling simulated results of OTA for rail-to-rail input common-mode voltage (Step = 50 mV).

The obtained total harmonic distortion of OTA with a 200 mV amplitude and 10 KHz input frequency sampled at 100 KHz is shown in Figure 10. The third harmonic is

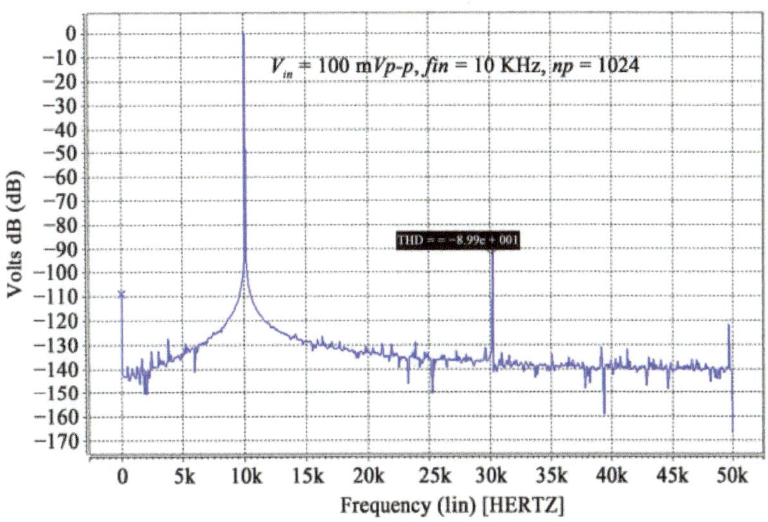

Figure 10. Total harmonic distortion of OTA with f_{in} = 10 KHz and V_{in} = 200 mVp–p.

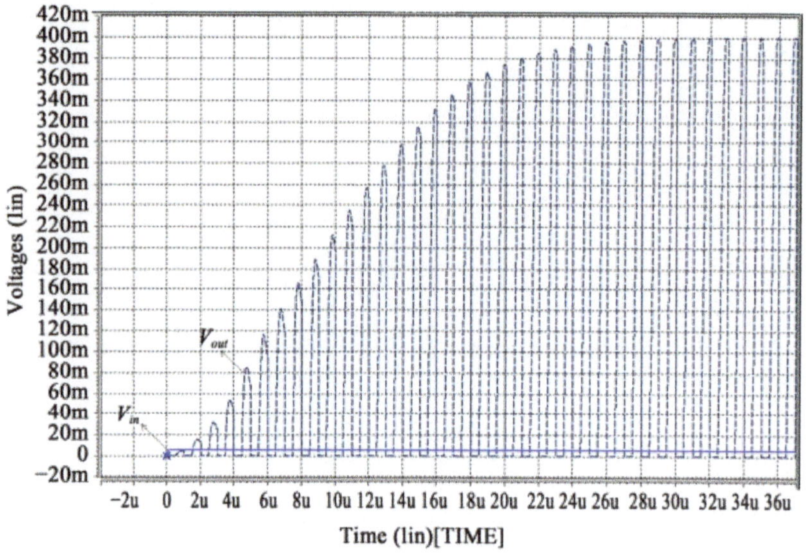

Figure 11. Integrated output voltage of OTA with f_s = 1 MHz and f_{in} = 1 KHz.

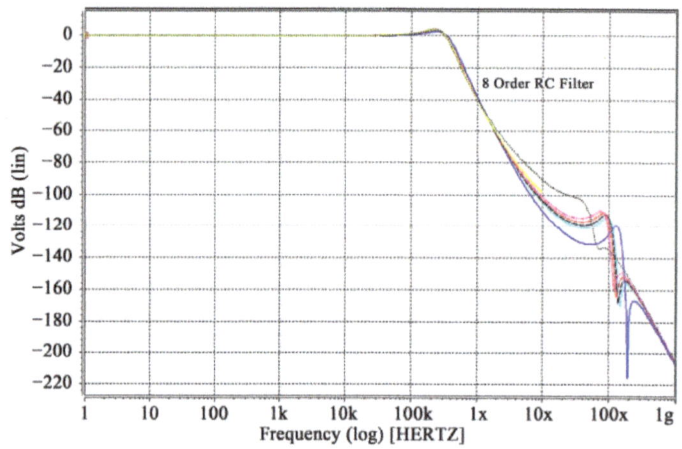

about 90 dB below the fundamental. Also, for 400 mV amplitude, the third harmonic is about −74.4 dB. As can be seen, the extra harmonics except main harmonic has been eliminated in this design.

CONCLUSIONS

In this paper a novel two-stage configuration of the ultra low-voltage and low-power bulk-input CMOS OTA in 90 nm CMOS process which is able to operate with a single supply voltage as low as 0.5 V has been presented. A new merged cross-coupled self-cascode pair was used and higher DC gain was achieved. In addition, a MFF compensation was utilized which has been improved the GBW and phase margin of the OTA. Also, a new bulkmode CMFB circuit which no longer loads OTA has been discussed. A large linear signal swing has been achieved due to the well controlled CM behavior. Finally this structure was checked by a typical SC integrator and was tested in an 8th order active Biquadrate RC filter. Correct functionality for this configuration is verified from −30°C to 70°C. In addition, this structure can be used for modern ultra low-voltage and low-power mixed- signal SoCs.

ACKNOWLEDGEMENTS

The work described in this paper is financially supported by a grant from the Research department of the Islamic Azad University-Lahijan Branch. Also, the corresponding author wishes to thank Reviewers for their useful comments and suggestions.

REFERENCES

1. E. Sanchez-Sinencio and A. G. Andreou, "Low-Voltage and Low Power Integrated Circuits and Systems," Low-Voltage Mixed-Signal Circuits, IEEE Press, New York, 1998.
2. A. N. Mohieldin, "High Performance Continuous-Time Filters for Information Transfer Systems," Ph.D. Dissertation, Department of Electrical Engineering, Texas A & M University, College Station, 2003.
3. S. Yan and Sanchez-Sinencio, "Low-Voltage Analog Circuit Design Techniques: A Tutorial," IEICE Transactions on Analog Integrated Circuits and Systems, Vol. E00-A, No. 2, 2000, pp. 179-196.

4. S. S. Rajput and S. S. Jamuar, "Low-Voltage Analog Circuit Design Techniques," IEEE Circuit and System Magazine, Vol. 2, No. 1, 2002, pp. 24-42.

5. S. Chatterjee, Y. Tsvidis and P. Kinget, "0.5 V Analog Circuit Techniques and Their Application to OTA and Filter Design," IEEE Journal of Solid State Circuits, Vol. 40, No. 12, 2005, pp. 2373-2387. doi:10.1109/JSSC.2005.856280

6. M. Trakimas and S. Sonkusale, "A 0.5 V Bulk-Input OTA with Improved Common-Mode Feedback for Low-Frequency Filtering Applications," Analog Integrated Circuits and Signal Processing, Springer Press, New York, 2009, pp. 83-89.

7. A. Ahmadpour and R. Fouladi, "A Modified High-Performance Structure of Low-Voltage CMOS Op-Amp," Proceeding of the 2008 IEEE International Conference on Electron Devices and Solid-State Circuits, Hong Kong, 8-10 December 2008, pp. 1-5. doi:10.1109/EDSSC.2008.4760687

8. A. Ahmadpour, "An Ultra Low-Voltage and High-Speed Amplifier for Switched-Capacitor Applications," 6th International Conference on Electrical Engineering/Electronics, Computer, Telecommunications and Information Technology, Vol. 1, Pattaya, 6-9 May 2009, pp. 552-555.

9. M. Dessouky and A. Kaiser, "Very Low-Voltage Digital-Audio Modulator with 88-dB Dynamic Range Using Local Switch Bootstrapping," IEEE Journal of Solid-State Circuits, Vol. 36, No. 3, March 2001, pp. 349-355. doi:10.1109/4.910473

10. R. Schaumann and M. E. Van Valkenburg, "Design of Analog Filters," Oxford University Press, Oxford, 2001.

CHAPTER 10

PERFORMANCE IMPROVEMENT OF PHASE-CHANGE MEMORY CELL USING ALSB$_3$TE AND ATOMIC LAYER DEPOSITION TIO$_2$ BUFFER LAYER

Sannian Song, Zhitang Song, Cheng Peng, Lina Gao, Yifeng Gu, Zhonghua Zhang, Yegang Lv, Dongning Yao , Liangcai Wu and Bo Liu

Keywords: Phase change memory, atomic layer depositionTiO$_2$ buffer layer, reset voltageAlSb$_3$Te

BACKGROUND

Phase change memory (PCM) has been regarded as the one of the most promising nonvolatile memories for the next generation because of the advantages of high speed, low power, good endurance, high scalability, and fabrication compatibility with complementary metal-oxide-semiconductor (CMOS) process [1–4]. PCM uses the reversible phase change between the crystalline and amorphous states of chalcogenide materials brought about by Joule heating. $Ge_2Sb_2Te_5$ (GST) is the most widely used due to its relatively good trade-off between thermal stability and crystallization speed. However, with low crystallization temperature (around 140°C), GST is susceptible to the issue of thermal cross-talk by the proximity effect [5]. The high reset current (mA) results in high power consumption for GST-based PCM [6]. The switching speed, which is limited by its nucleation-dominated crystallization mechanism, is insufficient to satisfy the requirement of dynamic random access memory (around 10 ns) is also not satisfactory [7]. These issues stimulate us to explore novel material system in order to improve the storing media characteristics. Compared with GST, Sb-rich Sb-Te materials have many advantages such as low melting point and fast crystallization [8]. However, it is difficult to guarantee a satisfactory data-retention time at 80°C due to its relatively low crystallization temperature [9]. Recently, the Al-Sb-Te (AST) ternary system has been proposed for application in electric memory [10, 11]. Compared with GST, Al-Sb-Te exhibits a high crystallization temperature, good data retention, and high switching speed.

It was reported that merely 0.2% to 1.4% of the total applied energy is effectively used for phase changing, and nearly 60% to 70% of the energy transfers back along the columnar tungsten (W) bottom electrode, having not participated in the heating process of the phase change material (for a T-shaped PCM cell) [12]. Such a low thermal efficiency inevitably leads to a large operating bias/current during the phase change processes. Consequently, one of the effective solutions that has been tried to enhance the thermal efficiency is using an appropriate heating layer between the phase change material layer and the underlying W electrode, or replacing the W plug with some other suitable material. There are some qualified materials that have already been applied in reducing the programming current, such as TiON [13], Ta_2O_5[14], SiGe [15], TiO_2[16, 17], $SiTaN_x$[18], C60 [19], and WO_3[20]. All these materials have the common physical characteristics of high electrical resistivity and low thermal conductivity. Indeed, a heater material with a large electrical resistivity (>0.1 Ω cm) but low thermal conductivity is most favorable for heat generation and restriction in a PCM cell.

Titanium oxide (TiO_2) is an n-type semiconductor and has very low thermal conductivity (approximately 0.7 to 1.7 W m^{-1} K^{-1} for 150- to 300-nm thick film) [21]. Note that the thermal conductivity will be even less for a thinner TiO_2 film. The electrical resistivity of a crystalline TiO_2 film measured by the van der Pauw method in this study is about 1.2 Ω cm, which is close to the result reported by Xu et al. [17]. In addition,

Citation: Sannian Song, Zhitang Song, Cheng Peng, Lina Gao, Yifeng Gu, Zhonghua Zhang, Yegang Lv, Dongning Yao , Liangcai Wu and Bo Liu, Performance improvement of phasechange memory cell using AlSb3Te and atomic layer deposition TiO_2 buffer layer. DOI: 10.1186/1556-276X-8-77.

TiO_2 has a high melting point (approximately 2116 K) and will be thermally stable under high temperature (approximately 900 K) during the reset operation. Generally speaking, with the suitable electrical resistivity, thermal conductivity and thermal stability, a crystalline TiO_2 layer should hopefully serve as the bottom heating layer in PCM cells to improve the thermal efficiency and, therefore, reduce the power requirement during phase transitions. In this study, the atomic layer deposition (ALD) TiO_2 was used as a buffer layer which was expected to improve the thermal efficiency and reduce the reset voltage of PCM.

METHODS

The PCM cells in this study are fabricated using 0.18 μm CMOS technology. Figure 1a shows a cross-section transmission electron microscopy (TEM) image of the fabricated cell without TiO_2buffer layer. The diameter and height of the columnar W electrode are 260 and 700 nm, respectively. Figure 1b shows a schematic diagram of the cross-section structure of the fabricated cell with TiO_2 buffer layer. The thin TiO_2 layer was interposed between the phase change layer (PCL) and W plug. A 2-, 4-, and 8-nm thick TiO_2 buffer layer was deposited by ALD at 400°C using Beneq TFS 500 ALD system (Beneq, Vantaa, Finland). One deposition cycle was composed of Ti precursor ($TiCl_4$) pulse (250 ms), 200 sccm N_2 purge (2 s), water (H_2O) pulse (250 ms), and 200 sccm N_2 purge (s2 s). The deposition rate is 0.5 A/cycle. The as-deposited films were crystallized with rutile structure measured by X-ray diffraction. Then, 100-nm thick AST PCL was deposited by magnetron sputtering. The background pressure and Ar gas pressure were 2.0×10^{-4} and 0.18 Pa, respectively. The stoichiometry of the deposited films was confirmed by electron dispersive spectroscopy. The Al/Sb/Te ratio was 1:3:1. Then, 20 nm TiN and 200 nm Al were deposited by sputtering as top electrode. For comparison, sputter-deposited AST film without the interposed TiO_2 layer was also fabricated with the same structure. The electric property tests of PCM were carried out by a Tektronix AWG5012b arbitrary waveform generator (Tektronix, Inc., Shanghai, China) and a Keithley 2602A parameter analyzer (Keithley Instruments, Inc., OH, USA).

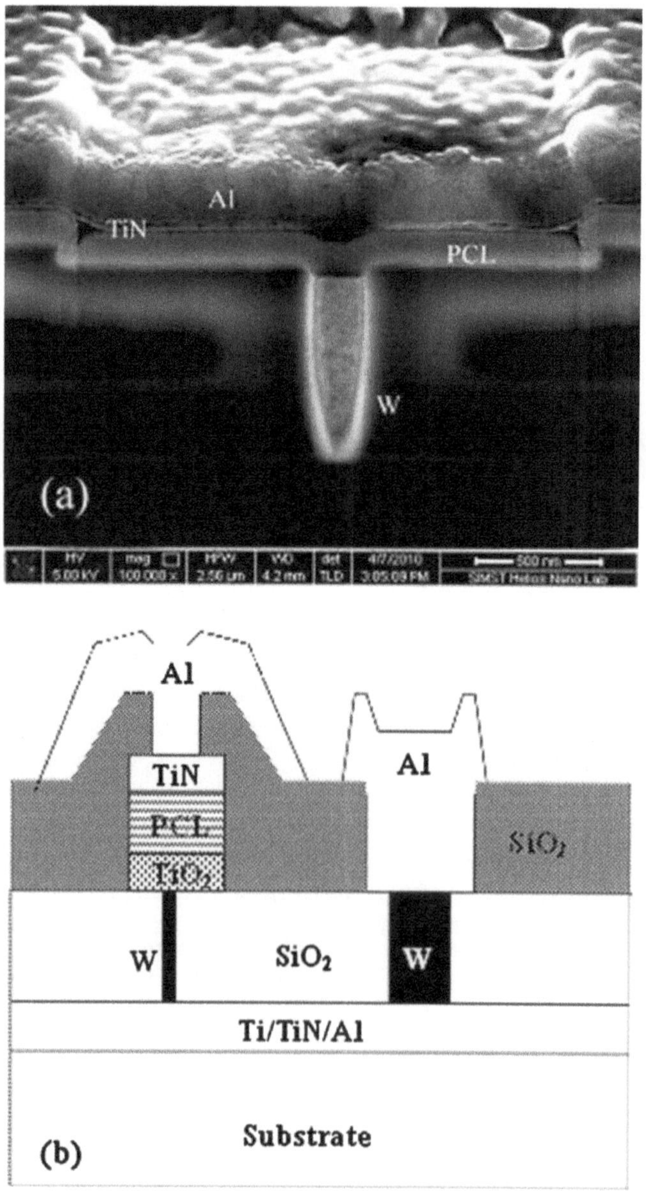

Figure 1: **Cross-sectional structures of PCM cells.** (**a**) Cross-sectional structure of PCM cell without TiO_2 buffer layer and (**b**) schematic diagram of the cross-section structure of the fabricated cell with TiO_2 buffer layer.

RESULTS AND DISCUSSION

Figure 2a shows the sheet resistance change of AST films as a function of temperature. The sample with a thickness of 100 nm was prepared on the SiO_2/Si(100) by sputtering at room temperature. Upon heating, the sheet resistance of AST films decreased with a rapid drop at the crystallization temperature (T_c). The T_c, defined by the temperature corresponding to minimum of the first derivative of *RT* curve, was about 198°C which is higher than that of GST films (approximately 145°C) [22]. To check the crystallization kinetics, electrical resistivity was *in situ* measured with increasing temperature with various heating rates dT/dt. Applying Kissinger's analysis which relates the transition temperature T_c, the rate of heating (dT/dt), and the activation energy (E_a) for crystallization by the formula below:

$$\ln\left[(dT/dt)/T_c^2\right] = C - (E_a/k_B T_c), \quad (1)$$

where C is a constant, k_B is the Boltzmann constant, a plot of $\ln[(dT/dt)/T_c^2$ against $1/T_c$ yields a straight line with slope, $-E_a/k_B$. From the Kissinger plot shown in Figure 2b, the activation energy for crystallization of AST was determined to be about 3.55 eV which is higher than that of GST films (approximately 2.01 eV) [22]. It has to be noted that the high crystallization temperature and high activation energy of AST offer a large benefit for a stable operation of the PCM device because the cells in the amorphous state tend to switch to the crystalline state due to cross talk, i.e., the heat dissipation from other cells.

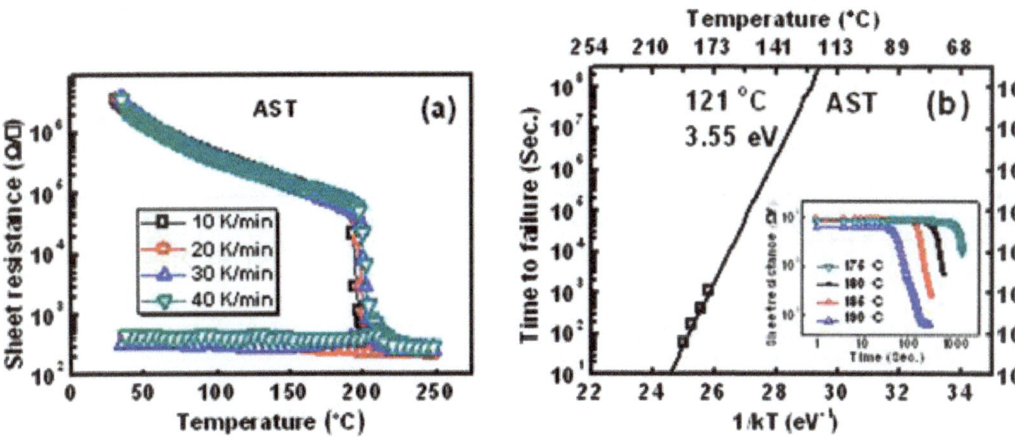

Figure 2: **Sheet resistance change and Kissinger plot.** (**a**) Temperature dependence of the sheet resistance of AST films and (**b**) Kissinger plot from which the E_a of the amorphous to crystalline transition at T_c of AST films are determined.

The bright-field TEM was used to study the structure of thin films. Figure 3 shows the TEM image of AST film after a 2-min heating at 400°C in Ar atmosphere; nanocrystals (dark spots) were observed. Peng et al. reported that an embedded crystal structure of hexagonal (Sb_2Te) and monoclinic (Al_2Te_3) phases can be found in AST materials [10]. The black area in the image results from an overlap of Sb_2Te and Al_2Te_3 crystalline grains. The overlap of grains will lead to a larger local density, and the incident electrons will be more scattered by these areas.

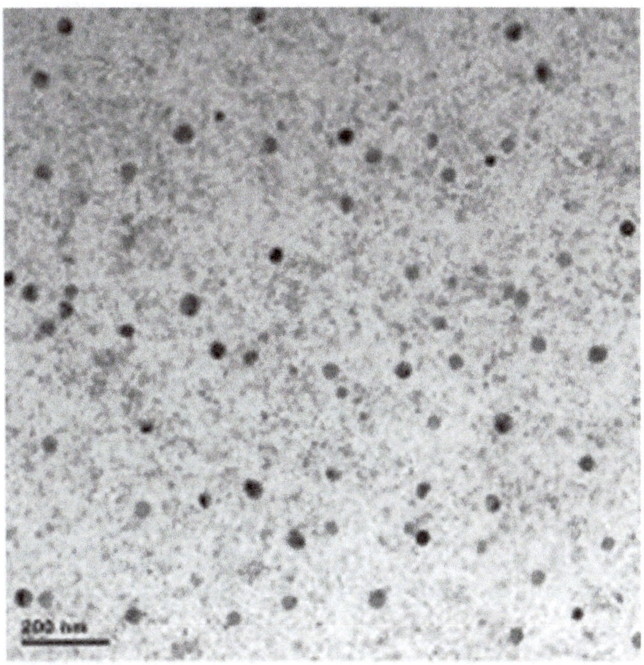

Figure 3: TEM image of AST film after a 2-min heating at 400°C.

The phase transition of PCM cell can be characterized from the relation between the cell resistance and the corresponding amplitude of voltage pulse or current pulse (so called *R-V* or *R-I* curve). The measured *R-V* curves for AST PCM cells with different pulse width are shown in Figure 4a. Reversible phase-change process has been observed. As revealed, once the programming voltage increases beyond the threshold voltage, the cell resistance starts to drop due to the crystallization of AST alloy and then reaches a minimum, which is corresponding to the set resistance. When the voltage is further increased, the resistance again rises and then returns to the reset state. It is clear that the set resistance decreases with the pulse width. The higher set resistance resulted from a shorter pulse implies that incomplete crystallization states are formed after set programming. It can be seen from Figure 4a the resistance of the AST devices dramatically increased by two orders of magnitude at a reset voltage of around 4.1 V (at 50 ns).

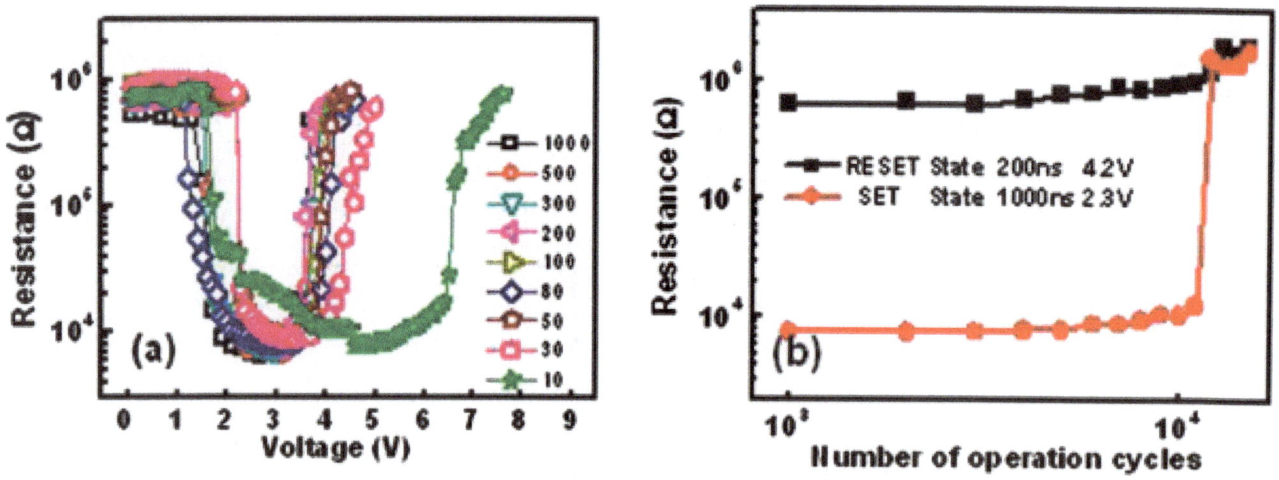

Figure 4: Resistance voltage and endurance characteristics. (a) Resistance voltage characteristics of PCM cell with AST films by different voltage pulse widths. (**b**) Endurance characteristics of the PCM cell with AST film.

Figure 5a,c,e shows the variations in cell resistance with the 2-, 4-, and 8-nm thick TiO_2 buffer layer as a function of the voltage for the set and reset operations, respectively. For the device with 2 nm TiO_2, as shown in Figure 5a, a 100-ns width pulse fails to set the cell and a pulse width of 100 ns is insufficient for a complete reset programming, suggesting that 2 nm TiO_2layer indeed leads to a slower crystallization process, thus longer write time for the set operation. For a device with 8 nm TiO_2, as shown in Figure 5e, a 5-ns pulse can trigger reversible phase-change of the cell, and the reset voltage of approximately 3.8 V (at 50 ns) of the cell is clearly lower than that of the AST cells (about 4.1 V) without TiO_2 layer. With 50-ns, pulse reset voltage of 2.4 V was achieved for the device with 4 nmTiO_2 layer (in Figure 5c), which is only about half of the voltage required by the device without TiO_2 buffer layer. The voltage reduction could be understood from the high Joule heating efficiency and the good thermal confinement. The oxide interfacial layer prevents heat generated in the programming volume of the AST from diffusing to the plug, which has high thermal conductivity, resulting in low power set/reset operation. Similar improvement has been reported on other kinds of oxide interfacial heater layers [23, 24]. Besides that, both of the resistances in amorphous and crystalline states retained at the same levels after inserting the TiO_2 layer. These results prove a fact that the inserted TiO_2 layer will not drift the resistance but can sharply diminish the operation voltage, which will be helpful to solve the difficult problem in the compatibility with the continuing scaling down dimension in CMOS process. It is worthy to point out that the set resistance is very stable for the cells with TiO_2 layer at different pulse widths, suggesting that the TiO_2 layer helps to raise the temperature profile within the phase change film and, thereby, enhances the heat-induced phase transition process. Furthermore, there are some other advantages of TiO_2such as easily fabricated, no pollution, fully compatible with CMOS process, and avoids the diffusion between phase change material and bottom electrode.

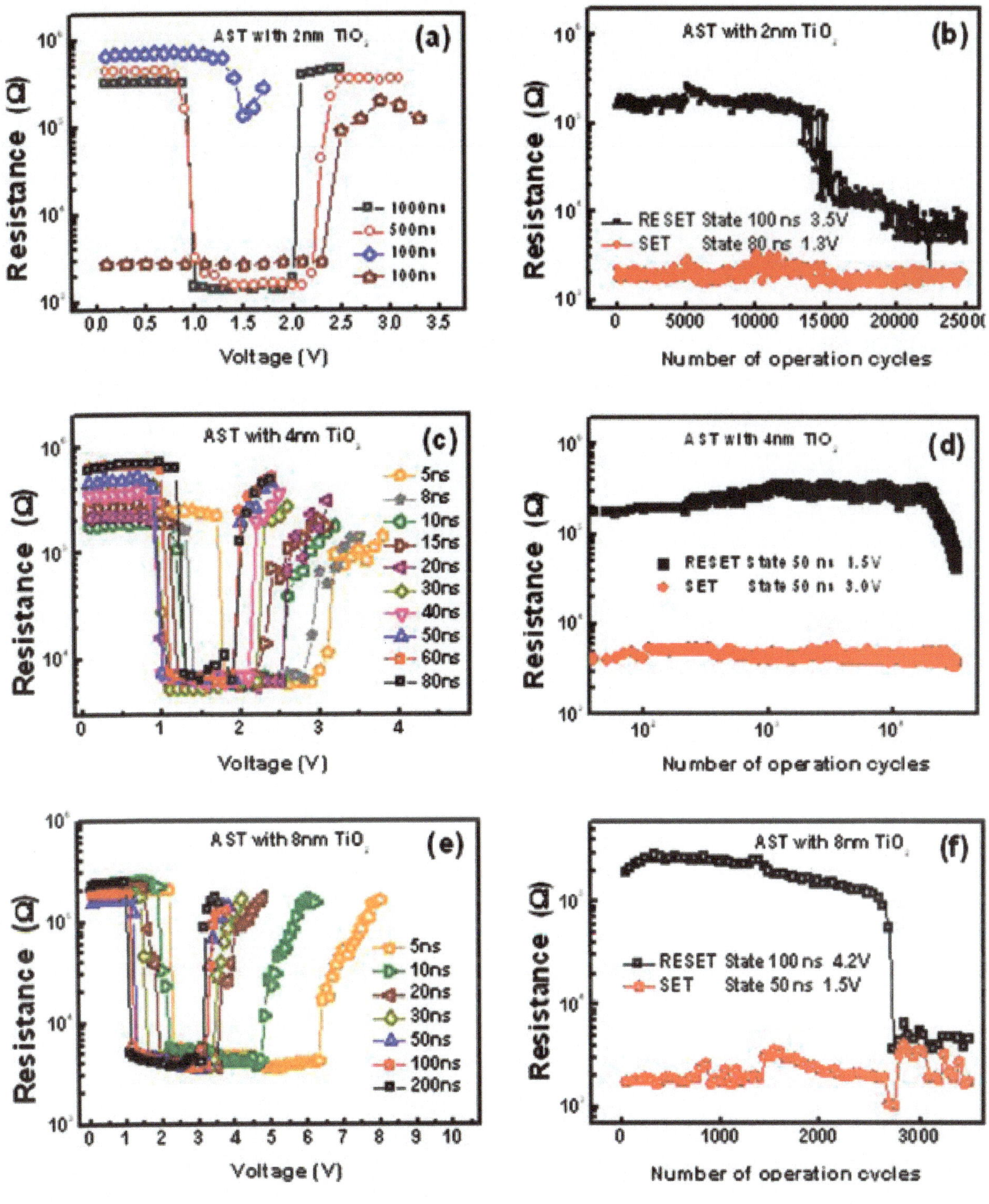

Figure 5: Resistance voltage characteristics of PCM cell at different pulse widths. (a) 2, (c) 4, and (e) 8 nm TiO_2. Endurance characteristics of the PCM cell (b) with 2, (d) 4, and (f) 8 nm TiO_2.

Figure 4b and Figure 5b,d,e show the repeatable resistance switching between the set and reset states of the cells without and with TiO_2 layer, respectively. For the device without TiO_2, as shown in Figure 4b, the endurance capability keeps about 20,000 cycles before the presence of resistance disorder with a set stuck failure mechanism. For the device with 2 nm TiO_2, as shown in Figure 5b, the reset resistance reduced gradually during the cycling. After 14,000 cycles, the reset resistance dropped rapidly, leading to the endurance failure by losing the set and reset resistance window. For the device with 8 nm TiO_2, as shown in Figure 5f, the endurance capability keeps about 2,700 cycles before the presence of resistance disorder with a reset stuck failure mechanism. Good endurance characteristics ($>10^4$ cycles) was found in the cell with 4-nm TiO_2 buffer layer. The low resistance state maintained around 10^3 Ω magnitude, and the high resistance state kept on 10^5 Ω level, indicating a satisfactory data resolution capability for random access memory application. The difference cyclic operation behavior shown in Figure4b and Figure 5b,d,e suggested the different performance degradation processes for the device with and without TiO_2 layer, which is currently under investigation. Among the various thicknesses of the TiO_2 buffer layer, 4 nm was the most appropriate thickness that maximized the improvement with negligible sacrifice of the other device performances, such as the reset/set resistance ratio, voltage window, and endurance.

CONCLUSIONS

This paper reports an efficient method for reducing the reset voltage and power of the conventional T-shaped PCRAM, which has the potential to replace the current nonvolatile memories. We inserted TiO_2 layer between phase change memory and bottom electrode to increase the utilization of the Joule heat and reduce the heat dissipation. Due to the suitable electrical resistivity and the low thermal conductivity of TiO_2 film, the overall set resistance of the PCM cell will not be greatly increased, while the remarkably increased overall thermal resistance helps to reduce the reset voltage.

ACKNOWLEDGMENTS

This work was supported by the National Key Basic Research Program of China (2010CB934300, 2011CB9328004, and 2011CBA00607), the National Integrate Circuit Research Program of China (2009ZX02023-003), the National Natural Science Foundation of China (61006087, 61076121, 61176122, and 61106001), the Science and Technology Council of Shanghai (11DZ2261000 and 1052nm07000), and the Chinese Academy of Sciences (20110490761).

Authors' contributions

SS and ZS conceived the study and revised the manuscript. CP and LG carried out the XRD and TEM characterizations. YG and ZZ participated in the sample preparation. YL and DY participated in the fabrication of the device. LW and BL read the manuscript and contributed to its improvement. All the authors discussed the results and contributed to the final version of the manuscript. All the authors read and approved the final manuscript.

REFERENCES

1. Ovshinsky SR: Reversible electrical switching phenomena in disordered structures. *Phys Rev Lett* 1968, 21: 1450–1453. 10.1103/PhysRevLett.21.1450
2. Wuttig M, Yamada N: Phase-change materials for rewriteable data storage. *Nat Mater*2007, 6: 824–832. 10.1038/nmat2009
3. Kolobov AV, Fons P, Frenkel AI, Ankudinov AL, Tominaga J, Uruga T: Understanding the phase-change mechanism of rewritable optical media. *Nat Mater* 2004, 3: 703–708. 10.1038/nmat1215
4. Lai S: Current status of the phase change memory and its future. In *Electron Devices Meeting: December 8–10 2003, Santa Clara*. New York: IEEE; 2003:10.1.1–10.1.4.
5. Kao KF, Chang CC, Chen FT, Tsai MJ, Chin TS: Antimony alloys for phase-change memory with high thermal stability. *Scr Mater* 2010, 63: 855–858. 10.1016/j.scriptamat.2010.06.036
6. Jung Y, Agarwal R, Yang CY, Agarwal R: Chalcogenide phase-change memory nanotubes for lower writing current operation. *Nanotechnology* 2011, 22: 254012. 10.1088/0957-4484/22/25/254012
7. Wong HSP, Raoux S, Kim S, Liang JL, Reifenberg JP, Rajendran B, Asheghi M, Goodson KE: Phase change memory. *Proc IEEE* 2010, 98: 2201–2227.
8. Lee ML, Miao XS, Ting LH, Shi LP: Ultrafast crystallization and thermal stability of In-Ge doped eutectic Sb70Te30 phase change material. *J Appl Phys* 2008, 103: 043501. 10.1063/1.2841724
9. Wang F, Zhang T, Song ZT, Liu C, Wu LC, Liu B, Feng SL, Chen B: Temperature influence on electrical properties of Sb-Te phase-change material. *Jpn J Appl Phys* 2008, 47: 843–846. 10.1143/JJAP.47.843
10. Peng C, Song ZT, Rao F, Wu LC, Zhu M, Song HJ, Liu B, Zhou XL, Yao DN, Yang PX, Chu JH: Al1.3Sb3Te material for phase change memory application. *Appl Phys Lett* 2011, 99: 043105. 10.1063/1.3616146
11. Ren K, Rao F, Song ZT, Lv SL, Cheng Y, Wu LC, Peng C, Zhou XL, Xia MJ, Liu B, Feng SL: Pseudobinary Al2Te3-Sb2Te3 material for high speed phase change memory application.*Appl Phys Lett* 2012, 100: 052105. 10.1063/1.3680580
12. 1Sadeghipour SM, Pileggi L, Asheghi M: Phase change random access memory, thermal analysis. In *The Tenth Intersociety Conference on Thermal and Thermomechanical Phenomena and Emerging Technologies in Electronic Systems, ITherm 2006: May 30–June 2 206; San Diego, California*. New York: IEEE; 2006:660–665.
13. Kang DH, Kim IH, Jeong JH, Cheong BK, Ahn DH, Lee D, Kim HM, Kim KB, Kim SH: An experimental investigation on the switching reliability of a phase change memory device with an oxidized TiN electrode. *J Appl Phys* 2006, 100: 054506. 10.1063/1.2338130
14. Matsui Y, Kurotsuchi K, Tonomura O, Morikawa T, Kinoshita M, Fujisaki Y, Matsuzaki N, Hanzawa S, Terao M, Takaura N, Moriya H, Iwasaki T, Moniwa M, Koga T: Ta2O5 interfacial layer between GST and W plug enabling low power operation of phase change memories. In *Electron Devices Meeting: December 11–13 2006; San Francisco, CA*. New York: IEEE; 2006:1–4.
15. Lee SY, Choi J, Ryu SO, Yoon SM, Lee NY, Park YS, Kim SH, Lee SH, Yu BG: Polycrystalline silicon-germanium heating layer for phase-change memory applications. *Appl Phys Lett* 2006, 89: 053517. 10.1063/1.2335363
16. Choi BJ, Oh SH, Choi S, Eom T, Shin YC, Kim KM, Yi KW, Hwang CS, Kim YJ, Park HC, Baek TS, Hong SK: Switching power reduction in phase change memory cell using CVD Ge2Sb2Te5 and ultrathin TiO2 films. *J Electrochem Soc* 2009, 156: 59–63.
17. Xu C, Song ZT, Liu B, Feng SL, Chen B: Lower current operation of phase change memory cell with a thin TiO2 layer. *Appl Phys Lett* 2008, 92: 062103. 10.1063/1.2841655
18. Cheng HY, Chen YC, Lee CM, Chung RJ, Chin TS: Thermal stability and electrical resistivity of SiTaNx heating layer for phase-change memories. *J Electrochem Soc* 2006, 153: 685–691. 10.1149/1.2196727
19. Kim C, Suh DS, Kim KHP, Kang YS, Lee TY, Khang Y, Cahill DG: Fullerene thermal insulation for phase change memory. *Appl Phys Lett* 2008, 92: 013109. 10.1063/1.2830002

20. Rao F, Song ZT, Gong YF, Wu LC, Feng SL, Chen B: Programming voltage reduction in phase change memory cells with tungsten trioxide bottom heating layer/electrode.*Nanotechnology* 2008, 19: 445706. 10.1088/0957-4484/19/44/445706

21. Mun J, Kim SW, Kato R, Hatta I, Lee SH, Kang KH: Measurement of the thermal conductivity of TiO2 thin films by using the thermo-reflectance method. *Thermochim Acta*2007, 455: 55–59. 10.1016/j.tca.2006.11.018

22. Song SN, Song ZT, Liu B, Wu LC, Feng SL: Stress reduction and performance improvement of phase change memory cell by using Ge2Sb2Te5–TaOx composite films. *J Appl Phys*2011, 109: 034503. 10.1063/1.3544432

23. Rao F, Song ZT, Gong YF, Wu LC, Liu B, Feng SL, Chen B: Phase change memory cell using tungsten trioxide bottom heating layer. *Appl Phys Lett* 2008, 92: 223507. 10.1063/1.2939218

24. Li MH, Zhao R, Law LT, Lim KG, Shi LP: TiWOx interfacial layer for current reduction and cyclability enhancement of phase change memory. *Appl Phys Lett* 2012, 101: 073502. 10.1063/1.4746286

CHAPTER 11 EMERGING NONVOLATILE MEMORY TECHNOLOGIES

Jagan Singh Meena, Simon Min Sze, Umesh Chand and Tseung-Yuen Tseng

Keywords: Emerging nonvolatile memory technologies, Magnetic storage, Market memory technologies, Memristors Phase change memories, Random-access storage, Flash memory technologies, Three-dimensional memory, Transparent memory, Unified memory

REVIEW

Background

General Overview

The idea of using a floating gate (FG) device to obtain a nonvolatile memory device was suggested for the first time in 1967 by Kahng D and Sze SM at Bell Labs [1]. This was also the first time that the possibility of nonvolatile MOS memory device was recognized. From that day, semiconductor memory has made tremendous contributions to the revolutionary growth of digital electronics since a 64-bit bipolar RAM chip to be used in the cache memory of an IBM computer was reported in 1969 [2]. Semiconductor memory has always been an indispensable component and backbone of modern electronic systems. All familiar computing platforms ranging from handheld devices to large supercomputers use storage systems for storing data temporarily or permanently [3]. Beginning with punch card which stores a few bytes of data, storage systems have reached to multiterabytes of capacities in comparatively less space and power consumption. Regarding application aspects, the speed of storage systems needs to be as fast as possible [4]. Since Flash memory has become a common component of solid-state disks (SSDs), the falling prices and increased densities have made it more cost-effective for many other applications [5]. Memory devices and most SSDs that use Flash memory are likely to serve very different markets and purposes. Each has a number of different attributes which are optimized and adjusted to best meet the needs of particular users. Because of natural inherent limitations, the long-established memory devices have been shorted out according to their inventions to match with portable electronic data storage systems. Today, the most prominent one is the limited capacity for continued scaling of the electronic device structure. Research is moving along the following paths for embedded Flash devices: (i) scaling down the cell size of device memory, (ii) lowering voltage operation, and (iii) increasing the density of state per memory cell by using a multilevel cell. To sustain the continuous scaling, conventional Flash devices may have to undergo revolutionary changes. Basically, it is expected that an entire DVD collection be in the palm of a hand. Novel device concepts with new physical operationing principles are needed. It is worthwhile to take a look at semiconductor memories against the background of digital systems. The way semiconductor devices are used in a systems environment determines what is required of them in terms of density, speed/power, and functions. It is also worthwhile to look into the economic significance of semiconductor memories and the relative importance of their various types. For the past three and a half decades in existence, the family of semiconductor memories has expanded greatly and achieved higher densities, higher speeds, lower power, more functionality, and lower costs [3, 6, 7]. At the same time, some of the limitations within each type of memory are also becoming more realized. As such, there are several emerging

Citation: Jagan Singh Meena, Simon Min Sze, Umesh Chand, Tseung-Yuen Tseng, Emerging nonvolatile memory technologies, DOI: 10.1186/1556-276X-9-526.

technologies aiming to go beyond those limitations and potentially replace all or most of the existing semiconductor memory technologies to become a universal semiconductor memory (USM). In addition, the rewards for achieving such a device would be to gain control of an enormous market, which has expanded from computer applications to all of consumer electronic products. Looking forward to the future, there are wide ranges of emerging memory applications for automation and information technology to health care. The specification of nonvolatile memory (NVM) is based on the floating gate configuration, which is the feature of an erased gate put into many cells to facilitate block erasure. Among them, designed Flash memories such as NOR and NAND Flash have been developed and then proposed as commercial products into bulk market. They have been considered as the most important products. NOR has high operation speed for both code and data storage applications; on the other hand, NAND has high density for large data storage applications [8]. Since the inception of Flash memory, there has been an exponential growth in its market driven primarily by cell phones and other types of consumer electronic equipment. While, today, integration of a silicon chip is not economical, toys, cards, labels, badges, value paper, and medical disposables could be imagined to be equipped with flexible electronics and memory. With growing demands for high-density digital information storage, memory density with arriving technology has been increased dramatically from the past couple of years. The main drive to develop organic nonvolatile memory is currently for applications of thin-film, flexible, or even printed electronics. One needs a technology to tag everything to electronic functionality which can be foreseen in a very large quantity and at a very low cost on substrates such as plastic and paper. Accessible popularization of roll-to-roll memory commercialization is a way to make an encounter interesting and challenging to have charge storage devices of choice for applications with enormous flexibility and strength. Recently, polymer (plastic memory) and organic memory devices have significant consideration because of their simple processes, fast operating speed, and excellent switching ability [9, 10]. One significant advantage polymer memory has over conventional memory designs is that it can be stacked vertically, yielding a three-dimensional (3-D) use of space [11]. This means that in terabyte solid-state devices with extremely low transistor counts such as drives about the size of a matchbook, the data persists even after power is removed. The NAND Flash market is continually growing by the successive introduction of innovative devices and applications. To meet the market trend, 3-D NVMs are expected to replace the planar ones, especially for 10-nm nodes and beyond. Moreover, simple-structure organic bistable memory exhibiting superior memory features has been realized by employing various nanoparticles (NPs) blended into a single-layered organic material sandwiched between two metal electrodes [12, 13]. The NPs act as traps that can be charged and discharged by suitable voltage pulses. NP blends show promising data retention times, switching speed, and cycling endurance, but the on-state current is too low to permit scaling to nanometer dimensions [10, 14]. A lot of these great ideas tend to die before reaching this point of development, but that is not to say that we will be seeing plastic memory on store shelves next year. There are still many hurdles to get over; software alone is a big task, as is the manufacturing process, but it does bring this technology one step closer to reality [15]. It is not an exaggeration to say that the equivalent of 400,000 CDs, 60,000 DVDs, or 126 years of MPG music may be stored on a polymer memory chip the size of a credit card.

The Vision of This Review

In this review, we focus on electrically programmable nonvolatile memory changes from silicon nanocrystal memory scaling to organic and metallic NP memory devices. Further, the scaling trend move towards the emerging NVM to flexible and transparent redox-based resistive switching memory technologies. This review is intended to give an overview to the reader of storage systems and components from conventional memory devices that have been proposed in the past years of recent progress in current NVM devices based on nanostructured materials to redox-based resistive random-access memory (RRAM) to 3-D and transparent memory devices. We describe the basics of Flash memory and then highlight the present problems with the issue of scaling tunnel dielectric in these devices. We briefly describe a historical change, how the conventional FG nonvolatile memory suffers from a charge loss problem as the feature size of the device continues to shrink. A discrete polysilicon-oxide-nitride-oxide-silicon (SONOS) memory is then proposed as a replacement of the conventional FG memory. The NC memory is expected to efficiently preserve the trapped charge due to the discrete charge storage node while also demonstrating excellent features such as fast program/erase speeds, low programming potentials, and high endurance. We also discuss current ongoing research in this field and the solutions proposed to solve the scaling problems by discussing a specific solution in detail which would be the centerpiece in recent memory work progress. Moreover, this review makes distinct emerging memory concepts with more recent molecular and quantum dot programmable nonvolatile memory concepts, specifically using charge trapping in conjugated polymers and metal NPs. We classify several possible devices, according to their operating principle, and critically review the role of π-conjugated materials in the data storage device operation. We describe specifications for applications of emerging NVM devices as well as already existing NAND memory and review the state of the art with respect to these target specifications in the future. Conclusions are drawn regarding further work on materials and upcoming memory devices and architectures.

Classification of Solid-State Memory Technologies

Data storage devices can be classified based on many functional criteria. Of them, silicon-based semiconductor memories are categorized into two: volatile and nonvolatile [3, 16]. In volatile memories, the information eventually fades while power supply is turned off unless the devices used to store data will be periodically refreshed. On the other hand, nonvolatile memories retain the stored information even when the power supply is turned off. Volatile memories, such as static random-access memory (SRAM) and dynamic random-access memory (DRAM), need voltage supply to hold their information while nonvolatile memories, namely Flash memories, hold their information without one. DRAM (dynamic stands for the periodical refresh) is needed for data integrity in contrast to SRAM. The basic circuit structures of DRAM, SRAM, and Flash memories are shown in Figure 1. DRAM, SRAM, and Flash are today's dominant solid-state memory technologies, which have been around for a long time, with Flash the youngest, at 25 years. DRAM is built using only one transistor and one capacitor component, and SRAM is usually built in CMOS technology with six transistors. Two cross-coupled inverters are used to store the information like in a flip-flop. For the access control, two further transistors are needed. If the write line is enabled, then data can be read and set with the bit lines. The Flash memory circuit works with the FG component. The FG is between the gate and the source-drain area and isolated by an oxide layer. If the FG is uncharged, then the gate can control the source-drain current. The FG gets filled (tunnel effect) with electrons when a high voltage at the gate is supplied, and the negative potential on the FG works against the gate and no current is possible. The FG can be erased with a high voltage in reverse direction of the gate. DRAM has an advantage over SRAM and Flash of only needing one MOSFET with a capacitor. It also has the advantage of cheap production as well as lower power consumption as compared to SRAM but slower than SRAM. On the other hand, SRAM is usually built in CMOS technology with six transistors and two cross-coupled inverters, and for the access control, two further transistors are needed. SRAM has the advantage of being quick, easy to control, integrated in the chip, as well as fast because no bus is needed like in DRAM. But SRAM has the disadvantages of needing many transistors and hence expensive, higher power consumption than DRAM. In comparison to DRAM and SRAM, Flash memory has FG between the gate and the source-drain area and isolated with an oxide layer. Flash memory does not require power to store information but is slower than SRAM and DRAM.

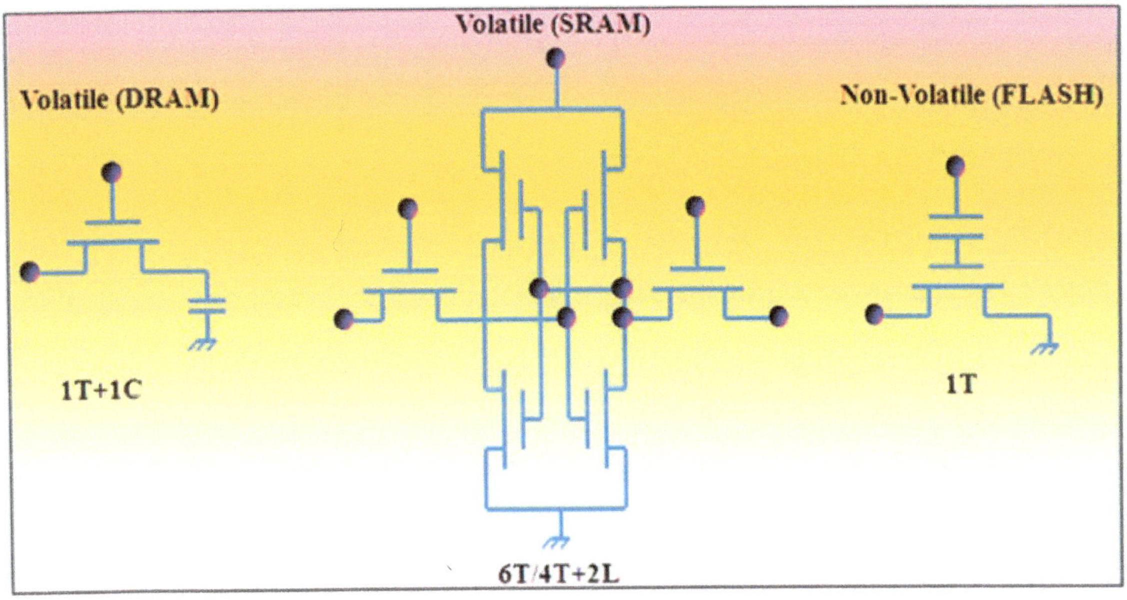

Figure 1: The circuitry structures of DRAM, SRAM, and Flash memories.

Both types of memories can be further classified based on the memory technology that they use and based on data volatility as shown in the classification flow chart depicted in Figure 2. Volatile memories consist mostly of DRAM [17], which can be further classified into SDRAM and mobile RAM which only retain information when current is constantly supplied to the device [18]. Another small but very important memory device is SRAM. The market for DRAM devices far exceeds the market for SRAM devices, although a small amount of SRAM devices is used in almost all logic and memory chips. However, DRAM uses only one transistor and one capacitor per bit, allowing it to reach much higher densities and, with more bits on a memory chip, be much cheaper per bit. SRAM is not worthwhile for desktop system memory, where DRAM dominates, but is used for its cache memories. SRAM is commonplace in small embedded systems, which might only need tens of kilobytes or less. Forthcoming volatile memory technologies that hope to replace or compete with SRAM and DRAM include Z-RAM, TTRAM, A-RAM, and ETA RAM. In the industry, new universal and stable memory technologies will appear as real contenders to displace either or both NAND Flash and DRAM. Flash memory is presently the most suitable choice for nonvolatile applications for the following reasons: Semiconductor nonvolatile memories consist mostly of the so-called 'Flash' devices and retain their information even when the power is turned off. Other nonvolatile semiconductor memories include mask read-only memory (MROM), antifuse-based one-time programmable (OTP) memory, and electrically erasable read-only memory (EEPROM). Flash is further divided into two categories: NOR, characterized by a direct write and a large cell size, and NAND, characterized by a page write and small cell size. Nonvolatile memory is a computer memory that can retain the stored information even when not powered [3, 19, 20]. Nonvolatile semiconductor memories are generally classified according to their functional properties with respect to the programming and erasing operations, as shown in the flow chart described in Figure 2. These are floating gate, nitride, ROM and fuse, Flash, emerging, and other new next-generation memory technologies. Today, these nonvolatile memories are highly reliable and can be programmed using a simple microcomputer and virtually in every modern electronic equipment, which are expected to replace existing memories.

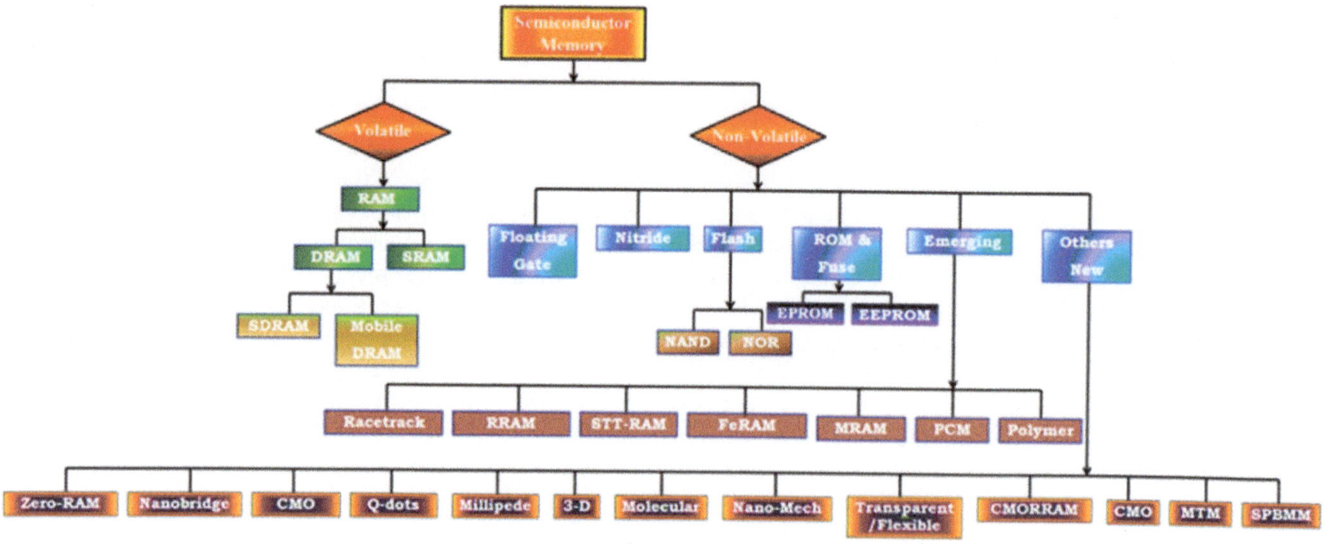

Figure 2: Flow chart for the semiconductor memory classification according to their functional criteria.

Among them, emerging nonvolatile memories are now very captivating. The next-generation memory market will cover up these emerging memory technologies [21]. There are mainly five types of nonvolatile memory technology: Flash memory, ferroelectric random-access memory (FeRAM), magnetic random-access memory (MRAM), phase-change memory (PCM), and RRAM. Nonvolatile memory, specifically 'Flash' memory, which is characterized by a large-block (or 'sector') erasing mechanism, has been the fastest growing segment of the semiconductor business for the last 10 years. Some of these newer emerging technologies include MRAM, FeRAM, PCM, spin-transfer torque random-access memory (STT-RAM), RRAM and memristor. MRAM is a nonvolatile memory [10, 22]. Unlike DRAM, the data is not stored in an electric charge flow, but by magnetic storage elements. The storage elements are formed by two ferromagnetic plates, each of which can hold a magnetic field, separated by a thin insulating layer. One of the two plates is a permanent magnet set to a particular polarity; the other's field can be changed to match that of an external field to store memory. STT-RAM is an MRAM (nonvolatile) but with better scalability over traditional MRAM. The STT is an effect in which the orientation of a magnetic layer in a magnetic tunnel junction or spin valve can be modified using a spin-polarized current. Spin-transfer torque technology has the potential to make MRAM devices combining low current requirements and reduced cost possible; however, the amount of current needed to reorient the magnetization is at present too high for most commercial applications. PCM is a nonvolatile random-access memory, which is also called ovonic unified memory (OUM), based on reversible phase conversion between the amorphous and the crystalline state of a chalcogenide glass, which is accomplished by heating and cooling of the glass. It utilizes the unique behavior of chalcogenide (a material that has been used to manufacture CDs), whereby the heat produced by the passage of an electric current switches this material between two states. The different states have different electrical resistance which can be used to store data. The ideal memory device or the so-called unified memory would satisfy simultaneously three requirements: high speed, high density, and nonvolatility (retention). At the present time, such memory has not been developed. The floating gate nonvolatile semiconductor memory (NVSM) has high density and retention, but its program/erase speed is low. DRAM has high speed (approximately 10 ns) and high density, but it is volatile. On the other hand, SRAM has very high speed (approximately 5 ns) but limited from very low density and volatility. It is expected that PCM will have better scalability than other emerging technologies. RRAM is a nonvolatile memory that is similar to PCM. The technology concept is that a dielectric, which is normally insulating, can be made to conduct through a filament or conduction path formed after application of a sufficiently high voltage. Arguably, this is a memristor technology and should be considered as potentially a strong candidate to challenge NAND Flash. Currently, FRAM, MRAM, and PCM are in commercial production but still, relative to DRAM and NAND Flash, remain limited to niche applications. There is a view that MRAM, STT-RAM, and RRAM are the most promising emerging technologies, but they are still many years away from competing for industry adoption [23]. Any new technology must be able to deliver most, if not all, of the following attributes in order to drive industry adoption on a mass scale: scalability of the technology, speed of the device, and power consumption to be better than existing memories. The NVSM is in inspiring search of novel nonvolatile memories, which will successfully lead to the realization and commercialization of the unified memory.

In progress, another new class of nonvolatile memory technologies will offer a large increase in flexibility compared to disks, particularly in their ability to perform fast, random accesses. Unlike Flash memory, these new technologies will support in-place updates, avoiding the extra overhead of a translation layer. Further, these new nonvolatile memory devices based on deoxyribonucleic acid (DNA) biopolymer and organic and polymer materials are one of the key devices for the next-generation memory technology with low cost. Nonvolatile memory based on metallic NPs embedded in a polymer host has been suggested as one of these new cross-point memory structures. In this system, trap levels situated within the bandgap of the polymer are introduced by the NPs [24, 25]. Memory devices play a massive role in all emerging technologies; as such, efforts to fabricate new organic memories to be utilized in flexible electronics are essential. Flexibility is particularly important for future electronic applications such as affordable and wearable electronics. Much research has been done to apply the flexible electronics technology to practical device areas such as solar cells, thin-film transistors, photodiodes, light-emitting diodes, and displays [26–28]. Research on flexible memory was also initiated for these future electronic applications. In particular, organic-based flexible memories have merits such as a simple, low-temperature, and low-cost manufacturing process. Several fabrication results of organic resistive memory devices on flexible substrates have been reported [29, 30]. In addition, with growing demand for high-density digital information storage, NAND Flash memory density has been increased dramatically for the past couple of decades. On the other hand, device dimension scaling to increase memory density is expected to be more and more difficult in a bit-cost scalable manner due to various physical and electrical limitations. As a solution to the problems, NAND Flash memories having stacked layers are under developing extensions [31, 32]. In 3-D memories, cost can be reduced by building multiple stacked cells in vertical direction without device size scaling. As a breakthrough for the scaling limitations, various 3-D stacked memory architectures are under development and expecting the huge market of 3-D memories in the near future. With lots of expectation, future-generation memories have potential to replace most of the existing memory technologies. The new and emerging memory technologies are also named to be a universal memory; this may give rise to a huge market for computer applications to all the consumer electronic products.

Market Memory Technologies by Applications

The semiconductor industry has experienced many changes since Flash memory first appeared in the early 1980s. The growth of consumer electronics market urges the demand of Flash memory and helps to make it a prominent segment within the semiconductor industry. The Flash memories were commercially introduced in the early 1990s, and since that time, they have been able to follow Moore's law and the scaling rules imposed by the market. There are expected massive changes in the memory market over the next couple of years, with more density and reliable technologies challenging the dominant NAND Flash memory now used in SSDs and embedded in mobile products. Server, storage, and application vendors are now working on new specifications to optimize the way their products interact with NVM - moves that could lead to the replacement of DRAM and hard drives alike for many applications, according to a storage networking industry association (SNIA) technical working group [33,34]. The Flash memory marketplace is one of the most vibrant and exciting in the semiconductor industry, not to mention one of the most competitive. The continuous invention of new memory technologies and their applications in the memory market also increase performance demands. These new classes of memories with the latest technology increase the vertical demand in the future memory market. In the next coming years, cumulative price reductions should become disruptive to DVDs and hard disk drives (HDDs), stimulate huge demand, and create new Flash markets.

The nonvolatile memories offer the system a different opportunity and cover a wide range of applications, from consumer and automotive to computer and communication. Figure 3 shows NVSM memory consumption by various applications in the electronics industry by market in 2010 extending upwards from computers and communication to consumer products [22]. It is noticed that there is a faster growth rate of the digital cellular phone since 1990; the volume of production has increased by 300 times, e.g., from 5 million units per year to about 1.5 billion units per year. Nowadays, flexibility and transparency are particularly of great significance for future electronic applications such as affordable and wearable electronics. Many advanced research technologies are applied to flexible technology to be used in a real electronics area [35]. Although silicon-based semiconductor memories have played significant roles in memory storage applications and communication in consumer electronics, now, the recent focus is turning from rigid silicon-based memory technology into a soft nonvolatile memory technology for low-cost, large-area, and low-power flexible electronic applications. Further, the memory market for the long term is continuously growing, even if with some ups and downs, and this is expected to continue in the coming years [36]. Since innovation drives the semiconductor industry, a new trend with transparency as well as flexibility and 3-D technologies will be attractive and move towards continuous growth in the near future.

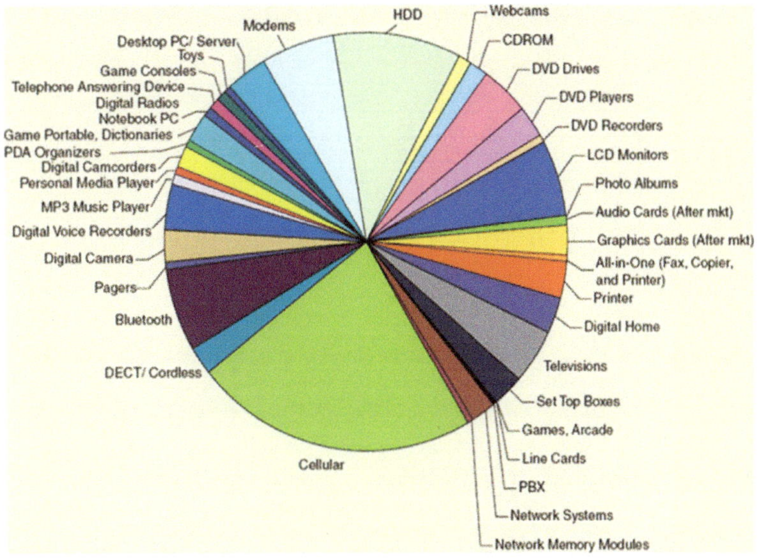

Figure 3: Various NVSM applications in the electronics industry by market size in 2010. Reprinted from ref. [22].

Successive creation of new mobile devices leads to the continual growth of NAND products as shown in Figure 4. To meet this market demand, early this year, 30-nm node technologies are in ramping-up phase, 20-nm node technologies are in the phase of transition to mass production, and a 10-nm node technology is under development. In addition, the future market requires high-speed operation even up to approximately 1,500 MB/s in order to satisfy a large amount of data correspondence [37]. However, high-speed operations cause high power consumption and chip temperature increase, which can deteriorate NAND reliability. Hence, reduction of operating voltage is inevitable to achieve the future NAND. Opportunities for the use of 3-D as well as polymer memory design in modern electronic circuits are rapidly expanding, based on the very high performance and unique functionality. However, their practical implementation in electronic applications will ultimately be decided by the ability to produce devices and circuits at a cost that is significantly below that needed to manufacture conventional electronic circuits based on, for example, silicon. If successful, these low-cost fabrication processes will ultimately result in the printing of large-area organic electronic circuits on a sheet of plastic paper using a roll-to-roll method, where low-temperature deposition of organics is followed by metal deposition and patterning in a continuous, high-speed process analogous, perhaps, to processes used in the printing of documents or fabrics.

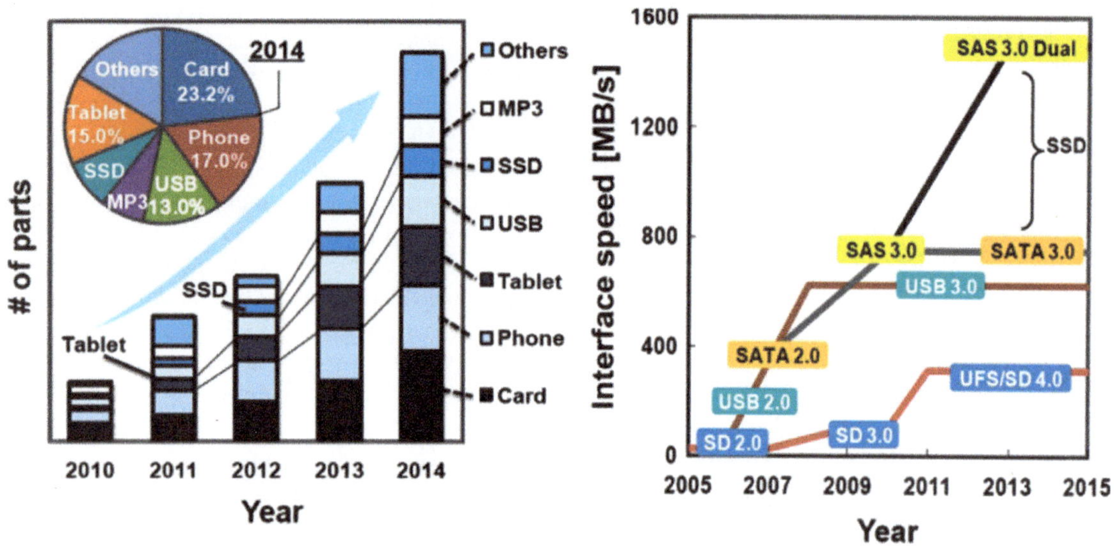

Figure 4: Growth of NAND Flash market up to 2014 (iSuppli) and the interface speed of various NAND applications. Reproduced from ref. [37].

In recent years, IDTechEx finds that the total market for printed, flexible, and organic electronics will grow from $16.04 billion in 2013 to $76.79 billion in 2023 and this growing trend is expected to continue in the coming years (see Figure 5a). The majority of that is OLEDs (only organic, not printed) and conductive ink used for a wide range of applications. On the other hand, stretchable electronics, logic and memory, and thin-film sensors are much smaller ingredients but having huge growth potential as they emerge from R&D [38]. The report specifically addresses the big picture that over 3,000 organizations are pursuing printed, organic, flexible electronics, including printing, electronics, materials, and packaging companies. While some of these technologies are in use now - indeed there are main sectors of business which have created billion-dollar markets - others are commercially embryonic.

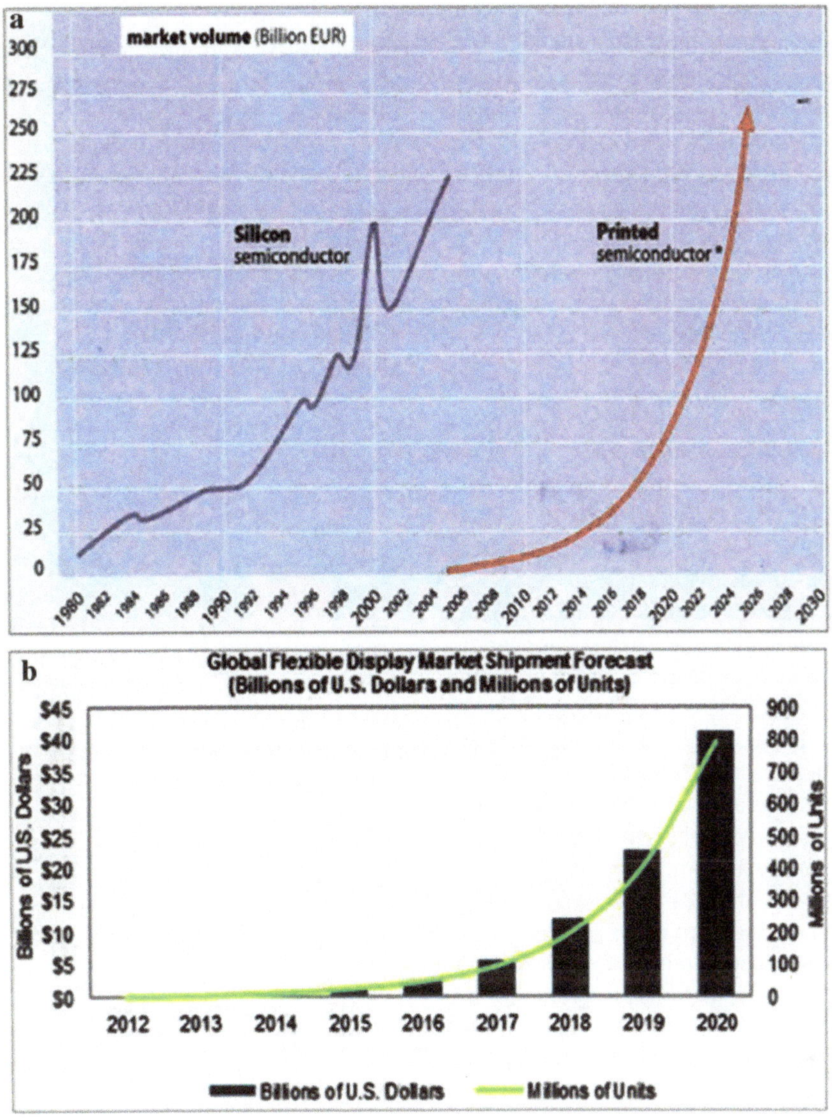

Figure 5: Market volume (a) and global flexible display market shipment forecast (b).Reproduced from refs. [38, 39].

Another key potential market for printed/flexible electronics is next-generation transparent conductive film to replace brittle and expensive indium tin oxide (ITO) in touch screens and displays, lighting, and photovoltaics. Touch display research says that the market for non-ITO transparent conductors will be about $206 million this year and grow to some $4 billion by 2020 as shown in Figure 5b. 'High demand for touchscreens for notebook and PC size displays has created a shortage of ITO touch sensors since the end of last year to drive more interest in these technologies, and the more flexible and potentially cheaper replacement technologies are getting more mature, notes Jennifer Colegrove, president and analyst, who will speak at the FlexTech workshop on transparent conductors. She notes that Atmel, Fujifilm, Unipixel and Cambrios are all in some phase of production' [39]. A large amount of the semiconductor market (approximately 20%) is given by the semiconductor memories; thus, the market for chips will develop in the next few years. This study reports that there is an analysis of the production process and the subsequent value chain, which comprises a benchmark analysis of the main segments of the semiconductor industry.

Recently, the 3-D nonvolatile memory structure has also attracted considerable attention due to its potential to replace conventional Flash memory in next-generation NVM applications [37,40]. 3-D memories are gathering increasing attention as future ultra-high-density memory technologies to keep a trend of increasing bit density and reducing bit cost. The NAND Flash market is continuously growing by the successive introduction of innovative devices and applications. To meet the market trend, 3-D NVMs are expected to replace the planar one, especially for 10-nm nodes and beyond. Therefore, the fundamentals and current status of the 3-D NAND Flash memory are reviewed and future directions are discussed [41]. 3-D integration promises to be an excellent replacement of current technologies for the development of NAND Flash memory. Time is running out for planar NAND technology. It will not be long that planar NAND will be completely replaced by 3-D NAND. 3-D NAND promises to satisfy the growing need of NAND memory [37].

Finally, NVM technologies have a bright future since every end-use application needs to store some parameters or some amount of an application program in the on-board NVM to enable it to function. The upcoming NVMs are the big hope for a semiconductor memory market, which provides memories for systems to run with flexibility, reliability, high performance, and low power consumption in a tiny footprint in nearly every electronic application. Recent market trends have indicated that commercialized or near-commercialized circuits are optimized across speed, density, power efficiency, and manufacturability. Flash memory is not suited to all applications, having its own problems with random-access time, bit alterability, and write cycles. With the increasing need to lower power consumption with zero-power standby systems, observers are predicting that the time has come for alternative technologies to capture at least some share in specific markets such as automotive smart airbags, high-end mobile phones, and RFID tags. An embedded nonvolatile memory with superior performance to Flash could see widespread adoption in system-on-chip (SoC) applications such as smart cards and microcontrollers.

Emerging NVM Technologies for Applications

The new emerging nonvolatile random-access memory products address the urgent need in some specific and small-form devices. Therefore, *iRAP* felt a need to do a detailed technology update and market analysis in this industry [42]. Recently, Yole Développement reports describe that emerging memory technologies have great potential to improve future memory devices to be increasingly used in various markets of industry and transportation, enterprise storage, mobile phones, mass storage, and smart cards [43]. Emerging NVM applications in various markets are shown in Figure 6. But there are numerous opportunities existing for novel architectures and applications that these emerging memory technologies can enable. These new emerging NVM products address the urgent need in some specific and small-form devices. Therefore, emerging nonvolatile memory products provide market data about the size of growth of the application segments and the developments of business opportunities. Until now, only FeRAM, PCM, and MRAM were industrially produced and available in low-density chips to only a few players. Thus, the market was quite limited and considerably smaller than the volatile DRAM- and nonvolatile Flash NAND-dominant markets (which enjoyed combined revenues of $50+ billion in 2012). However, in the next 5 years, the scalability and chip density of those memories will be greatly improved and will spark many new applications with NVM market drivers explained in more detail.

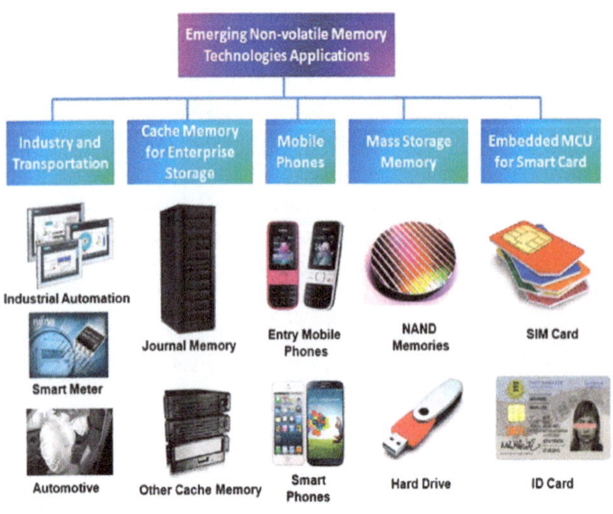

Figure 6: Emerging NVM applications in various markets.

Accompanied by the adoption of STT-MRAM and PCM cache memory, enterprise storage will be the largest emerging NVM market. NVM will greatly improve the input/output performance of enterprise storage systems whose requirements will intensify with the growing need for web-based data supported by floating mass servers. In addition, mobile phones will increase their adoption of PCM as a substitute to Flash NOR memory in MCP packages to 1-gigabyte (GB) chips made available by Micron in 2012. Higher-density chips, expected in 2015, will allow access to smart phone applications that are quickly replacing entry-level phones. STT-MRAM is expected to replace SRAM in SoC applications, thanks to lower power consumption and better scalability. Smart cards and microcontrollers (MCU) will likely adopt MRAM/STT-MRAM and PCM as a substitute to embed Flash. Indeed, Flash memory cell size reduction is limited in the future. The NVM could reduce the cell size by 50% and thus be more cost-competitive. Additional features like increased security, lower power consumption, and higher endurance are also appealing NVM attributes. The mass storage markets served by Flash NAND could begin using 3-D RRAM in 2017 to 2018, when 3-D NAND will slow down its scalability as predicted by all of the main memory players. If this happens, then a massive RRAM ramp-up will commence in the next decade that will replace NAND; conditional 3-D RRAM cost-competitiveness and chip density are available. It is expected surely that the emerging NVM business will be very dynamic over the next 5 years, thanks to improvements in scalability/cost and density of emerging NVM chips [44].According to a recently published report from Yole Développement, Emerging Non-volatile Memory Technologies, Industry Trends and Market Analysis, the global market for emerging nonvolatile random-access memory products was projected to have reached $200 million in 2012. This market is expected to increase to $2,500 million by 2018 at an average annual growth at a CAGR of +46% through the forecast period with mobile phones, smart cards, and enterprise storage as main growth drivers (Figure 7). Market adoption of memory is strongly dependent on its scalability. This Yole Développement report provides a precise memory roadmap in terms of technological nodes, cell size, and chip density for each emerging NVM such as FeRAM, MRAM/STT-MRAM, PCM, and RRAM. A market forecast is provided for each technology by application, units, revenues, and also market growth as given a detailed account of emerging NVM market forecast (Figure 7). PCM devices, the densest NVM in 2012 at 1 GB, will reach 8 GB by 2018, which are expected to replace NOR Flash memory in mobile phones and will also be used as a storage class memory in enterprise storage. MRAM/STT-MRAM chips will reach 8 to 16 GB in 2018. They will be widely sold as a storage class memory and possibly as a DRAM successor in enterprise storage after 2018. By 2018, MRAM/STT-MRAM and PCM will surely be the top two NVM on the market. Combined, they will represent a $1.6 billion business by 2018, and their sales will almost double each year, with double-density chips launched every 2 years. FeRAM will be more stable in terms of scalability, with 8- to 16-MB chips available by 2018; the development of a new FRAM material could raise scalability, but we do not expect it to be widely industrialized and commercialized before 2018. FeRAM will grow at a steady growth rate (10% per year) and will focus on industrial and transportation applications because of the low-density availability, whereas RRAM revenues would not really surge by 2018, with the availability of high-density chips of several tens of gigabytes that could replace NAND technology. Meanwhile, it has also been considered by memory technologist experts that for large-volume markets like mass storage NAND, only one technology will be adopted in order to reduce production cost and RRAM seems to be the best candidate. But the real massive adoption of emerging NVM as a replacement for NAND and DRAM will happen after 2020.

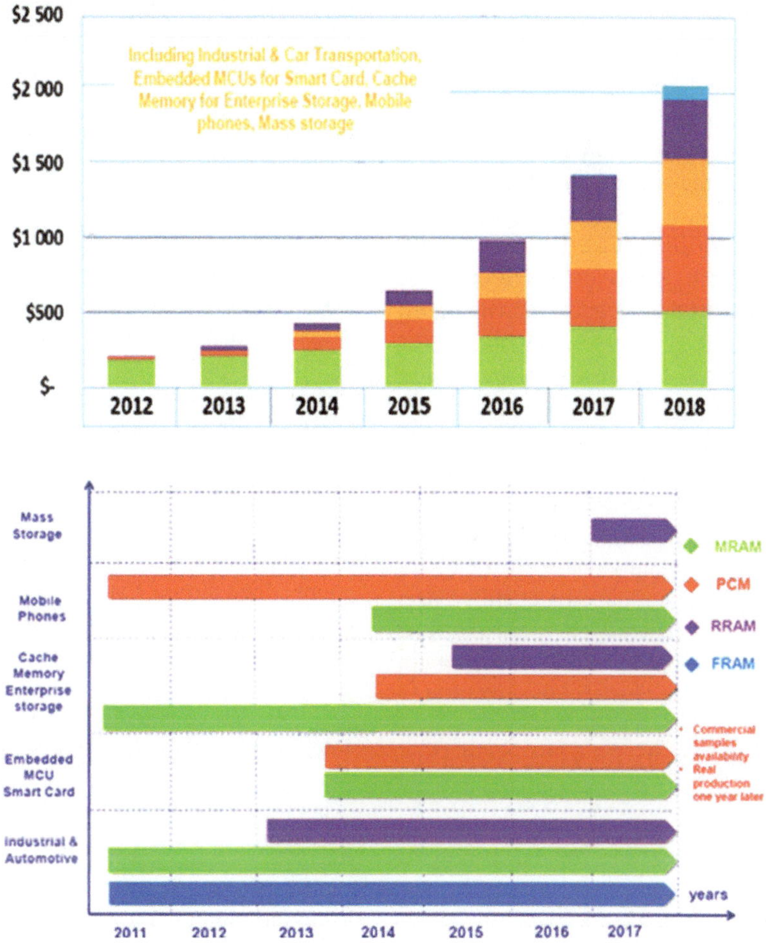

Figure 7: Emerging NVM market forecast by applications from 2012 to 2018 (in M$).Reproduced from ref. [43].

Advances in Flash Memory Technologies

Flash memory is basically a MOSFET nonvolatile device that can be electrically erased and reprogrammed [3, 45]. It is a technology that is primarily used in memory cards and Flash drives for general storage and transfer of data between computers and other digital products. Since the invention of the transistor, NVSM had been the most important invention in the electron device field. The floating gate memory was used to store the information and a tunneling current for programming and erasing operations. The charge is injected into or removed from the floating gate and the floating gate remains in that state, even after power is removed, which means that Flash memory is nonvolatile. The invention of NVSM further gave rise to a new class of memory devices and hence broadened its applications to become ubiquitous. There are a large number of products in the market now which use Flash devices exclusively as secondary storage. Few examples of their applications include medical diagnostic systems, notebook computers, digital audio players, digital cameras, mobile phones, personal digital assistants, digital televisions, universal serial bus (USB) Flash personal disks, Global Positioning Systems, and many more. Semiconductor storage devices store data in tiny memory cells made of very small transistors and capacitors made of semiconductor materials such as silicon. Each cell can hold 1 bit of information and an array of cells stores a large chunk of information. Flash devices are gaining popularity over conventional secondary storage devices like hard disks. The Flash memory fabrication process is compatible with the current CMOS process and is a suitable solution for embedded memory applications. A Flash memory cell is simply a MOSFET cell, except that a polysilicon floating gate [46] (or a silicon nitride charge trap layer) is sandwiched between a tunnel oxide and an inter-polyoxide to form a charge storage layer [47]. Although Flash memory is likely the standard charge storage device for the next generation, scaling may eventually be limited by the tunnel oxide limit [8]. In terms of the operation speed of program and erase, Flash memory requires a thin tunnel oxide to enhance the carrier transport between the floating gate and the silicon substrate. However, the very thin tunnel oxide suffers from many reliability issues like reduction in operation voltage, and after a considerable number of program and erase cycles, the tunnel oxide undergoes deterioration loss [48]. Thus, researchers have focused on possible solutions and proposed alternate technologies, including nitride-based memory, nanocrystal memory, and switching memory. All other nonvolatile memories require integration of new materials that are not as compatible as the conventional CMOS process.

NOR and NAND Flash Memory Technologies

NOR and NAND Flash, two major Flash types, are dominant in the memory market. NOR Flash has lower density but a random-access interface, while NAND Flash has higher density and interface access through a command sequence [49]. Their corresponding structures are shown in Figure 8. NOR and NAND Flash come from the structure used for the interconnections between memory cells. Intel is the first company to introduce a commercial (NOR type) Flash chip in 1988, and Toshiba released the world's first NAND Flash in 1989 [50]. Depending on how the cells are organized in the matrix, it is possible to distinguish between NAND Flash memories and NOR Flash memories. In NOR Flash, cells are connected in parallel to the bit lines, which notably allow the cells to be read and programmed individually. The parallel connection of NOR Flash cells resemble the parallel connection of transistors in a CMOS NOR gate architecture. On the other hand, in NAND Flash, the cells are connected in series, resembling a NAND gate. The series connections consume less space than the parallel ones, reducing the cost of NAND Flash. It does not, by itself, prevent NAND cells from being read and programmed individually. Most of the engineers and scientists are not so familiar with the differences between these two technologies. Generally, they usually refer to the NOR architecture as 'Flash' and are unaware of the NAND Flash technology and its many benefits over NOR [51]. This could be due to the fact that most Flash devices are used to store and run codes (usually small), for which NOR Flash is the default choice, although we are providing some major differences between NOR and NAND Flash technologies by their architecture and the internal characteristic features of the individual Flash.

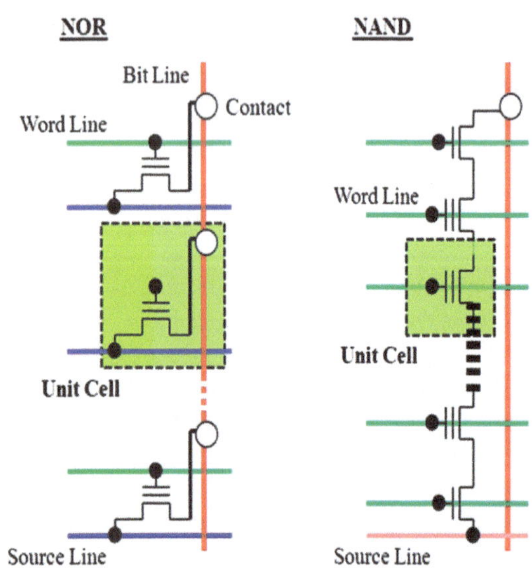

Figure 8: Comparison of NOR Flash array and NAND Flash array architectures.

NOR Flash is slower in erase operation and write operation compared to NAND Flash [52]. This means that NAND Flash has faster erase and write times. Moreover, NAND Flash has smaller erase units, so fewer erases are needed. NOR Flash can read data slightly faster than NAND Flash. NOR Flash offers complete address and data buses to randomly access any of its memory locations (addressable to every byte). This makes it a suitable replacement for older ROM BIOS/firmware chips, which rarely needs to be updated. Its endurance is 10,000 to 1,000,000 erase cycles. NOR Flash is highly suitable for storing codes in embedded systems. Most of today's microcontrollers come with built-in Flash memory [53].

NAND Flash occupies a smaller chip area per cell. This makes NAND Flash available in greater storage densities and at lower costs per bit than NOR Flash. It also has up to ten times the endurance of NOR Flash. NAND is more fit as storage media for large files including video and audio. USB thumb drives, SD cards, and MMC cards are of NAND type [54]. NAND's advantages are fast write (program) and erase operations, while NOR's advantages are random access and byte write capability. NOR's random access ability allows for execute in place (XiP) capability, which is often a requirement in embedded applications. NAND is slow random accessible, while NOR is hampered by having slow write and erase performance. NAND is better suited for filing applications. However, more processors include a direct NAND interface and can boot directly from NAND (without NOR). However, NAND cannot perform read and write operations simultaneously; it can accomplish these at a system level using a method called shadowing, which has been used on PCs for years by loading the BIOS from the slower ROM into the high-speed RAM.

Table 1 highlights the major differences between NOR and NAND. It shows that NAND is ideal for high-capacity data storage while NOR is best used for code storage and execution, usually in small capacities. There are many other differences between these two technologies which will be further discussed individually. However, those listed in the table are enough to strongly differentiate the types of applications using them: NOR is typically used for code storage and execution. This, mainly in capacities up to 4 MB, is common in applications such as simple consumer appliances, low-end cell phones, and embedded applications, while raw NAND is used for data storage in applications such as MP3 players, digital cameras, and memory cards [55–57]. The codes for raw NAND-based applications are stored in NOR devices.

Table 1: Comparison between NOR and NAND Flash memories[55–57]

Features	NOR	NAND
Memory size	≤512 Mbit	1 to 8 Gbit
Sector size	Approximately 1 Mbit	Approximately 1 Mbit

Program time	9 µs/word	400 µs/page
Erase time	1 s/sector	1 ms/sector
Read access time	<80 ns	20 µs
Write parallelism	8 to 16 words	2 Kbyte
Output parallelism	Byte/word/dword	Byte/word
Read parallelism	8 to 16 words	2 Kbyte
Access method	Random	Sequential
Price	High	Very low
Reliability	Standard	Low

Scaling and Challenges of Flash Memory Technologies

Currently, there have been increasing demands on reducing the feature size in microelectronic products and more interest in the development of Flash memory devices to meet the growing worldwide demand. A conventional FG memory device must have a tunnel oxide layer thickness of 8 nm to prevent charge loss and to make 10 years' data retention certain. This necessity will limit scalability for Flash memory devices [8, 58]. Thus, in order to meet technology scaling in the field of memory and data storage devices, mainstream transistor-based Flash technologies will be developed gradually to incorporate material and structural innovations [59]. Dielectric scaling in nonvolatile memories has been reached near to the point where new approaches will be required to meet the scaling requirements while simultaneously meeting the reliability and performance requirements for future products. High-dielectric-constant materials are being explored as possible candidates to replace both the traditional SiO_2 and oxide/nitride/oxide (ONO) films used in Flash memory cells. Flash cell scaling has been demonstrated to be really possible and to be able to follow Moore's law down to the 90-nm technology generations. The technology development and the consolidated know-how are expected to sustain the scaling trend down to the 50-nm technology node and below as forecasted by the International Technology Roadmap for Semiconductors (ITRS) in Figure 9, which indicates that the silicon MOSFET was already in the nanoscale. The minimum feature size of an individual CMOSFET has shrunk to 15 nm with an equivalent gate oxide thickness (EOT) of 0.8 nm in 2001 [13]. However, semiconductor Flash memory scaling is far behind CMOS logic device scaling. For example, the EOT of the gate stack in semiconductor Flash memory is still more than 10 nm. Moreover, semiconductor Flash memory still requires operation voltages of more than 10 V, which is still far from the operation voltage of CMOS logic devices. It is important to scale the EOT of the gate stack to achieve a small memory cell size and also prolong battery life.

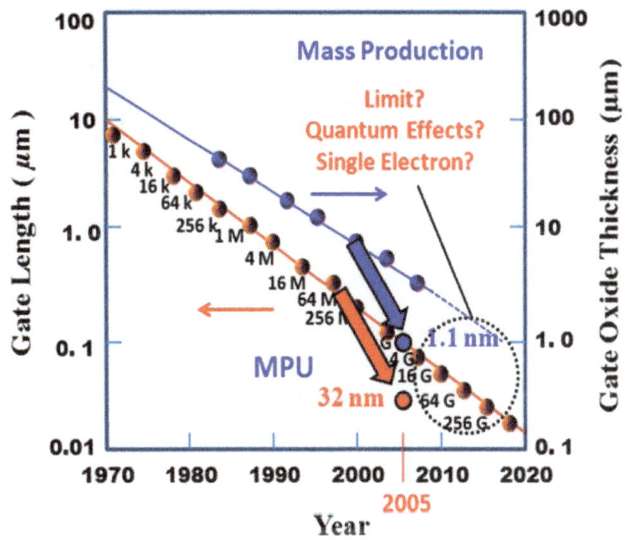

Figure 9: The trend of MOSFET scaling from ITRS. Reproduced from ITRS Corp.

Another limitation of FG technology is that tunnel oxide scaling is limited by stress-induced leakage current (SILC) related to charge transfer problem as indicated in Figure 10[60, 61]. The SILK increases with decreasing oxide thickness. This can be attributed to tunneling assisted by the traps in the bulk of the dielectric. Trap-assisted tunneling can take place at very low electric fields. If the density of traps is increased, the leakage will also increase. Electrical stress can increase the number of these traps. So it becomes an important limitation of scaling down the memory device [62]. For EOT<8 nm, a single oxide trap will cause to complete the charge loss in the FG Flash cell. The scaling of the gate stacks and operation voltages are often related to each other. A tunnel oxide thickness of more than 8 nm is currently used in the commercial Flash memory chip to meet the 10 years' data retention time requirement. If the tunnel oxide were to be scaled below 2 nm, the operation voltage could be reduced from more than 10 V to below 4 V [63]. Unfortunately, the retention time would also be reduced, from 10 years to several seconds. This physical damage to the tunnel oxide during the cycling process causes data retention problems, program disturbance, read disturbance, and erratic characteristic behavior of the FG memory cell. Such problems severely limit the reliability and multilevel cell operation. This basic limitation of the tunnel oxide thickness becomes increasingly important with scaling. New storage node concepts are also becoming attractive as an alternative approach to address some of the dielectric scaling limitations. Flash memory adopts a charge stored in a silicon nitride as the trapping layer, which exhibits significantly reduced defect-related leakage current and very low SILC as compared to SiO_2 with a similar EOT [64]. Such a relentless reduction of device dimensions has many challenges like retention, endurance, reduction in the number of electrons in the FG, dielectric leakage, cell-to-cell cross talk, threshold voltage shift, and reduction in memory window margins [65, 66]. The key concept of real scaling issues such as material and structural changes in Flash memory technologies is provided in detail in the next distinct part.

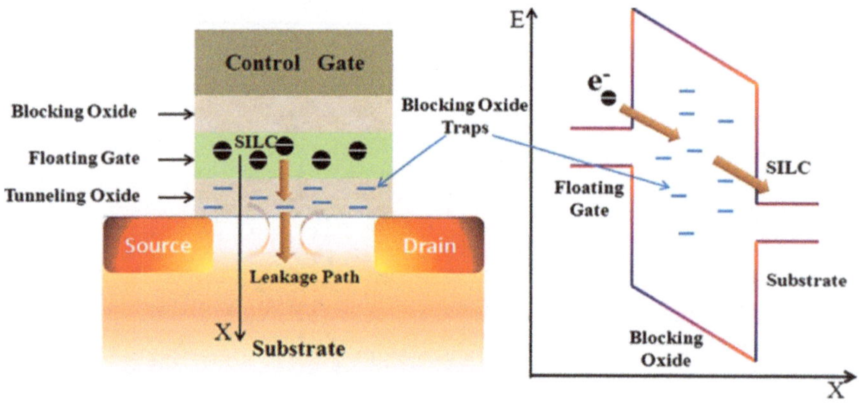

Figure 10: Schematic plots of a Flash memory cell and the degradation of its tunnel oxide. The degradation leads to the formation of percolation paths responsible for the FG charge loss, hence the loss of the stored information. The presence of traps in the energy barrier yields the trap-assisted tunneling mechanism and originates the stress-induced leakage current (SILC).

FG Flash Memory Technology

The FGNV memory is a basic building block of Flash memory, which is based on FG thin-film storage (TFS) memories that have been developed with the addition of an erase gate configuration. The conventional FG memory (Figure 11a) consists of a MOSFET configuration that is modified to include polysilicon as a charge storage layer surrounded by an insulated inner gate (floating gate) and an external gate (control gate). This what makes Flash memory nonvolatile and all floating gate memories to have the same generic cell structure. Charge is transferred to or from the floating gate through a thin (8 to 10 nm) oxide [1, 67]. Because the floating gate is electrically isolated by the oxide layer, any electrons placed on it are trapped there. Flash memory works by adding (charging) or removing (discharging) electrons to and from a floating gate. A bit's 0 or 1 state depends upon whether or not the floating gate is charged or discharged. When electrons are present on the floating gate, current cannot flow through the transistor and the bit state is '0'. This is the normal state for a floating gate. When electrons are removed from the floating gate, current is allowed to flow and the bit state is '1'. The FG memory has achieved high density, good program/erase speed, good reliability, and low operating voltage and promotes endurance for Flash memory application.

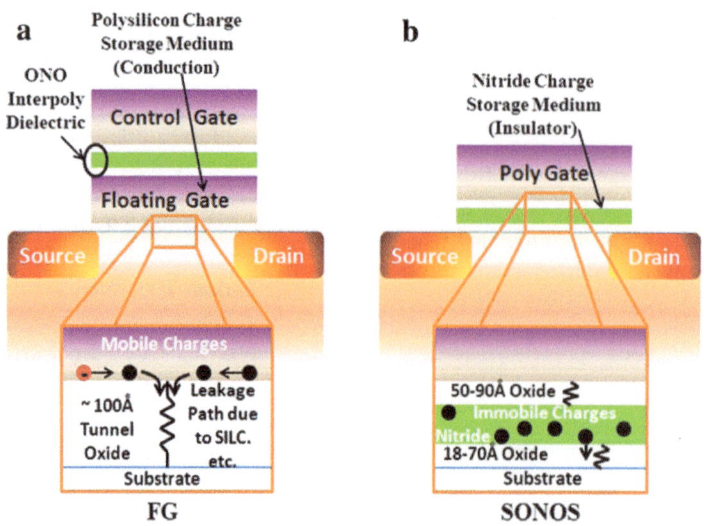

Figure 11: Schematics of the conventional FG memory and SONOS. Schematics of **(a)**floating gate and thin-film storage-based embedded nonvolatile memory bit cells, depending on the charge stored inside the gate dielectric of a MOSFET, and **(b)** the nitride traps (SONOS), embedded into the gate oxide of a MOSFET.

SONOS Memory Technology

In order to solve the scaling issue of the FG memory, the SONOS memory has been proposed as a Flash technology since the 1980s [68, 69]. The acronym SONOS is derived from the structure of the device as shown in Figure 11b. The SONOS device is basically a MOSFET, where the gate has been replaced by an ONO dielectric. The SONOS memory has a better charge retention than the FG memory when the FG bit cell's tunneling oxide layer is below 10 nm [70]. Moreover, the SONOS memory exhibits many advantages, e.g., easy to fabricate, high program/erase (P/E) speed, low programming voltage and power consumption, and better potential for scalability below the 70-nm node, according to the ITRS [71]. The charge, holes or electrons, are injected into the nitride layer using direct tunneling through the tunnel oxide layer. The nitride layer is electrically isolated from the surrounding transistor, although charges stored on the nitride directly affect the conductivity of the underlying transistor channel. Since the SONOS memory possesses spatially isolated deep-level traps, a single defect in the tunneling oxide will not cause discharge of the memory cell. The thickness of the top oxide is important to prevent the Fowler-Nordheim tunneling of electrons from the gate during erase. When the polysilicon control gate is biased positively, electrons from the transistor source and drain regions tunnel through the oxide layer and get trapped in the silicon nitride. This results in an energy barrier between the drain and the source, raising the threshold voltage V_{th} (the gate-source voltage necessary for current to flow through the transistor). Moreover, the nitride layer is electrically isolated from the surrounding transistor, although charges stored on the nitride directly affect the conductivity of the underlying transistor channel. The oxide/nitride sandwich typically consists of a 2-nm-thick oxide lower layer, a 5-nm-thick silicon nitride middle layer, and a 5- to 10-nm-thick oxide upper layer [72, 73]. However, SONOS-type Flash memories have several drawbacks such as shallow trap energy level, erase saturation, and vertical stored charge migration [74]. The programming speed and operating voltage problems can be solved by reducing the tunnel oxide thickness. At low tunnel oxide thickness, the issues that impact SONOS-type memories include erase saturation and vertical charge migration, which seriously degrade the retention capability of the memory [75]. Thus, many concerns still remain for the SONOS type of memories, which will be discussed in the next section.

Limitations of FG and SONOS Memory Technologies

Scaling demands very thin gate insulators in order to keep short channel effects and control the shrinkage of the device size and maximize the performance. When the tunneling oxide thickness is below 10 nm, the storaged charge in the FG is easy to leak due to a defect in the tunneling oxide formed by repeated write/erase cycles or direct tunneling current.

The tunneling gate oxide thickness in a conventional Flash memory cannot be scaled down to sub-7 nm because of charge retention [76]. The SONOS Flash memory can relieve the problem but still has a relatively thick gate dielectric thickness of about 7 nm. Therefore, conventional SONOS Flash memory also has a scaling-down problem. Many studies have shown that the charge retention characteristics in scaled SONOS nonvolatile memory devices with a low gate oxide thickness and at high temperature are problematic with shallow-level traps [48, 77, 78]. For the conventional SONOS memory, erase saturation and vertical stored charge migration [79, 80] are the two major drawbacks; the most challenging tasks are how to maintain an acceptable charge capability of the discrete storage nodes and how to fabricate nanocrystals with constant size, high density, and uniform distributions [81]. When the trap energy level is shallow, erase saturation and vertical migration occur and the electron charge decay rate increases due to low tunnel oxide thickness, issues that impact SONOS-type memories as shown in Figure 12. This erase saturation makes SONOS erase less as the erase voltage or the tunnel oxide thickness is increased. Since the SONOS memory uses silicon nitride as a charge trapping layer, the electrons in the Si sub-conduction band will tunnel through the tunneling oxide and a portion of the nitride, and this consequently degrades the program speed. Besides this, the conduction band offset of nitride is only 1.05 eV and back-tunneling of the trapped electron may also occur. Although applying a very high electric field may accelerate the de-trapping rate, the gate electron injection current exceeds the de-trapping but resulting in practically an increase in charge and no erasing. Using an ultra-thin (<2 nm) tunnel oxide offers an efficient charge direct tunneling erase and opens a memory window. However, the direct tunneling cannot be turned off at a low electric field, leading to poor retention and read disturb. Thus, the SONOS memory cannot be used for NAND Flash without further innovation of new memory technologies. The main reason for the growth of emerging NVM technologies is that scaling has now become a serious issue for the memory industry. Not only are many of these new technologies inherently more scalable, but also they seem well suited to the next generation of mobile computing and communications that will demand high-capacity memories capable of storing and rapidly accessing video and a large database without overburdening battery power sources.

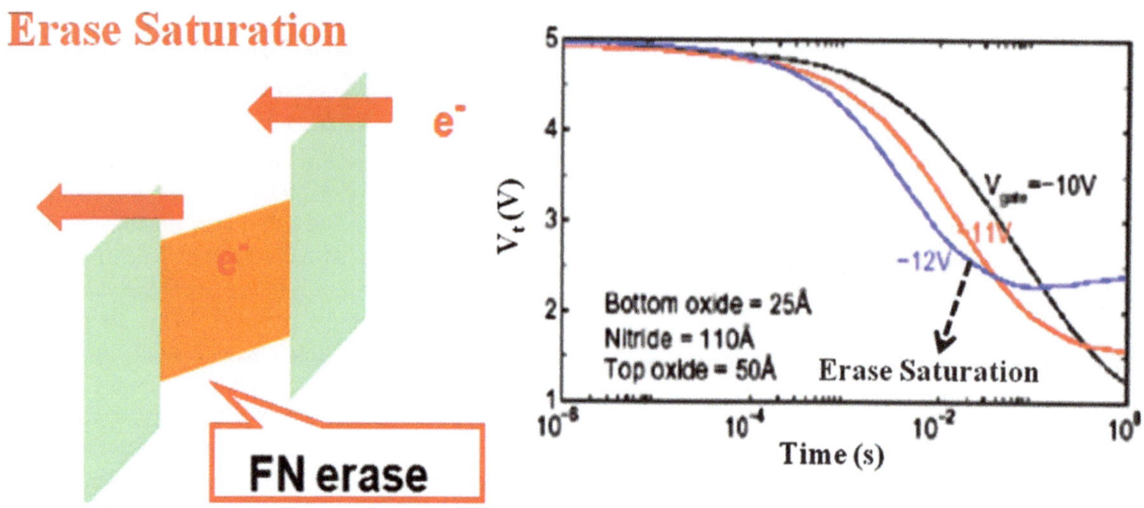

Figure 12: Fowler-Nordheim (FN) tunneling of electrons from the gate during erase and erase saturation in SONOS nonvolatile memory. This indicates the reduced memory window as the erase voltage is increased. Reproduced from ref. [74].

Many alternate device structures are proposed to hopefully circumvent these scaling challenges and to improve the device performance. In an effort to continue Moore's law and overcome the ultimate limitations of MOS-based memory devices, other storage concepts have been proposed in search of the 'unified memory'. The ideal memory device or the so-called 'unified memory' would satisfy simultaneously three requirements: high speed, high density, and nonvolatility. At the present time, such an ideal memory has not been developed. FGNVSM has high density and nonvolatility, but its P/E speed is low. DRAM has high speed (approximately 10 ns) and relatively high density, but it is volatile. SRAM has very high speed (approximately 5 ns), but it suffers from very low density and volatility. Many nonvolatile memory devices have been proposed on the basis of changing charge storage materials and new device concepts for the 'unified memory'. These structures will be considered in the next sections. In light of such issues, emerging memory solutions seem to be a key technology.

Current Emerging Memory Technologies

Recent studies have revealed that there is a close correlation among existing and emerging memory technologies in view of scalability. The scaling trend of memory transition leads to smaller and smaller memory devices, which have been routinely observed. To further support this assertion, another set of current progress in memory technology is described to the increasing importance of memory to users' experience and the importance of memory to system performance. There are many emerging memory technologies which are trying to replace existing memory technologies in the market. These new memory devices such as RRAM, PCM, and STT-RAM have read/write/retention/endurance characteristics different from those of conventional SRAM, DRAM, and Flash [82]. But the ideal characteristics of new emerging memory technologies have to be meeting the performance of SRAM and the density of NAND Flash in terms of stability, scalability, and switching speed. Thus, going beyond the traditional bistable memory, the possibilities of multilevel, high-performance memory devices suitable for market must be explored. Currently, there are several technologies that show some promise; some of these new emerging technologies are MRAM, FeRAM, PCM, STT-RAM, nano-random-access memory (NRAM), racetrack memory, RRAM and memristor, molecular memory, and many others [10, 83]. Each of these memory technologies will be briefly outlined and discussed in the following sections. In view of the commercial production, currently, MRAM, FeRAM, and PCM are in commercial production but still remain limited to niche applications relative to DRAM and NAND Flash. There is a prospect that among the emerging memory technologies, MRAM, STT-RAM, and RRAM are the most promising ones, but they are still many years away from competing for industry adoption [84]. It is necessary for any new technology to be able to deliver most for industry adoption. For industry adoption on a mass scale, some parameters must be matched with existing memory technologies. In consideration of new technology for industry application, the scalability of the technology, speed of the device, power consumption to be better than existing memories, endurance, densities, better than existing technologies and most importantly the cost; if the emerging technology can only run one or two of these attributes, then, at most desirable, it is likely to be resigned to niche applications.

MRAM

MRAM or magnetic RAM is a nonvolatile RAM technology under development since the 1990s. RRAM methods of storing data bits use magnetic charges instead of the electrical charges used by DRAM and SRAM technologies. MRAM, first developed by IBM in the 1970s [85], is expected to replace DRAM as the memory standard in electronics. MRAM is basically based on memory cells having two magnetic storage elements, one with a fixed magnetic polarity and another with a switchable polarity. These magnetic elements are positioned on top of each other but separated by a thin insulating tunnel barrier as shown in the cell structure in Figure 13. Moreover, scientists define a metal as magnetoresistive if it shows a slight change in electrical resistance when placed in a magnetic field. By combining the high speed of static RAM and the high density of DRAM, proponents say that MRAM could be used to significantly improve electronic products by storing greater amounts of data, enabling it to be accessed faster while consuming less battery power than existing electronic memories. Technically, it works with the state of the cell, which is sensed by measuring the electrical resistance while passing a current through the cell. Because of the magnetic tunnel effect [86], if both magnetic moments are parallel to each other, then the electrons will be able to tunnel and the cell is in the low resistance 'ON' state. However, if the magnetic moments are antiparallel, the cell resistance will be high. The memory characteristics of MRAM of writing and erasing are fulfilled by passing a current through the write line to induce a magnetic field across the cell. MRAM has been slowly getting off the ground but has now entered the market and will become increasingly available for mass production in the couple of years and beyond. Currently, it has reached some level of commercial success in niche applications [87]. Various companies such as Samsung, IBM, Hitachi and Toshiba, and TSMC are actively developing variant technologies of MRAM chips. In view of power consumption and speed, MRAM competes favorably than other existing memories such as DRAM and Flash, with an access time of a few nanoseconds [88–90]. Although it has some limitation during the 'write' operation, the smaller cell size could be limited by the spread of the magnetic field into neighboring cells and need an amendment to compete completely as a universal memory. The price of MRAM is also another issue and considered a limiting factor, with prices far in excess of all the currently established memories at approximately £2 to £3 ($3 to $5) per megabyte [91]. According to this price level, MRAM is in excess of 1,000 times the price of Flash memory and over 10,000 times the price of hard disk drives. It is expected that of the next-generation memory technologies, MRAM, in the future, will have the biggest market, followed by FeRAM, PCRAM, and memristors.

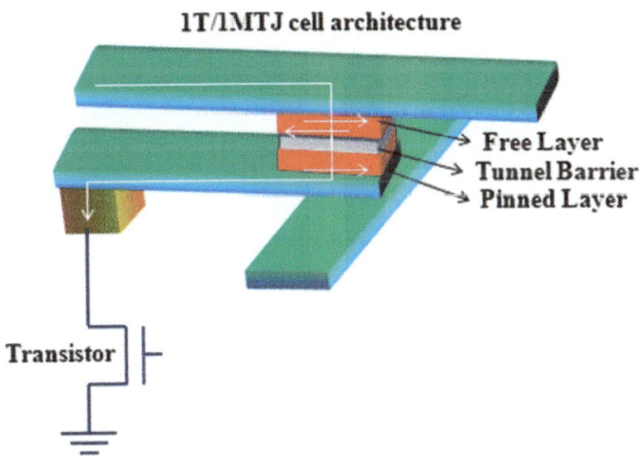

Figure 13: Basic MRAM cell structure.

STT-MRAM

STT-MRAM is a magnetic memory technology that exerts the base platform established by an existing memory called MRAM to enable a scalable nonvolatile memory solution for advanced process nodes [92, 93]. It is a new kind of magnetic RAM with the following features: fast read and write times, small cell sizes, potentially even smaller, and compatibility with existing DRAM and SRAM. As we have discussed in the previous section, MRAM stores data according to the magnetization direction of each bit and the nanoscopic magnetic fields set the bits in conventional MRAM. On the other hand, STT-MRAM uses spin-polarized currents, enabling smaller and less energy-consuming bits. The basic cell structure of STT-RAM is depicted in Figure 14. In addition, STT-RAM writing is a technology in which an electric current is polarized by aligning the spin direction of the electrons flowing through a magnetic tunnel junction (MTJ) element. Data writing is performed by using the spin-polarized current to change the magnetic orientation of the information storage layer in the MTJ element [94]. The resultant resistance difference of the MTJ element is used for information readout. STT-RAM is a more appropriate technology for future MRAM produced using ultra-fine processes and can be efficiently embedded in subsequent generations of such semiconductor devices as FPGAs, microprocessors, microcontrollers, and SoC. A special bonus for embedded designers is the fact that the internal voltage STT-RAM requires is only 1.2 V. The difference between STT-MRAM and a conventional MRAM is only in the writing operation mechanism; the read system is the same. The memory cell of STT-MRAM is composed of a transistor, an MTJ, a word line (WL), a bit line (BL), and a source line (SL) [95]. Currently, STT-RAM is being developed in companies including Everspin, Grandis, Hynix, IBM, Samsung, TDK, and Toshiba. However, for STT-RAM to be adopted as a universal mainstream semiconductor memory, some key challenges should be resolved: the simultaneous achievement of low switching current and high thermal stability. It must be dense (approximately 10 F^2), fast (below 10 ns of read and write speeds), and operating at low power [96].

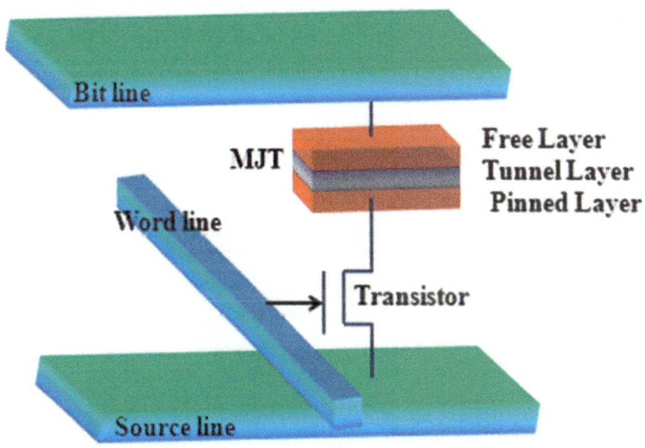

Figure 14: Basic STT-RAM cell structure.

FeRAM

FeRAM is a nonvolatile RAM that combines the fast read and write access of DRAM cells, consisting of a capacitor and transistor structure as shown in Figure 15. The cell is then accessed via the transistor, which enables the ferroelectric state of the capacitor dielectric to be sensed. In spite of its name, FeRAM does not contain iron. The polarization properties of a ferroelectric substance are used as a memory device. Today's FeRAM uses lead zirconate titanate (PZT); other materials are being considered. The main developer of FeRAM is Ramtron International. FeRAM is the most common kind of personal computer memory with the ability to retain data when power is turned off as do other nonvolatile memory devices such as ROM and Flash memory [97]. In a DRAM cell, the data periodically need refreshing due to the discharging of the capacitor, whereas

FeRAM maintains the data without any external power supply. It achieves this by using a ferroelectric material in the place of a conventional dielectric material between the plates of the capacitor. When an electric field is applied across dielectric or ferroelectric materials, it will polarize, and while that field is removed, it will depolarize. But the ferroelectric material exhibits hysteresis in a plot of polarization versus electric field, and it will retain its polarization. One disadvantage of FeRAM is that has a destructive *read* cycle. The read method involves writing a bit to each cell; if the state of the cell changes, then a small current pulse is detected by indicating that the cell was in the OFF state. However, it is a fast memory that can endure a high number of cycles (e.g., 10^{14}) [98], meaning that the requirement for a *write* cycle for every *read* cycle will not result in short product lives with a very low power requirement. It is expected to have many applications in small consumer devices such as personal digital assistants (PDAs), handheld phones, power meters, and smart cards, and in security systems. FeRAM is faster than Flash memory. It is also expected to replace EEPROM and SRAM for some applications and to become a key component in future wireless products. Even after FeRAM has achieved a level of commercial success, with the first devices released in 1993 [99, 100], current FeRAM chips offer performance that is either comparable to or exceeding current Flash memories [98, 101], but still slower than DRAM.

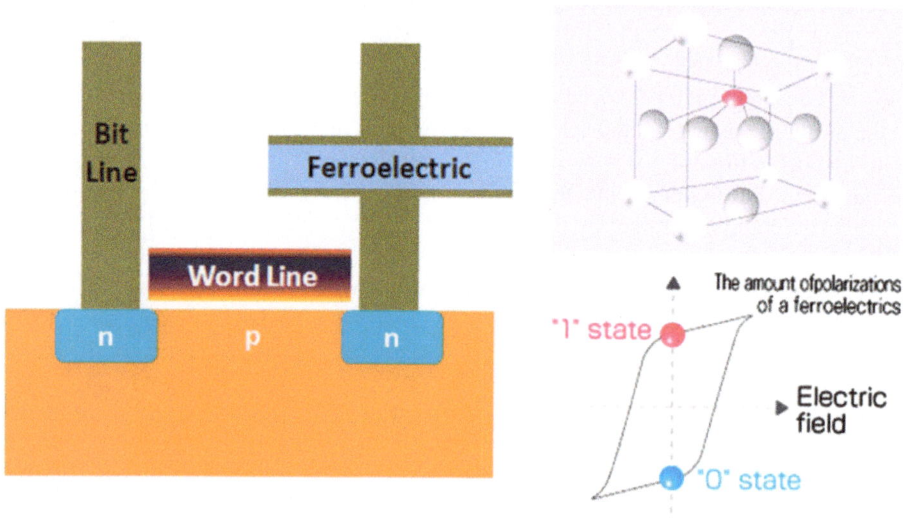

Figure 15: Basic structure of a FeRAM cell. The crystal structure of a ferroelectric and an electric polarization-electric field hysteresis curve are also shown.

PCRAM

PCRAM, also known as PCM, perfect RAM (PRAM), OUM, and chalcogenide RAM (CRAM), is a type of nonvolatile RAM based on a class of material called chalcogenide glasses that can exist in two different phase states (e.g., crystalline and amorphous) [102, 103]. The basic PCRAM cell structure is depicted in Figure 16. Most phase-change materials contain at least one element from group 6 of the periodic table, and the choice of available materials can be further widened by doping these materials [104–107]. In particular, the most promising are the GeSbTe alloys which follow a pseudobinary composition (between GeTe and Sb_2Te_3), referred to as GST. These materials are in fact commonly used as the data layer in rewritable compact disks and digital versatile disks (CD-RW and DVD-RW) where the change in optical properties is exploited to store data. The structure of the material can change rapidly back and forth between amorphous and crystalline on a microscopic scale. The material has low electrical resistance in the crystalline or ordered phase and high electrical resistance in the amorphous or disordered phase. This allows electrical currents to be switched *ON* and *OFF*, representing digital high and low states. This process has been demonstrated to be on the order of a few tens of nanoseconds [108], which potentially makes it compatible with Flash for the *read* operation, but several orders of magnitude faster for the *write* cycle. This makes it possible for PCM to function many times faster than conventional Flash memory while using less power. In addition, PCM technology has the potential to provide inexpensive, high-speed, high-density, high-volume nonvolatile storage on an unprecedented scale. The physical structure is three-dimensional, maximizing the number of transistors that can exist in a chip of fixed size. PCM is sometimes called perfect RAM because data can be overwritten without having to erase it first. Possible problems facing PCRAM concern the high current density needed to erase the memory; however, as cell sizes decrease, the current needed will also decrease. PCM chips are expected to last several times as long as currently available Flash memory chips and may prove cheaper for mass production. Working prototypes of PCM chips have been tested by IBM, Infineon, Samsung, Macronix, and others. Also, the production of PCM has been announced recently by both collaborations between Intel and STMicroelectronics as well as with Samsung [109, 110].

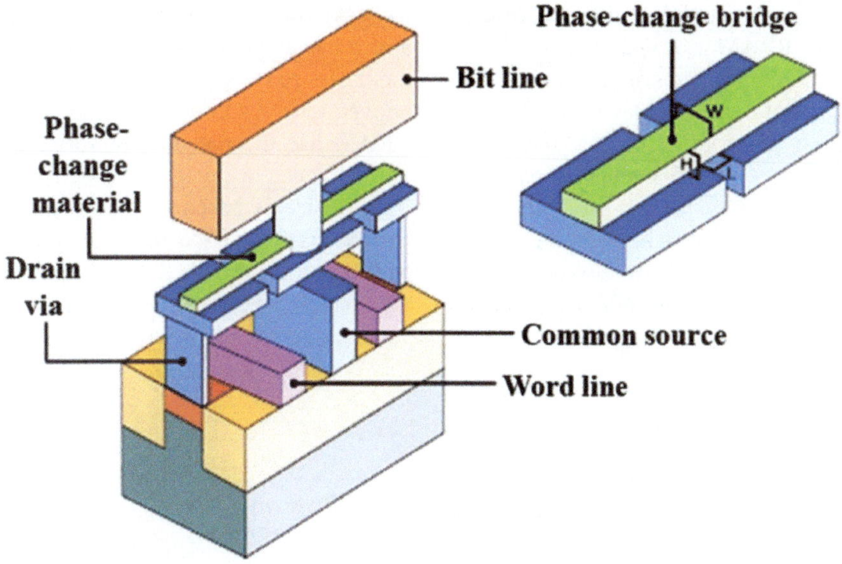

Figure 16: Basic PCRAM cell structure. Reproduced from IBM-Macronix-Qimonda.

Comparison of Primary Contenders for MRAM, STT-RAM, FeRAM, and PCM Technologies

Before going to other emerging memories, we herein provide a comparison among MRAM, FeRAM, and PCM. The specific features of these memory devices are provided in Table 2. Relatively mature, new-material memories such as MRAM, STT-RAM, FeRAM, and PCM can offer a variety of features that have potential to be the candidates for next-generation nonvolatile memory devices. Brand-new concepts such as RRAM, molecular, organic/polymer, and other nanowire-based memory technologies have also been proposed. These are discussed in detail in the following section.

Table 2: Summary of primary contenders for MRAM, FeRAM, STT-RAM, and PCM technologies

Features	FeRAM	MRAM	STT-RAM	PCM
Cell size (F^2)	Large, approximately 40 to 20	Large, approximately 25	Small, approximately 6 to 20	Small, approximately 8
Storage mechanism	Permanent polarization of a ferroelectric material (PZT or SBT)	Permanent magnetization of a ferromagnetic material in a MTJ	Spin-polarized current applies torque on the magnetic moment	Amorphous/polycrystal phases of chalcogenide alloy
Read time (ns)	20 to 80	3 to 20	2 to 20	20 to 50
Write/erase time (ns)	50/50	3 to 20	2 to 20	20/30
Endurance	10^{12}	$>10^{15}$	$>10^{16}$	10^{12}
Write power	Mid	Mid to high	Low	Low
Nonvolatility	Yes	Yes	Yes	Yes
Maturity	Limited production	Test chips	Test chips	Test chips
Applications	Low density	Low density	High density	High density

RRAM

RRAM is a disruptive technology that can revolutionize the performance of products in many areas, from consumer electronics and personal computers to automotive, medical, military, and space. Among all the current memory technologies, RRAM is attracting much attention since it is compatible with the conventional semiconductor processes. Memristor-based RRAM is one of the most promising emerging memory technologies and has the potential of being a universal memory technology [111]. It offers the potential for a cheap, simple memory that could compete across the whole spectrum of digital memories, from low-cost, low-performance applications up to universal memories capable of replacing all current market-leading technologies, such as hard disk drives, random-access memories, and Flash memories [112]. RRAM is a simple, two-terminal metal-insulator-metal (MIM) bistable device as shown in the basic configuration in Figure 17. It can exist in two distinct conductivity states, with each state being induced by applying different voltages across the device terminals. RRAM uses materials that can be switched between two or more distinct resistance states. Many companies are investing metal oxide nanolayers switched by voltage pulses. Researchers generally think that the pulses' electric fields produce conducting filaments through the insulating oxide. HP Labs plans to release prototype chips this year based on 'memristors' in which migrating oxygen atoms change resistance [113]. Xu et al. have also defined that among all the technology candidates, RRAM is considered to be the most promising as it operates faster than PCRAM and it has a simpler and smaller cell structure than magnetic memories (e.g., MRAM or STT-RAM) [114]. In contrast to a conventional MOS-accessed memory cell, a memristor-based RRAM has the potential of forming a cross-point structure without using access devices, achieving an ultra-high density. This device is based on the bistable resistance state found for almost any oxide material, including NiO, ZrO_2, HfO_2, $SrZrO_3$, and $BaTiO_3$[115–119]. Currently, Samsung and IBM are actively investigating RRAM.

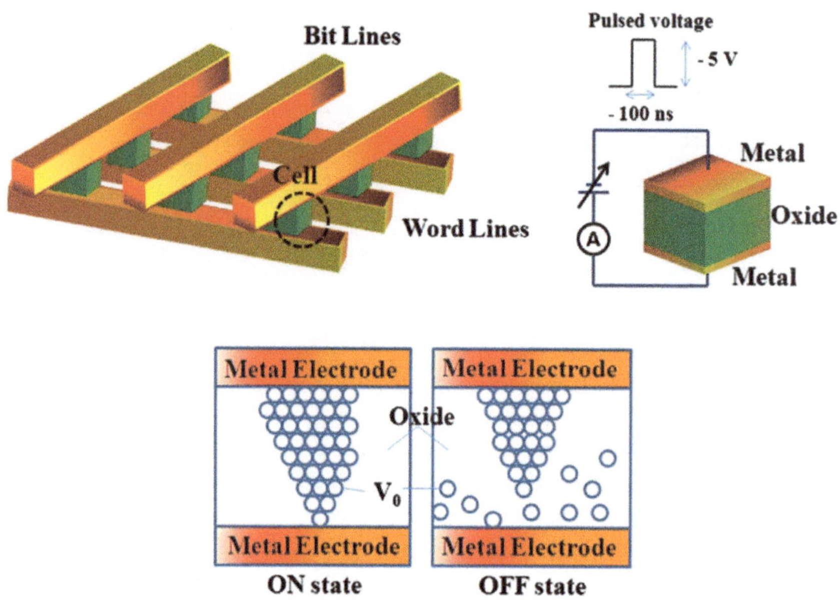

Figure 17: Basic RRAM cell structure. A schematic diagram of the mechanism of the resistive switching in a metal/oxide/metal-structured memory cell is also shown. Reproduced from ref. [123].

Kamiya et al. have revealed by a theoretical mechanism that RRAM shows filamentary-type resistive switching, where the oxygen vacancy is considered to form conductive filaments in the resistive material as shown in Figure 17[120]. The formation and disruption of these filaments are thus the mechanisms responsible for the ON-OFF switching in RRAM devices. The key issue is, therefore, to reveal electronic roles in the formation and disruption of the vacancy filaments. RRAM can be switched between the low resistance state (LRS) and the high resistance state (HRS) of the resistive material by applying voltages to the electrodes. Lee has explained that during the SET process, the current level increases from HRS to LRS as the voltage increases from 0 V to the critical point which is called the set voltage (*V* set), while the current level abruptly decreases from LRS to HRS at the reset voltage (*V* reset) under the RESET process. The SET and RESET processes are repeatedly carried out by sweeping the gate voltage with the binary states LRS and HRS [121]. Wang and Tseng and Lin et al. have indicated that the interface plays an important role in enhancing the performances of RRAM [122, 123]. Recently, Goux et al. have explained that using a stacked RRAM structure has been shown to be one of the most promising methods to improve the memory characteristics [124]. Although being a most promising memory element, critical issues for the future development of RRAM devices are reliable, such as data retention and memory endurance [125]. A data retention time of over 10 years can be extrapolated from retention characteristics measured at high temperatures and a memory endurance of over 10^6 cycles [126]. Therefore, a statistical study of reliability, availability, and maintainability is essential for the future development of RRAM.

Polymer Memory

Throughout the last few years, polymers have found growing interest as a result of the rise of a new class of nonvolatile memories. In a polymer memory, a layer consists of molecules and/or nanoparticles in an organic polymer matrix is sandwiched between an array of top and bottom electrodes as illustrated in Figure 18. Moreover, polymer memory has the advantage of a simple fabrication process and good controllability of materials [127]. Polymer memory could be called digital memory with the latest technology. It is not possible for a silicon-based memory to be established in less space, but it is possible for polymer memory. Ling et al. explained that polymer materials have simplicity in structure, free *read* and *write* capability, better scalability, 3-D stacking ability, low-cost potential, and huge capacity of data storage [128]. They revealed that a polymer memory stores information in a manner that is entirely different from that of silicon-based memory devices. Rather than encoding '0' and '1' from the number of charges stored in a cell, a polymer memory stores data on the basis of high and low conductivity while responding to an applied voltage. Among the large number of emerging memory technologies, polymer memory is the leading technology. It is mainly because of its expansion capability in 3-D space [129] since most polymers are organic materials consisting of long chains of single molecules. Prior to polymer memory fabrication, deposition of an organic layer is usually done by the sol-gel spin coating technique. All the other necessary constituent materials are dissolved in a solvent which is then spin-coated over a substrate. When the solvent is evaporated, a thin film of material with 10- to 100-nm thickness is successfully deposited at bottom electrodes. Top electrodes are deposited as the final step. The conductivity of the organic layer is then changed by applying a voltage across the memory cell, allowing bits of data to be stored in the polymer memory cell. When the polymer memory cell becomes electrically conductive, the electrons are introduced and removed. Even the polymer is considered as a 'smart' material to the extent that functionality is built into the material itself of switchability and charge store. This will open up tremendous opportunities in the electronics world, where tailor-made memory materials represent an unknown territory. The nonvolatileness and other features are inbuilt at the molecular level and offers very high advantages in terms of cost. But turning polymer memory into a commercial product would not be easy. Memory technologies compete not only on storage capacity but on speed, energy consumption, and reliability. 'The difficulty is in meeting all the requirements of current silicon memory chips,' says Thomas, the Director of Physical Sciences at IBM's Watson Research Center in Yorktown Heights, NY. They are likely to be limited to niche applications [130].

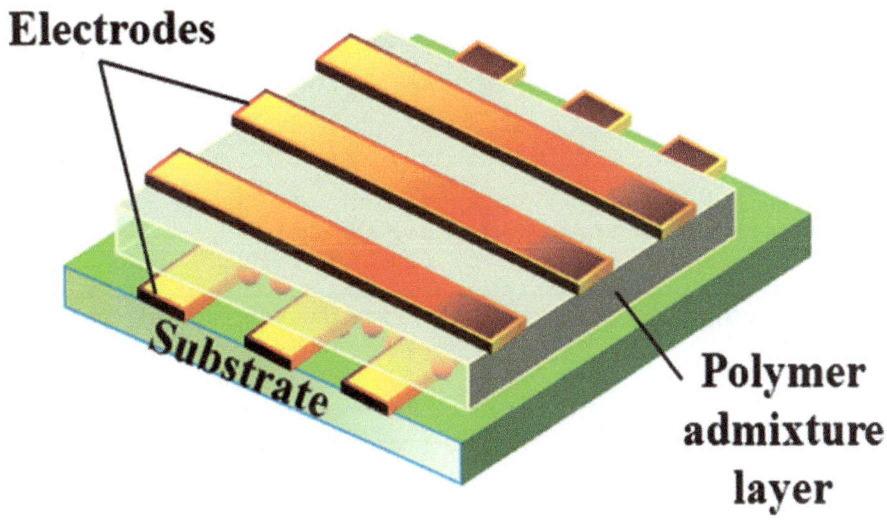

Figure 18: Structure of a polymer memory device.

Racetrack Memory

In a racetrack memory, information is stored on a U-shaped nanowire as a pattern of magnetic regions with different polarities. The U-shaped magnetic nanowire is an array of keys, which are arranged vertically like trees in a forest as shown in Figure 19. Achieving capacities comparable to vertical RM or hard drives would require stacks of these arrays. The nanowires have regions with different magnetic polarities, and the boundaries between the regions represent 1 or 0 s, depending on the polarities of the regions on either side [131, 132]. The magnetic information itself is then pushed along the wire, past the write and read heads by applying voltage pulses to the wire ends. The magnetic pattern to speed along the nanowire, while applying a spin-polarized current, causes the data to be moved in either direction, depending on the direction of the current. A separate nanowire perpendicular to the U-shaped 'racetrack' writes data by changing the polarity of the magnetic regions. A second device at the base of the track reads the data. Data can be written and read in less than a nanosecond. A racetrack memory using hundreds of millions of nanowires would have the potential to store vast amounts of data [133,134]. In this way, the memory requires no mechanical moving of parts and it has a greater reliability and higher performance than HDDs, with theoretical nanosecond operating speeds. For a device configuration where data storage wires are fabricated in rows on the substrate, conventional manufacturing techniques are adequate. However, for the maximum possible memory density, the storage wires are proposed to be configured rising from the substrate in a 'U' shape, giving rise to a 3-D forest of nanowires. While this layout does allow high data storage densities, it also has the disadvantage of complex fabrication methods, with so far, only 3-bit operation of the devices demonstrated [133]. As the access time of the data is also dependent on the position of the data on the wire, these would also be performance losses if long wires are used to increase the storage density further. The speed of operation of the devices has also been an issue during development, with much slower movement of the magnetic domains than originally predicted. This has been attributed to crystal imperfections in the permalloy wire, which inhibit the movement of the magnetic domains. By eliminating these imperfections, a data movement speed of 110 m/s has been demonstrated [133].

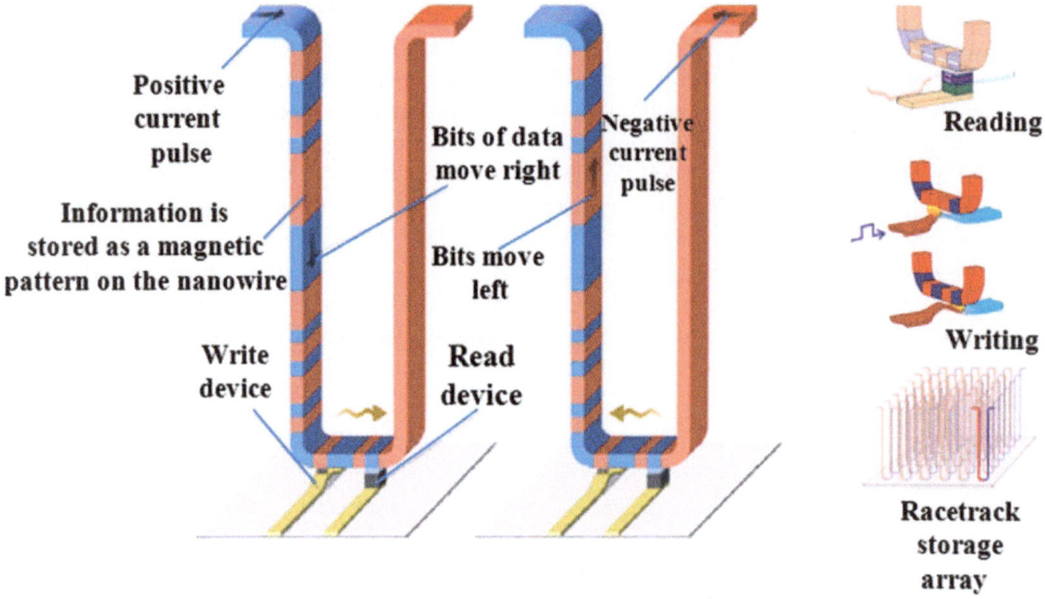

Figure 19: Racetrack memory diagram showing an array of U-shaped magnetic nanowires. The nanowires are arranged vertically like trees in a forest and a pair of tiny devices that read and write the data. Adopted from IBM.

Other New Memory Technologies

Researchers are already working hard on several emerging technologies, as discussed in previous sections, to pursue storage-class memories with a more traditional design than that of the racetrack memory, which places the bits in horizontal arrays.

Molecular Memory

A molecular memory is a nonvolatile data storage memory technology that uses molecular species as the data storage element, rather than, e.g., circuits, magnetics, inorganic materials, or physical shapes [135]. In a molecular memory, a monolayer of molecules is sandwiched between a cross-point array of top and bottom electrodes as shown in Figure 20. The molecules are packed in a highly ordered way, with one end of the molecule electrically connected to the bottom electrode and the other end of the molecule connected to the top electrode, and this molecular component is described as a molecular switch [136]. Langmuir-Blodgett (LB) deposition is ideally suited for depositing the molecular layer for the fabrication of molecular memory devices [137, 138]. Then, regarding the molecular memory operation, by applying a voltage between the electrodes, the conductivity of the molecules is altered, enabling data to be stored in a nonvolatile way. This process can then be reversed, and the data can be erased by applying a voltage to the opposite polarity of the memory cell. The increasing demand for nonvolatile electronic memories will grow rapidly in order to keep pace with the requirements for subsystems involved in flight demonstration projects and deep space operations. At the same time, mass, volume, and power must be minimized for mission affordability concerning these requirements; molecular memory could be a very promising candidate to fill this need.

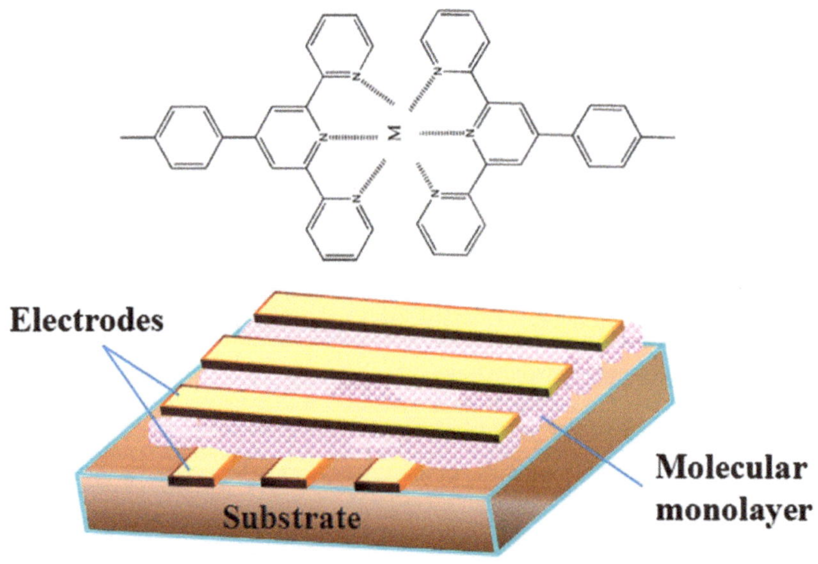

Figure 20: Cell structure of a molecular memory device.

Recently, Plafke has revealed clearly via an article that like most experimental technology that sounds so amazing that we want it right now, the molecular memory cell does not provide enough power for a commercial device [139]. This is currently only able to produce a 20% jump in conductivity. However, the area of molecular switching memory is promising, having eliminated the need for near-absolute zero temperatures and removed some of the constraints of the shape and number of layers of the molecule sheets which intend to convey that two of the biggest barriers are taken away. Thus, molecular memory requires strong attention to work over such issues and needs immediate amendment to see the possibility of a universal memory in the future.

MNW

In the last two decades, an increasing interest is observed for electronics-related devices and the search for a universal memory data storage device that combines rapid *read* and *write* speeds, high storage density, and nonvolatility is driving the investigation of new materials in the nanostructured form [140]. As an alternative to the current Flash memory technology, a novel transistor architecture using molecular-scale nanowire memory cells holds the promise of unprecedently compact data storage. The molecular nanowire array (MNW) memory is fundamentally different from other semiconductor memories; information storage is achieved through the channel of a nanowire transistor that is functionalized with redox-active molecules rather than through manipulation of small amounts of charge. It is relatively slow and lacks the random access capability, wherein data that can be randomly read and written at every byte are being actively pursued. Figure 21 shows the schematic design of a MNW memory cell. Lieber, and Agarwal and Lieber have revealed that the nanowire-based memory technology is a powerful approach to assemble electronic/photonic devices at ultra-small scales owing to their sub-lithographic size, defect-free single-crystalline structure, and unique geometry [141, 142]. Nanowires synthesized by chemical or physical processes are nearly perfect single-crystal structures with a small geometry and perfect surface. The channel of a nanowire transistor is functionalized with redox-active molecules. During programming, control of the voltage acting on the substrate is possible to change the oxidation and reduction states of the active molecules. Finally, by measurement of the conductance of the nanowire with the gate bias fixed at 0 V or a small voltage and from the hysteresis, the two states can be defined as a high-conductance ON state and a low-conductance OFF state. The MNW memory has advantages of low power dissipation, ultra-high density, simple fabrication process, 3-D structure, and multilevel storage, and it functions at the nanoscale with a few electrons but limited by low retention time parameter [143, 144]. Moreover, the deposition of metals onto a monolayer of molecular wires can lead to low device yield, and this problem remains a major challenge [145]. However, mentioning the term emerging class memory, it could be expected that the MNW memory represents an important step towards the creation of molecular computers that are much smaller and could be more powerful than today's silicon-based computers.

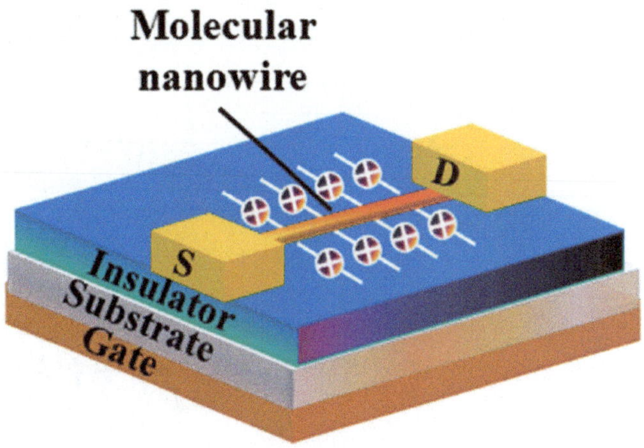

Figure 21: A MNW memory cell structure.

SNW

Semiconductor memory is essential for information processing as a key part of silicon technology; semiconductor memory has been continuously scaled to achieve a higher density and better performance in accordance with Moore's law [146]. Flash memory may reach fundamental scaling limits, however, because a thick tunneling oxide is required to prevent charge leakage and achieve 10 years' retention. As Flash memory approaches its scaling limit, several alternative strategies have been proposed to extend or replace the current Flash memory technology [147]. These approaches are revolutionary, but major challenges must be overcome to achieve small memory size and aggressive technology design architecture. In addition to the engineering of trapping layers, the device performance can also be improved by using innovative nonplanar channel geometries. Among the various nanostructure materials, semiconductor nanowire memory (SNW) has induced great scientific interest as possible building blocks for future nanoelectronic circuitry. In a SNW memory device, nanowires are integrated with SONOS technology. The basic schematic design of SNW is depicted in Figure 22. The SNW memory shows high mobility, less power dissipation, and high performance. Moreover, being 3-D-stacked, the SNW memory enhances cell density and data capacity without relying on advances in process technology. The nanowire-based memory device can store data electrically and is nonvolatile, meaning it retains data when the power is turned off, like the silicon-based Flash memory found in smart phones and memory cards [148], with minimal increase in chip size. In addition, the SNW device exhibits reliable write/read/erase operations with a large memory window and high on-to-off current ratio, which are highly advantageous for applications in nonvolatile memory [149]. The SNW memory cannot hold data as long as the existing Flash, but it is slower and has fewer rewrite cycles and it could potentially be made smaller and packed together more densely. And its main advantage is that it can be made using simple processes at room temperature, which means that it can be deposited even on top of flexible plastic substrates [150]. The SNW could, for instance, be built into a flexible display and could be packed into smaller spaces inside cell phones, MP3 players, plastic RFID tags, and credit cards.

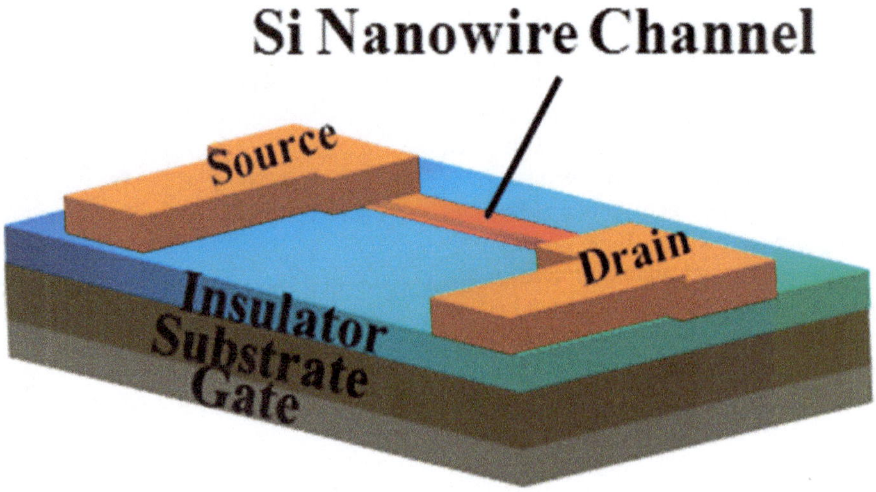

Figure 22: A bottom-gate FET-based nonvolatile SNW memory device.

NRAM

NRAM is a carbon nanotube (CNT)-based memory, which works on a nanomechanical principle, rather than a change in material properties [151]. NRAM uses carbon nanotubes for the bit cells, and the 0 or 1 is determined by the tube's physical state: up with high resistance, or down and grounded. NRAM is expected to be faster and denser than DRAM and also very scalable, able to handle 5-nm bit cells whenever CMOS fabrication advances to that level. It is also very stable in its 0 or 1 state. Produced by Nantero, these memories consist of the structure shown in Figure 23a with an array of bottom electrodes covered by a thin insulating spacer layer [152]. CNTs are then deposited on the spacer layer, leaving them freestanding above the bottom electrodes. Unwanted CNTs are removed from the areas around the electrode, with top contacts and interconnects deposited on top of the patterned CNT layer. During the time that the CNTs are freestanding, there is no conduction path between the bottom and top electrodes and hence the memory cell is in the *OFF* state. However, if a large enough voltage is applied over the cell, the nanotubes are attracted to the bottom electrode where they are held in place by van der Waals forces [153]. Due to the conductive nature of the CNTs, the electrodes are now connected and the cell reads the low conductivity *ON* state as shown in Figure 23b. The *OFF* state can be returned by repelling the nanotubes with the opposite electrode polarity. Nonvolatility is achieved due to the strength of the van der Waals forces overcoming the mechanical strain of the bent nanotubes, hence holding the cell in the *ON* state. NRAM offers the possibility of a simple cell architecture, which could operate at much higher speeds than the conventional Flash and with low power use. Cui et al. reported CNT memory devices exhibiting an extraordinarily high charge storage stability of more than 12 days at room temperature [154]. However, as NRAM is based on CNTs, it suffers from fabrication problems that are inherent in carbon nanotube-based devices. The issues include the cost and fabrication complexity of producing the CNTs, ensuring uniform dispersions of nanotubes, and difficulties in removing nanotubes from the unwanted positions on the substrate.

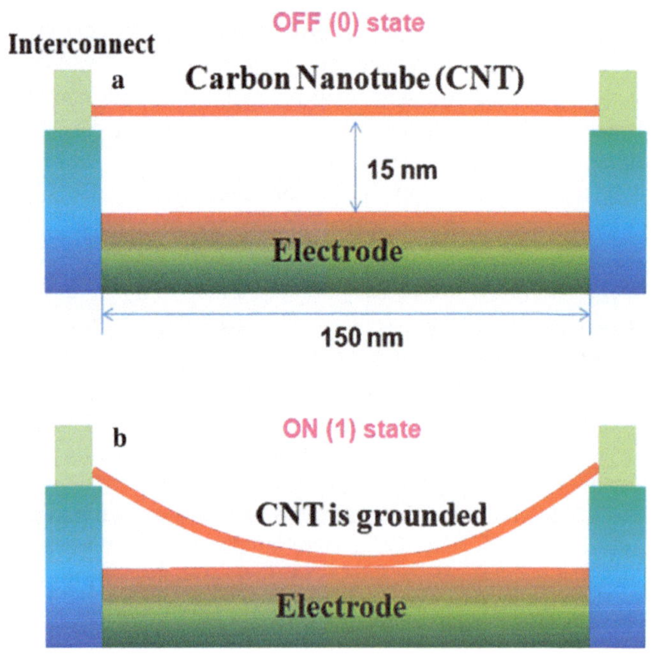

Figure 23: NRAM structure with (a) OFF state and (b) ON state.

Millipede Memory

In 2002, IBM developed a punch card system known as Millipede, which is a nonvolatile computer memory stored in a thin polymer sheet with nanoscopic holes to provide a simple way to store binary data [155]. It can store hundreds of gigabytes of data per square centimeter. However, the polymer reverts to its pre-punched form over time, losing data in the process. Millipede storage technology is being pursued as a potential replacement for magnetic recording in hard drives, at the same time reducing the form factor to that of Flash media. The prototype's capacity would enable the storage of 25 DVDs or 25 million pages of text on a postage stamp-sized surface and could enable 10 GB of storage capacity on a cell phone. Millipede uses thousands of tiny sharp points (hence the name) to punch holes in a thin plastic film. Each of the 10-nm holes represents a single bit. The pattern of indentations is a digitized version of the data. The layout of the millipede cantilever/tip in contact with the data storage medium is shown in Figure 24. According to IBM, Millipede can be thought of as a nanotechnology version of the punch card data processing technology developed in the late nineteenth century [156]. However, there are significant differences: Millipede is rewritable, and it may eventually enable storage of over 1.5 GB of data in a space no larger than a single hole in the punch card. Storage devices based on IBM's technology can be made with existing manufacturing techniques, so they will not be expensive to make. According to P. Vettiger, head of the Millipede project, there is not a single step in fabrication that needs to be invented. Vettiger predicts that a nanostorage device based on IBM's technology could be available as early as 2005 [155]. Now, researchers at IBM's Zurich Research Laboratory in Switzerland have clocked the rate of data loss. They have calculated that at 85°C, a temperature often used to assess data retention, it would lose just 10% to 20% of information over a decade, comparable to Flash memory [157].

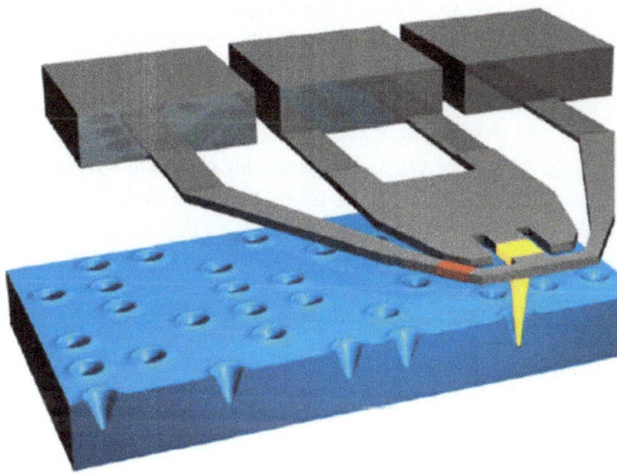

Figure 24: Schematic layout of the millipede cantilever/tip in contact with the data storage medium. Adopted from ref. [157].

WORM Memory Based on DNA Biopolymer Nanocomposite

The use of DNA is well known as a good model for metal NP synthesis due to its affinity to the metal ions [158]. In recent years, DNA has also been shown to be a promising optical material with the material processing fully compatible with conventional polymer for thin-film optoelectronic applications [159, 160]. Researchers from National Tsing Hua University in Taiwan and the Karlsruhe Institute of Technology in Germany have created a DNA-based memory device, that is, write-once-read-many-times (WORM), that uses ultraviolet (UV) light to encode information [161]. The device consists of a single biopolymer layer sandwiched between electrodes, in which electrical bistability is activated by *in situ* formation of silver nanoparticles embedded in a biopolymer upon light irradiation (Figure 25). The device functionally works when shining UV light on the system, which enables a light-triggered synthesis process that causes the silver atoms to cluster into nanosized particles and readies the system for data encoding. For some particular instance, the team has found that using DNA may be less expensive to process into storage devices than using traditional, inorganic materials like silicon, the researchers say [161, 162]. They said that when no voltage or low voltage is applied through the electrodes to the UV-irradiated DNA, only a low current is able to pass through the composite; this corresponds to the 'OFF' state of the device. But the UV irradiation makes the composite unable to hold a charge under a high electric field, so when the applied voltage exceeds a certain threshold, an increased amount of charge is able to pass through. This higher state of conductivity corresponds to the 'ON' state of the device. The team found that this change from low conductivity ('OFF') to high conductivity ('ON') was irreversible: once the system had been turned on, it stayed on, no matter what voltage the team applied to the system. Once information is written, the device appears to retain that information indefinitely. The researchers hope that the technique will be useful in the design of optical storage devices and suggest that it may have plasmonic applications as well. Consequently, WORM memories based on DNA a biopolymer nanocomposite have emerged as an excellent candidate for next-generation information storage media because of their potential application in flexible memory devices. This work combines new advances in DNA nanotechnology with a conventional polymer fabrication platform to realize a new emerging class of DNA-based memory.

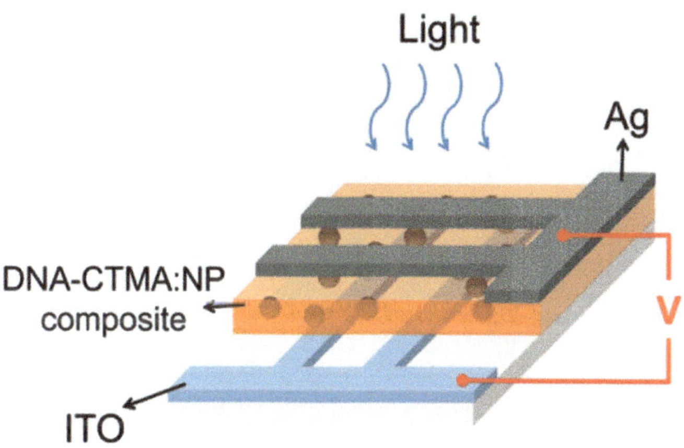

Figure 25: Schematic design of a memory device consisting of a thin DNA biopolymer film sandwiched between electrodes. The memory switching effect is activated upon light irradiation. Adopted from ref. [161].

QD Memory

Memory made from tiny islands of semiconductors - known as quantum dots - could fill a gap left by today's computer memory, allowing storage that is fast as well as long lasting. Researchers have shown that they can write information into quantum dot memory in just nanoseconds. Memory is divided into two forms: DRRAM and Flash [163, 164]. Computers use DRAM, for short-term memory, but data does not persist for long and must be refreshed over 100 times per second to maintain its memory. On the other hand, Flash memory, like that used in memory cards, can store data for years without refreshing but writes information about 1,000 times slower than DRAM. New research shows that memory based on quantum dots can provide the best of both: long-term storage with write speeds nearly as fast as DRAM. A tightly packed array of tiny islands, each around 15 nm across, could store 1 terabyte (1,000 GB) of data per square inch, the researchers say. Dieter Bimberg and colleagues at the Technical University of Berlin, Germany, with collaborators at Istanbul University, Turkey, demonstrated that it is possible to write information to the quantum dots in just 6 ns [165, 166]. The key advantages of quantum dot (QD) NVMs are the high read/write speed, small size, low operating voltage, and, most importantly, multibit storage per device. However, these features have not been realized due to variations in dot size and lack of uniform insulator cladding layers on the dots [167]. Incorporating QDs into the floating gate results in a reduction in charge leakage and power dissipation with enhanced programming speed. Researchers in India and Germany have now unveiled the memory characteristics of silicon and silicon-germanium QDs embedded in epitaxial rare-earth oxide gadolinium oxide (Gd_2O_3) grown on Si (111) substrates as shown in the DQM structure in Figure 26. Multilayer Si as well as single-layer $Si_{1-x}Ge_x$ (where x=0.6) QDs show excellent memory characteristics, making them attractive for next-generation Flash-floating-gate memory devices [168, 169].

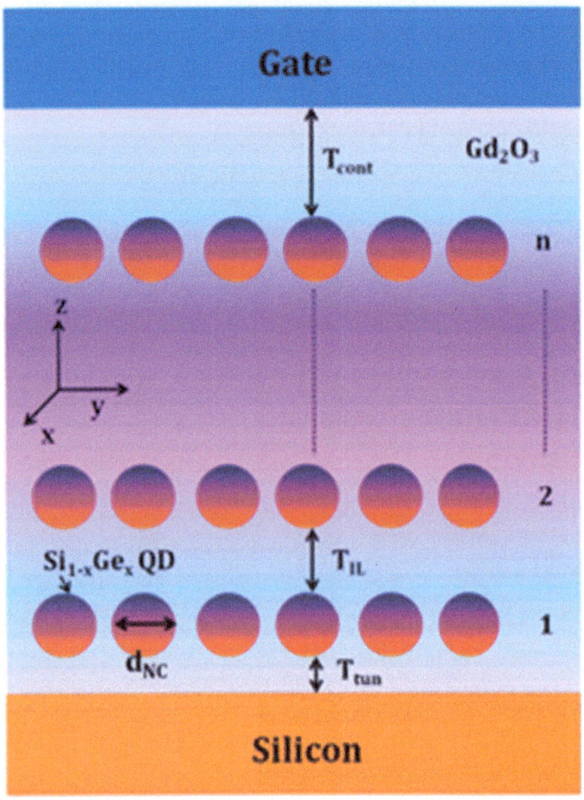

Figure 26: Structure of quantum dot memory. Adopted from ref. [168].

3-D Cross-Point Memory

Memory producers are also trying to develop alternative technologies that may be scalable beyond 20-nm lithography. For true scalability beyond 20-nm technology nodes, it is necessary to design a cross-point memory array which does not require diodes for access elements [170]. The cross-point memory architecture could be designed such that it can be easily fabricated in multiple layers to form a stacked 3-D memory [171]. The 3-D technology has brought to high volume an NVM where arrays of memory cells are stacked above control logic circuitry in the third dimension, and stacking 3-D memory directly over CMOS allows for high array efficiency and very small die size [172]. The 3-D technology uses no new materials, processes, or fabrication equipment, which control logic circuitry composed of typical CMOS. The memory construction uses typical back-end processing tools, and each memory layer is a repeat of the layers below it. The basic design of the 3-D cell consists of a vertical diode in series with a memory element as shown in Figure 27. Building integrated circuits vertically allows for a reduced chip footprint when compared to a traditional 2-D design, by an approximate factor of the number of layers used. This offers significant advantages in terms of reduced interconnect delay when routing to blocks that otherwise would have been placed laterally. The process for the 2-D cross-point array can be built into a multilayer 3-D architecture. Traditionally, a 3-D integrated circuit (3-D-IC) has used more than one active device layer. While resistance-change memory cells are not active devices, they function as rectifying devices in design. Further characterization of the resistance-change material is also necessary in order to guarantee that the 3-D cross-point memory will be practical for data storage. Also, the scalability of metal-oxide resistance-change materials beyond 20-nm technology nodes still needs to be studied. Moreover, the programming operation is expected to be competitive with both NAND and NOR Flash in terms of speed because of the relatively low voltage requirements of resistance-change materials. If the peripheral circuitry for accommodating the write operation can be made sufficiently compact, then the 3-D cross-point memory will indeed be a viable replacement for NAND and NOR Flash in future process generations.

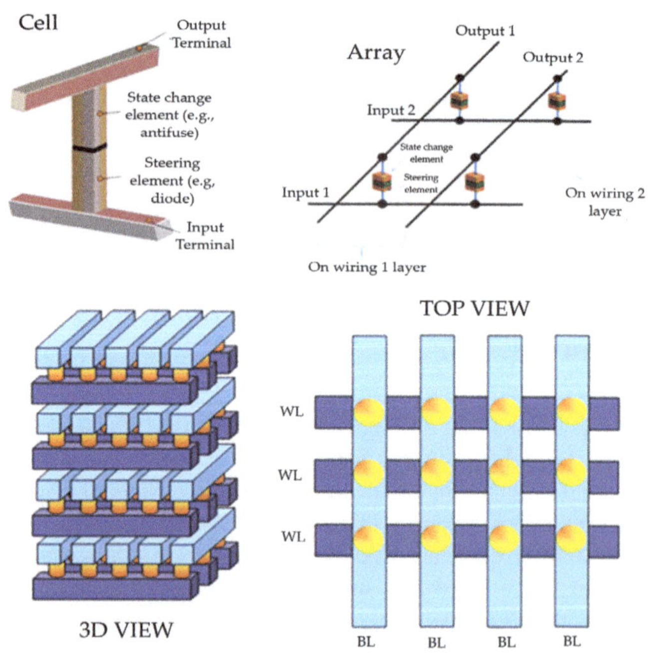

Figure 27: The basic design of a 3-D cell that consists of a vertical diode in series. (top) Side view, (bottom right) top view, and (bottom left) 3-D view.

TFM

Transparent and flexible electronics (TFE) is, today, one of the most advanced topics for a wide range of device applications, where the key component is transparent conducting oxides (TCOs), which are unique materials that oxides of different origin play an important role, not only as a passive component but also as an active component [173]. TFE is an emerging technology that employs materials (including oxides, nitrides, and carbides) and a device for the realization of invisible circuits for implementing next-generation transparent conducting oxides in an invisible memory generation [174]. In general, the TF-RRAM device is based on a capacitor-like structure (e.g., ITO/transparent resistive material/ITO/transparent and flexible substrate), which provides transmittance in the visible region [175]. For such new class of memory technology, data retention is expected to be about 10 years. The basic structural design of the new memory chips is configured, namely with two terminals per bit of information on a transparent and flexible substrate rather than the standard three terminals per bit on a rigid and opaque substrate (Figure 28). They are much better suited for the next revolution in electronic 3-D memory than Flash memory. These new memory chips that are transparent, are flexible enough to be folded like a sheet of paper, shrug off 1,000°F temperatures twice as hot as the max in a kitchen oven, and survive other hostile conditions could usher in the development of next-generation Flash-competitive memory for tomorrow's keychain drives, cell phones, and computers, a scientist reported today. Speaking at the 243rd National Meeting and Exposition of the American Chemical Society, the world's largest scientific society, he said that devices with these chips could retain data despite an accidental trip through the drier or even a voyage to Mars. And with a unique 3-D internal architecture, the new chips could pack extra gigabytes of data while taking up less space [176]. Despite the recent progress in TF-RRAM, it needs lots of work to satisfy the dual requirements of resistance to repeated bending stress and transparent properties. Thus, it is supposed that an achievement of such TF-RRAM device will be the next step towards the realization of transparent and flexible electronic systems. We hope that FT-RRAM devices will mark a milestone in the current progress of such unique and invisible electronic systems in the near future.

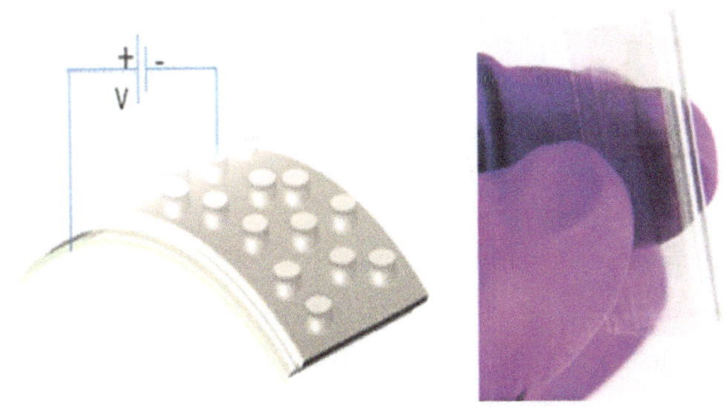

Figure 28: A schematic design of FT-RRAM and a flexible, transparent memory chip image created by researchers at Rice University. Reproduced from Tour Lab/Rice University.

1T1R-RRAM

One-transistor one-resistor (1T1R)-RRAM is also one class of emerging memory technology with impressive characteristics. It meets the demands for next-generation memory systems. It is expected that 1T1R-RRAM could be able to meet the demand of high-speed (e.g., performance) memory technology. The 1T1R structure is chosen because the transistor isolates current to cells, which are selected from cells which do not. The basic cell structure of 1T1R is depicted in Figure 29. 1T1R-RRAM consists of an access transistor and a resistor as a storage element. Zangeneh and Joshi have also mentioned that the 1T1R cell structure is similar to that of a DRAM cell in that the data is stored as the resistance of the resistor and the transistor serves as an access switch for reading and writing data [177, 178]. In reference to this, they revealed the 1T1R cell as the basic building block of a NVRRAM array as it avoids sneak path problem to ensure reliable operation. Moreover, the 1T1R structure is more compact and may enable vertically stacking memory layers, ideally suited for mass storage devices. But, in the absence of any transistor, the isolation must be provided by a 'selector' device, such as a diode, in series with the memory element, or by the memory element itself. Such kinds of isolation capabilities have been inferior to the use of transistors, limiting the ability to operate very large RRAM arrays in 1T1R architecture. 1T1R memory polarity can be either binary or unary. Bipolar effects cause polarity to reverse reset operation to set operation. Unipolar switching leaves polarity unaffected but uses different voltages.

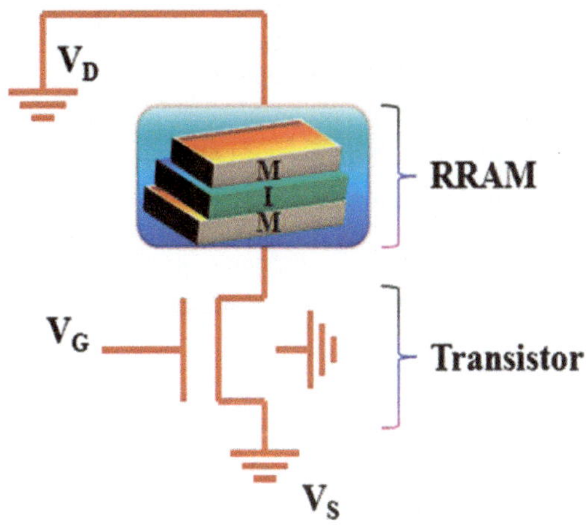

Figure 29: The basic cell structure of 1T1R-RRAM.

MTM, PFRAM, SPBMM, and CMORRAM - future alternate NVMs

Other potential emerging classes of memory technologies, we are describing in short, are molecular tunnel memory (MTM), polymeric ferroelectric RAM (PFRAM), spin-polarized beam magnetic memory (SPBMM), light memory, and complex metal-oxide RRAM (CMORRAM). We can say that these are sister memory technologies of molecular memory, ferroelectric/polymer memory, magnetic memory, and metal-oxide RRAM, respectively. Although these new technologies will almost certainly result in more complex memory hierarchies than their family memories, they are likely to allow the construction of memory chips that are nonvolatile, have low energy, and have density and development close to or better than those of DRAM chips, with improved performance and allowing memory systems to continue to scale up.

CONCLUSIONS

This article reviewed the historical development to the recent advancement on memory architecture and scaling trend of several conventional types of Flash within the MOS family and projected their future trends. With great progress being made in the emerging memory technologies, current trends and limitations were discussed before leading to some insight into the next generation of memory products. For the past three and a half decades in existence, the family of semiconductor memories has expanded greatly and achieved higher densities, higher speeds, lower power, more functionality, and lower costs. In the past 40 to 50 years, NVSM has grown from the FG concept to FAMOS, SAMOS, Flash memory, multilevel cells, RRAM, 3-D structures, and TF-RRAM. Since 1990, NVSM is an inspired technology, which has ushered in the digital age, enabled the development of all modern electronic systems, and brought unprecedented benefit to humankind. At the same time, some of the limitations within each type of memory are also becoming more realized. As the device dimension is reduced to the deca-nanometer regime, NVSM faces many serious scaling challenges such as the interface of neighboring cells, reduction of stored charges, and random telegraph noise. As such, we hope and are confident that there are several emerging technologies aiming to go beyond those limitations and potentially replace all or most of the existing semiconductor memory technologies to become a USM. Despite these limitations, the field of conventional semiconductor memories would continue to flourish and memory device scientists will find the way to meet these challenges and may even develop a 'unified memory' with low cost, high performance, and high reliability for future electronic systems. Progress towards a viable new resistive memory technology relies on fully understanding the mechanisms responsible for switching and charge transport, the failure mechanisms, and the factors associated with materials reliability. Moreover, the development of current redox-based resistive switching will help to improve our old technologies, and further research will produce more impressive results that will benefit industries and society to improve the quality of life for billions of people around the world.

ACKNOWLEDGEMENTS

First of all, the authors would like to thank and gratefully acknowledge all corresponding publishers for the kind permission to reproduce their figures and related description, used in this review article. This work was supported by the National Science Council, Taiwan, under Project No. NSC 102-2221-E-009-134-MY3.

Competing interests

The authors declare that they have no competing interests.

AUTHORS' CONTRIBUTIONS

JSM designed the structure and modified the manuscript. SMS and TYT participated in the sequence alignment and editing the manuscript. UC participated in its design and coordination and helped to draft the manuscript. All authors read and approved the final manuscript.

REFERENCES

1. Kahng D, Sze SM: A floating gate and its application to memory devices. *Bell Syst Tech J*1967, 46(6):1288–1295.
2. Ayling JK, Moore RD, Tu GK: A high-performance monolithic store. In *Dig Tech Papers IEEE Int Solid-State Circuits Conf: February 19–21 1969; Philadelphia*. Piscataway: IEEE; 1969:36–37.
3. Bez R, Camerlenghi E, Modelli A, Visconti A: Introduction to flash memory. *Proc IEEE*2003, 91(4):489–502.
4. Aritome S: Advanced flash memory technology and trends for files storage application. In*IEEE IEDM Tech Dig: December 10–13 2000; San Francisco*. Piscataway: IEEE; 2002:763–766.
5. SNIA: *Solid State Storage 101: An Introduction to Solid State Storage*. San Francisco: SNIA; 2009.
6. Kim K, Koh G: The prospect on semiconductor memory in nano era. *IEEE 7th Inter Conf Solid-State Integer Circuits Technol* 2004, 1: 662–667.
7. Lam CH: The quest for the universal semiconductor memory. In *IEEE Conf Electron Devices and Solid-State Circuits: December 19–21 2005; Hong Kong*. Piscataway: IEEE; 2005:327–333.
8. Tseng TY, Sze SM: *Nonvolatile Memories: Materials, Devices and Applications*. Lewis Way: American Scientific Publishers; 2012:25650.
9. Lin CW, Wang DY, Tai Y, Jiang YT, Chen MC, Chen CC, Yang YJ, Chen YF: Type-II heterojunction organic/inorganic hybrid non-volatile memory based on FeS_2 nanocrystals embedded in poly(3-hexylthiophene). *J Phys D Appl Phys* 2011, 44(29):292002.
10. Burr GW, Kurdi BN, Scott JC, Lam CH, Gopalakrishnan K, Shenoy RS: Overview of candidate device technologies for storage-class memory. *IBM J Res & Dev* 2008, 524: 449–464.
11. Intel is hot on plastic memory [http://www.geek.com/chips/intel-is-hot-on-plastic-memory-548187]
12. Lin HT, Pei Z, Chen JR, Hwang GW, Fan JF, Chan YJC: A new nonvolatile bistable polymer-nanoparticle memory device. *IEEE Electron Device Lett* 2007, 28(11):951–953.
13. Cho SH, Lee DI, Jung JH, Kim TW: Electrical bistabilities and memory stabilities of nonvolatile bistable devices fabricated utilizing C_{60} molecules embedded in a polymethyl methacrylate layer. *Nanotechnology* 2009, 20(34):345204.
14. Tang W, Shi HZ, Xu G, Ong BS, Popoyic ZD, Deng JC, Zhao J, Rao GH: Memory effect and negative differential resistance by electrode-induced two-dimensional single-electron tunneling in molecular and organic electronic devices. *Adv Mater* 2005, 17(19):2307–2311.
15. Intel brings plastic memory one step closer to reality [http://www.geek.com/chips/intel-brings-plastic-memory-one-step-closer-to-reality-553620]
16. Wang X: Electrical characterization of novel charge-trapping non-volatile semiconductor memory devices. In *Theses and Dissertations. Volume 932*. Lehigh University; 2006.
17. Lankhorst MHR, Ketelaars BWSMM, Wolters RAM: Low-cost and nanoscale non-volatile memory concept for future silicon chips. *Nat Mater* 2005, 4: 347–352.
18. Viking Technology: *The Expanding Role of Non-volatile Memory in High-Performance Embedded Architecture.*

Foothill Ranch: Viking Technology; 2012:643–7255.

19. Coburn J, Caulfield AM, Akel A, Grupp LM, Gupta RK, Jhala R, Swanson S: NV-heaps: making persistent objects fast and safe with next-generation, non-volatile memories. In*Proceedings of the Sixteenth International Conference on Architectural Support for Programming Languages and Operating Systems (ASPLOS'11): March 5–11 2011; Newport Beach*. California: ACM; 2011:1–13.

20. Brewer JE: Lest we forget: NVSM from origins to the "beyond CMOS" era. In *IEEE Non-volatile Memory Technology Symposium (NVMTS): November 15–17 2004; Orlando*. Florida: IEEE; 2004.

21. Global next generation memory markets (2012–2017), by Technology (DRAM, SRAM, Flash Memory, Memristor, Magneto Resistive RAM (MRAM), Phase Change RAM (PCRAM), Ferroelectric RAM) by Technology Application & Geography; 2013 [http://www.marketsandmarkets.com/Market-Reports/Memristor-Memory-Market-632.html]

22. Lu CY: *International Symposium on Non-volatile Memory: the Technology Driver of the Electronics Industry. Volume 2. 1st edition*. Singapore: World Scientific; 2012.

23. Proctor A: Non-volatile memory & its use in enterprise applications. In *Viking Technology Understanding Non-volatile Memory Technology Whitepaper*. Irvine: Viking Technology; 2012:1–8.

24. 2Ouyang J, Chu CW, Szmanda CR, Ma L, Yang Y: Programmable polymer thin film and non-volatile memory device. *Nat Mater* 2004, 3: 918–922.

25. Bozano LD, Kean BW, Deline VR, Salem JR, Scott JC: Mechanism for bistability in organic memory elements. *Appl Phys Lett* 2004, 84(4):607–609.

26. Tung VC, Chen LM, Allen MJ, Wassei JK, Nelson K, Kaner RB, Yang Y: Low-temperature solution processing of graphene-carbon nanotube hybrid materials for high-performance transparent conductors. *Nano Lett* 2009, 9(5):1949–1955.

27. Zirkl M, Haase A, Fian A, Schön H, Sommer C, Jakopic G, Leising G, Stadlober B, Graz I, Gaar N, Schwödiauer Gogonea RSB, Bauer S: Low-voltage organic thin-film transistors with high-k nanocomposite gate dielectrics for flexible electronics and optothermal sensors. *Adv Mater* 2007, 19(17):2241–2245.

28. Koeppe R, Bartu P, Bauer S, Sariciftci NS: Light- and touch-point localization using flexible large area organic photodiodes and elastomer waveguides. *Adv Mater* 2009, 21(34):3510–3514.

29. Lin HT, Pei Z, Chen JR, Kung CP, Lin YC, Tseng CM, Chan YJ: A 16-byte nonvolatile bistable polymer memory array on plastic substrates. *IEEE Int Eletron Devices Meet Tech Dig* 2007, 1/2: 233–236.

30. Kinoshita K, Okutani T, Tanaka H, Hinoki T, Agura H, Yazawa K, Ohmi K, Kishida S: Flexible and transparent ReRAM with GZO memory layer and GZO-electrodes on large PEN sheet. *Solid-State Electronics* 2011, 58(1):48–53.

31. Hubert A, Nowak E, Tachi K, Maffini-Alvaro V, Vizioz C, Arvet C, Colonna JP, Hartmann JM, Loup V, Baud L, Pauliac S, Delaye V, Carabasse C, Molas G, Ghibaudo G, De Salvo B, Faynot O, Ernst T: A stacked SONOS technology, up to 4 levels and 6 nm crystalline nanowires, with Gate-All-Around or independent gates (φ-Flash), suitable for full 3D integration. In *IEDM Tech Dig: December 7–9 2009; Baltimore*. Piscataway: IEEE; 2009:637–640.

32. Yun JG, Lee JD, Park BG: 3D NAND flash memory with laterally-recessed channel (LRC) and connection gate architecture. *Solid-State Electron* 2011, 55(1):37–43.

33. Mearian L: *Non-volatile memory's future is in software: new memory technology to serve dual roles of mass storage and system memory*. Computerworld: Framingham; 2012.

34. Eleftheriou E, Haas R, Jelitto J, Lantz MA, Pozidis H: Trends in storage technologies. *IEEE Data Eng Bull* 2010, 33(4):4–13.

35. Ji Y, Cho B, Song , Kim TW, Choe M, Kahng YH, Lee T: Stable switching characteristics of organic nonvolatile memory on a bent flexible substrate. *Adv Mater* 2010, 22(28):3071–3075.

36. Ballhaus W, Pagella A, Vogel C: *A Change of Pace for the Semiconductor Industry?*. Düsseldorf: Price Waterhouse Coopers; 2009.

37. Choi J, Seol KS: 3D approaches for non-volatile memory. In *IEEE Symp VLSI Tech Digest: June 14–16 2011, Honolulu*. Piscataway: IEEE; 2011:178–179.

38. Das R, Harrop P: *Printed, Organic & Flexible Electronics: Forecasts, Players & Opportunities 2013–2023*. Cambridge: IDTechEx; 2013.

39. Doe P: Options for adding memory and logic to printed or flexible electronics. [http://electroiq.com/blog/2013/07/options-for-adding-memory-and-logic-to-printed-or-flexible-electronics/]

40. Park KT, Kim D, Hwang S, Kang M, Cho H, Jeong Y, Seo Y, Jang J, Kim HS, Jung SM, Lee YT, Kim C, Lee WS: A 45nm 4Gb 3-dimensional double-stacked multi-level NAND flash memory with shared bitline structure. In *Digest of Technical Papers. IEEE International Solid-State Circuits Conference: February 3–7 2008; San Francisco*. California: IEEE; 2008:510–632.

41. Elliott J, Jung ES: Ushering in the 3D memory era with V-NAND. In *Flash Memory Summit: August 12–15 2013*. Santa Clara: Santa Clara Convention Center; 2013.

42. Innovative Research and Products Inc: *Advanced Solid-State Memory Systems and Products: Emerging Non-Volatile Memory Technologies, Industry Trends and Market Analysis*. Stamford: Innovative Research and Products Inc; 2010:146.

43. Yole Développement: *Emerging Non-volatile Memory*. Lyon: Yole Développement; 2013:275.

44. Global Industry Analysts, Inc: *Advanced Solid-State Memory Systems - Global Strategic Business Report*. San Jose: Global Industry Analysts, Inc; 2012:731.

45. Groeseneken G, Maes HE, Van Houdt J, Witters JS: Basics of nonvolatile semiconductor memory devices.

In *Nonvolatile Semiconductor Memory Technology*. Edited by: Brown WD, Brewer JE. New York: IEEE; 1998:1–88.

46. Nonvolatile memories for 90 nm SoC and beyond. Electronic Engineering (EE) Times 2003 [http://www.eetimes.com/document.asp?doc_id=1201899]

47. Arreghini A, Driussi F, Vianello E, Esseni D, van Duuren MJ, Golubovi'c DS, Akil N, van Schaijk R: Experimental characterization of the vertical position of the trapped charge in Si nitride-based nonvolatile memory cells. *IEEE Trans Electron Devices* 2008, 55(5):1211–1219.

48. Salvo BD, Gerardi C, Schaijk RV, Lombardo SA, Corso D, Plantamura C, Serafino T, Ammendola G, Duuren MV, Goarin P, Mei WY, Jeugd KVD, Baron H, Gély M, Mur P, Deleonibus S: Performance and reliability features of advanced nonvolatile memories based on discrete traps. *IEEE Trans Device Mater Rel* 2004, 4(3):77–389.

49. Toshiba NAND vs. NOR flash memory technology overview. Toshiba America Electronic Components Inc; 1–4 [http://umcs.maine.edu/~cmeadow/courses/cos335/Toshiba%20NAND_vs_NOR_Flash_Memory_Technology_Overviewt.pdf]

50. Micron Technology Inc: *Technical Note: NAND Flash 101: An Introduction to NAND Flash and How to Design It in to Your Next Product*. Boise: Micron Technology Inc; 2006:1–27.

51. Yan Y, Chen W, Fan R, Guo X, Guo H, Zhang F, Ding L, Zhang K, Lin D, Wang Y: Analysis of the TID induced failure modes in NOR and NAND flash memories. *IEEE Trans Nucl Sci*2013, 60(1):224–229.

52. Cooke J: *Flash Memory 101: An Introduction to NAND Flash*. Micron Technology Inc: Boise; 2006.

53. Lai SK: Brief history of ETOX NOR flash memory. *J Nanosci Nanotech* 2012, 12(10):7597–7603.

54. Jose ST, Pradeep C: Design of a multichannel NAND Flash memory controller for efficient utilization of bandwidth in SSDs. In *Proc Intern Multi-Conf Autom Computing Commun Control and Comp Sensing (IMAC4S): March 22–23 2013; Kottayam*. Piscataway: IEEE; 2013:235–923.

55. Zheng M, Tucek J, Qin F, Lillibridge M: Understanding the robustness of SSDs under power fault. In *11th USENIX Conference on File and Storage Technologies (FAST'13): February 12–15 2013: San Jose*. Berkeley: USENIX Association; 2013:271–284.

56. Crippa L, Micheloni R, Motta I, Sangalli M: *Nonvolatile memories: NOR vs. NAND Architectures: Memories in Wireless Systems. Signal Communication Technology*. Berlin, Heidelberg: Springer; 2008:29.

57. Tal A: *Two Flash Technologies Compared: NOR vs NAND. 91-SR-012–04–8 L REV. 1.0*. M-Systems Flash Disk Pioneers: Newark; 2002.

58. Hu CW, Chang TC, Tu CH, Chiang CN, Lin CC, Sze SM, Tseng TY: NiSiGe nanocrystals for nonvolatile memory devices. *Appl Phys Lett* 2009, 94(6):062102.

59. Fazio A: Flash memory scaling. *MRS Bull* 2004, 29: 814–817.

60. Tu CH: *Electrical Switching and Memory Behaviors in Organic-based Devices*. Ann Arbor: ProQuest LLC; 2008.

61. Prall K: Scaling non-volatile memory below 30 nm. In *22nd IEEE Non-volatile Semiconductor Memory Workshop: August 26–30 2007; Monterey*. Piscataway: IEEE; 2007:5–10.

62. Nicollian PE, Rodder M, Grider DT, Chen P, Wallace RM, Hattangady SV: Low voltage stress-induced-leakage-current in ultrathin gate oxides. In *37th IEEE Annual Intern Reliab Phys Symp: March 23–25 1999: San Diego*. Piscataway: IEE; 1999:400.

63. King Y: Thin dielectric technology and memory devices. In *PhD dissertation*. University of California, Berkeley; 1999:60–05. 2265 2265

64. Brewer JE, Gill M: *Nonvolatile Memory Technologies with Emphasis on Flash: A Comprehensive Guide to Understanding and Using Flash Memory Devices*. Hoboken: Wiley; 2008:750.

65. Zhao W, Dong G, Sun H, Zheng N, Zhang T: Reducing latency overhead caused by using LDPC codes in NAND flash memory. *EURASIP J Adv Signal Proc* 2012, 203: 1–9.

66. Ielmimi D: Reliability issues and modeling of flash and post-flash memory. *Microelectron Enger* 2009, 86: 1870–1875.

67. Cappelletti P, Golla C, Olivo P, Zanoni E: *Flash Memories*. New York: Springer; 1999.

68. Ramkumar K, Prabhakar V, Kouznetsov I, Geha S: *Cypress SONOS Technology*. San Jose: Cypress Semiconductor; 2013:1–13.

69. White MH, Adams DA, Bu J: On the go with SONOS. *IEEE Circuit Devices Magazine*2000, 16(4):22–31.

70. You HC, Hsu TH, Ko FH, Huang JW, Yang WL, Lei TF: SONOS-type flash memory using an HfO_2 as a charge trapping layer deposited by the sol–gel spin-coating method. *IEEE Electron Device Lett* 2006, 27(8):653–655.

71. Sematech International: *Test equipment. The International Technology Roadmap for Semiconductors (ITRS)*. Austin: Sematech International; 2001:27–28.

72. Bu J, White MH: Design considerations in scaled SONOS nonvolatile memory devices.*Solid State Electron* 2001, 45: 113–120.

73. French ML, Chen CY, Sathianathan H, White MH: Design and scaling of SONOS multidielectric device for NVSM application. *IEEE Trans Comp Pack Manu Tech Part A*1994, 17(3):390–397.

74. Steimle RF, Muralidhar R, Rao R, Sadd M, Swift CT, Yater J, Hradsky B, Straub S, Gasquet H, Vishnubhotla L, Prinz EJ, Merchant T, Acred B, Chang K Jr, White BE Jr: Silicon nanocrystal non-volatile memory for embedded memory scaling. *Microelectron Reliab*2007, 47(4):585–592.

75. Hsu H, Lee JY: Electrical characterization of metal-oxide-high-k dielectric-oxide-semiconductor (MOHOS)

structures for memory applications. *Microelectron Reliab* 2007, 47(4):606–609.

76. Watanabe H, Aritome S, Hemink GJ, Maruyama T, Shirota R: Scaling of tunnel oxide thickness for flash EEPROMs realizing stress-induced leakage current reduction. In *Symp VLSI Tech Dig Tech Papers: June 7–9 1994; Honolulu*. Piscataway: IEEE; 1994:47–48.

77. Chowdhury NA, Misra D: Charge trapping at deep states in Hf–silicate based high-k gate dielectrics. *J Electrochem Soc* 2007, 154(2):G30-G37.

78. Yang Y, White MH: Charge retention of scaled SONOS nonvolatile memory devices at elevated temperatures. *Solid-State Electron* 2000, 44(6):949–958.

79. Xuan P, She M, Harteneck B, Liddle A, Bokor J, King TJ: FinFET SONOS flash memory for embedded applications. In *IEDM Tech Dig: December 8–10 2003; Washington, DC*. Piscataway: IEEE; 2003:609–613.

80. Sugizaki T, Kobayashi M, Ishidao M, Minakata H, Yamaguchi M, Tamura Y, Sugiyama Y, Nakanishi T, Tanaka H: Novel multi-bit SONOS type flash memory using a high-k charge trapping layer. In *VLSI Symp Tech Dig: June 10–12 2003; Kyoto*. Piscataway: IEEE; 2003:27–28.

81. Ostraat ML, De Blauwe JW, Green ML, Bell LD, Brongersma ML, Casperson J, Flagan RC, Atwater HA: Synthesis and characterization of aerosol silicon nanocrystal nonvolatile floating-gate memory devices. *Appl Phys Lett* 2001, 79(3):433–435.

82. Wong HSP: Emerging memory devices. In *Intern Semicond Dev Res Symp (ISDRS): December 7–9 2011; College Park*. Piscataway: IEEE; 2011:1.

83. Kryder MH, Kim CS: After hard drives—what comes next? *IEEE Trans Magn* 2009, 45(10):3406–3413.

84. Wang B: *Emerging Technology Analysis: The Future and Opportunities for Next-Generation Memory*. Gartner, Inc: Stamford; 2011.

85. Gallagher WJ, Parkin SSP: Development of the magnetic tunnel junction MRAM at IBM: from first junctions to a 16-Mb MRAM demonstrator chip. *IBM J Res & Dev* 2006, 50(1):5–23.

86. Julliere M: Tunneling between ferromagnetic films. *Physics Letters A* 1975, 54(3):225–226.

87. Wang D, Tondra M, Pohm AV, Nordman C, Anderson J, Daughton JM, Black WC: Spin dependent tunneling devices fabricated for magnetic random access memory applications using latching mode. *J Appl Phys* 2000, 87(9):6385–6387.

88. Daughton JM: Magnetoresistive memory technology. *Thin Solid Films* 1992, 216(1):162–168.

89. Naji PK, Durlam M, Tehrani S, Calder J, Deherrera MF: A 256 kb 3.0 V 1T1MTJ nonvolatile magnetoresistive RAM. In *IEEE Intern Solid-State Circ Conf Dig Tech Papers: February 7 2001; San Francisco. Volume 438*. Piscataway: IEEE; 2001:122–123.

90. Tehrani S, Engel B, Slaughter JM, Chen E, Deherrera M, Durlam M, Naji P, Whig R, Jenesky J, Calder J: Recent developments in magnetic tunnel junction MRAM. *IEEE Trans Magn* 2000, 36(5):2752–2757.

91. Everspin Technologies-MR2A16AYS35-MRAM 4MB. 2009 [http://www.everspin.com/PDF/EST_MR2A16A_prod.pdf]

92. Chen E, Apalkov D, Diao Z, Driskill-Smith A, Druist D, Lottis D, Nikitin V, Tang X, Watts S, Wolf SA, Ghosh AW, Lu JW, Poon SJ, Stan M, Butler WH, Gupta S, Mewes CKA, Mewes T, Visscher PB: Advances and future prospects of spin-transfer torque random access memory. *IEEE Trans Magn* 2010, 46(6):1873–1878.

93. Khvalkovskiy AV, Apalkov D, Watts S, Chepulskii R, Beach RS, Ong A, Tang X, Driskill-Smith A, Butler WH, Visscher PB, Lottis D, Chen E, Nikitin V, Krounbi M: Basic principles of STT-MRAM cell operation in memory arrays. *J Phys D Appl Phys* 2013, 46(7):074001.

94. Kawahara T, Ito K, Takemura R, Ohno H: Spin-transfer torque RAM technology: review and prospect. *Microelectron Reliab* 2012, 52(4):613–627.

95. Hosomi M, Yamagishi H, Yamamoto T, Bessho K, Higo Y, Yamane K, Yamada H, Shoji M, Hachino H, Fukumoto C, Nagao HH, Kano H: A novel nonvolatile memory with spin torque transfer magnetization switching: spin-ram. In *Int Elec Dev Meet (IEDM) Tech Dig: December 5 2005; Washington, DC*. Piscataway: IEEE; 2005:459–462.

96. Apalkov D, Khvalkovskiy A, Watts S, Nikitin V, Tang X, Lottis D, Moon K, Luo X, Chen E, Ong A, Driskill-Smith A, Krounbi M: Spin-transfer torque magnetic random access memory (STT-MRAM). *ACM J Emerg Tech Comp Sys (JETC)* 2013, 9(2):13:2–13:35.

97. Auciello O, Scott JF, Ramesh R: The physics of ferroelectric memories. *Phys Today* 1998, 51: 22–27.

98. Ramtron: *Ramtron Announces 8-Megabit Parallel Nonvolatile F-RAM Memory*. Colorado Springs: Ramtron International Corporation; 2009.

99. Ramtron: *F-RAM Technology Brief*. Colorado Springs: Ramtron International Corporation; 2007.

100. Moise TS, Summerfelt SR, McAdams H, Aggarwal S, Udayakumar KR, Celii FG, Martin JS, Xing G, Hall L, Taylor KJ, Hurd T, Rodriguez J, Remack K, Khan MD, Boku K, Stacey G, Yao M, Albrecht MG, Zielinski E, Thakre M, Kuchimanchi S, Thomas A, McKee B, Rickes J, Wang A, Grace J, Fong J, Lee D, Pietrzyk C, Lanham R, *et al.*: Demonstration of a 4 Mb, high density ferroelectric memory embedded within a 130 nm, 5 LM Cu/FSG logic process. In *Int Elec Dev Meet (IEDM) Tech Dig: December 8–11 2002; San Francisco*. Piscataway: IEEE; 2002:535–538.

101. Toshiba develops world's highest bandwidth, highest density non-volatile RAM. 2009 [https://www.toshiba.co.jp/about/press/2009_02/pr0902.htm]

102. Hudgens S, Johnson B: Overview of phase-change chalcogenide nonvolatile memory technology. *MRS Bull* 2004, 29(11):829–832.

103. Pirovano A, Lacaita AL, Benvenuti A, Pellizzer F, Hudgens S, Bez R: Scaling analysis of phase-change memory technology. In *Int Elec Dev Meet (IEDM) Tech Dig: December 8–10 2003; Washington, DC*. Piscataway: IEEE; 2003:29.6.1–29.6.4.

104. Ahn SJ, Song YJ, Jeong CW, Shin JM, Fai Y, Hwang YN, Lee SH, Ryoo KC, Lee SY, Park JH, Horii H, Ha YH, Yi JH, Kuh BJ, Koh GH, Jeon GT, Jeong HS, Kim K, Ryu BI: Highly manufacturable high density phase change memory of 64 Mb and beyond. In *Int Elec Dev Meet (IEDM) Tech Dig: December 13–15 2004; San Francisco*. Piscataway: IEEE; 2004:907–910.

105. Raoux S, Burr GW, Breitwisch MJ, Rettner CT, Chen YC, Shelby RM, Salinga M, Krebs D, Chen SH, Lung HL, Lam CH: Phase-change random access memory: a scalable technology.*IBM J Res & Dev* 2008, 52(4/5):465–479.

106. Ryu SW, Oh JH, Choi BJ, Hwang SY, Hong SK, Hwang CS, Kim HJ: SiO incorporation effect in $Ge_2Sb_2Te_5$ films prepared by magnetron sputtering for phase change random access memory devices. *Electrochem Solid-State Lett* 2006, 9(8):G259-G261.

107. Raoux S, Salinga M, Jordan-Sweet J, Kellock A: Effect of Al and Cu doping on the crystallization properties of the phase change materials SbTe and GeSb. *J Appl Phys* 2007, 101(4):44909–44915.

108. Atwood G, Bez R: Current status of chalcogenide phase change memory. In *63rd Device Research Conf Dig (DRC '05): June 22 1005; Santa Barbara. Volume 1.* Piscataway: IEEE; 2005:29–33.

109. Intel N: *Intel News Release: STMicroelectronics Deliver Industry's First Phase Change Memory Prototypes.* Intel: Santa Clara; 2008.

110. Wire B: *Samsung Electronics and Numonyx Join Forces on Phase Change Memory*. San Francisco: Business Wire; 2009.

111. Chua L: Memristor—the missing circuit element. *IEEE Trans Circuit Theory* 1971, 18(5):507–519.

112. Kim S: Resistive RAM (ReRAM) Technology for High Density Memory Applications. In *4th Workshop Innovative Memory Technol MINATEC 2012; June 21–24 2012*. Grenoble: MINATEC; 2012:4.

113. Strukov DB, Snider GS, Stewart DR, Williams RS: The missing memristor found. *Nature*2008, 453: 80–83.

114. Xu C, Dong X, Jouppi NP, Xie Y: Design implications of memristor-based RRAM cross-point structures. In *Desig Auto Test Eur Conf Exhib (DATE): March 14–18 2011; Grenoble*. Piscataway: IEEE; 2011:1–6.

115. Seo S, Lee MJ, Seo DH, Jeoung EJ, Suh DS, Joung YS, Yoo IK, Hwang IR, Kim SH, Byun IS, Kim JS, Choi JS, Park BH: Reproducible resistance switching in polycrystalline NiO films. *Appl Phys Lett* 2004, 85(23):5655–5657.

116. Wu MC, Lin YW, Lin CH, Tseng TY: Low-power and highly reliable multilevel operation in ZrO_2 1T1R RRAM. *IEEE Electron Dev Lett* 2011, 32(8):1026–1028.

117. Huang CY, Huang CY, Tsai TL, Lin CA, Tseng TY: Switching mechanism of double forming process phenomenon in ZrO_x/HfO_y bilayer resistive switching structure with large endurance. *Appl Phys Lett* 2014, 104(6):062901(4).

118. Lin MH, Wu MC, Lin CH, Tseng TY: Effects of vanadium doping on resistive switching characteristics and mechanisms of $SrZrO_3$-based memory films. *IEEE Trans Electron Devices* 2010, 57(8):1801–1808.

119. Yang JJ, Pickett MD, Li X, Ohlberg DAA, Stewart DR, Williams RS: Memristive switching mechanism for metal/oxide/metal nanodevices. *Nat Nanotech* 2008, 3: 429–433.

120. Kamiya K, Yang MY, Park SG, Magyari-Köpe B, Nishi Y, Niwa M, Shiraishi K: ON-OFF switching mechanism of resistive–random–access–memories based on the formation and disruption of oxygen vacancy conducting channels. *Appl Phys Lett* 2012, 100: 073502.

121. Lee JS: Progress in non-volatile memory devices based on nanostructured materials and nanofabrication. *J Mater Chem* 2011, 21: 14097–14112.

122. Wang SY, Tseng TY: Interface engineering in resistive switching memory. *J Adv Dielectrics*2011, 1(2):141–162.

123. Lin CY, Wu CY, Wu CY, Hu C, Tseng TY: Modified resistive switching behavior of ZrO_2memory films based on the interface layer formed by using Ti top electrode. *J Appl Phys*2007, 102(9):094101–094105.

124. Goux L, Fantini A, Kar G, Chen YY, Jossart N, Degraeve R, Clima S, Govoreanu B, Lorenzo G, Pourtois G, Wouters DJ, Kittl JA, Altimime L, Jurczak M: Ultralow sub-500nA operating current high-performance $TiNAl_2O_3HfO_2HfTiN$ bipolar RRAM achieved through understanding-based stack-engineering. In *VLSI Tech Dig: June 12–14 2002; Honolulu*. Piscataway: IEEE; 2012:159–160.

125. Sawa A: Resistive switching in transition metal oxides. *Mater Today* 2008, 11(6):28–36.

126. Seo YJ, An HM, Kim HD, Kim TG: Improved performance in charge-trap-type flash memories with an Al_2O_3 dielectric by using bandgap engineering of charge-trapping layers.*J Korean Phys Soc* 2009, 55(6):2679–2692.

127. Prakash A, Ouyang J, Lin JL, Yang Y: Polymer memory device based on conjugated polymer and gold nanoparticles. *J Appl Phys* 2006, 100(5):054309–054314.

128. Ling QD, Lim SL, Song Y, Zhu CX, Chan DSH, Kang ET, Noeh KG: Nonvolatile polymer memory device based on bistable electrical switching in a thin film of poly(N-vinylcarbazole) with covalently bonded C_{60}. *Langmuir* 2007, 23(1):312–319.

129. Moller S, Perlov C, Jackson W, Taussing C, Forrest SR: A polymer/semiconductor write-once read-many-times memory. *Nature* 2003, 426: 166–169.

130. Stikeman A: Polymer memory: computer memory could soon earn the ultimate commercial validation: the cheap plastic knock-off. *MIT Techno Rev* 2002.

131. Hubert A, Schäfer R: *Magnetic Domains: The Analysis of Magnetic Microstructures*. Heidelberg, Berlin: Springer; 1998.

132. Parkin SSP, Jiang X, Kaiser C, Panchula A, Roche K, Samant M: Magnetically engineered spintronic sensors and memory. *Proc IEEE* 2003, 91(5):661–680.

133. Parkin SSP, Hayashi M, Thomas L: Magnetic domain-wall racetrack memory. *Science*2008, 320(5873):190–194.

134. Parkin SSP: *(2004–2007) U.S. Patents 6,834,005, 6,898,132, 6,920,062, 7,031,178, and 7,236,386.*

135. Shipway AN, Katz E, Willner I: Molecular memory and processing devices in solution and on surfaces. *Structure and Bonding* 2001, 99: 237–281.

136. Nonvolatile molecular memory. Ames Research Center, National Aeronautics and Space Administration; 2008 [http://www.nasa.gov/centers/ames/research/technology-onepagers/nonvolatile_memory.html]

137. Islam MS, Li Z, Chang SC, Ohlberg DAA, Stewart DR, Wang SY, Williams RS, Stanley R: Dramatically improved yields in molecular scale electronic devices using ultra-smooth platinum electrodes prepared by chemical mechanical polishing. In *Proc 5th IEEE Conf Nanotech: July 11–15 2005; Nagoya. Volume 1.* Piscataway: IEEE; 2005:80–83.

138. Oliveira ON Jr: Langmuir-Blodgett films - properties and possible applications. *Braz J Phys* 1992, 22(2):60–69.

139. Plafke J: Individual molecules: storage devices of the future. [http://www.extremetech.com/computing/146690-individual-molecules-storage-devices-of-the-future]

140. Lee SH, Jung Y, Agarwal R: Highly scalable non-volatile and ultra-low-power phase-change nanowire memory. *Nat Nanotech* 2007, 2: 626–630.

141. Lieber CM: Nanoscale science and technology: building a big future from small things.*MRS Bull* 2003, 28(7):486–491.

142. Agarwal R, Lieber CM: Semiconductor nanowires: optics and optoelectronics. *Appl Phys A Mater Sci Proc* 2006, 85(3):209–215.

143. Reed M, Chen J, Rawlett AW, Price DW, Tour JM: Molecular random access memory cell.*Appl Phys Lett* 2001, 78(23):3735–3737.

144. Li C, Zhang D, Liu X, Han S, Tang T, Zhou C, Fan W, Koehne J, Han J, Meyyappan M, Rawlett AW, Price DW, Tour JM: Fabrication approach for molecular memory arrays. *Appl Phys Lett* 2003, 82(4):645–647.

145. Li C, Lei B, Fan W, Zhang D, Meyyappan M, Zhou C: Molecular memory based on nanowire–molecular wire heterostructures. *J Nanosci Nanotechnol* 2007, 7(1):138–150.

146. Atwood A: Future directions and challenges for ETox flash memory scaling. *IEEE Trans Device Mater Reliab* 2004, 4(3):301–305.

147. Govoreanu B, Brunco DP, Van Houdt J: Scaling down the interpoly dielectric for next generation Flash memory: challenges and opportunities. *Solid State Electron* 2005, 49(11):1841–1848.

148. Patel P: Computing News. Bendable memory made from nanowire transistors. *MIT Technol Rev Rep* 2010.

149. Li Q, Zhu X, Xiong HD, Koo SM, Ioannou DE, Kopanski JJ, Suehle JS, Richter CA: Silicon nanowire on oxide/nitride/oxide for memory application. *Nanotechnology* 2007, 18: 235204.

150. Sohn JI, Choi SS, Morris SM, Bendall JS, Coles HJ, Hong WK, Jo G, Lee T, Welland ME: Novel nonvolatile memory with multibit storage based on a ZnO nanowire transistor. *Nano Lett* 2010, 10(11):4316–4320.

151. Kang JW, Lee JH, Lee JH, Hwang HJ: A study on carbon nanotube bridge as a electromechanical memory device. *Physica E* 2005, 27(3):332–340.

152. Nantero, imec collaborate on development of carbon-nanotube-based memory. Nantero: Nantero Inc; 2012 [http://www.azom.com/news.aspx?newsID=34644#]

153. Rueckes T, Kim K, Joselevich E, Tseng GY, Cheung CL, Lieber CM: Carbon nanotube-based nonvolatile random access memory for molecular computing. *Science* 2000, 289(5476):94–97.

154. Cui JB, Sordan R, Burghard M, Kern K: Carbon nanotube memory devices of high charge storage stability. *Appl Phys Lett* 2002, 81(17):3260–3262.

155. Vettiger P, Cross G, Despont M, Drechsler U, Dürig U, Gotsmann B, Häberle W, Lantz MA, Rothuizen HE, Stutz R, Binnig GK: The "millipede"—nanotechnology entering data storage.*IEEE Trans Nanotechnol* 2002, 1(1):39–55.

156. Walsh C: IBM's Millipede. [http://www.ewp.rpi.edu/hartford/~walshc2/FWM/Research%20Project/Final/IBM%27s%20Millipede.pdf]

157. Gotsmann B, Knoll AW, Pratt R, Frommer J, Hedrick JL, Duerig U: Designing polymers to enable nanoscale thermomechanical data storage. *Adv Funct Mater* 2010, 20(8):1276–1284.

158. Jones MR, Osberg KD, Macfarlane RJ, Langille MR, Mirkin CA: Templated techniques for the synthesis and assembly of plasmonic nanostructures. *Chem Rev* 2011, 111(6):3736–3827.

159. Hagen JA, Li W, Steckl AJ, Grote JG: Enhanced emission efficiency in organic light-emitting diodes using deoxyribonucleic acid complex as an electron blocking layer. *Appl Phys Lett* 2006, 88: 171109.

160. Yu Z, Li W, Hagen JA, Zhou Y, Klotzkin D, Grote JG, Steckl AJ: Photoluminescence and lasing from deoxyribonucleic acid (DNA) thin films doped with sulforhodamine. *Appl Opt*2007, 46(9):1507–1513.

161. Hung YC, Hsu WT, Lin TY, Fruk L: Photoinduced write-once read-many-times memory device based on DNA biopolymer nanocomposite. *Appl Phys Lett* 2011, 99: 253301.

162. Monson CF, Woolley AT: DNA-templated construction of copper nanowires. *Nano Lett*2003, 3(3):359–363.

163. Pashley RD, Lai SK: Flash memories: the best of two worlds. *IEEE Spectrum* 1989, 26: 30–33.

164. Pavan P, Bez R, Olivo P, Zanoni E: Flash memory cells-an overview. *Proc IEEE* 1997, 85(8):1248–1271.

165. Geller M, Marent A, Nowozin T, Bimberg D, Akcay N, Öncan N: A write time of 6 ns for quantum dot-based memory structures. *Appl Phys Lett* 2008, 92(9):092108.

166. Cui K, Ma W, Zhang Y, Huang J, Wei Y, Cao Y, Guo X, Li Q: 540-meV hole activation energy for GaSb/GaAs quantum dot memory structure using AlGaAs barrier. *IEEE Elect Dev Lett* 2013, 34(6):759–761.

167. Tiwari S, Rana F, Hanafi H, Hartstein A, Crabbe EF, Chan K: A silicon nanocrystals based memory. *Appl Phys Lett* 1996, 68(10):1377–1379.

168. 1Manna S, Aluguri R, Katiyar A, Das S, Laha A, Osten HJ, Ray SK: MBE-grown Si and $Si_{1-x}Ge_x$ quantum dots embedded within epitaxial Gd_2O_3 on substrate for floating gate memory device. *Nanotechnology* 2013, 24: 505709.

169. 16Geller MP: Investigation of carrier dynamics in self-organized quantum dots for memory devices. *PhD thesis*. Technical University of Berlin; 2007.

170. Jung SM, Jang J, Cho W, Cho H, Jeong J, Chang Y, Kim J, Rah Y, Son Y, Park J, Song MS, Kim KH, Lim JS, Kim K: Three dimensionally stacked NAND Flash memory technology using stacking single crystal Si layers on ILD and TANOS structure for beyond 30 nm node. In *Proc Intern Elect Devices Meet (IEDM): December 11–13 2006; San Francisco*. Piscataway: IEEE; 2006.

171. Wong S, El-Gamal A, Griffin P, Nishi Y, Pease F, Plummer J: Monolithic 3D integrated circuits. In *IEEE Inter Symp VLSI Technol Systems and Applications (VLSI-TSA): April 23–25 2007; Hsinchu*. Piscataway: IEEE; 2007:1–4.

172. Vyyoda MA: *3-Dimensional Monolithic Nonvolatile Memories and the Future of Solid-State Data Storage*. Milpitas: SanDisk Corporation; 2008.

173. Wager JF: Transparent electronics. *Science* 2003, 300(5623):1245–1246.

174. Kiriakidis G, Nathan A, Keszler DA, Fortunato E, Nomura K: E-MRS 2011 Fall Symposium I: Advances in Transparent Electronics, from Materials to Devices III. *IOP Conf Ser Mater Sci Eng* 2012, 34: 011001.

175. Seo JW, Park JW, Lim KS, Kang SJ, Hong YH, Yang JH, Fang L, Sang GY, Kim HK: Transparent flexible resistive random access memory fabricated at room temperature. *Appl Phys Lett* 2009, 95: 133508.

176. Transparent, flexible 3-D memory chips may be the next big thing in small memory devices. Featured Research. American Chemical Society 2012 [http://www.kurzweilai.net/transparent-flexible-3d-memory-chips-may-be-the-next-big-thing-in-small-memory-devices]

177. Zangeneh M, Joshi A: Performance and energy models for memristor-based 1T1R RRAM cell. In *Proc. Great Lakes Symposium on Very Large Scale Integration (GLSVLSI): May 3–4 2012; Salt Lake City*. Utah: ACM; 2012:9–14.

178. Zangeneh M, Joshi A: Design and optimization of nonvolatile multibit 1T1R resistive RAM.*IEEE Trans VLSI System* 2014, 22(8):1815–1822.

CHAPTER 12 PROBE-BASED STORAGE TECHNOLOGIES

Lei Wang, Ci Hui Yang, Jing Wen, Si Di Gong and Yuan Xiu Peng

INTRODUCTION

Today, the prosperity of social networks such as Youtube, Twitter, and Facebook, in conjunction with the digitalization of the daily service of worldwide citizens, has triggered a radical increase on the total amount of global digital data. According to IDC's report, the amount of digital data has already surpassed 4.4 ZettaBytes (ZB) in 2013, and is predicted to reach 44 ZettaBytes in 2020 [1]. Under this situation, the capacities of various data storage devices where the digital data can be recorded and replayed are conversely limited in TeraBytes (TB) regime, lagging far behind the phenomenal growth of the digital data. The conventional mass storage devices can be categorized into hard disk, optical disc, and flash memory. Today, the hard disk drive (HDD) is undoubtedly the most famous mainstream storage device mainly due to its high capacity, fast data rate, and low cost. Nevertheless, in order to satisfy the current storage demand, the magnetic grain inside the HDD needs to be further shrunk so as to enhance the resulting density. However, the reduction on the volume of the magnetic grain would deteriorate the thermal stability of the magnetic written bits whose magnetization may be reversed without magnetic field due to the thermal fluctuation. This is well-known as superparamagnetic limit [2, 3]. The evolution of the optical disc has experienced three generations ranging from compact disc (CD) to Blu-ray disc. The major hurdle to prevent optical disc from booming its capacity results from the fact that the diameter of the laser spot focused on the optical disc is approximately proportional to the wavelength of the laser beam and inversely proportional to the numerical aperture (NA) of the objective lens [4, 5]. In this case, the storage capacity of the optical disc can seemingly be enhanced by either reducing the beam wavelength or increasing the NA. However, as there is no breakthrough progress recently made in the technologies of solid state laser and objective lens, the shortest wavelength and the largest NA to date are still limited to 405 and 0.85 nm for optical recording [6], which is usually referred to optical diffraction. Flash memory, represented by universal serial bus (USB) drive and iPhone series products, has recently received numerous attentions because of its high capacity, short data access time, non-volatility, portability, and low expense, which is expected to replace HDD in the near future. The capacity improvement of the flash memory strongly depends on the downscale of the flash cell size. In this case, further scaling down the cell size would result in a decrease on the thickness of the tunnel oxide layer usually sandwiched between the floating gate and wafer. A tunnel oxide layer with a very thin thickness is unable to prevent electrons from escaping from the floating gate, thus causing the loss of data, which is known as scaling limit [7,8]. Consequently, as HDD, optical disc, and flash memory are suffering from superparamagnetic limit, diffraction limit, and scaling limit, it is timely to explore more innovative mass storage devices with higher storage capacity, faster data rate, lower energy consumption, and longer data retention time than conventional counterparts, giving rise to the advent of probe-based storage devices.

A probe-based storage device can be defined as a storage memory that utilizes nanoscale probe to record and read bits in the storage medium. Prior to the probe-based storage device, nanoscale probe has been widely used for

Citation: Lei Wang, Ci Hui Yang, Jing Wen, Si Di Gong and Yuan Xiu Peng, Probe-based Storage Technologies, DOI 10.1186/s11671-016-1556-9.

scanning probe microscope (SPM) applications to map out the tomography image of the sample [9]. To achieve this, nanoscale probe usually makes use of a cantilever with a very sharp tip integrated on its end to perform the raster scanning in which tip is brought closely to or in contact with sample. During the scanning, the tip can be deflected up or down due to the variation of the atomic force between tip apex and sample surface, and tomography image of sample surface can therefore be constructed by measuring the position of the tip, as schematically shown in Fig. 1. For probe-based storage device, an external excitation that is initially applied to the nanoscale probe can be employed to change the physical properties of the storage medium when in contact with the probe tip during the scanning, and thereby the region subjected to the external excitation exhibits different physical properties from the region without suffering from the excitation. As a result, these two regions with different characteristics can be regarded as the binary bits "1" and "0", respectively, to realize the storage function. The principle of the probe-based storage device is illustrated in Fig. 2. According to the descriptions above, the size of the recorded bit is primarily determined by the size of the tip apex used for recording process, implying a fact that the implementation of tip apex on the order of nanometers would allow for an areal density over multi Terabit per square inch (Tbit/in^2). In addition, the small size of recorded bit can lead to an aggressive reduction on the required energy consumption. Thanks to these merits, worldwide researchers have been dedicated to exploring probe-based storage technologies using various storage materials and system architectures, and this gives rise to an emergence of a variety of innovative probe memories such as thermo-mechanical probe memory, magnetic probe memory, ferroelectric probe memory, and phase-change probe memory. In this case, it is essential to have a thorough review on the physical principles and current status of various probe-based storage devices to help scientists to understand the inherent physics of these devices. However, such review articles are rarely found according to Web of Science by the time of writing and the latest review article about probe storage technologies has been dated back to 2011 [10]. As a result, a comprehensive review concerning probe-based storage devices is urgently required so as to reveal the physical mechanisms of these devices, and thus to ignite the research enthusiasm to further mitigate their respective storage performances.

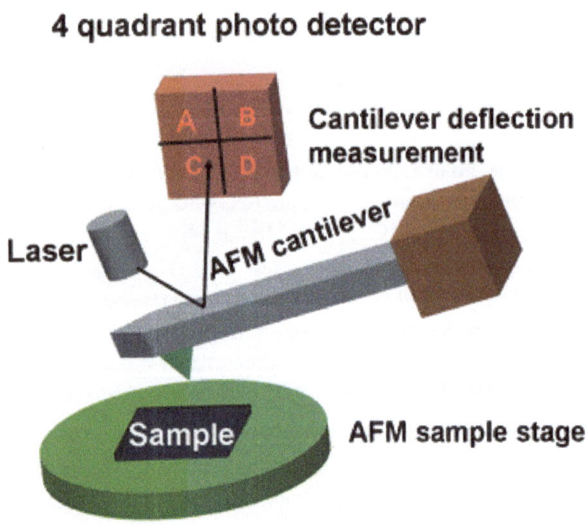

Figure 1: Schematic of AFM probe.

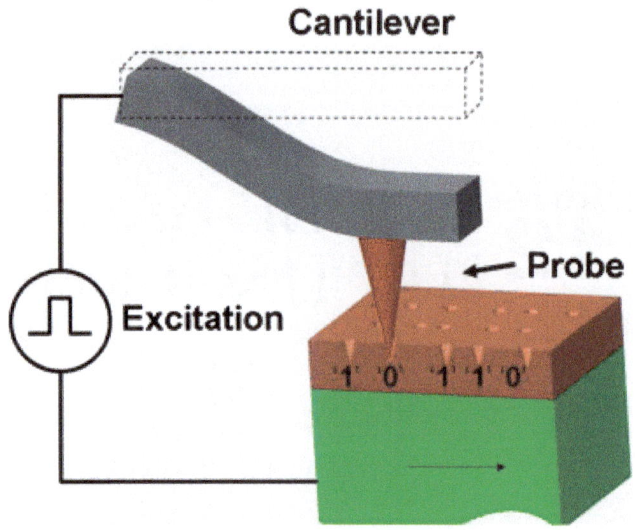

Figure 2: Principle of probe-based storage memory.

Thermo-Mechanical Probe Memory

The principle of thermo-mechanical probe memory is to pull a heated tip into the polymer medium to generate topographic change that can be denoted as binary digit "1" on medium surface [11]. The driving force behind the development of thermo-mechanical probe memory can be ascribed to the presence of IBM's Millipede system that consists of an array of SPM tips with a moving storage medium underneath [11], as shown in Fig. 3. A localized heater, usually made of low-doped resistive silicon is integrated at the tip base, acted as heating element. During the writing process, an electrostatic force is applied to the cantilever to push the tip that is heated around polymer's glass transition temperature (400 °C) into the polymer medium to soften it. The heated polymer medium would melt and allow tip to penetrate into it, thus generating an indentation on medium surface. This indentation and surrounding continuous flat surface can be used to represent "1" and "0", respectively. The erasing process can be achieved by removing the applied force to pull the tip up back from the indentation. This would result in a surface tension to flatten the previously written indentation [12]. During the readout process, a read resistor whose resistance strongly depends on the temperature is integrated in one of the side-arms of the three-legged cantilever design, which is heated to around 300 °C. When tip falls into an indentation, the distance between tip and substrate is reduced, thereby accelerating heat leaking through the substrate. In this case, the temperature of the read resistor is changed due to the cooling process, and this would change the resistance of the read sensor, which can be detected as readout signals. The writing, readout, and erasing modes of thermo-mechanical probe memory are schematically shown in Fig. 4.

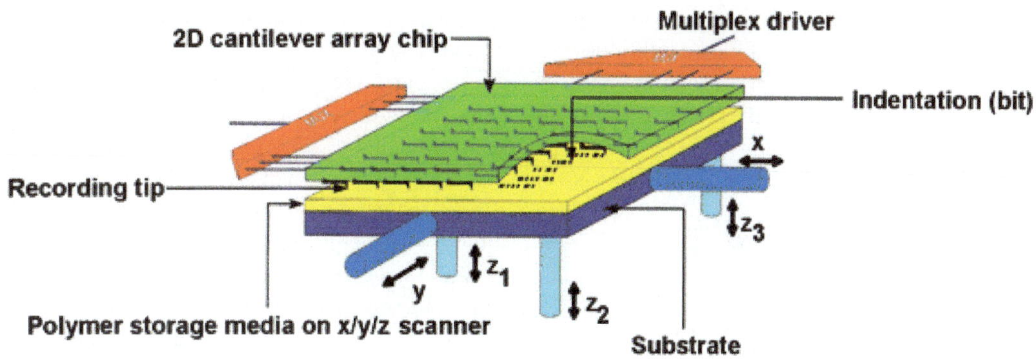

Figure 3: Millipede system. Reprinted with permission from [11].

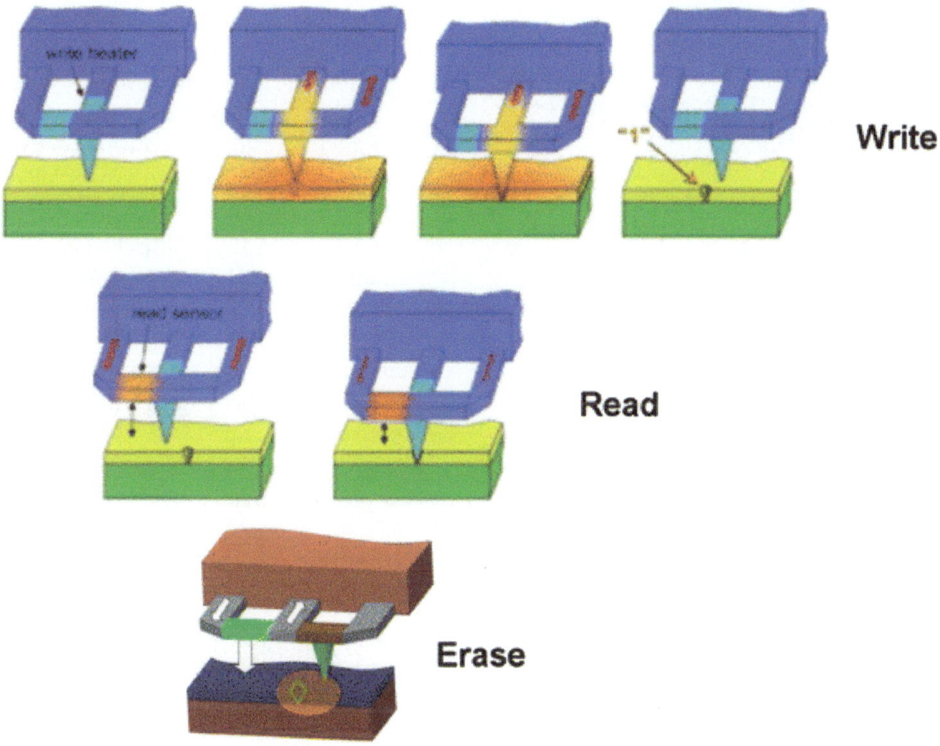

Figure 4: Millipede memory when operated in write mode, read mode, and erase mode. Reprinted with permission from [13].

The ability to simultaneously record and read bits using tip arrays obviously endows thermo-mechanical probe memory with some advantageous traits such as ultra-high recording density and high data/write speed when in comparison with conventional data storage devices [13–16]. In addition, the resulting density can be further boomed by reducing the bit pitch that is determined by the rim size of the written bit [17]. It has been found that either using a tip with small contact radius or reducing the indentation depth can effectively compact the rim size, and therefore allow for higher recording density. These exciting findings and the inherent advantages of device itself have recently led to some remarkable technological advances on thermo-mechanical probe memory. In perspective of the storage medium, a Diels-Adler (DA) polymer that has a highly cross-linked, high-molecular weight at low temperature was proposed [18]. This polymer has a high wear resistance with good thermal stability below the transition temperature, whereas it turns out to be easily deformable and gentle on the tip above the transition temperature. The feasibility of using DA polymer to achieve 1 Tb/in^2 has been demonstrated [18]. In addition to DA polymer, a polyaryletherketone (PAEK) polymer that unites phenyl-ethynyl groups in the backbone with diresorcinol units in the backbone for control of the glass-transition temperature was considered as another promising storage medium for thermo-mechanical probe memory due to its low glass-transition temperature less than 160 °C in the cross-linked state and superb thermal stability up to 450 °C [17]. These exceptional characteristics would give rise to a writing of indentation on a microsecond time scale as well as an effective minimization of thermal degradation for indentation with a hot tip. The capability of using a PAEK polymer templated on a cleaved mica surface to provide an areal density of 4 Tb/in^2, the endurance of 108 write cycles, and data retention time of 10 years at 85 °C has been demonstrated [19–21], as illustrated in Fig. 5. The possibility of achieving 9 Tb/in^2 based on the PAEK polymer was also theoretically predicted [22]. Another prospective material for thermo-mechanical recording as an alternative to polymer is NiTi shape-memory alloys (SMAs) [23]. NiTi is more immune to the normal environment effects at the shape-memory effect's normal operation temperature than the poly (methyl methacrylate) (PMMA) polymer that is apt to charring at the temperatures around 350 °C. Besides, the low operating temperature and high thermal conductivity of SMAs allows the device using SMAs to have a less energy consumption for writing and more rapid erasing operation than that with PMMA polymer. The practicality of using SMAs as the storage medium for thermo-mechanical probe storage to achieve 0.5 Tbit/in^2 has already been proven [23].

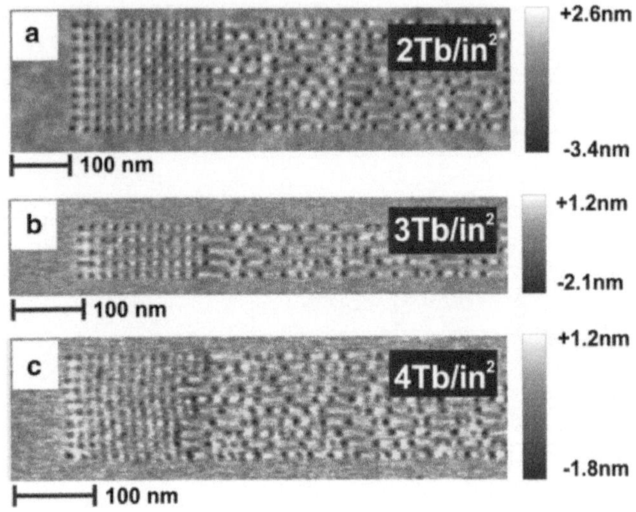

Figure 5: Topographic images of random data patterns used in the recording tests showing (**a**) 2 Tbit/in^2 on a normal spin coated samples, and (**b**) 3 Tbit/in^2, and (**c**) 4 Tbit/in^2 on an ultraflat templated sample. Reprinted with permission from [17].

As thousands of cantilevers scan repeatedly back and forth across the medium surface, the tip and sample wear become more evident than other storage devices, thus exacerbating the lifetime of the thermo-mechanical probe device as well as the resulting density. In order to suppress the wear from media's point of view, one possible approach is to introduce a photo-resistant layer between the silicon substrate and the PMMA polymer [12]. From the perspective of tip, tip wear can be aggressively mitigated by coating the tip with some hard materials. Recently, a moulded DLC tip designed based on atom by atom attrition mechanism was reported to give approximately four orders of magnitude improvement on wear compared to the carbon nanotube tip [24]. Such an improvement can be further enhanced by the use of SiC tip [25]. Another possibility to alleviate the tip wear is to actuate the tip with a periodic force at frequencies at or above the natural resonant frequency of the cantilever [26, 27]. In practice, it is preferable to operate AFM at intermittent-contact mode that can effectively reduce tip wear. However, the intermittent-contact mode that usually requires high cantilever stiffness is usually contradictory with the feeble cantilever used in thermo-mechanical storage to allow easy electrostatic actuation [28]. This can be solved by either utilizing amplitude modulation of the cantilever through electrostatic actuation [26] or slightly modulating the force on the tip-sample contact [28], allowing for an areal density of 1 Tb/in^2 attained even after a sliding distance of 140 m.

Magnetic Probe Memory

Magnetic probe memory exhibits an analogous recording/readout mechanism to the conventional HDD. A probe tip from magnetic force microscope (MFM) scans over the surface of the magnetic medium whose magnetization can be reversed by the magnetic field emanated from the magnetic tip, to accomplish the recording operation, as illustrated in Fig. 6. During readout mode, the MFM tip is brought in close to the previously written magnetic dots and thus deflected up and down due to the magnetic force existing between the tip and magnetic dot. Such a variation of the position of MFM tip can be detected as the readout signal. It is well-known from the Neel-Arrhenius relation that the ratio of the product of the crystalline anisotropy and the volume of the bit to the product of the Boltzmann constant and the absolute temperature should be greater than 50 so as to maintain the thermal stability of the magnetic dots [29]. In this case, the reduction on the bit volume for higher density would require a larger crystalline anisotropy, thereby requiring a much higher writing field from magnetic tip. Nevertheless, due to the low field emanated from the magnetic tip, the crystalline anisotropy of the storage medium for magnetic probe memory needs to remain low to ensure the successful magnetization reversal while at the expense of the resulting recording density. For this reason, the magnetic probe memory has initially shown a relatively low recording density down to 200 Gb/in^2, resulting from a vacuum MFM on Co/Pt multi-layered dots with perpendicular anisotropy [30, 31]. Fortunately, several methods to strongly enhance the writing field of MFM probes have been proposed recently, one of which is to applying an external field on a small coil mounted below the magnetic medium [32–38]. As can be seen from Fig. 7, the magnetization of the recorded film can be reversed by the superimposition of the writing field directly from MFM tip and the external field from the coil, leading to a density possibly up to 1.2 Tb/in^2 by mean of a 30 nm tip radius [37, 38]. An alternative to achieve the magnetization reversal in the absence of the external field is to locally heat the medium using a tunneling current from scanning tunneling microscope (STM) probe to lower its crystalline anisotropy [39–43], as illustrated in Fig. 8. Thanks to this, the magnetization of the heated region can be switched by the demagnetization field of the surrounding film. This technique results in a variety of bit size on Co/Pt multi layers ranging from 800 to 170 nm probably due to the variations on medium itself and tip radius [44–48]. However, if the writing is performed using STM tip, the subsequent readout of the written bits can only be read by spin-polarized tunneling. In this case, MFM tip that is operated in field-emission mode is implemented instead of STM, by which a minimum bit size with 80 nm is secured with the assist of a pulse background field [32]. In addition to the methods above, such as heat-assisted magnetic probe recording technology can be realized by an incorporation of a heated AFM tip with a Co/Pt multilayer patterned medium prepared by sputtering on pillars of 90 nm diameter [49], whereas the readout manner for this technique remain unreported. Such a nano-indentation technique was recently applied to a ferromagnetic shape memory alloy (FSMA) to achieve four magnetic-based memory states using MFM probe due to magnetic field or stress-induced twin rearrangement along two crystal orientations [50].

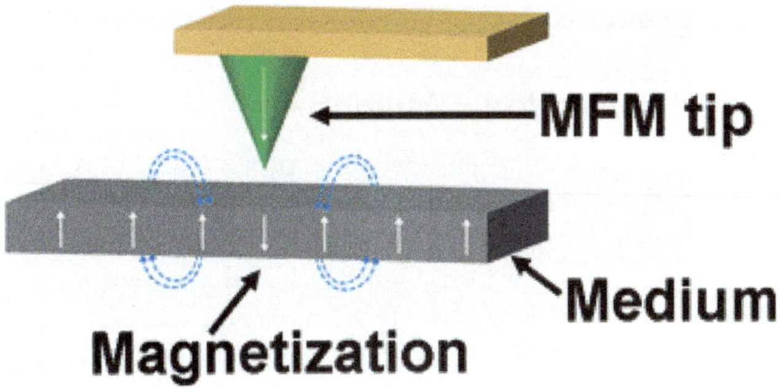

Figure 6: Magnetic probe memory system.

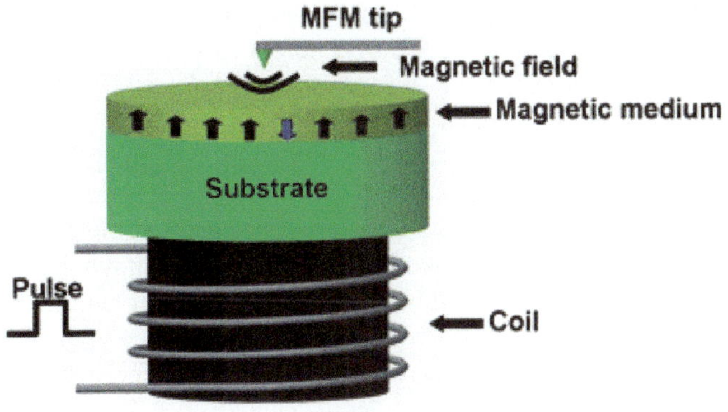

Figure 7: Magnetic probe memory using an external field.

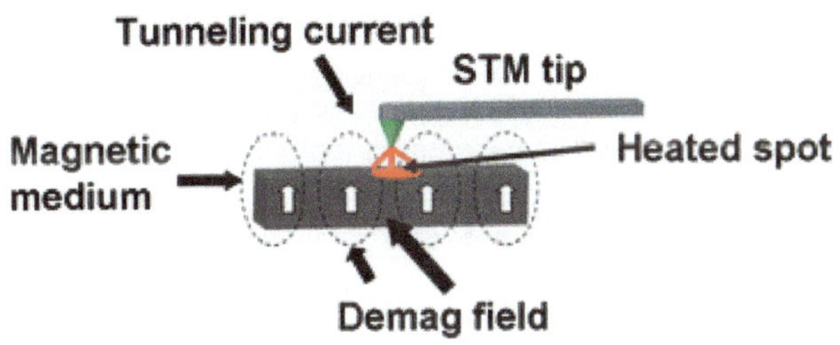

Figure 8: Magnetic probe memory using STM.

In terms of readout process, a short tip-sample distance on the order of 10 nm or less is required in order to obtain a readout image with high resolution. However, for MFM tip operating within such a small distance, non-magnetic tip-sample interactions play a more important role and would introduce unwanted topographic interference. This drawback can be overcome by integrating a magneto-resistive sensor on the magnetic tip [51–55]. A cantilever integrated with a spin-valve sensor was recently reported, and leads to a resolution around 1 μm, which is somewhat inadequate for probe-based storage [56]. However, it is possible to incorporate the magnetic field sensor used for modern HDDs that can readily give a resolution below 20 nm with the magnetic probe memory. In addition to this, another novel approach making use of dual tips for both topography and magnetic field imaging has been recently proposed [57]. According to this method, a cantilever is cut in half using the focused ion beam technique, whereby two different tips, one magnetized tip and one non-magnetized tip, are implemented to tackle the topography imaging and magnetic imaging, respectively. This dual-tip technique can radically suppress the perturbations of the magnetic tip as compared to the standard MFM methods.

Ferroelectric Probe Memory

The storage medium of ferroelectric probe memory is ferroelectric material showing a non-linear relationship between the applied electric field and the store charges. As illustrated in Fig. 9, applying an external electric field to the ferroelectric material can align its dipole with the field direction due to the small shifts in the position of atoms associated with shifts in the distributions of electronic charge in the crystal structure, whereas removing the electric field would drive dipole back to its original polarization state. Such a relationship can be simply interpreted by a hysteretic loop, as shown in Fig. 10. Therefore, these two distinct polarizations of dipole can be used to represent binary digits. During the recording process, a conductive probe where an external field is applied is brought in contact with the ferroelectric medium whose polarization can be switched according to the direction of the applied field. The readout process of ferroelectric probe memory adopts a very different mechanism from other probe storage memories, as its readout operation can only be conducted under "0" state. Consequently, no readout signal is detected for the case of reading "0" state. In contrast, to read "1" state, a current pulse is introduced to induce the polarization re-orientation, which is regarded as the readout signal. The readout process of ferroelectric probe memory apparently requires an erasing of the previously written data, which is known as "destructive readout" [58]. It should be kept in mind that ferroelectric materials, exemplified by lead zirconate titanate (PZT), has exhibited much higher energy density than the highest ever reported energy density for magnetic materials [59, 60]. Thanks to this, an unwanted thermal fluctuation effect induced by ambient thermal energy for ferroelectric materials would occur at a much higher density limit than magnetic materials. The density up to 3.6 Tb/in^2 on an atomically smooth PZT medium has been achieved [61]. The contrast in the positive and negative poling region can be significantly mitigated by either carefully controlling the deposition temperature or using the PZT films exposed to an ion beam using an electron cyclotron resonance (ECR) ion source with an argon and oxygen gas mixture [62]. The minimum single-digit domain with good stability in PZT film was recently reported to 4 nm in diameter, suggesting for a recording density of 40 Tbit/in^2 [63]. Another promising ferroelectric material in addition to PZT is $LiTaO_3$ that shows unique electro-optical properties as well as good mechanical and chemical stability [64–69]. The demonstration of accomplishing 10 Tbit/in^2 and 13 Tbit/in^2 using thin $LiTaO_3$ single-crystal films and a background field has been recently reported [70, 71]. For the latter case, data retention was measured by investigating the readback signals at elevated temperatures, and an activation energy of 0.8 eV at an attempt frequency of 200 kHz was found, sufficient for a data retention of 10 years [72].

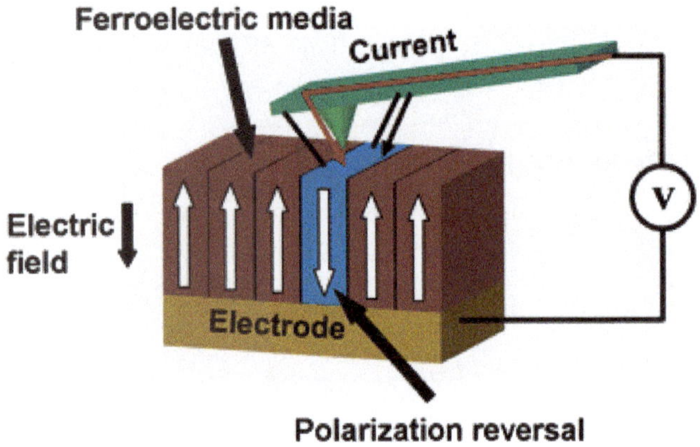

Figure 9: Ferroelectric probe memory system.

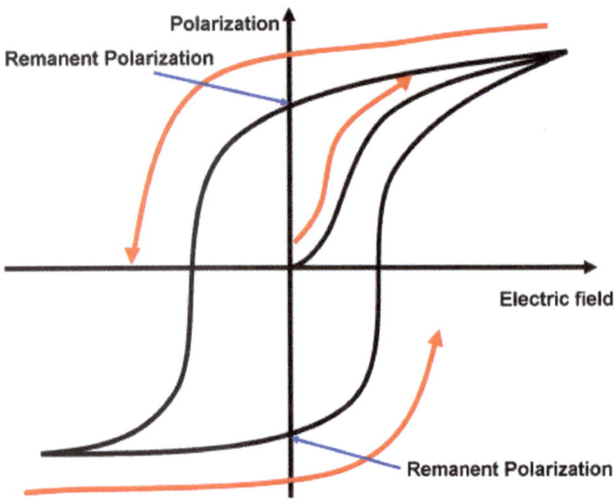

Figure 10: Hysteretic loop of FeRAM.

Although the ferroelectric probe memory exhibits a capability of providing densities of multi-Tbit/in^2, its destructive readout mechanism requires data to be refreshed after readout and this would cumulate read/write cycles and eventually causes severe fatigue. In order to overcome the adverse effects of the destructive readout, several non-destructive readout methods have been recently proposed including piezoelectric force microscopy (PFM) readout [73–75] and scanning nonlinear dielectric microscopy (SNDM) readout [64, 76–78]. For PFM readout, a conductive probe where a small AC tip-sample voltage is applied is brought in contact with the sample, and the response of the probe related to the polarization state of ferroelectric domains is detected by PFM as a readout signal at a frequency below the cantilever resonance. It should be pointed out that the resulting electric field would possibly change the permittivity of the ferroelectric domain, thus resulting in an advent of second harmonics [73]. An alternative method to operate probe in non-contact mode is SNDM readout that possesses from a fact that the reversal of the ferroelectric polarization can slightly change the storage medium's capacitance due to the non-linearity in the permittivity tensor [79]. The variation on the capacitance would slightly change the resonance conditions that can be detected by monitoring the cantilever vibration if the cantilever is excited with a fixed AC voltage by means of a lock-in technique. Another method reported the direct piezoelectric effect to build up charge on the tip as a result of the tip-sample load force [78]. The resulting current is proportional to the load force, leading to a trade-off with endurance, as tip wear increases with the load force.

Phase-Change Probe Memory

Although the history of phase-change probe memory is fairly shorter than its compatriots, it has been considered as one of the most promising mass storage devices for the next generation due to its potential for ultra-high density, high data rate, long data retention, and low cost. Its advantageous characteristics arise from the combination of phase-change materials with nanoscale-conductive probe. Phase-change materials, mainly represented by chalcogenide alloy, are capable of exhibiting two distinct physical phases depending on the arrangement of the atomic structure [80], i.e., amorphous phase and crystalline phase. The atomic structure of amorphous phase presents a short range translational order, while for crystalline phase, the atoms are distributed in a repeating or periodic array over large atomic distances. Such a structural variation makes the electrical and optical properties of amorphous phase significantly different from the crystalline phase, as the phase-change materials in crystalline phase normally shows much lower electrical resistivity and higher optical reflectivity than those of amorphous phase. Another intriguing feature of phase-change materials is that its two distinct phases can be switched between each other [81]. In order to generate crystalline phase, phase-change materials under amorphous phase needs to be heated above glass transition temperature but below the melting point, followed by a slow cooling to drive materials back to the energetically more favorable crystalline phase. To recover amorphous phase, phase-change materials in crystalline phase is required to be heated above melting temperature, subsequently followed by a fast quench to prevent the re-crystallization. Such a reversible transition is illustrated in Fig. 11. In addition to the thermal switching, a unique electronic switching phenomenon, known as "threshold switching", is also found on phase-change materials [82], which is depicted in Fig. 12. As can be seen from Fig. 12, a high-resistance state (OFF state) is initially presented in the phase-change materials along with the increase of the bias voltage, while a sudden "snap back" takes place when the bias voltage reaches a so-called threshold voltage. Under this circumstance, a sharp increase on the resulting current is monitored, leading to a low resistance state (ON state). Threshold switching is highly important for phase-change materials, as it allows for the phase transition at relatively low voltage. Otherwise, phase transition process can only be induced using very high voltage, thereby causing unnecessary energy consumption. Several theoretical models have been proposed to interpret the threshold switching from different scientific perspective. The threshold switching was first ascribed to the thermal breakdown effect in the amorphous film [83], whereas later, more evidence revealed that threshold switching was mainly dominated by electronic mechanism [84, 85]. According to this, two different electronic models based on trap-limited theory [86] and a balance between a strong Shockley Reed Hall (SRH) recombination through trap levels and a generation phenomenon [87] were, respectively, speculated to be the origin of the threshold switching. Differing from above models, a crystallization model owes the origin of threshold switching to the field induced nucleation of conductive cylindrical crystallites [88]. To date, the physical mechanism that can explicitly explain the threshold switching behavior still remains unclear.

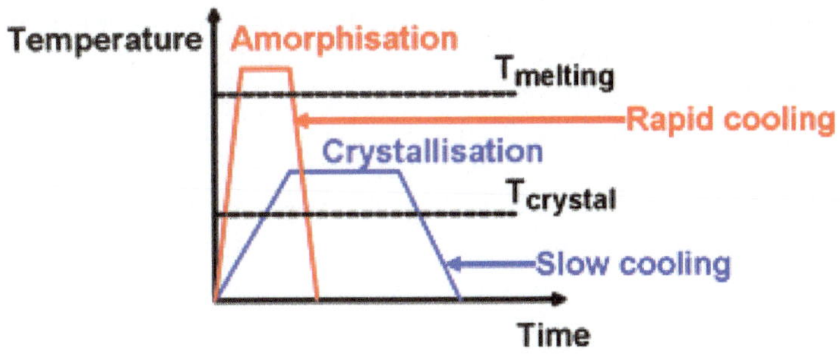

Figure 11: Thermal switching of phase-change materials between amorphous and crystalline phases. T_m is the melting temperature and $T_{crystal}$ is the crystalline transition temperature.

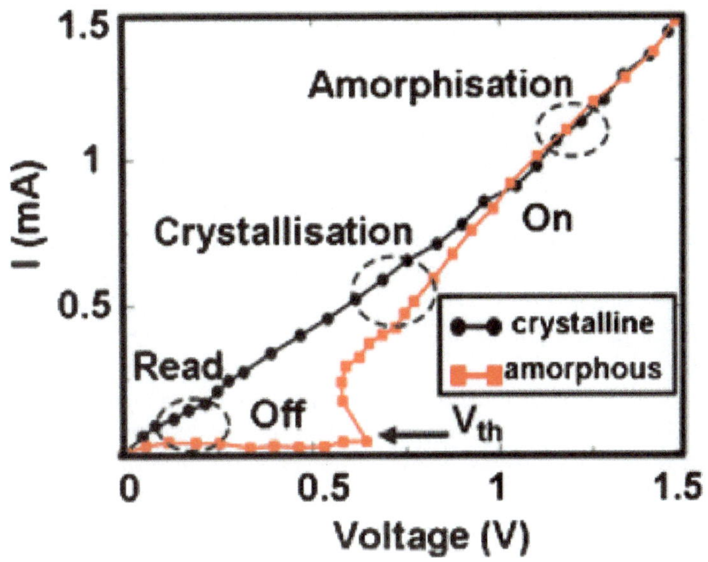

Figure 12: Threshold switching phenomenon observed in phase-change materials. V_{th} is the threshold-switching voltage, and *I* represents the resulting current. Reprinted with permission from [82].

As phase-change materials in amorphous phase exhibit distinct electrical/optical properties from crystalline phase, it is viable for phase-change probe memory to utilize such properties difference to denote the binary digits "1" and "0". In terms of phase-change probe memory, during the recording process, a write current is injected to phase-change media through the conductive probe tip to either heat the media to crystalline temperature for crystallization, or to the melting temperature followed by a rapid cooling to generate amorphization. The readout process is achieved by detecting the difference on the sensing current between highly conductive crystalline state and highly resistive amorphous state. The operations of phase-change probe memory are schematically shown in Fig. 13. As the conductive probe is employed as the conductive path for writing current, the dimension of the resulting phase-transition region would strongly depend on the radius of tip apex. In other words, it is viable to achieve multi-Tbit/in^2 if the radius of the tip apex can be downscaled below around 20 nm [89]. In addition, the most commonly used storage medium in phase-change probe memory is $Ge_2Sb_2Te_5$, usually referred to GST, which has been extensively adopted by rewritable optical disc and phase-change random access memory due to its capability of offering fast transition speed, long data retention time, and great endurance [90–93]. Given that the nanoscale probe tip and the GST medium are two key components of the phase-change probe memory, it is natural to infer that phase change probe-memory is capable of inheriting the performance superiorities of the nanoscale probe and the GST media.

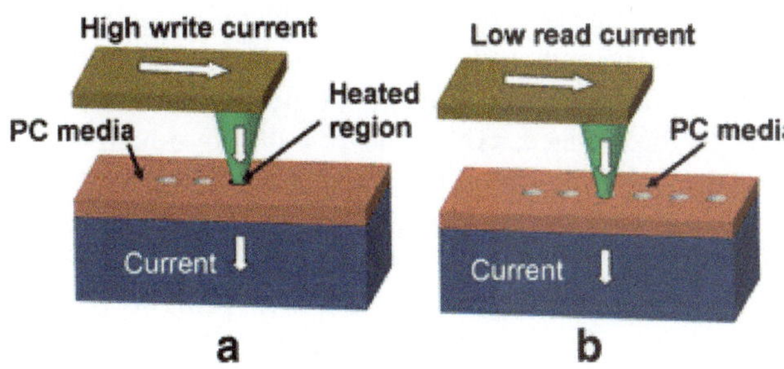

Figure 13: Phase-change probe memory when operated in **a** write mode, and **b** read mode.

Depending on the adopted architecture of phase-change probe memory, a variety of resulting bit size from 80 to 20 nm has been reported [94–100]. By the time of writing, the maximum areal density of phase-change probe memory that can be achieved experimentally is up to 1 Tbit/in^2 [94] with corresponding power consumption of 100 PJ per bit [95], much lower than the thermo-mechanical probe memory.

This is expected as phase-change probe memory requires the heating of bit volume only. Nevertheless, a new paradigm to induce the phase transition by a heated tip has recently been introduced [101], in which the AFM probe is heated to crystalline temperature either by a resistive heater or by a pulsed laser diode, and subsequently brought in contact with the GST to induce the phase transition. An areal density up to 3.3 Tb/in^2 was obtained with the desired write speed of 50 Mb/s for one probe when using a spinning disk to position the medium [101]. In addition, it is viable to achieve high density recording using a novel recording scenario that the probe moves a small distance over the media and thus gives rise to a recorded mark with "arbitrary" length by making use of the inter-symbol-interference (ISI) effect, also known as mark-length recording. The advantage of mark-length recording over the conventional mark-position recording arises from the independence of the resulting density on the size of the tip apex, allowing for an increase of user densities by at least 50 % and potentially 100 % with an implementation of relatively blunt tip [102]. A typical phase-change probe memory usually consists of a nanoscale conductive probe and a phase-change media stack that comprises a GST storage layer sandwiched by a capping layer and a under layer, which are all deposited on silicon (Si) or silicon dioxide (SiO_2) substrate. The role of capping layer is not only to protect GST layer from wear and oxidization, but also to build an effective electrically and thermally conductive path between probe tip and GST layer [103]. To satisfy the latter, the electrical conductivity of the capping layer needs to be relatively high in order to generate sufficient Joule heating inside the GST layer, while its thermal conductivity is required to be low to prevent heat from dissipating across the capping layer. However, a capping layer with high electrical conductivity is usually not desired as this would cause the presence of the short circuit during the readout process [104]. Compared with the capping layer, the under layer plays a less important role in determining the writing and readout performance of the phase-change probe memory, as it primarily acts as a bottom electrode. In this case, the electrical conductivity of the bottom layer is required to be much higher than capping layer so as to collect adequate write/readout current, while a relatively low thermal conductivity is preferable for the bottom layer to maintain sufficient Joule heat inside the storage layer [105, 106]. In addition, as both capping and bottom layers show thickness-dependent electrical/thermal conductivities [107, 108], their thicknesses also need to be carefully determined. Given the fact that threshold voltage is approximately equal to the product of layer thickness and threshold field, a thin GST is usually preferable so as to reduce the threshold voltage and thus the written power for a given low writing pulse. Therefore, an optimized design of the electrical probe memory device has been proposed by taking into account aforementioned factors as well as the underlying issues from practical fabrication process [109], as shown in Fig. 14. As can be seen from Fig. 14, this optimized phase-change media stack adopts a diamond-like-carbon (DLC) capping layer with thickness, electrical conductivity, and thermal conductivity of 2 nm, 100 $\Omega^{-1}m^{-1}$, 0.2 $Wm^{-1}K^{-1}$, and a titanium nitride (TiN) bottom layer with thickness, electrical conductivity, and thermal conductivity of 40 nm, 0.5×10^7 $\Omega^{-1}m^{-1}$, 12 $Wm^{-1}K^{-1}$, respectively. The thickness of GST layer is fixed to be 10 nm, which is the typical thickness for phase-change probe memory [89, 96]. The write and readout performances of the designed probe memory was assessed by a previously developed electro-thermal model that is, however, supplemented with several advanced modeling techniques including threshold switching, electrical contact resistance at PtSi/DLC interface, and thermal boundary resistance (TBR) at the interfaces of DLC/GST and GST/TiN [109]. The crystallization process in this model is determined by simultaneously solving the Laplace equation, classical heat transfer equation, and the nucleation-growth equation [6, 58, 103]. Solving the Laplace equation results in an electric field distribution inside the GST active layer, which is subsequently adopted as the Joule heating source for the classical heat transfer equation. Then, the temperature distribution inside the GST layer can be calculated by solving the heat transfer equation, whereby the temperature dependent crystal fraction of the GST layer can be obtained according to the nucleation-growth equation. Note that writing amorphous bit using electrical probe usually requires much higher temperature inside the capping layer than crystallization, and this may strongly harm the thermal stability of the capping layer. Because of this, the experimental writing of amorphous bits using electrical probe seems more difficult than writing crystalline bit, and was reported to be realized only by the stacks without capping layer [95, 110]. Therefore, in this paper, we are only taking into account the writing of crystalline bits. Figure 15 shows a production of a spherical shaped bit with a diameter of around 10 nm by a 6 V pulse of 100 ns, corresponding to the areal density of 10 Tb/inch^2 and the data rate of >Mb/s. Besides, the resultant writing power and writing energy are 0.016 mW and 1.6 pJ, respectively. This implies that the total writing energy required for using 100 and 1000 tips in parallel are 0.16 and 1.6 nJ, respectively, much lower than any other probe-based memories [11, 41, 67]. Most importantly, the crystalline bit shown in Fig. 15 is reported to be thermodynamically stable at the temperature below 550 °C, demonstrating its long retention time at archival temperature [89]. Meanwhile, the ability to achieve multi-level recording using phase-change probe memory was also demonstrated recently [111]. It was found that the resistance of the GST thin film can be toggled among three different states (high, intermediate, and low resistances) by carefully controlling the voltage between the conductive probe and the GST layer, as shown in Fig. 16. Therefore, three data values "0", "1", and "2" can be assigned to these three resistance states, respectively, allowing for multi-level operations in phase-change probe memory. In addition to the use of conductive AFM, scanning thermal microscopy (SThM) was recently reported to have the capability of measuring the thermal conductivities and thermal boundary resistances of GeTe (GT) and GST thin films [112, 113]. During the measurement, the SThM probe is repeatedly brought into and out of contact with material surface and the thermal properties of the probed materials can be determined by measuring the comparison of the heat flow from the SThM probe before and after the contact. According to such technique, it is possible to distinguish the amorphous phase from the crystalline phase by sensing the change in the thermal properties between amorphous and crystalline region, as revealed in Fig. 17. This finding also makes SThM as an alternative tool to provide bit write/read functions with spatial resolution down to the nanometer scale.

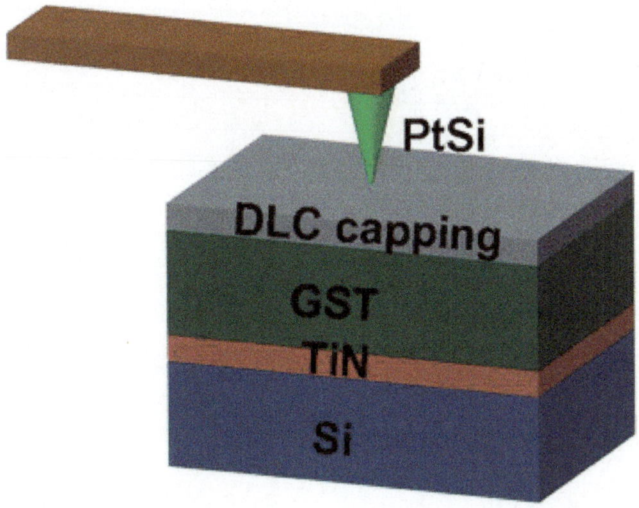

Figure 14: An optimized phase-change probe memory architecture.

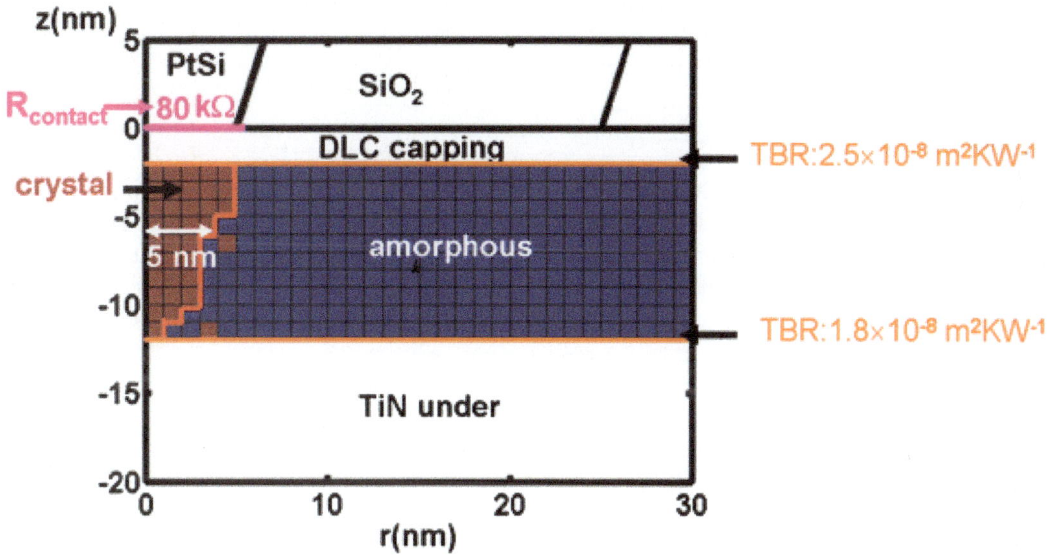

Figure 15: The resulting crystalline bit from the designed optimal electrical probe memory. The written bit exhibits a radius of approximately 5 nm, corresponding to 10 Tbit/in^2.

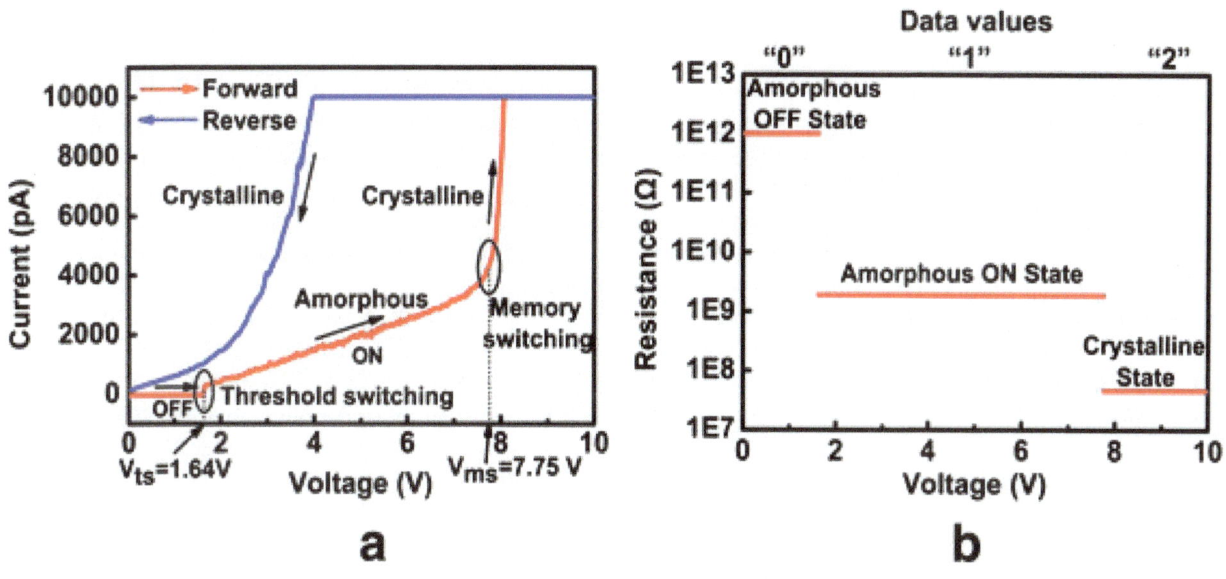

Figure 16: **a** Current-voltage spectrum of amorphous GST thin film measured in the range of 0–10 V with C-AFM; **b** the corresponding resistance of the amorphous OFF state, amorphous ON state, and crystalline state. Reprinted with permission from [111].

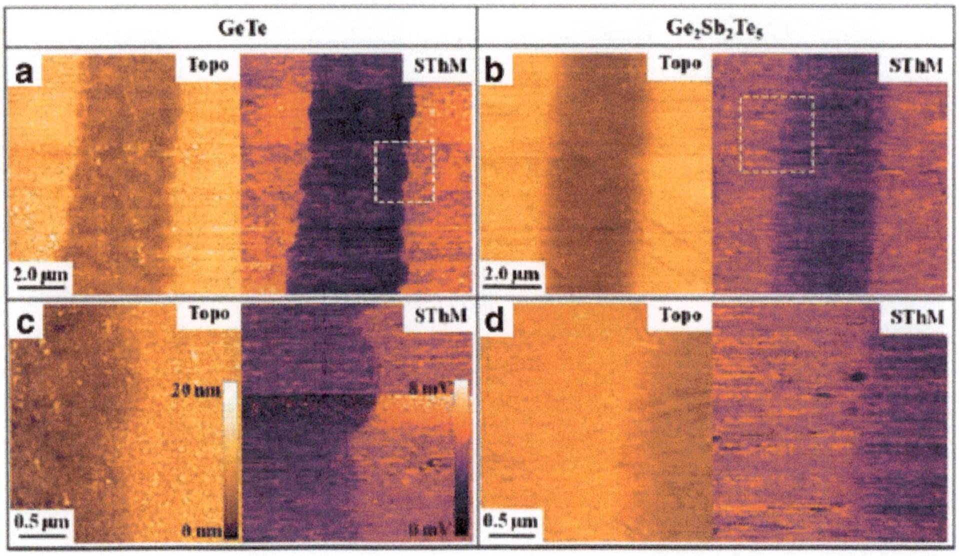

Figure 17: **a** Topography (*left sub-panel*) and SthM (*right sub-panel*) images with 10 μm and **b**8 μm scan sizes, revealing a crystalline line written into 200 nm GT and GST amorphous thin films by a focused laser beam. **c** The 2.5-μm images for GT and (**d**) GST taken from the spatial locations marked by the insets in **a** and **b**. Reprinted with permission from [112].

CONCLUSIONS

Today, the maximum achievable density resulting from conventional storage forms is currently limited below 1 Tb/in^2, and further density enhancement on these devices does not seem to be very optimistic provided that there are no innovative storage technologies emerging in the near future to break through their respective physical limits. Compared with the conventional storage devices, the practicality of using probe-based storage memories to achieve multi-Tbit/in^2 has been demonstrated on different probe storage devices. The characteristic comparison for different probe-based storage memories is shown in Table 1.

Table 1: Performance comparison of various probe memories [121]

	Phase-change	Magnetic	Thermomech	Ferroelectric
Density	3.3 Tb/in^2	60 Gb/in^2	4.0 Tb/in^2	4.0 Tb/in^2
Est. Max. density	≈10 Tb/in^2	≈100 Tb/in^2	≈10 Tb/in^2	>10 Tb/in^2
Read speed per probe	50 Mb s^{-1}	<10 b/s	40 kb s^{-1}	2 Mb s^{-1}
Write speed per probe	50 Mb s^{-1}	<10 b/s	1 Mb s^{-1}	50 kb s^{-1}
Travel per probe	2.5 m	0.5 m	750 m	5000 m

The advent of thermal-mechanical probe memory has pioneered a novel application of probe on data storage field and has been considered as the most mature probe technology in comparison to its juniors. However, the energy consumption caused by thermo-mechanical probe memory seems to be problematic as the entire tip needs to be heated to perform write and readout process. This can be alleviated by a non-thermal writing mechanism that the AFM tip at room temperature is pressed into the polymer to generate the indentation [114]. This innovative writing strategy gives rise to a density up to 1 Tbit/in^2 or more with an exemption from heating the tip [114]. Due to the superparamagnetic limit, the maximum areal density obtained from magnetic probe memory in the laboratory level is limited to a few of Gbit/in^2, much lower than that of other probe storage memories. Moreover, the necessity to integrate a compliant cantilever with a sensitive force sensor for the readout of magnetic probe memory actually makes the array design more complicated. As a consequence, the practicality of magnetic probe memory is being strongly challenged, which may accelerate its demise in the future. Ferroelectric probe memory has recently received more attention than magnetic probe memory because of its higher density, faster readout speed, and more endurance cycles. In addition, several approaches aiming to reduce the tip wear have been recently examined. One possible way is to use a platinum iridium tip in the readout of ferroelectric media to increase the endurance of the tip [61]. It was found that no loss in either the read or the write resolution took place after a sliding distance of 5 km at 5 mm/s, resulting in an extremely low wear volume down to 5.6×10^3 nm^3. An alternative method to overwhelm tip wear is to implement a dielectric-sheathed carbon nanotube (CNT) probe with micrometer-long tips that can be immune to significant wear and ensure a high resolution for both write and readout processes [115]. Recently, a hard HfB_2 coated tip was reported to extend the tip's endurance beyond 8 km of sliding [116]. Benefiting from the merge of the nanoscale conductive tip and phase-change medium, the capability of phase-change probe memory to offer ultra-high density, fast write/readout speed, long retention time, and great endurance cycles has been demonstrated both experimentally and theoretically. A more attractive feature of phase-change probe memory than other probe storage devices arises from the fact that the conductive tip used in phase-change probe memory is only required to be electrically sharp rather than physically sharp. As a result, it is viable to fabricate an electrically sharp tip but with a large physical diameter, thereby mitigating tip tribology and wear issues without the sacrifice of the resulting density. Such an idea has indeed been experimentally demonstrated by the fabrication of a SiO_2-encapsulated silicon tip with platinum silicide (PtSi) at the tip apex [117, 118], as shown in Fig. 18. Forming PtSi at the tip apex can significantly improve the conduction and wear properties of the conductive tip, due to its ability to provide superior anti-wear characteristics and very high electrical conductivity. In addition, the use of the dielectric encapsulation (SiO_2) can increase the physical diameter of the tip and thus contribute to lower levels of wear due to the reduced pressure between tip and sample. Therefore, the integration of highly conductive PtSi tip apex with a non-conductive SiO_2 encapsulation assigns such an encapsulated tip a capability to present a very small electrical contact area with the media even if the tip itself is not physically very sharp, and consequently enables a high density writing and reading simultaneously with an extremely good anti-wear characteristic. Nevertheless, phase-change probe memory is severely limited by the difficulty of amorphizing crystalline regions of the storage medium [119, 120].

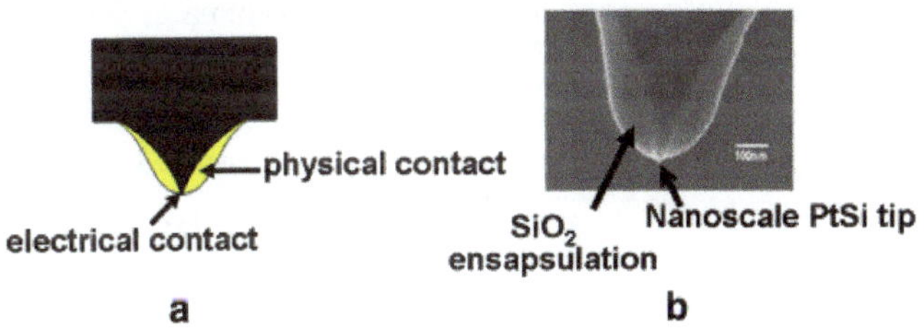

Figure 18: **a** Schematic of the "encapsulated" tip concept, and **b** the experimentally fabricated encapsulated tip. Reprinted with permission from [117].

ACKNOWLEDGEMENTS

The authors acknowledge the financial support of the National Natural Science Foundation of China (grant No. 61201439).

Competing Interests

The authors declare that they have no competing interests.

REFERENCES

1. Turner V (2014) Digital universe of opportunites. IDC Tech Report. 1
2. Wood R (2000) IEEE Trans Magn 36:36
3. Iwasaki S (2012) J Magn Magn Mater 324:244
4. Koide DC, Kajiyama T, Tokumaru H, Takano Y, Nabata Y, Ogata T, Miyazaki T, Ohishi K (2011) Jpn J Appl Phys 51:08JA04
5. Overton G (2012) Laser Focus World 48:39
6. Wang L, Gai S (2014) Contemp Phys 55:75
7. Lu CY (2012) J Nanosci Nanotechnol 12:7604
8. Lai SK (2012) J Nanosci Nanotechnol 12:7597
9. Binnig G, Quate CF (1986) Phys Rev Letts 56:930
10. Wright CD, Aziz MM, Shah P, Wang L (2011) Curr Appl Phys 11, e104
11. Vettiger P, Cross G, Despont M, Drechsler U, Durig U, Gotsmann B, Haberle W, Lantz MA, Rothuizen HE, Stutz R, Binnig GK (2002) IEEE Trans Nanotechnol 1:39
12. Binnig GK, Despont M, Drechsler U, Haberle W, Lutwyche M, Vettiger P, Mamin HJ, Chui BW, Kenny TW (1999) Appl Phys Lett 74:1329
13. Pozidis H, Haberle W, Wiesmann D, Drechsler U, Despont M, Albrecht TR, Eleftheriou E (2004) IEEE Trans Magn 40:2531
14. Vettiger P, Despont M, Drechsler U, Durig U, Haberle W, Lutwyche MI, Rothuizen HE, Stutz R, Widmer R, Binnig GK (2000) IBM J Res & Dev 44:323
15. Cannara RJ, Gotsmann B, Knoll A, Durig U (2008) Nanotechnology 19:395305
16. King WP, Santiago JG, Kenny TW, Goodson KE (1999) Proc ASME MEMS 1:583
17. Wiesmann D, Rawlings C, Vecchione R, Porro F, Gotsmann B, Knoll A, Pires D, Duerig U (2009) Nano Letts 9:3171
18. Gotsmann B, Duerig U, Frommer J, Hawker CJ (2006) Adv Funct Mater 16:1499
19. Pires D, Gotsmann B, Porro F, Wiesmann D, Durig U, Knoll A (2009) Langmuir 25:5141
20. Knoll A, Wiesmann D, Pires D, Gotsmann B, Durig U (2010) J Nanosci Nanotechnol 10:4538
21. Gotsmann B, Knoll AW, Pratt R, Frommer J, Hedrick JL, Durig U (2010) Adv Funct Mater 20:1276
22. H. J. M, J. S. Trethewey, A. D. Johnson, W. J. Drugan, and W. C. Crone, Adv. Maters. 17, 1123(2005).
23. Shaw GA, Trethewey JS, Johnson AD, Drugan WJ, Crone WC (2005) Adv Maters 17:1123
24. Bhaskaran H, Gotsmann B, Sebastian A, Drechsler U, Lantz MA, Despont M, Jaroenapibal P, Carpick RW, Chen Y, Sridharan K (2010) Nat Nanotechnol 5:181
25. Lantz MA, Gotsmann N, Jaroenapibal P, Jacobs TDB, O'Connor SD, Sridharan K, Carpick RW (2012) Adv Funct Mater 22:1639
26. Sahoo DR, Haberle W, Bachtold P, Sebastian A, Pozidis H, Eleftheriou E (2008) 2008 American Control Conference. ACC, Seattle, pp 2034–2039
27. Knoll A, Rothuizen H, Gotsmann B, Duerig U (2010) Nanotechnol 21:185701
28. Lantz MA, Wiesmann D, Gotsmann B (2009) Nat Nanotechnol 4:586
29. Shalaby M, Vicario C, Hauri CP (2016) New J Phys 18:013019
30. Mironov VL, Ermolaeva OL (2009) J Surf Invest 3:840
31. Mironov VL, Gribkov BA, Vdovichev SN, Gusev SA, Fraerman AA, Ermolaeva OL, Shubin AB, Alexeef AM, Zhdan PA, Binns C (2009) J Appl Phys 106:053911

32. Onoue T, Siekamn M, Abelmann L, Lodder JC (2004) J Magn Magn Mater 272:2317
33. Ohkubo T, Kishigami J, Yanagisawa K, Kaneko R (1991) IEEE Trans Magn 27:5286
34. Ohkubo T, Kishigami J, Yanagisawa K, Kaneko R (1993) NTT R&D 42:545
35. Ohkubo T, Maeda Y, Koshimoto Y (1995) IEICE Trans On Electron E78-C:1523
36. Manalis S, Babcock K, Massie J, Elings V, Dugas M (1995) Appl Phys Lett 66:2585
37. El-Sayed RT, Carley LR (2004) IEEE Trans Magn 40:2326
38. El-Sayed RT, Carley LR (2003) IEEE Trans Magn 39:3566
39. Zhang L, Bain JA, Zhu J-G, Abelmann L, Onoue T (2007) Physica B 387:328
40. Zhong ZY, Zhang L, Zhang H-W (2008) Curr Appl Phys 8:57
41. Zhang L, Bain JA, Zhu J-G, Abelmann L, Onoue T (2006) J Magn Magn Mater 305:16
42. Zhang L, Bain JA, Zhu J-G (2002) IEEE Trans Magn 38:1895
43. Zhang L, Bain JA, Zhu J-G, Abelmann L, Onoue T (2004) IEEE Trans Magn 40:2549
44. Watanuki O, Sonobe Y, Tsuji S, Sai F (1991) IEEE Trans Magn 27:5289
45. Hosaka S, Koyanagi H, Kikukawa A, Miyamoto M, Imura R, Ushiyama J (1995) J of Vac Sci &Technol B 13:1307
46. Nakamura J, Miyamoto M, Hosaka S, Koyanagi H (1995) J Appl Phys 77:779
47. Zhong ZY, Zhang L, Zhang HW (2007) EPJ Appl Phys 38:259
48. Onoue T, Siekman H, Abelmann L, Lodder JC (2008) J Appl Phys D 41:155008
49. Algre E, Gaudin G, Bsiesy A, Nozieres JP (2005) IEEE Trans Magn 41:2857
50. Watson CS, Hollar C, Anderson K, Knowlton WB, Mullerner P (2013) Adv Funct Mater 23:3995
51. E. Sarajlic, R. Vermeer, M. Y. Delalande, M. H. Siekman, R. Huijink, H. Fujita, and L. Abelmann. in: Oric, 23rd Ubt, Conf. on Micro Electro Mechanical Systems (MEMS 2006), Hong Kong, 2010, 328(2010).
52. Petit D, Atkinson D, Johnston S, Wood D, Cowburn RP (2004) IEE Proc: Science. Measurement and Technology 151:127
53. Oral A, Bending SJ, Henini M (1996) Appl Phys Lett 69:1324
54. Gregusova D, Martaus J, Fedor J, Kudela R, Kostic I, Cambel V (2009) Ultramicroscopy 109:1080
55. Hatakeyama K, Vermeer R, Siekman MH, Sarajlic E, Tas NR, Abelmann L (2014) In: INTERMAG 2014, Dresden. pp. 1741–1742. IEEE
56. Takezaki T, Yagisawa D, Sueoka K (2006) Jpn J Appl Phys 45:2251
57. Precner M, Fedor J, Soltys J, Cambel V (2015) Nanotechnol 26:055304
58. Wang L, Yang CH, Wen J (2015) Electron Mater Lett 11:505
59. Pertsev NA, Contreras JR, Kukhar VG, Hermanns B, Kohlstedt H, Waser R (2003) Appl Phys Lett 83:3356
60. Zybill CE, Li B, Koch F, Graf T (2000) Phys Status Solidi A 177:303
61. Tayebi N, Zhang Y, Chen RJ, Tran Q, Chen R, Nishi Y, Ma Q, Rao V (2010) ACS Nano 4:5713
62. Hiranaga Y, Cho Y (2014) Jpn J Appl Phys 53:09PA05
63. Tayebi N, Narui Y, Franklin N, Collier CP, Giapis KP, Nishi Y, Zhang Y (2010) Appl Phys Lett 96:023103
64. Cho Y, Fujimoto K, Hiranaga Y, Wagatsuma Y, Onoe A, Terabe K, Kitamura K (2002) Appl Phys Lett 81:4401
65. Cho Y, Hashimoto S, Odagawa N, Tanaka K, Hiranaga Y (2005) Appl Phys Lett 87:232907
66. Hiranaga Y, Fujimoto K, Cho Y, Wagatsuma Y, Onoe A, Terabe K, Kitamura K (2002) Integr Ferroelectr 49:203
67. Cho Y, Fujimoto K, Hiranaga Y, Wagatsuma Y, Onoe A, Terabe K, Kitamura K (2003) Nanotechnol 14:637
68. Hiranaga Y, Cho Y, Fujimoto K, Wagatsuma Y, Onoe A (2003) Jpn J Appl Phys 42:6050
69. Cho Y, Hiranaga Y, Fujimoto K, Wagatsuma Y, Onoe A (2004) Integr Ferroelectr 61:77
70. Cho Y, Fujimoto K, Hiranaga Y, Wagatsuma Y, Onoe A, Terabe K, Kitamura K (2003) Ferroelectrics 292:51
71. Hiranaga Y, Uda T, Kurihashi Y, Tanaka K, Cho Y (2007) IEEE Trans Ultrason Ferr Freq Control 54:2523
72. Tanaka K, Hiranaga Y, Cho Y (2008) Record of electrical and commun. Eng Conversazione Tohoku University 76:384
73. Franke K, Besold J, Haessler W, Seegebarth C (1994) Surf Sci 302:L283
74. Hidaka T, Maruyama T, Saitoh M, Mikoshiba N, Shimizu M, Shiosaki T, Wills LA, Hiskes R, Dicarolis SA, Amano J (1996) Appl Phys Lett 68:2358
75. Kalinin SV, Jesse S, Rodriquez BJ, Shin J, Baddorf AP, Lee HN, Borisevich A, Pennycook SJ (2006) Nanotechnol 17:3400
76. Saurenbach F, Terris BD (1990) Appl Phys Lett 56:1703
77. Saurenbach F, Terris BD (1992) IEEE Trans Ind Appl 28:256
78. Kim BM, Adams DE, Tran Q, Ma Q, Rao V (2009) Appl Phys Lett 94:063105
79. Tanaka K, Kurihashi Y, Uda T, Daimon Y, Odagawa N, Hirose R, Hiranaga Y, Cho Y (2008) Jpn J Appl Phys 47:3311
80. Wuttig M, Yamada N (2007) Nature 6:824
81. Matheswaran P, Sathyamoorthy R, Asokan K (2012) Electron Mater Lett 8:417
82. Welnic W, Wuttig M (2008) Materials Today 11:20
83. Adler D, Henisch HK, Mott N (1978) Rev Mod Phys 50:209

84. Adler D, Shur MS, Silver M, Ovshinsky SR (1980) J Appl Phys 51:3289
85. Adler D, Shur MS, Silver M, Ovshinsky SR (1984) J Appl Phys 56:579
86. Ielmini D, Zhang YG (2007) J Appl Phys 102:054517
87. Redaelli A, Pirovano A, Benvenuti A, Lacaita AL (2008) J Appl Phys 103:111101
88. Karpov VG, Kryukov YA, Savransky SD, Karpov IV (2007) Appl Phys Lett 90:123504
89. Wright CD, Marilyn M, Mustafa MM (2006) IEEE Trans Nanotechnol 5:50
90. Yamada N, Ohno E, Nishiuchi K, Akahira N, Takao M (1991) Jpn J Appl Phys 69:2849
91. Kolobov AV, Fons P, Tominaga J, Uruga T (2006) J Non-Cryst Solids 352:1612
92. Wong H-SP, Raoux S, Kim SB, Liang JL, Reifenberg JP, Rajendran B, Asheghi M, Goodson KE (2010) Proc IEEE 98:2201
93. Raoux S, Welnic W, Ielmni D (2010) Chem Rev 10:240
94. Gidon S, Lemonnier O, Rolland B, Bichet O, Dressler C, Samson Y (2004) Appl Phys Lett 85:6392
95. Satoh H, Sugawara K, Tanaka K (2006) J Appl Phys 99:024306
96. Bhaskaran H, Sebastian A, Pauza A, Pozidis H, Despont M (2009) Rev Sci Instrum 80:083701
97. Wang L, David CD, Shah P, Aziz MM, Sebastian A, Pozidis H, Pauza A (2011) Jpn J Appl Phys 50:09MD04
98. Wang L, Wright CD, Aziz MM, Shah P, Yang CH, Yang GW (2013) Mater Lett 112:51
99. Wang L, Wright CD, Aziz MM, Yang CH, Yang GW (2016) Electron Mater Lett 10:1045
100. Wang L, Gong SD, Wen J, Yang CH (2016) J Nanomater 2016:1
101. Hamann HF, O'Boyle M, Martin YC, Rooks M, Wickramasinghe HK (2006) Nat Mater 5:383
102. Wright CD, Shah P, Wang L, Aziz MM, Sebastian A, Pozidis H (2010) Appl Phys Lett 97:173104
103. Wang L, Yang CH, Wen J, Gai S, Peng YX (2015) J Mater Sci Mater Electron 26:4618
104. Wang L, Wright CD, Aziz MM, Yang CH, Yang GW (2014) Jpn J Appl Phys 53:028002
105. Wang L, Wright CD, Aziz MM, Ying J, Yang GW (2013) EPL 104:56007
106. Wang L, Wen J, Yang CH, Shan G, Miao XS (2015) Nanosci Nanotechnol Lett 7:870
107. Balandin AA, Shamsa ML, Liu WL, Casiraghi C, Ferrari AC (2008) Appl Phys Lett 93:043115
108. Shamsa M, Liu WL, Balandin AA, Casiraghi C, Milne WI, Ferrari AC (2006) Appl Phys Lett 89:161921
109. Wang L, Wen J, Yang CH, Gai S, Peng YX (2015) Nano 10:1550118
110. Wright CD, Wang L, Shah P, Aziz MM, Varesi E, Bez R, Moroni M, Cazzaniga F (2011) IEEE Trans Nanotechnol 10:900
111. Yang F, Xu L, Chen J, Xu J, Yu Y, Ma ZY, Chen K-J (2016) Nanotechnology 27:035706
112. Bo JL, Timofeeva M, Tovee PD, Robinson BJ, Huey BD, Kolosov OV (2014) J Appl Phys 116:134904
113. Velea A, Borca CN, Socol G, Galca AC, Grolimund D, Popescu M, van Bokhoven JA (2014) J Appl Phys 116:234306
114. Jo A, Joo W, Jin WH, Nam H, Kim JK (2009) Probe-based data storage. Nature Nanotechnol 4:727.
115. Tayebi N, Narui A, Chen RJ, Collier CP, Giapis KP, Zhang Y (2008) Appl Phys Lett 93:103112
116. Tayebi N, Yanguas-Gil A, Kumar N, Zhang Y, Abelson JR, Nishi Y, Ma Q, Rao VR (2012) Appl Phys Lett 101:091909
117. Bhaskaran H, Sebastian A, Drechsler U, Despont M (2009) Nanotechnol 20:105701
118. Bhaskaran H, Sebastian A, Despont M (2009) IEEE Trans Nanotechnol 8:128
119. Wang L, Yang CH, Wen J, Peng YX (2015) Current Nanoscience 11:676.
120. Tanaka K (2007) J Non-Crystals Solids 353:1899
121. W. W. Koelmans, J. B. C. Engelen, and L. Abelmann. arXiv. 1(2015).

CHAPTER 13 A THYRISTOR-ONLY INPUT ESD PROTECTION SCHEME FOR CMOS RF ICS

Jin Young Choi and Choongkoo Park

Keywords: ESD Protection, HBM, Thyristor, Mixed-Mode Simulation, RF ICs

INTRODUCTION

CMOS chips are vulnerable to electrostatic discharge (ESD) due to thin gate oxides used, and therefore protection devices such as NMOS transistors are required at input pads. A large size for the protection devices is needed to reduce discharge current density and thereby to protect them against thermal-related problems. However, using a large size tends to increase parasitic capacitances added to input nodes generating problems such as gain reduction and poor noise characteristics in RF ICs [1].

To reduce the added parasitics, the protection schemes utilizing thyristor or diode protection devices were suggested [2,3] and have been used as fundamental protection schemes in RF ICs. In the protection scheme utilizing thyristor or diode protection devices, it is conventional to include a V_{DD}-V_{SS} clamp NMOS device in the input pad structure to provide discharge paths for all possible human-body model (HBM) test modes [4]. A large size for the clamp NMOS device is essential to prevent thermal device failure. Even though the clamp NMOS device does not add parasitics to the input node since it is not connected to it, adopting a large size makes the design to consume an excessive area for the pad structure. This requirement can be a serious limitation in a chip where a pad size is a critical issue in chip design. In this paper, we suggest an input protection scheme utilizing low-voltage triggering thyristor devices only without using a clamp NMOS device in input pad structure. This scheme can be implemented into input pad structures of CMOS RF ICs to provide protection against HBM and MM (Machine mode) discharge events. We present a comparative analysis result of the proposed scheme and the conventional thyristor protection scheme incorporating a clamp NMOS device. The characteristics of the latter scheme are already presented in [4].

A 2-dimensional device simulator, together with a circuit simulator, is utilized as a tool for the comparative analysis. The analysis methodology utilizing a device simulator has been widely adopted with credibility [4-6] since it provides valuable information relating the mechanisms leading to device failure, which cannot be obtained by measurements only.

In Section 2, we suggest low-voltage triggering thyristor protection device structures assuming usage of standard CMOS processes, and introduce device characteristics based on DC device simulations. In Section 3, we briefly explain discharge modes in HBM tests and introduce two input protection scheme utilizing the suggested protection devices. In Section 4, we construct an equivalent circuit model for CMOS chips equipped with the input protection devices to simulate various input HBM test situations, and execute mixed-mode transient simulations. Based on the simulation results, we figure out weak modes in real discharge tests, and present indepth analysis results relating critical characteristics such as peak voltages developed across gate oxides in input buffers, locations of peak temperature inside protection devices, and so on. In Section 5, based on the simulation results, considerations relating device and circuit design are discussed.

Citation: J. Choi and C. Park, "A Thyristor-Only Input ESD Protection Scheme for CMOS RF ICs," Circuits and Systems, Vol. 2 No. 3, 2011, pp. 170-182. doi: 10.4236/cs.2011.23025.

PROTECTION DEVICE STRUCTURES AND DC CHARACTERISTICS

Figure 1 shows the NMOS protection device structure assumed in this work, which is utilized as a V_{DD}-V_{SS} clamp device. The structure is same with the one suggested in [4]. The p+ junctions located at the upper left/ right corners represent diffusions for substrate ground contacts. A series resistor of 1 MΩ·μm, which is not shown in Figure 1, is connected at the bottom substrate node considering distributed resistances leading to substrate contacts located far away. DC simulations were performed using a 2-dimensional device simulator ATLAS [7]. All necessary physical models including an impact ionization model and latticeheating models were included in the simulations. The source, the gate, and the substrate were grounded, and the drain bias was varied for simulation.

Figure 2 shows simulated drain current vs. voltage characteristics of the NMOS device in Figure 1 in a semi-log scale. The underlying physics relating the characteristics in Figure 2 are fully explained in [4] previously. The device shows an n^+ -drain/p-sub junction breakdown when the drain voltage is increased above 9.3 V. A generated hole current by avalanche flows to the substrate terminal to increase the body potential. With a sufficient hole current flowing, a parasitic lateral npn bipolar transistor is triggered. The source, the body, and the drain act as an emitter, a base, and a collector, respectively. As the BJT is being triggered, the required drain-source voltage is reduced to show a snapback, as indicated as "BJT trigger" in Figure 2. After the snapback at about 9.4 V, the drain-source voltage drops to about 4.6 V of a bipolar holding voltage.

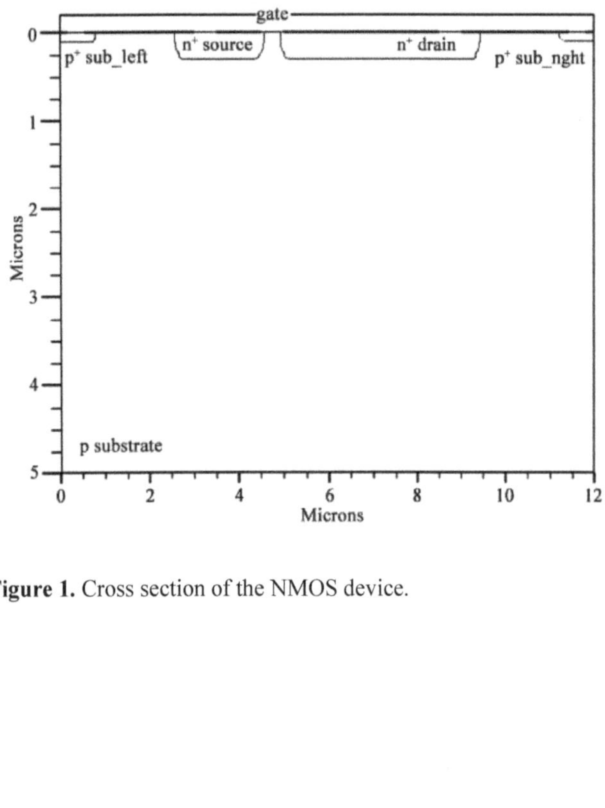

Figure 1. Cross section of the NMOS device.

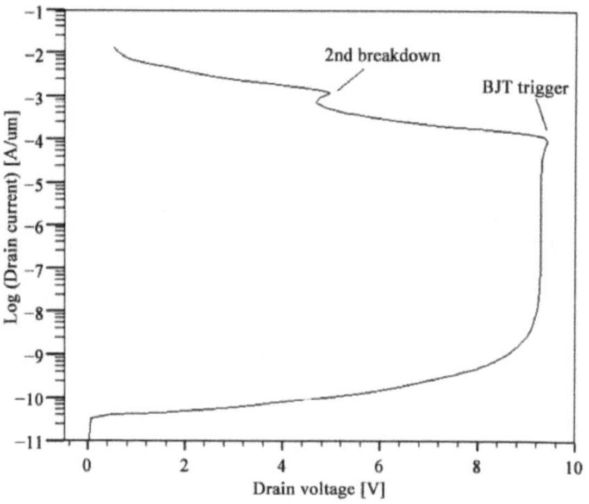

Figure 2. Drain I-V characteristics of the NMOS device.

Figure 3 shows the lvtr_thyristor_down device structure assumed in this work, which is used as a protection device between an input pad and a V_{SS} node. An lvtr_ thyristor device is a pnpn-type device suggested to lower the snapback voltage by incorporating an NMOS transistor into it [2]. The drain/ gate/source NMOS structure composed of the n^+ well (n^+ region at the right-hand corner of the n well), the gate, and the n+ cathode is similar to the NMOS structure in Figure 1. However, it does not incorporate ESD implant steps, which is implied by the relatively shallow junctions. The structure shown in Figure 3 is same with the one of the lvtr_thyristor device suggested in [4]. The NMOS gate oxide thickness, the gate length, the effective channel length, and the channel peak doping are same with those in Figure 1. A series resistor is also connected at the bottom substrate node as in the NMOS device in Figure 1.

The n^+ and p^+ anodes in Figure 3 are tied together to serve as an anode. The cathode, the gate, and the substrate were grounded, and the anode bias was varied for simulation. Figure 4 shows the simulated DC anode current vs. voltage characteristics of the lvtr_thyristor_down device in Figure 3. The underlying physics relating the characteristics in Figure 4 are fully explained in [4] previously. The device shows an n^+ -drain/p-sub junction breakdown when the anode voltage is increased to above 8.8 V. With a sufficient hole current flowing to the substrate terminal, the p-sub/n^+ -cathode junction is forward biased triggering a lateral npn bipolar transistor. The n^+ well, the p substrate, and the cathode act as a collector, a base, and an emitter, respectively. At this situation, a snapback is monitored as shown in Figure 4. The collector current from the n+ anode flows through the n well to decrease the potential of the region under the p^+ anode by an ohmic drop. When the collector current is large enough, the p^+ -anode/n-well junction is forward biased to trigger a pnpn (p^+ -anode/n-well/p-sub/ n^+ -cathode) thyristor, which causes another decrease in the anode voltage, as indicated as "pnpn trigger" in Figure 4. The resulting holding voltage drops to about 1 V, which is much smaller compared to 4.6 V of the NMOS device in Figure 2. We note here that the device shown in Figure 3 includes a well conducting thyristor by virtue of the larger n^+-p^+ anode space. This is verified by the fact that the current level for the pnpn trigger is not much higher than that for the bipolar trigger in Figure 4. As the anode space decreases, the current level for the pnpn trigger increases [8].

If we want to use the lvtr_thyristor_down device in Figure 3 as a protection device between a V_{DD} node and a pad in a thyristor-only protection scheme, we should

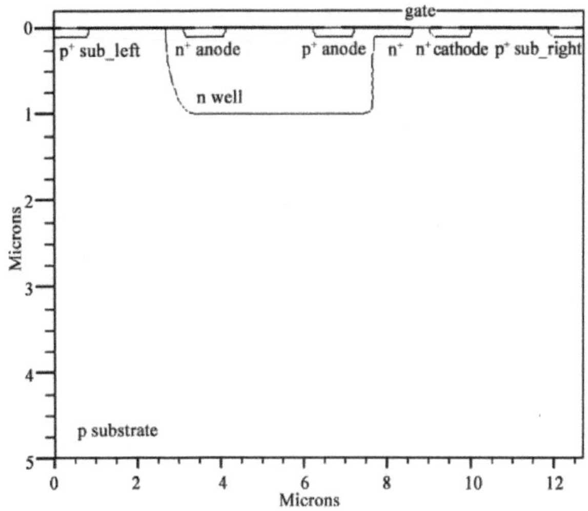

Figure 3. Cross section of the lvtr_thyristor device.

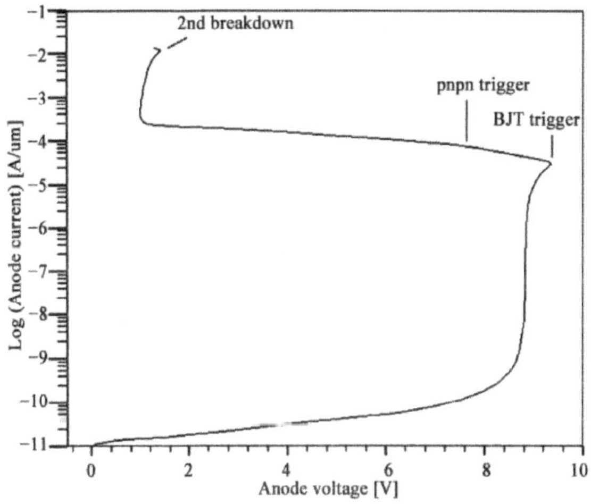

Figure 4. Anode I-V characteristics of the lvtr_thyristor device.

connect the p^+ anode and the n^+ anode to a V_{DD} node, and n^+ cathode to a pad, and the common p substrate to a VSS node. In this case, however, we found that a serious problem occurs in a PD mode (a positive ESD voltage applied to a pad with a V_{DD} pin grounded) since there is no forward diode path from the pad to the V_{DD} node. Without the forward diode path provided, an npn (n^+ -cathode/p-sub/n^+ -anode) bipolar transistor conducts from the pad to the V_{DD} node to cause a thermal-related problem with a larger holding voltage. Figure 5 shows the lvtr_thyristor_up device, which is suggested in this work to solve the problem. The device apparently includes a forward diode path from the cathode to the anode.

As we can see in Figure 5, the device needs addition of a p base region, which is provided in most of standard CMOS processes below 0.35 μm to allow forming simple bipolar transistors. Depth of the p-base region was assumed as 0.5 μm, and the base is assumed to have a channel peak doping of 2.35×10^{17} cm^{-3}, which is similar to that of the NMOS structure in the lvtr_thyristor_ down device. Doping profiles of the n well, the n^+ and p^+ junctions are same with those in the the lvtr_thyristor_down device. The NMOS transistor, which is incorporated in the device to lower the snapback voltage, is located inside the p base. The n^+ well (n^+ region at the left-hand corner of the p base), the gate, and the n^+ cathode play the role of the drain, the gate, and the source, respectively. The NMOS structure in the device is very similar to that in the lvtr_thyristor_down device except a small difference in the acceptor doping profile in the body region. We note that, in deep n-well processes, a p well separated from a p substrate can replace the p base.

The n^+ and p^+ anodes in Figure 5 are tied together to serve as an anode. The n^+ and p^+ cathodes, the gate, and the substrate were grounded, and the anode bias was varied for simulation. Figure 6 shows the simulated DC anode current vs. voltage characteristics of the lvtr_thyristor_up device in Figure 5. The characteristics and the underlying physics are very similar to those of the lvtr_thyristor_down device except that an npn bipolar transistor is formed by the n^+-well/p-base/n^+ -cathode structure and a pnpn thyristor is formed by the p^+ -anode/n-well/p-base/n^+ -cathode structure. The breakdown voltage in this device corresponds to that of the n^+ -well/p-base junction. Table 1 summarizes the principal DC characteristics of the three protection devices.

ESD PROTECTION SCHEMES

Since parasitics added to an input pad should be minimized, it is desired to connect fewer number of protection devices to an input pad. An effective way to reduce the number is to use a V_{DD}-V_{SS} clamp device since it provides discharge paths without adding parasitics to an input pad. Figure 7 shows a popular ESD protection scheme utilizing a thyristor device, which minimizes the added parasitics and is chosen for a comparison study with the thyristor-only protection scheme in this work. A CMOS inverter was assumed as an input buffer.

The lvtr_thyristor_down device in Figure 3 is used for T1 in Figure 7. The NMOS device in Figure 1 is used for M_2. In M_2, the drain is connected to V_{DD}, and the gate, the source, and the substrate are connected to V_{SS}. In T_1, the p+ and n^+ anodes are connected to the pad, and the p substrate and the n^+ cathode are connected to V_{SS}. The NMOS gate (G_1) in T_1 is also connected to V_{SS} to maintain an off state in normal operations. It is important to locate all the protection devices close to the pad to minimize variation of the gate voltage in the input buffer when an ESD voltage is applied to the pad. Even though same discharge paths can be provided with V_{DD}-V_{SS} clamp devices located in other places, there exists an enhanced danger of oxide failure since the voltage applied to the gate oxide of the input buffer may increase due to added high voltage drops in power bus lines with a large discharge current flowing.

Figure 8 shows the thyristor-only protection scheme suggested in this work. The lvtr_thyristor_down device in Figure 3 is used for T_1, and the lvtr_thyristor_up device in Figure 5 is used for T_2. Connection of T_1 is same as that in Figure 7. In T_2, the p+ and n+ anodes are connected to V_{DD}, and the p^+ and n^+ cathodes are connected to the pad. The NMOS gate (G_2) in T_2 is connected to the pad to maintain an off state in normal operations. Al though it is not shown in Figure 8, the p substrate in T_2 is connected to V_{SS}.

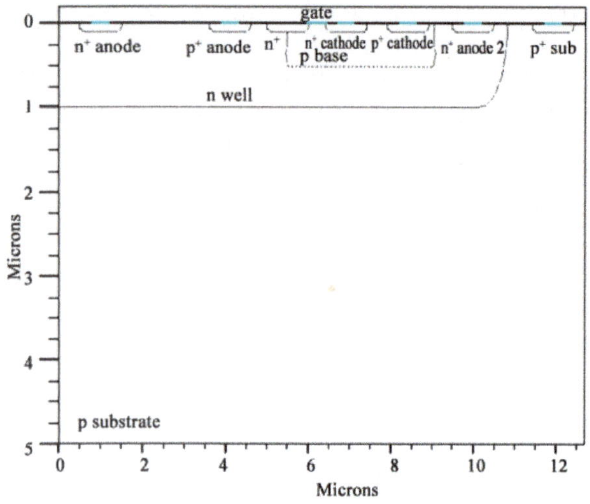

Figure 5. Cross section of the diode device.

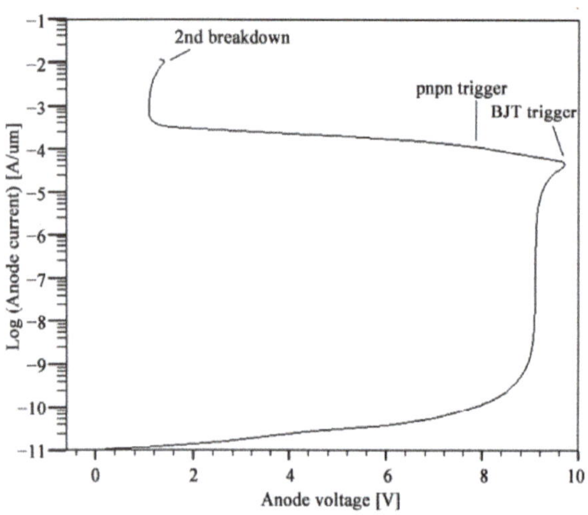

Figure 6. Anode current-voltage characteristics of the lvtr_ thyristor_up device.

Table 1. Principal principal DC characteristics of the protection devices.

Protection device	Holding Voltage	Breakdown voltagee	Snapback voltagee
NMOS	4.6 V	9.3 V	9.4 V
lvtr_thyristor_down	1.0 V	8.8 V	9.4 V
lvtr_thyristor_up	1.1 V	9.1 V	9.7 V

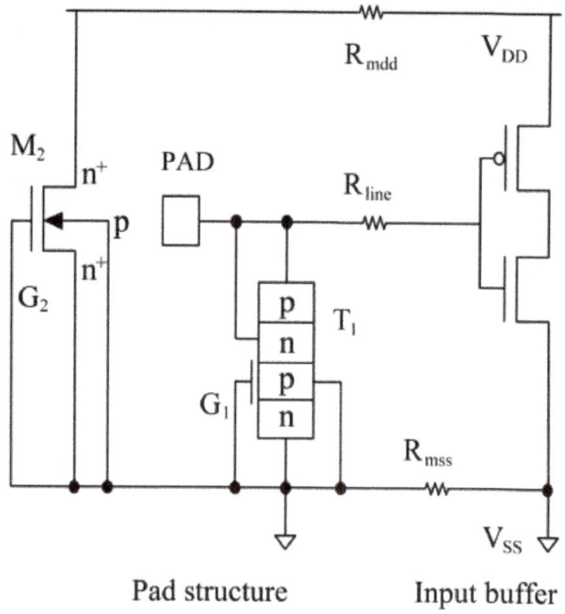

Figure 7. Protection scheme (1) utilizing a thyristor device.

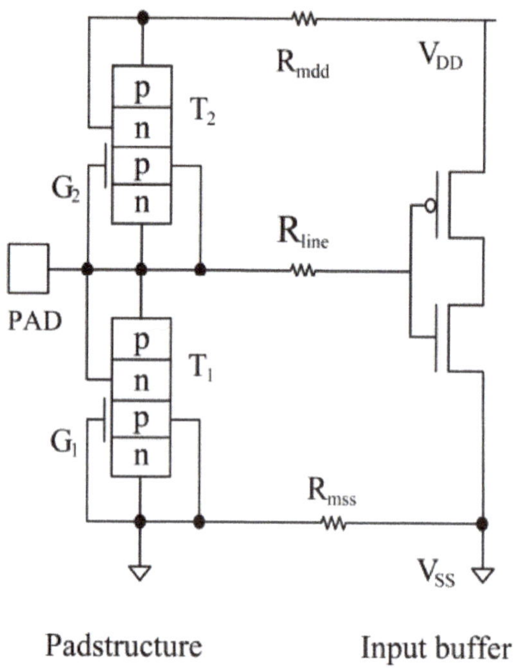

Figure 8. Protection scheme (2) utilizing thyristor devices only.

Input protection schemes utilizing thyristor devices only had been suggested [9,10]; however the trigger voltages of the suggested devices are too high [9] to be used in recent technologies, or the trigger voltages for the two LVTSCR thyristor devices are uneven [10] due to a difference in hole and electron mobilities resulting uneven trigger voltage for different HBM test modes, which is certainly not beneficial, and the PMOS-triggered thyristor structure does not provide a forward diode path from a pad to a V_{DD} node [10] to make the complementary-LVTSCR structure hard to be optimized. While the amount of added capacitance to an input pad is expected to increase to about twice of that in case of using the protection scheme (1), the area consumed by a pad structure is expected to be reduced a lot by eliminating the clamp NMOS device, and the peak voltages applied to gate oxides of input buffers can be reduced somewhat since the series clamp NMOS device disappears in a discharge path. Since HBM tests for input pins should include all possible discharge modes, tests are performed for five modes defined as PS, NS, PD, ND, and PTP modes [4]. Main discharge paths for test modes in each protection scheme are shown in Figures 9 and 10.

Figure 9 shows main discharge paths for each test mode when using the protection scheme (1). In a PS mode, a pnpn thyristor in T_1 provides a main discharge path, and in an NS mode, a forward-biased pn (p^+ -sub/ n^+ -anode) diode in T_1 provides it. In a PD mode, a pnpn thyristor in T_1 and a forward-biased pn (p^+ -sub/n^+ -drain) diode in M_2 in series provide a main discharge path, and in an ND mode, a parasitic npn bipolar transistor in M_2 and a forward-biased pn (p^+ -sub/n^+ -anode) diode in T_1 in series provides it. In a PTP mode, a pnpn thyristor in T_1 and a forward-biased pn (p^+ -sub/n^+ -anode) diode in T_3, which is located in the other pad, in series provide a main discharge path.

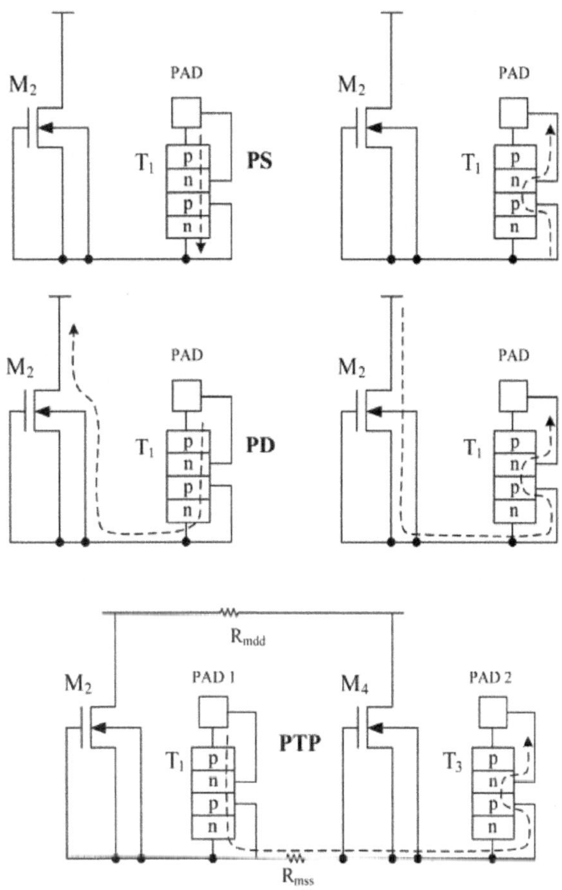

Figure 9. Main discharge paths for each test mode in the protection scheme (1).

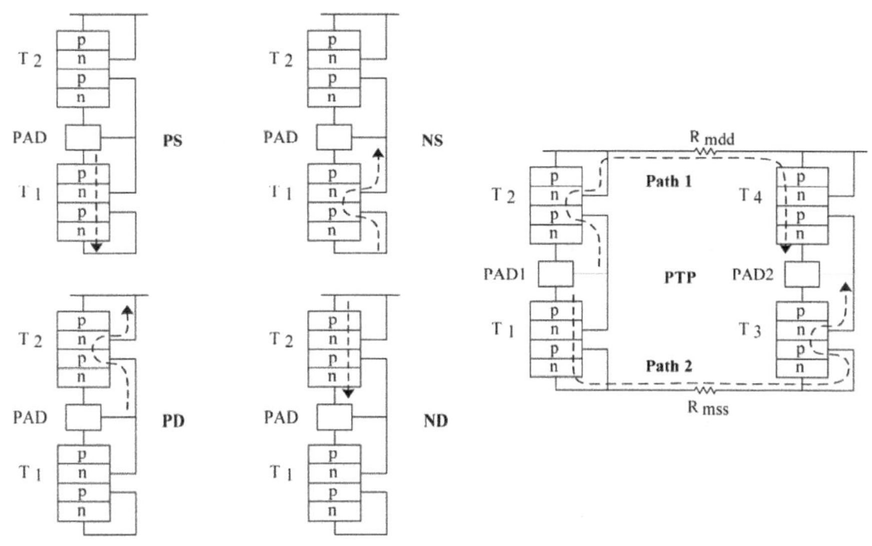

Figure 10. Main discharge paths for each test mode in the protection scheme (2).

Local lattice heating is proportional to a product of current density and electric field intensity, and therefore temperature-related problems in the protection devices can occur in the parasitic npn bipolar transistor rather than in the forward-biased diode or in the pnpn thyristor since the holding voltage of the bipolar transistor is much larger. Therefore the width of the lvtr_thyristor_down device can be small, however, we should assign a sufficient device width to M_2 considering an ND mode. Figure 10 shows main discharge paths for each test mode when using the protection scheme (2). In a PS mode, a pnpn thyristor in T1 provides a main discharge path, and in an NS mode, a forward-biased pn (p^+ -sub/ n_+ -anode) diode in T1 provides it. In a PD mode, a forward-biased pn (p^+ -cathode/ n^+ -anode) diode in T_2 provides a main discharge path, and in an ND mode, a pnpn thyristor in T_2 provides it. In a PTP mode, there exist two main discharge paths. One is a series path composed of a forward-biased pn (p^+ -cathode/n^+ -anode) diode in T_2 and a pnpn thyristor in T_4, and the other is a series path composed of a pnpn thyristor in T_1 and a forward-biased pn (p^+ -sub/n^+ -anode) diode in T_3. Since the holding voltages of the thyristor devices are not large, widths of all the thyristor devices don't need to be large.

MIXED-MODE TRANSIENT SIMULATIONS

Figure 11 shows an equivalent circuit of an input HBM test situation assuming a PS mode, which is fully explained in [4] previously. V_{ESD} is a HBM test voltage, and a switch S_1 charges C_{ESD} and then a switch S_2 initiates discharge. By utilizing time-varying resistors for the switches, the switching times of S_1 and S_2 were set short as 0.15 ns. In Figure 11, a V_{DD}-V_{SS} clamp NMOS protection device M_2 and a protection device T_1 form a representative protection circuit in the input pad, assuming usage of the protection scheme (1). In case of using the protection scheme (2), M_2 is eliminated, and the additional protection device T_2 should be inserted between the V_{DD} node and the pad.

A CMOS inverter is assumed as an input buffer inside a chip, which is modeled by a capacitive network. C_{ngate} and C_{pgate} represent gate-oxide capacitances of an NMOS transistor and a PMOS transistor, respectively. C_{ds} represents an n-well/p-sub junction capacitance. 0.1 pF, 0.1 pF, and 0.01 pF were assumed for C_{ngate}, C_{pgate}, and C_{ds}, respectively.

Using ATLAS, we performed mixed-mode transient simulations utilizing the equivalent circuit in Figure 11 equipped with two different input protection circuits shown in Figures 7 and 8. When a mixed-mode simulation is performed, the active protection devices are solved by device and circuit simulations simultaneously. For all the mixed-mode transient simulations performed for each test mode, $V_{ESD} = \pm 2000$ V was assumed. To have fair comparison on ESD robustness of the different protection schemes, widths of the protection devices were adjusted to maintain utmost peak lattice temperature inside the protection devices below 500 K in all the mixed-mode simulations, resulting 250 μm, 20 μm, and 20 μm for M2, T_1, and T_2, respectively.

As an example of the mixed-mode simulation results, Figure 12 shows variation of the anode current of T_1 as a function of time in a PS mode in case of using the protection circuit (1) in Figure 7. The anode current peaks up to 1.37A, and shows decaying characteristics with a time constant of roughly RESDCESD = 0.15 μs, which can be expected from the equivalent circuit in Figure 11. Figure 13 shows variations of the voltages developed across C_{ngate} and C_{pgate} in the input buffer from the same simulation result. In Figure 13, the pad voltage is not shown since it is almost same with the voltage developed across C_{ngate}.

From the DC simulation result in Figure 4, we can estimate transient discharge characteristics of T_1, which lies in the main discharge path in this case. When a positive ESD voltage is applied, the developed voltage across the device will increases at least up to the snapback voltage (9.4 V). As the bipolar transistor and the pnpn thyristor are triggered in order, the developed voltage will drop down to the holding voltage (1 V) and main discharge will proceed. In the later stage of discharge when the discharge current decreases below the holding current for the thyristor action, the developed voltage will increase again at most up to the snapback voltage (9.4 V) and will remain constant around the breakdown voltage (8.8 V) for some duration even though the discharge current decreases. As the discharge current decreases further, the device will go out of the breakdown mode and the developed voltage will decrease towards zero to end the discharge.

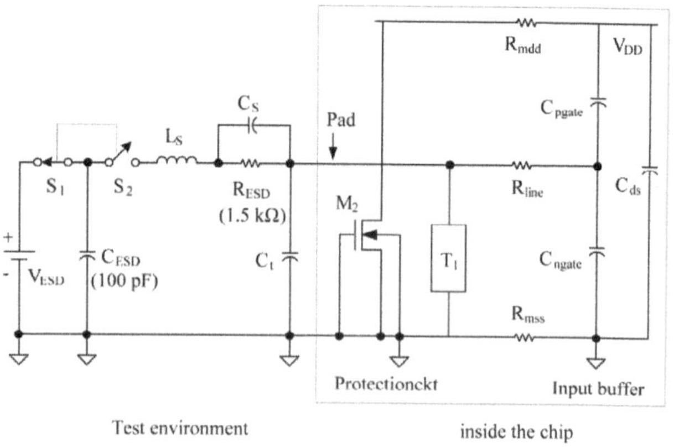

Figure 11. Equivalent circuit of an input-pin HBM test situation.

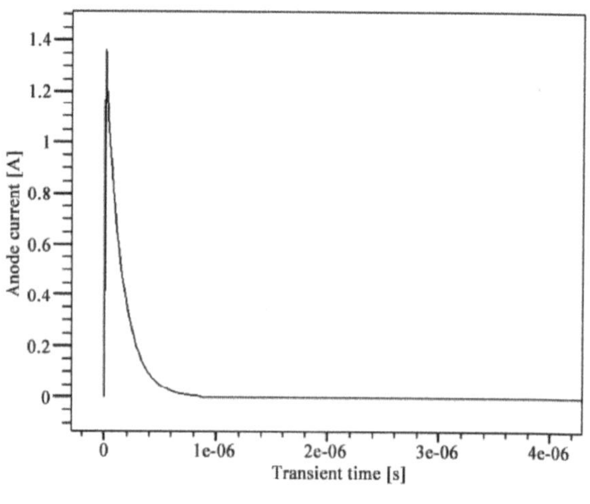

Figure 12. Variation of the anode current of T1 in a PS mode when using the protection circuit (1).

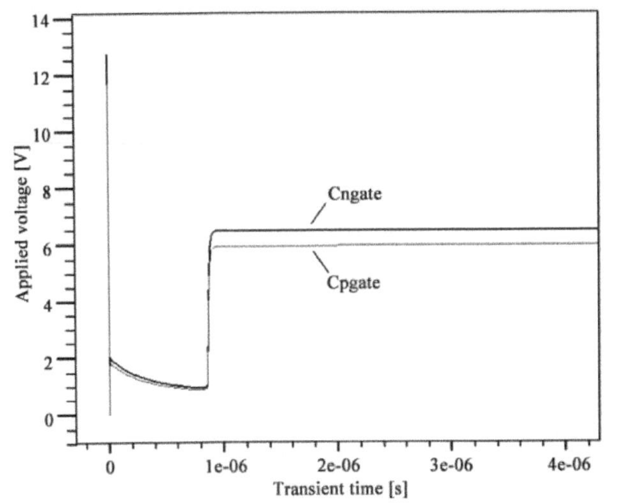

Figure 13. Variations of the voltages developed on C_{ngate} and C_{pgate} in a PS mode when using the protection circuit (1).

With the expectation above in mind, let's examine the results shown in Figures 12 and 13. From the results shown in Figures 12 and 13, we can see that the parasitic bipolar transistor in T_1 is triggered when the pad voltage in the early stage of discharge increases to about 12.8 V at 0.77 ns after S2 in Figure 11 is closed. The trigger voltage is lager than the expected DC value probably due to the time needed for charge redistribution. Main discharge through the pnpn thyristor proceeds as the pad voltage, which is equal to the anode-cathode voltage of T_1, drops to the holding voltage of about 2 V. The pad voltage decreases down to 1 V as the discharge current decreases with time. We can also see that the pad voltage increases again and reaches to 6.5 V at about 0.9 μs, when the anode current is reduced below the holding current for the pnpn thyristor action, and decreases very slowly thereafter. We confirmed from additional simulations that it takes 510 ms for the pad voltage to decrease down to 3 V. We also confirmed that main components of the anode current at 0.9 μs are the leakage current through the n-well/p-sub junction and the weak-inversion MOS current. The developed peak voltage in this later stage of discharge is smaller than the breakdown voltage (8.8 V) of the lvtr_thyristor device shown in Figure 4. This seems to be caused by the long duration (0.9 μs) of the main discharge by virtue of the excellent conducting pnpn thyristor. We confirmed that, with the sufficient discharge through the pnpn thyristor, the remaining discharge current level in this stage of discharge is only about 40 nA, which is too low for T_1 to conduct in a breakdown mode. We also confirmed from an additional simulation that, if we decrease the n^+ /p^+ anode contact space down to 0.7 μm, the pnpn thyristor turns off earlier at 0.815 μs and the developed peak voltage increases up to 9 V with the remaining discharge current level of about 0.1 mA, which is certainly high enough for the device to conduct in a breakdown mode.

Figure 13 shows that, in overall, a lower voltage by about 1 V is developed on C_{pgate} since the V_{DD} node does not lie in the main discharge path. Figure 14 shows variations of the voltages developed on C_{ngate} and C_{pgate} in a PS mode in case of using the protection circuit (2) in Figure 8. We confirmed that the variation of the anode current of T_1 is similar to that in Figure 12.

As we can see from Figure 14, the variation of the pad voltage, which is again almost same with the voltage developed on C_{ngate}, is similar to that in case of using the protection circuits (1). The parasitic bipolar transistor in T_1 is triggered when the pad voltage in the early stage of discharge increases to about 12.8 V at 0.77 ns after S_2 is closed. Main discharge through the pnpn thyristor proceeds as the pad voltage drops to the holding voltage of about 2 V.

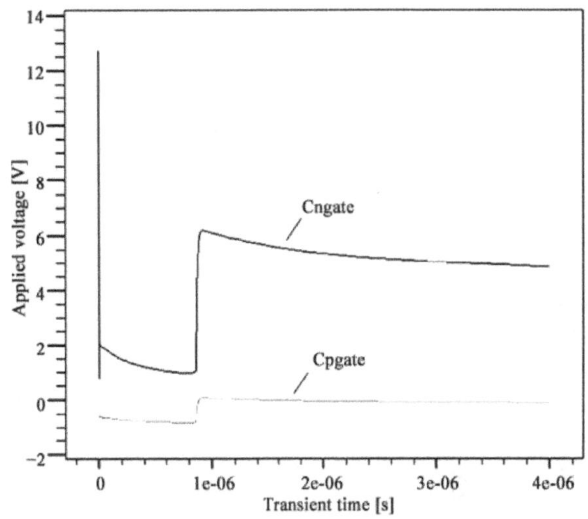

Figure 14. Variations of the voltages developed on C_{ngate} and C_{pgate} in a PS mode when using the protection circuit (2).

The pad voltage increases again and reaches to 6.2 V at about 0.9 μs, and decreases very slowly thereafter. We confirmed that the pad voltage (6.2 V) in this case is slightly lower than that in case of using the protection circuit (1) due to a difference in the current components. We confirmed that, when the pad voltage increases to 6.2 V, a pnp (p^+ -cathode/n-well/p^+ -sub) bipolar transistor in T_2 is triggered and the pad voltage is limited by the pnp bipolar holding voltage. Since the holding voltage is somewhat large, thermal heating may cause a problem. However, we confirmed that the bipolar current level at this moment is too low to cause thermal heating. The components of the anode current in T1 in this later stage of discharge also include the leakage current through the n-well/p-sub junction and the weak-inversion MOS current, however the bipolar current through T_2 is a major discharge current for some duration. Due to this current component, the pad voltage in this later stage of discharge decreases faster, compared to the case using the protection circuit (1). We confirmed that it takes 23 ms for the pad voltage to decrease down to 3 V.

In Figure 14, we can see that the voltage developed across C_{pgate} remains low all the time. This is because the p^+ -cathode/n^+ -anode diode is conducting if the V_{DD}-pad voltage becomes larger than the forward diode drop. In the later stage, there is almost no conduction through T_2, and the voltage stays close to zero.

Voltages across the Gate Oxides in the Early Stage of Discharge

For the two PS modes analyzed above, the trigger time for T_1 is 0.77 ns. Due to this trigger time, the voltage (12.8 V) larger than the DC snapback voltage is developed across T_1 right after S_2 is closed, resulting the high voltage developed on C_{ngate} in the early stage of discharge in Figures 13 and 14. Depending on test modes, larger peak voltages across the gate oxides in the input buffer appear at C_{ngate} or C_{pgate}. If we define the test modes, which produce larger peak voltages in the mixed-mode transient simulations performed for 5 test modes, as weak modes, the results can be summarized as shown in Table 2.

In Table 2, 13.3 V on Cpgate in the PD mode in case of using the protection scheme (1) corresponds to a sum of the voltages applied on the pnpn structure in T_1 and the forward diode in M_2, which can be easily expected from Figures 7 and 9. The voltage applied on the pnpn structure peaks up to 12 V in this case. 13.5 V on Cpgate in the ND mode in case of using the protection scheme (1) corresponds to a sum of the voltages applied on the npn structure in M_2 and the forward diode in T_1. The voltage applied on the npn structure peaks up to 10.8 V. 12.3 V on Cpgate in the ND mode in case of using the protection scheme (2) corresponds to the voltage applied on the pnpn structure in T_2, which can be easily expected from Figures 8 and 10.

The peak voltages in Table 2 can be regarded as excessive; however, durations of the peak voltages applied are very short. We confirmed that, for example, the durations for which the voltages exceed 10 V are at most 0.2 ns. Therefore it may be inferred that the gate oxides in the input buffer won't be damaged in the early stage of discharge [11]. Notice that the peak voltages can be suppressed by reducing the bipolar trigger voltage of the NMOS transistor in the NMOS protection device or the thyristor protection devices. To make the bipolar trigger voltage of the NMOS transistor even lower than the off-state DC breakdown voltage, the gate-coupled NMOS (gcNMOS) structure [12] can be adopted. It is possible to obtain a similar result by simply inserting a series resistor between the gate (G_2) and V_{SS} nodes of M_2 in Figure 7 since the gate-drain overlap capacitance (C_{gd}) already exists in the NMOS structure [4]. For the lvtr_thyristor_down device, the same technique can be applied since it includes the same NMOS structure in it [4]. For the lvtr_thyristor_up device, the same technique can be also applied. A series resistor inserted between the gate (G_2) and the input node in Figure 8 will do the role.

We performed addition simulations to confirm that the early peaking can be suppressed by adding the series resistor to the gate node. The results are summarized in Table 3. For the 250 μm M_2, a 10 kΩ resistor was inserted between the gate and V_{SS} nodes. For the 20 μm T_1, a 125 kΩ resistor was inserted between the gate and V_{SS} nodes. For the 20 μm T_2, a 125 kΩ resistor was inserted between the gate and the input pad nodes. It is certain that the early peaking can be suppressed if needed.

Voltages across the Gate Oxides in the Later Stage of Discharge

Depending on test modes, larger peak voltages across the gate oxides also appears at C_{ngate} or C_{pgate} in the later stage of discharge. If we define the test modes, which produce larger peak voltages, as weak modes, the results can be summarized as shown in Table 4. We confirmed that use of the gate-coupling techniques does not affect the peak voltages in the later stage of discharge at all.

Table 2. Peak voltages developed across the gate oxides in the early stage of discharge.

Protection scheme	Weak mode	Peak voltage [V] C_{ngate}	Peak voltage [V] C_{pgate}	Time [ns]
(1)	PS	12.8		0.77
	PD		13.3	0.66
	ND		13.5	0.82
(2)	PS	12.8		0.77
	ND		12.3	0.80

Table 3. Peak voltages developed across the gate oxides in the early stage of discharge when adopting the gate-coupling technique.

Protection scheme	Weak mode	Peak voltage [V] C_{ngate}	Peak voltage [V] C_{pgate}	Time [ns]
(1)	PS	8.7		0.65
	PD		9.7	0.57
	ND		10.4	0.79
(2)	PS	8.7		0.65
	ND		8.8	0.69

Table 4. Peak voltage developed across the gate oxides in the later stage of discharge.

Protection scheme	Weak mode	Peak voltage [V] C_{ngate}	Peak voltage [V] C_{pgate}	Time [μs]
(1)	PD	7.6		0.92
	ND		10.7	0.50
(2)	PS	6.2		0.92
	ND		8.8	0.89
	PTP	7.4	7.9	0.89

From Table 4, we can see that, in case of using the protection scheme (1), the ND mode is the weakest one. 10.7V developed on C_{pgate} corresponds to a sum of the breakdown voltage of M2 (9.6 V) and the forward diode drop in T_1 (1.1 V) right after the main discharge through M_2 and T_1 is finished, which can be easily expected from Figures 9 and 7. Differently from the results of the thyristor devices in Figures 13 and 14, the peak voltage developed across M_2 in this case is about same with the DC breakdown voltage (9.3 V). This is because the bipolar transistor in M_2 is not as excellent conducting as the pnpn thyristor, and the main discharge through the bipolar transistor ends earlier (at 0.5 μs) compared to that through the pnpn thyristor as shown in Table 4. In case of using the protection scheme (2), the ND mode is also the weakest one. 8.8 V developed on C_{pgate} corresponds to the voltage developed across T_2, which is somewhat smaller than the DC breakdown voltage (9.1 V) of T2. In a PTP mode, 7.4 V and 7.9 V are developed on C_{ngate} and C_{pgate}, respectively, which correspond to the voltages developed across T_1 and T_4 in Figure 10.

We note that high voltages in the later stage of discharge can damage gate oxides in input buffers since they last for long time. When judging from the peak voltages developed across the gate oxides in the later stage of discharge in Table 4, the weakest modes in case of using the protection scheme (1) is an ND mode, and the PMOS gate oxide is more vulnerable to HBM ESD damages if the gate-oxide thicknesses of the NMOS and the PMOS are same. In case of using the protection scheme (2), the weakest mode is also an ND mode and the PMOS gate oxide is also more vulnerable. When judging from the peak voltages developed, the advantage of the protection scheme (2) over the protection scheme (1) is expected to stand out more as the gate oxide thickness shrinks with advanced process technology used.

Location of Peak Temperature and Weak Modes

In case of using the protection scheme (1), the utmost peak temperature in a PS mode appears at T_1, and Figure 15 shows the variation of peak temperature inside T_1. Peak temperature increases up to 473 K at about 0.9 ns just before the pnpn thryristor trigger, but decreases down to 330 K as soon as the pnpn thryristor is triggered since the holding voltage decreases. It peaks again up to 421 K at about 45 ns with increasing discharge current, and decreases slowly with the discharge current decreasing. By examining 2-dimensional temperature distributions, we confirmed that peak temperature at 0.9 ns appears at the n^+ well junction, where the electric field intensity is highest, and that at 45 ns appears at the n^+ cathode junction, where the current density is highest. If we define the test modes, which produce larger temperature increase inside any protection device, as weak modes, the results can be summarized as shown in Table 5.

In case of using the protection scheme (1) incorporateing the 250 μm NMOS device and 20 μm lvtr_thyristor_down device, the weakest mode is the ND mode, and peak temperature appears in M_2, which conducts as an npn bipolar transistor. Peak temperature inside M_2 appears at the gate-side n^+ drain junction. This is the reason for assigning a large spacing between the gate and the drain contact in Figure 1 to avoid drain contact melting. The second weakest mode is the PS mode, and the 1st peak temperature appears in T_1, which happens just before the pnpn thyristor is triggered. At this point, peak temperature inside T_1 appears at the n^+ well junction

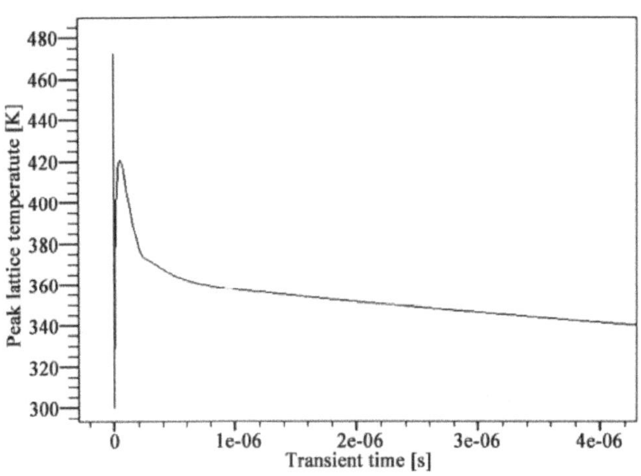

Figure 15. Peak temperature variation inside T_1 in a PS mode when using the protection scheme (1).

Table 5. Peak temperature locations and times.

Protection scheme	Weak mode	Peak temp.[°K]	Peak temperature: Location	Peak temperature: Time [ns]
(1)	PS	473	n^+ well junction in T_1	0.9
		421	n^+ cathode junction in T_1	45
	ND	495	gate-side drain junction in M_2	33
(2)	PS	471	n^+ well junction in T_1	0.8
		421	n^+ cathode junction in T_1	52
	ND	471	n^+ well junction in T_2	0.9
		473	n^+ cathode junction in T_2	44

at the right-hand corner of the n well. However, a problem with contact melting will not occur in this junction since there is no contact on it. The 2nd peak temperature in T_1 appears at the n^+ cathode junction when it conducts as a pnpn thyristor. Junction engineering such as increasing the junction area or adopting ESD ion implantation may be required to restrain temperature increase. However, it will not add parasitics to the input pad since the junction is not connected to it.

In case of using the protection scheme (2) incorporating the 20 μm lvtr_thyristor devices, the weakest mode is the ND mode, and the 1st peak temperature appears in T_2, which happens just before the pnpn thyristor is triggered. Peak temperature inside T_2 appears at the n^+ well junction at the left-hand corner of the p base. However, a problem with contact melting will not occur in this junction since there is no contact on it. The 2nd peak temperature in T_2 appears at the n^+ cathode junction when it conducts as a pnpn thyristor. Junction engineering such as increasing the junction area or adopting ESD ion implantation may be required to restrain temperature increase. This will not add parasitics to the input pad as long as the p-base region is not widened since the n^+ cathode is located inside the p base. As shown in Table 5, lattice heating characteristics in a PS mode are very similar to those in case of using the protection scheme (1). We confirmed from additional simulations incorporating the gate-coupling technique that all the temperature peaking prior to 1ns in Table 5 are suppressed below 380 K by virtue of the reduced bipolar trigger voltages. This can be easily expected from the results shown in Table 3.

From the result shown in Table 5, we can see that the 20 μm lvtr_thyristor devices are superior to the 250 μm NMOS device in ESD robustness in terms of thermal heating. Therefore we can save a lot of area consumed by a pad structure by eliminating the large clamp NMOS device.

DISCUSSIONS

Considerations in Designing the Lvtr_Thyristor_Down Device

By performing additional simulations, we figured out that a serious problem could occur if the p-type substrate contacts are not located close to the lvtr_thyristor_down device as shown in

Figure 3. When the p^+ -sub/n^+ -anode forward diodes in T_1 and T_3 in Figure 10 turn on in the early stage of discharge in the NS and PTP modes, respectively, the parasitic npn (n^+ -cath-ode/p-sub/n^+ -anode) bipolar transistor inside this small-sized device can be triggered to increase temperature around the n^+ cathode junction a lot, where electric field intensity is high. Therefore it is very important to locate the p^+ -sub contacts close as shown in Figure 3.

Considerations in Designing the Lvtr_Thyristor_Up Device

By performing additional simulations, we also figured out that a similar problem could occur if the n^+ anode2 contact at the right-hand side of the p base is not located in the lvtr_thyristor_up device as shown in Figure 5. When the p^+ -cathode/n^+ -anode forward diode in T_2 in Figure 10 gets on in the early stage of discharge in PD and PTP modes, a parasitic npn (n^+ -cathode/p-base/ n^+ -anode) bipolar transistor inside this small-sized device can be triggered to increase temperature around the n^+ cathode junction a lot. This can be completely solved by providing an additional p^+ -cathode/n^+ -anode2 forward diode path with the n+ anode2 contact as shown in Figure 5.

We also figured out that there is an important consideration to take care in connecting the gate node (G_2) in T_2 in Figure 8. Even though G_2 can be connected to the V_{SS} node without increasing an off-state leakage in normal operations, this may cause a problem by making the pnpn thyristor in T_1 never triggered in the PTP mode shown in Figure 10. This is because the voltage developed between the pad (connected to the n^+ / p^+ cathodes of T_2) and the V_{SS} node is restrained below 9 V, which is much smaller than the pnpn trigger voltage of 12.8 V, as a result of capacitive coupling between G_2 (connected to the VSS node) and the n^+ /p^+ cathodes of T_2. As a result, the discharge current flows mainly through the upper discharge path (Path 1) consisting of T_2 and T_4 in Figure 10. At the same time, the pnp (p^+ -cathode/n-well/p^+ -sub) bipolar transistor in T2 is triggered to provide another discharge path by way of the forward biased pn (p^+ -cathode/n^+ -anode) diode in T3. This causes a thermal heating problem by increasing lattice temperature near the p^+ sub junction at the right-hand corner of T_2 a lot. This pnp (p^+ -cathode/n-well/p^+ -sub) bipolar transistor is easily triggered since the p^+ -cathode/n-well diode is already forward biased due to the conduction through Path 1 in Figure 10. This problem is completely solved by making the pnpn thyristor in T_1 easy to be triggered by connecting G2 to the input pad as shown in Figure 8.

Providing Discharge Paths for VDD-VSS HBM Discharge

We have to check that a chip adopting the thyristor-only protection scheme can provide safe discharge paths when V_{DD}-V_{SS} HBM tests are performed. We note that large clamp devices such as the NMOS device shown in Figure 1 should be located between V_{DD} and V_{SS} buses in V_{DD} and V_{SS} pad structures and also anywhere a space is available to provide discharge paths for V_{DD}-V_{SS} ESD events and also to reduce a V_{DD} bounce during normal operation by increasing the capacitance between the two buses. We note that the bipolar trigger voltage of the clamp NMOS device without the gate-coupling technique was confirmed as less than 11 V relating the result shown in Table 3.

Using a single lvtr_thyristor protection device formed by assuming the lvtr_thyristor_down device in Figure 3 (T_1) and the lvtr_thyristor_up device in Figure 5 (T_2) located side by side on a same substrate, we confirmed by a mixed-mode simulation that the V_{DD}-V_{SS} peak voltage of the protection scheme (2) in a V_{DD}-V_{SS} HBM test is 17.6 V. The V_{DD}-V_{SS} voltage decreases down to 4 V (2 V each across T_1 and T_2) with both of the pnpn thyristors in T_1 and T2 being triggered. Therefore, in a V_{DD}-V_{SS} HBM ESD test, it is certain that the clamp NMOS devices will provide discharge paths before the pnpn path through T_1 and T_2 in any of input pad structures is triggered. Also when a surge voltage appears between V_{DD} and V_{SS} buses, the clamp NMOS device will constrain the rail voltage below 11 V to suppress the possibility of latch-up through T_1 and T_2. Also the latchup cannot persist since the conduction through T_1 and T_2 in series can be maintained only if the V_{DD}-V_{SS} voltage is higher than 4 V, which is higher than the normal supply voltage in recent technologies.

SUMMARY

We proposed an input protection scheme composed of thyristor devices only to minimize the size of an input pad structure. For this purpose, we suggested the lowvoltage triggering thyristor protection device structures assuming usage of standard CMOS processes, and attempted an in-depth comparison study with a conventional thyristor protection scheme incorporating a clamp NMOS device based on DC and mixed-mode transient analyses utilizing a 2-dimensional device simulator. We analyzed in detail the problems which can occur in real HBM tests to provide useful findings regarding the proposed protection scheme as follows.

1) We figured out weak modes in terms of peak voltages developed across gate oxides in input buffers.
2) We figured out weak modes in terms of temperature increase inside the protection devices, and also figured out locations of peak temperature inside the protection devices.
3) We suggested design guidelines for each protection device to minimize temperature increase inside it and to minimize voltages developed across gate oxides in input buffers.
4) We showed how we can incorporate the gate coupling technique into the suggested protection devices.
5) We showed that the suggested thyristor-only protection scheme can be made free from CMOS latch-up.

REFERENCES

1. P. Leroux and M. Steyaert, "High-Performance 5.2 GHz LNA with on-Chip Inductor to Provide ESD Protection," Electronics Letters, Vol. 37, No. 7, 2001, pp. 467-469. doi:10.1049/el:20010271
2. A. Chatterjee and T. Polgreen, "A Low-Voltage Triggering SCR for on-Chip ESD Protection at Output and Input Pads," IEEE Electron Devices Letters, Vol. 12, No. 1, 1991, pp. 21-22. doi:10.1109/55.75685

3. E. R. Worley, R. Gupta, B. Jones, R. Kjar, C. Nguyen and M. Tennyson, "Sub-micron Chip ESD Protection Schemes Which Avoid Avalanching Junctions," Electrical Overstress/Electrostatic Discharge Symposium Proceedings, Phoenix, 12 September 1995, pp. 13-20.

4. J. Y. Choi, "A Comparison Study of Input ESD Protection Schemes Utilizing NMOS, Thyristor, and Diode Devices," Communications and Network, Vol. 2, No. 1, 2010, pp. 11-25. doi:10.4236/cn.2010.21002

5. H. Feng, G. Chen, R. Zhan, Q. Wu, X. Guan, H. Xie and A. Z. H. Wang, "A Mixed-Mode ESD Protection Circuit Simulation-Design Methodology," IEEE Journal of Soilid-State Circuits, Vol. 38, No. 6, June 2003, pp. 995- 1006. doi:10.1109/JSSC.2003.811978

6. B. Fankhauser and B. Deutschmann, "Using Device Simulations to Optimize ESD Protection Circuits," International Symposium on Electromagnetic Compatibility, Santa Clara, 9-13 August 2004, pp. 963-968.

7. Silvaco International, "ATLAS II Framework," Version 5.10.2.R, Silvaco International, Austin, 2005.

8. J.-Y. Choi, W. S. Yang, D. Kim and Y. Kim, "Thyristor Input-Protection Device Suitable for CMOS RF ICs," Analog Integrated Circuits and Signal Processing, Vol. 43, No. 1, April 2005, pp. 5-14. doi:10.1007/s10470-005-6566-y

9. H. Feng, K. Gong and A. Wang, "A Comparison Study of ESD Protection for RFIC's: Performance vs. Parasitic," 2000 IEEE Radio Frequency Integrated Circuits Symposium, Boston, 11-13 June 2000, pp. 143-146.

10. M.-D. Ker, C.-Y. Wu and H.-H. Chang, "Complementary-LVTSCR ESD Protection Circuit for Submicron CMOS VLSI/ULSI," IEEE Transactions on Electron Devices, Vol. 43, No. 4, 1996, pp. 588-598. doi:10.1109/TED.1996.1210725

11. Z. H. Liu, E. Rosenbaum, P. K. Ko, C. Hu, Y. C. Cheng, C. G. Sodini, B. J. Gross and T. P. Ma, "A Comparative Study of the Effect of Dynamic Stressing on High-Field Endurance and Stability of Reoxidized-Nitrided, Fluorinated and Conventional Oxides," International Electron Devices Meeting, Washington, 8-11 February1991, pp. 723-726.

12. G. Chen, H. Fang and A. Wang, "A Systematic Study of ESD Protection Structures for RF ICs," 2003 IEEE Radio Frequency Integrated Circuits Symposium, Philadelphia, 8-10 June 2003, Vol. 46, pp. 347-350.

CHAPTER 14

A CMOS 3.1 - 10.6 GHZ UWB LNA EMPLOYING MODIFIED DERIVATIVE SUPERPOSITION METHOD

Amir Homaee

Keywords: Broadband; Low-Noise Amplifier (LNA); Noise Figure; Ultra-Wideband (UWB); Modified Derivative Superposition Method

INTRODUCTION

Development of the high-speed wireless communication systems puts increasing request on integrated low-cost RF devices with multi-GHz bandwidth operating at the lowest power consumption and supply voltage. Ultra wide band (IEEE 802.15.3a) appears as a new technology capable for high data transfer rates (up to 1 Gb/s) within short distances (10 m) at low power. This technology uses for some application such as wireless personal area networks (WPANs), providing an environment for transmission of audio, video, and other high-bandwidth data [1]. The amplifier that is used for this application must meet several requirements. For example to interface with the preselect filter and antenna, the amplifier input impedance should be close to 50 over the desired UWB band. However sufficient gain with wide band width to overtop the noise of a mixer, low noise figure to improve receiver sensitivity, low power consumption to increase battery life, small die area to reduce the cost, unconditional stability and good linearity are important parameters. There is a close trade-off between them. There are some proposed solutions and circuits for each parameter [2-8]. However, some parameters would be ruined by improving the others [4]. In this research a new circuit has achieved via modifying these methods [1,9]. The main parameter in this research is noise figure which has noticeably improved in comparison with the other references. It is 2.66 - 3 dB over 3.1 - 10.6-GHz band width.

INPUT STAGE

Common-gate and Cascode configurations are two kinds of methods usually used to design the input stage of LNA in CMOS circuits, while the Common-Gate and Cascode structure provides a wide-band and narrow-band input matching respectively. However Common-gate stage has an intrinsically high noise figure versus Cascode stage and the noise-canceling techniques must be used. In the narrow band application, a shunt inductor is added in the input stage to resonate with C_{gs} to enhance impedance matching at the desired frequency. However in most CMOS narrow band applications, cascode LNA with inductive degeneration is preferable but for isolating from the input to the output and omitting of the C_{gd} path, the Common-Gate LNA performs better reverse isolation and stability versus Common-Source LNA. Numerical value for the lower bound is about 2.2 dB for long-channel devices and 4.8 dB for short channel devices.

Citation: A. Homaee, "A CMOS 3.1 - 10.6 GHz UWB LNA Employing Modified Derivative Superposition Method," Circuits and Systems, Vol. 4 No. 3, 2013, pp. 323-327. doi: 10.4236/cs.2013.43044.

CIRCUIT DESIGN AND ANALYSIS

The proposed wide-band LNA is shown in Figure 1.

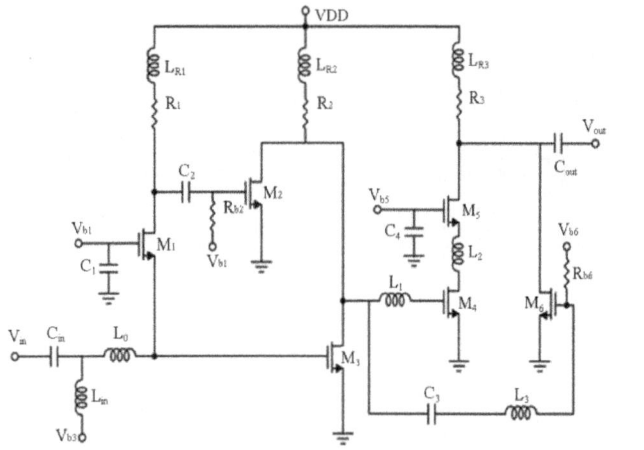

Figure 1. Proposed broadband noise-canceling LNA.

It consists of an input stage and a cascode second stage. An off-chip bias-T provides the gate bias of M_3 and the DC current path of M_1. The series inductors L_1 and L_3 further resonate with the input gate-source capacitance of M_4 and M_6 respectively, resulting in a larger bandwidth and some residual peaking on the frequency response [10]. The parasitic capacitances of M_1 and M_3 make an LC ladder structure with inductor L_0. The DC load resistors R_1 and R_2 are combined with shunt peaking inductors L_{R1} and L_{R2} respectively to extend circuit bandwidth effectively [11]. The series peaking inductor L_{R2} also resonate with the total parasitic capacitances C_{d2} and C_{d3} at the drain of M_2 and M_3. Since the load resistor, R_3, is added to reduce the Q factor of L_{R3} for flat gain and can be directly substitute for a switching quad to form a single-balanced mixer then the output 50 ohm matching is not demanded in an integrated receiver. The minimum channel length of 0.18 µm is considered for all the transistors in the proposed circuit to minimize parasitic capacitances and improve frequency performance. The Cascode stage extends bandwidth, provides better isolation and increases frequency gain. In fact the input stage and the Cascode stage support low-frequency power gain and high-frequency power gain, respectively. The combination of both frequency responses lead to a broadband power gain.

Table 1 shows the design values of the proposed CMOS LNA.

Table 1. Design values of the proposed CMOS LNA.

L_{in}	4 nH	**(W/L)3**	124/0.18
L_0	0.6 nH	**(W/L)4**	37.5/0.18
L_{R1}	4.5 nH	**(W/L)5**	55/0.18
L_{R2}	2.5 nH	**(W/L)6**	90/0.18
L_{R3}	1.2 nH	C_{in}, C_3, C_4	2PF
L_1	2.76 nH	C_{out}	7PF
L_2	0.7 nH	C_1, C_2	1PF
L_3	2 nH	R_1	320 Ω
(W/L)1	15/0.18	R_2	135 Ω
(W/L)2	24.3/0.18	R_3	80 Ω

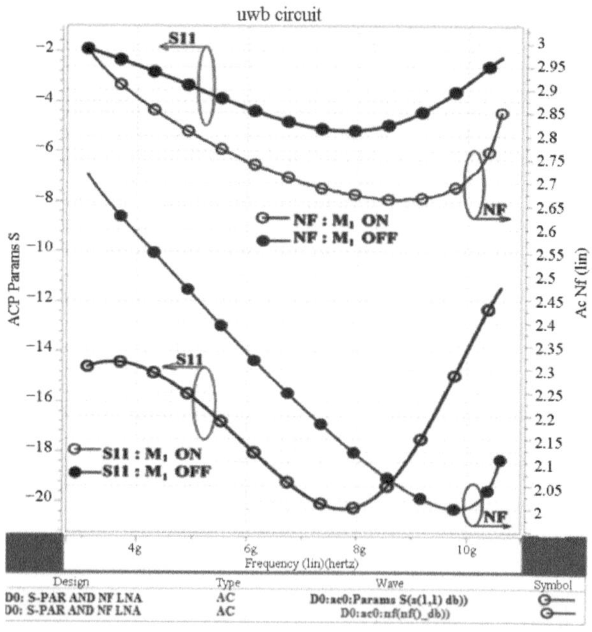

Figure 2. Simulated noise figure and input isolation with M_3 turned ON and OFF.

INPUT COMMON-GATE STAGE AND NOISE ISSUES

In Figure 2 the simulated NF and S11 parameter is compared to the case with M1 is turned OFF. There is a close tradeoff between NF and S11. When M_1 is turned on, the NF is increased by at least 0.6 dB and S21 parameter is decreased 2 dB with the same power dissipation and a similar bandwidth, but on the contrary an acceptable input matching will be achieved. Although the transistor M_1 provides a wide-Extra band matching, it has an intrinsically high noise figure. In order to investigate the noise performance, the MOS transistor noise model with the channel thermal noise is used. As shown in Figure 3, neglecting gate and flicker noises and assuming a perfect match in this analysis, the PSD of the channel thermal noise $\overline{i_{n,d}^2}$ is given as (1)

$$\overline{i_{n,d}^2} = 4KT\gamma g_{do}\Delta f = 4KT\frac{\gamma}{\alpha}g_m\Delta f \tag{1}$$

where k is the Boltzmann constant, T is the absolute temperature in Kelvin, g is the MOS transistor's coefficient of channel thermal noise, α is defined as the ratio of the transconductance g_m and the zero-bias drain conductance g_{ds} and Δf is the bandwidth over which the noise figure is measured respectively.

If the condition (2) is established the noise of the M_1 is omitted [1].

$$g_{m_2}R_1 = g_{m_3}R_s$$

The following equations describe the noise figure by R_1, M_2 and M_3 that they contribute to the overall noise figure.

$$F_{R_1} = \frac{4KTR_1 g_{m_2}^2}{KTR_s\left(g_{m_3} + \frac{g_{m_2} R_1}{R_s}\right)^2} = \frac{R_s}{R_1} \tag{3}$$

$$F_{M_2} = \frac{4KT\gamma/\alpha g_{m_2}}{KTR_s\left(g_{m_3} + g_{1m_1}\left(Z_{L_{R_1}} \| r_{o1}\right) g_{m_2}\right)^2} = \frac{\gamma}{\alpha} \frac{1}{g_{m_2} R_1} F_{R_1} \tag{4}$$

$$F_{M_3} = \frac{4KT\gamma/\alpha g_{m_3}}{KTR_s\left(g_{m_3} + g_{m_1}\left(Z_{L_{R_1}} \| r_{o1}\right) g_{m_2}\right)^2} = \frac{4\gamma/\alpha}{g_{m_3} R_s\left(1 + R_s g_{m_1}\right)^2} \tag{5}$$

Thus, the total noise figure can be approximated as (6)

$$F_{\text{total}} = \frac{R_s}{R_1}\left(1 + \frac{\gamma}{\alpha} \frac{1}{g_{m_2} R_1}\right) + \frac{4\gamma/\alpha}{g_{m_3} R_s\left(1 + R_s g_{m_1}\right)^2} \tag{6}$$

SIMULATION RESULT

The circuit was simulated with 0.18 μm TSMC library Hspice simulator. All simulations are done considering 50 Ω input and output terminals. In Figure 4 S parameter are simulated. S11 and S22 are approximately less than −11 dB. The average gain power is approximately 14 dB with 0.8 dB ripple over the frequency range and the reverse isolation is less than −33 dB. The measured noise figure is 2.66 - 3 dB over 3.1 - 10.6-GHz.

MODIFIED DERIVATIVE SUPERPOSITION METHOD FOR LINEARIZING

In this section by using the modified derivative superposition method [9], the linearity of LNA will be improved,

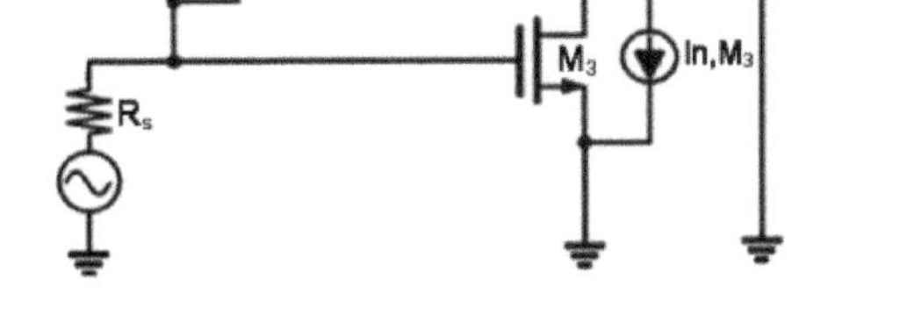

Figure 3. Principle of the noise schematic.

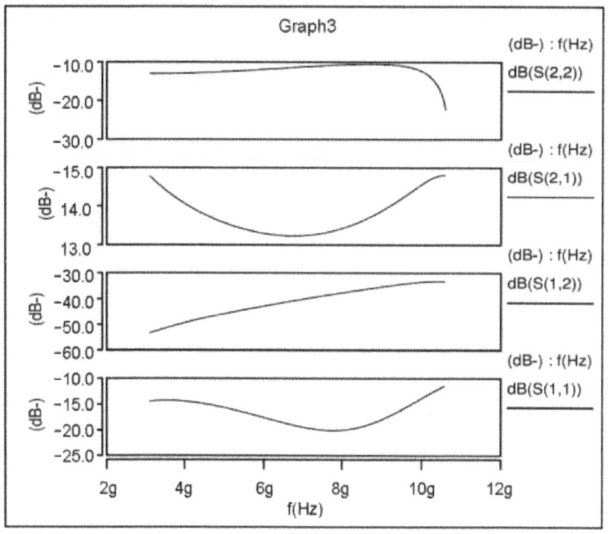

Figure 4. Simulated S parameter.

and IIP3 will be increased over the UWB band. The small-signal output current of a common-source biased in saturation region can be expressed as

$$i_d\left(v_{gs}\right) = g_1 v_{gs} + g_2 v_{gs}^2 + g_3 v_{gs}^3 + \cdots \tag{7}$$

where g_1 is the small-signal transconductance and the higher order coefficients (g_2, g_3, etc.) explain the strengths of the corresponding nonlinearities [9]. Among these coefficients, g_3 is the most important parameter because the third-order inter modulation distortion (IMD3) depends it and thus determines IIP3. The coefficients of (7) can be derived as (8)

$$g_1 = \frac{\partial I_D}{\partial V_{GS}}, g_2 = \frac{1}{2}\frac{\partial^2 I_D}{\partial V_{GS}}, g_3 = \frac{1}{6}\frac{\partial^3 I_D}{\partial V_{GS}} \tag{8}$$

when v_{gs} crosses from the weak and moderate inversion regions to the strong inversion (SI) region, g_3 changes from positive to negative [12]. If a positive g_3 with a specific g_3 (VGS) curvature of one MOSFET is aligned with a negative g3 with a similar, but mirror-image curvature of another MOSFET by offsetting their gate biases, and the g_3 magnitudes are equalized through a relative MOSFET scaling, the theoretical AIP3 will be efficiently improved in a wide range of the gate biases and the resulting composite g_3 will be close to zero [9].

As shown in Figure 5 at the optimum gate biases, when two FET are paralleled and one of them operates in the weak inversion (WI) region near the peak in its positive g3 and another works in the SI region near the dip in its negative g_3, the achieved AIP3 will be improved. Figure 6 presents the effect of modified deriva perposition method on the similar circuit [9]. By using this method IIP3 increases notably across the UWB band. Figure 7 shows the effect of using modified DS ethod on the IIP3 versus frequency respectively. If the M_6 is omitted the IIP3 change as Figure 7 but other parameters do not change considerably.

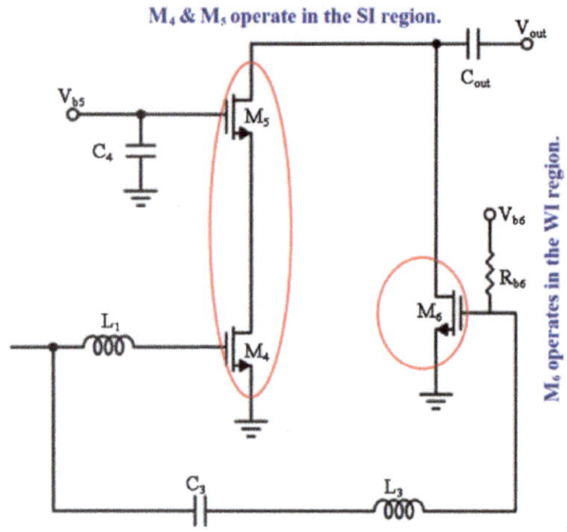

Figure 5. Modified derivative superposition meth d for o linearizing.

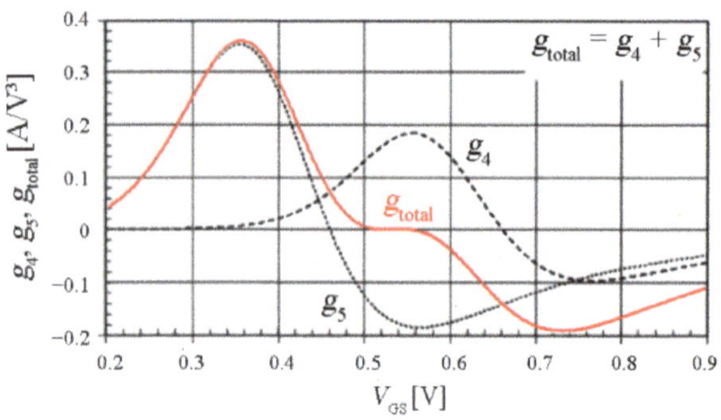

Figure 6. Third-order power series coefficients.

The results of this work are shown in Table 2 and are compared with recently published CMOS LNAs.

CONCLUSION

This paper presents a new design of an UWB LNA structure based on a standard RFCMOS technology. Satisfactory input matching and noise performance are obtained after regarding the tradeoff between the input impedance of the common-gate stage and its noise performance. The measured noise figure is 2.66 - 3 dB over 3.1 - 10.6-GHz that is noticeable in comparison with the other references. A flat gain is worth mentioning in all LNA design and the simulated power gain is 14 ± 0.8 dB.

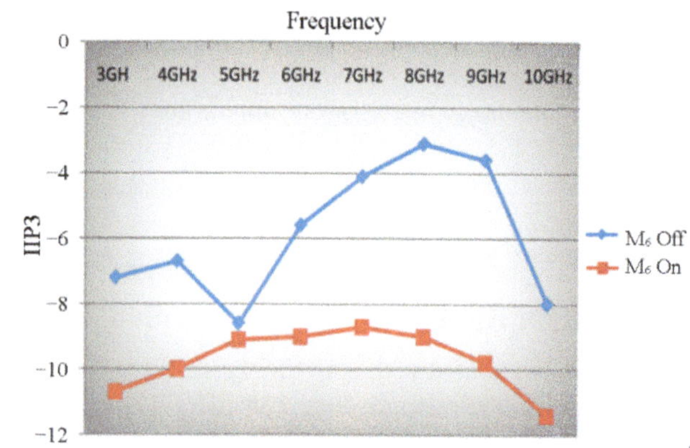

Figure 7. Measured IIP3 versus frequency.

Table 2: Performance summery.

Ref.	CMOS Technology	S11 (dB)	S22 (dB)	S12 (dB)	S21 (dB)	BW3-dB (GHz)	NF (dB)	Power (mw)	IIP3 (dbm)
This work	0.18 µm CMOS	<-11.5	<-10.5	<-33	13.2 - 14.8	3.1 - 10.6	<3	23.7	-3.1 - -8.6
[1]	0.18 µm CMOS	<-11	<-12	<-32	9.7	1.2 - 11.9	<5.4	20	-6.2
[2]	0.18 µm CMOS	<-9.4	<-8	<-40	10.9 - 13.9	3.1 - 10.6	<4.7	14.4	-8.5
[3]	0.18 µm CMOS	<-5.7	<-13.7	-	7.6 - 10.8	3.1 - 10.6	3.9 ~ 5.8	6.2	-5
[4]	0.18 µm CMOS	<-9	<-13	-	15.9 - 17.5	3.1 - 10.6	3.1 - 5.7	33.2	-
[5]	0.13 µm CMOS	<-11	-	-	13.5	2.6 - 10.7	2.7 - 4.2	13.5	+5

It consumes 23.7 mW from a 1.8 V supply. By employing modified derivation superposition method, the third-order intercept point, IIP3, is improved significantly.

ACKNOWLEDGEMENTS

The author would like to thank for support from research and development section of National Iranian Oil Company (NIOC) and Iranian Offshore Oil Company (IOOC).

REFERENCES

1. C.-F. Liao and S.-I. Liu, "A Broadband Noise-Canceling CMOS LNA for 3.1 10.6-GHz UWB Receivers," IEEE Journal of Solid-State Circuits, Vol. 42, No. 2, 2007, pp. 329-339. doi:10.1109/JSSC.2006.889356
2. K.-C. He, M.-T. Li, C.-M. Li and J.-H. Tarng, "Parallel-RC Feedback Low-Noise Amplifier for UWB Applications,"

IEEE Transactions on Circuit and System-II: Express Briefs, Vol. 57, No. 8, 2010, pp. 582-586.

3. Z.-Y. Huang, C.-C. Huang, C.-C. Chen, C.-C. Hung and C.-M. Chen, "An Inductor-Coupling Resonated CMOS Low Noise Amplifier for 3.1 10.6 GHz Ultra-Wideband System," IEEE International Symposium on Circuits and Systems, Taipei, 24-27 May 2009, pp. 221-224.

4. Y. Lu, K.-S. Yeo, A. Cabuk, J. Ma, M.-A. Do and Z. Lu, "A Novel CMOS Low-Noise Amplifier Design for 3.1-to 10.6-GHz Ultra-Wide-Band Wireless Receivers," IEEE Transactions on Circuit and System-I, Vol. 53, No. 8, 2006, pp. 1683-1692. doi:10.1109/TCSI.2006.879059

5. A. Mirvakili, M. Yavari and F. Raissi, "A Linear Current-Reuseed LNA for 1 10.6 GHz UWB Receivers," IEICE Electronics Express, Vol. 5, No. 21, 2008, pp. 908-914. doi:10.1587/elex.5.908

6. A. Mirvakili and M. Yavari, "A Noise-Canceling CMOS LNA Design for the Upper Band of UWB DS-CDMA Receivers," IEEE International Symposium on Circuits and Systems, Taipei, 24-27 May 2009, pp. 217-220.

7. C.-P. Liang, C.-W. Huang, Y.-K. Lin and S.-J. Chung, "3 10 GHz Ultra-Wideband Low-Noise Amplifier with New Matching Technique," Electronic Letters, Vol. 46, No. 16, 2010, pp. 1102-1103. doi:10.1049/el.2010.1279

8. H. Wang, L. Zhang and Z. Yu, "A Wideband Inductorless LNA with Local Feedback and Noise Cancelling for Low-Power Low-Voltage Applications," IEEE Transactions on Circuit and System-I: Regular Papers, Vol. 57, No. 8, 2010, pp. 1993-2005.

9. V. Aparin and L.-E. Larson, "Modified Derivative Superposition Method for Linearizing FET Low-Noise Amplifiers," IEEE Transactions on Microwave Theory and Techniques, Vol. 53, No. 2, 2005, pp. 571-581. doi:10.1109/TMTT.2004.840635

10. T.-H. Lee, "The Design of CMOS Radio-Frequency Integrated Circuits," Cambridge University Press, New York, 1998.

11. S.-S. Mohan, M.-D.-M. Hershenson, S.-P. Boyd and T.-H. Lee, "Bandwidth Extension in CMOS with Optimized On-Chip Inductors," IEEE Journal of Solid-State Circuits, Vol. 35, No. 3, 2000, pp. 346-355. doi:10.1109/4.826816

12. C. Xin and E.-S. Sinencio, "A Linearization Technique for RF Lownoise Amplifier," Proceedings of the 2004 International Symposium on Circuits and Systems, Vancouver, 23-26 May 2004, p. IV-313-16.

CHAPTER 15

CONDUCTIVE-BRIDGING RANDOM ACCESS MEMORY: CHALLENGES AND OPPORTUNITY FOR 3D ARCHITECTURE

Debanjan Jana, Sourav Roy, Rajeswar Panja, Mrinmoy Dutta, Sheikh Ziaur Rahaman, Rajat Mahapatra, Siddheswar Maikap

Keywords: CBRAM, Conductive bridge, Resistive switching, Chalcogenide, Solid electrolyte, Bilayer, Three-dimensional (3D)Memory

BACKGROUND

In recent days, resistive random access memory (RRAM) technology is one of the most promising and reliable alternative solutions to overcome the scaling bottleneck of FLASH [1]. The advantages of RRAM devices are their simple metal-insulator-metal structure, low fabrication cost, long endurance, and non-volatile properties with low power consumption, multi-level cell (MLC) operation, and especially excellent scaling below the <11 nm technology node [2-5]. One of the RRAM devices is conductive-bridging random access memory (CBRAM), which has high possibility to fulfill the requirements for next-generation non-volatile memory (NVM) technology as mentioned in the International Technology Roadmap for Semiconductors (ITRS) [1]. In addition, major benefits are high speed (few nanoseconds) and low voltage operation (±3 V). Basically, the structure of CBRAM devices consists of one metal electrode which is electrochemically active (i.e., anode) or oxidized under external positive bias, such as Ag or Cu, and another one which is electrochemically inert (i.e., cathode), such as platinum (Pt), iridium (Ir), gold (Au), tungsten (W), or titanium-nitride (TiN). These two electrodes are separated by a solid electrolyte or oxide materials. In 1976, Hirose et al. [6] reported memory switching using a Ag dendrite in a Ag-doped As_2S_3 film in a Ag/As_2S_3/Mo structure. In 1999, Kozicki et al. [7] reported a programmable metallization cell (PMC) device using metal-doped chalcogenide films. PMC devices have different names like electrochemical metallization cell (ECM), 'atom switch' [8], or CBRAM [9]. Generally, they are called 'CBRAM'. According to their concepts, when a positive voltage is applied on the anode, an electrochemical reaction occurs, which oxidizes the metal to form ions (Cu^{z+}, $z = 1$, 2, or Ag^+). The cations drift through the solid electrolyte switching layer under the electric field, and the metal ions are reduced on the inert electrodes. As this process continues, a metallic filament is established in between the two electrodes and the device switches from the high resistance state (HRS) to the low resistance state (LRS). By changing the polarity of the voltage, an electrochemical dissolution of the conductive bridges takes place, resetting the device from the LRS to the HRS. The resistance state changes by applying external bias, and the switching is reversible. It is true that device reliability including stable switching characteristics, program/erase endurance, and data retention at 85°C for 10 years at a low current of few microamperes is a very challenging issue for production. In addition, the switching mechanism in different structures including materials and electrodes is not understood clearly. Especially, the growth kinetics of the metallic filament in oxide- and non-oxide-based materials are still debated. Although the memory performances of a designed structure with low operation current have huge opportunities to fulfill the requirements of ITRS, they are not achieved yet. Even though many papers have been reported by several groups, a complete review on CBRAM performances including switching characteristics, reliability, mechanism, and so on for real production is needed, which is not reported yet.

Here, the challenges and opportunities of CBRAM devices using different switching materials such as chalcogenides, oxides, and bilayers in different structures have been reviewed. Memory performances such as switching characteristics, endurance, multi-level cell operation, and data retention have

Citation: Debanjan Jana, Sourav Roy, Rajeswar Panja, Mrinmoy Dutta, Sheikh Ziaur Rahaman, Rajat Mahapatra, Siddheswar Maikap, Conductive-bridging random access memory: challenges and opportunity for 3D architecture, DOI: 10.1186/s11671-015-0880-9.

been discussed. Memory devices can be operated with an operation current of 0.1 μA to 1 mA under an operation voltage of ±2 V. Filament growth dynamics in different switching materials have been discussed. Device reliabilities such as switching uniformity, endurance, and data retention under a few microamperes are very challenging issues. The device can be scaled down below the 11 nm technology node and has a very good opportunity for non-volatile crossbar memory for 3D architecture in the near future. Finally, CBRAM as a chip has been discussed for future production.

MATERIALS AND DEPOSITION METHODS

CBRAM performance depends on switching materials including different non-oxides and oxides. The non-oxides, i.e., chalcogenides, are $GeSe_x$ [10-12], GeS_2 [13,14], GeTe [15], Cu_2S [16], Ag_2S [8], and so on. The oxide-based materials are Ta_2O_5 [16,17], SiO_2 [18], ZrO_2 [19], GeO_x [20,21], and so on. Other materials such as amorphous Si [22] and Si_3N_4 [23] have been reported also. Some reported results show CBRAM characteristics using bilayer switching materials: one layer is used as a buffer or interfacial layer and the other one is used as a switching layer. Bilayers such as Cu-Te/GdO_x [24], Cu-Te/SiO_x [25], MoO_x/GdO_x [26], TiO_x/$TaSiO_y$ [27], $GeSe_x$/TaO_x [28], Ti/TaO_x [29], Cu-Te/Al_2O_3 [30], TiW/Al_2O_3 [31], and CuTe-C/Al_2O_3 [32] have been reported by many groups. Deposition process parameters of different switching materials are discussed below.

Memory devices using chalcogenide-based switching materials are reported below. Chalcogenide glasses are based on the chalcogen elements, and they have a wide range of properties such as optical and electrical ones, as reported in the literature. Several types of chalcogenide-based solid electrolytes have been reported for future resistive switching memory applications. Kozicki et al. [10] reported a Ag/$Ag_{33}Ge_{20}Se_{47}$/Ni PMC device. A 100-nm-thick Ni as a bottom electrode (BE) was fabricated on SiO_2/Si substrates. Then, 50-nm-thick $Ge_{70}Se_{30}$ was evaporated with the help of a Knudsen-type cell under 10^{-6} Torr vacuum. This approach will help in maintaining almost the same level of elementary content of the switching material. A very low deposition rate of 0.03 nm/s was maintained to enhance the step coverage and filled into the narrow via-holes. Immediately after that, a 30-nm-thick Ag film was deposited to form a Ag:GeSe solid electrolyte. In this case, Ag diffusion by exposure to 405-nm ultraviolet (UV) radiation was time consuming and cost effective. To avoid this issue, Schindler et al. [11] reported a Ag/Ag-GeSe/Pt structure. Their memory cells consisted of a 100-nm-thick continuous Pt BE on SiO_2/Si substrates. A Si_3N_4 layer was deposited on Pt by plasma-enhanced chemical vapor deposition (PECVD) method with a deposition rate of 0.2 nm/s. Via-holes with varying diameters from 2.5 to 50 μm were patterned by optical lithography. The solid electrolyte layer was 50-nm-thick $Ge_{0.3}Se_{0.7}$ and was deposited by radio-frequency sputtering or physical vapor deposition (PVD) at a rate of 0.2 nm/s. The applied power ranged from 5 to 25 W. The top electrodes were patterned by optical lithography. Then, the Ag layer was deposited by thermal evaporation at a deposition rate of 0.5 nm/s. After that, a lift-off process was carried out to obtain the final device. Due to the low melting point of Se (221°C), it is difficult to control the composition of GeSe after deposition by sputtering. To keep the same composition, the GeSe material was deposited by using an electron beam evaporator, which we have also reported previously [12]. The resistive switching material and electrodes were deposited by PVD processes such as sputtering, electron beam evaporation, and thermal evaporation. Initially, 8-in. p-type Si (100) wafers were used to deposit a SiO_2 layer with a thickness of approximately 200 nm. The SiO_2 layer was grown by thermal oxidation. Then, W or TiN metal was used as a BE on the SiO_2/Si substrates. The thickness of the BE was approximately 200 nm. To design the memory devices, a SiO_2 layer with a thickness of approximately 150 nm was deposited. Lithography was used to pattern via-holes with different nominal areas from 200×200 nm^2 to 8×8 μm^2 of the active switching region. After defining the via-hole regions, reactive-ion etching (RIE) process was used to etch the uncovered SiO_2 films and to open the W BE contact. Following this, a Ge_xSe_{1-x} solid electrolyte with a thickness of approximately 40 nm was deposited using an electron beam evaporator. Pure Ge_xSe_{1-x} ($x = 0.2$ to 0.5) granules (commercially available) with different compositions such as $Ge_{0.2}Se_{0.8}$, $Ge_{0.3}Se_{0.7}$, $Ge_{0.4}Se_{0.6}$, and $Ge_{0.5}Se_{0.5}$ were used for deposition. The chamber pressure was 5×10^{-6} Torr prior to deposition. The deposition rate was 1 to 2 Å/s. Then, a Cu top electrode (TE) with a thickness of approximately 40 nm was deposited using a thermal evaporator. A 160-nm-thick layer of Al was then deposited *in situ* using the same thermal evaporator. Otherwise, a 100- to 200-nm-thick Cu electrode was used. Finally, a lift-off process was performed to obtain the resistive switching memory device. To investigate the memory device and microstructure of the solid electrolyte films, transmission electron microscopy (TEM) was carried out using a FEI Tecnai G2F-20 field emission system with an energy of 200 kV. The memory devices for TEM observation were prepared using an FEI Helios-400s system with an operating voltage of 5 kV and using a Ga^+ion source. Figure 1a shows a TEM image and energy-dispersive X-ray spectroscopy (EDX) analysis of a Cu/$Ge_{0.4}Se_{0.6}$/W structure [33]. The thicknesses of the Cu and W electrodes are found to be 136 and 85 nm, respectively. The thickness of the $Ge_{0.2}Se_{0.8}$ solid electrolyte films is approximately 38 nm. A small via-hole size is about 140 nm in width, which is lower (slightly) than our design size (200 nm). The $Ge_{0.4}Se_{0.6}$ film is amorphous, as shown in the outside (Figure 1b) and inside (Figure 1c) of the via-hole regions. A layer-by-layer structure is observed from those images. All materials such as Cu, Ge, Se, and W in the Cu TE, GeSe switching, and W BE layers are confirmed by EDX analysis (Figure 1d). Figure 2a shows the X-ray photo-electron spectroscopy (XPS) spectra of the Ge_xSe_{1-x} solid electrolyte films. The peak binding energies of $Ge_{0.5}Se_{0.5}$, $Ge_{0.4}Se_{0.6}$, $Ge_{0.3}Se_{0.7}$, and $Ge_{0.2}Se_{0.8}$ films are found to be 30.1, 30.6, 31, and 31.1 eV, respectively. The peak binding energy of Ge *3d* electrons increased (30.1 to 31.1 = 1 eV) owing to increased Se content [34,35]. The chemical shifts of Ge *3d* (29 eV) electrons are 1.1, 1.6, 2.0, and 2.1 eV for $Ge_{0.5}Se_{0.5}$, $Ge_{0.4}Se_{0.6}$, $Ge_{0.3}Se_{0.7}$, and $Ge_{0.2}Se_{0.8}$ films, respectively. A similar chemical shift of Ge *3d* electrons was reported by Ueno et al. [35]. According to their reported results, the chemical shifts of Ge *3d* electrons are 2.2 ± 0.1 eV for $GeSe_2$ and 1.7 ± 0.1 eV for GeSe, while those of Se *3d* electrons are negative, which is similar to our results. The corresponding binding energies of Se *3d* core-level electrons for $Ge_{0.5}Se_{0.5}$, $Ge_{0.4}Se_{0.6}$, $Ge_{0.3}Se_{0.7}$, and $Ge_{0.2}Se_{0.8}$ films are found to be 54.2, 54.5, 54.7, and 54.8 eV, respectively (Figure 2b). A positive shift of binding energies from 54.2 to 54.8 = 0.6 eV is owing to the higher Se contents. The chemical shifts of Se *3d* (55.2 eV) electrons are found to be −1.0, −0.7, −0.5, and −0.4 eV for $Ge_{0.5}Se_{0.5}$, $Ge_{0.4}Se_{0.6}$, $Ge_{0.3}Se_{0.7}$, and $Ge_{0.2}Se_{0.8}$ films, respectively. However, the chemical shifts of Se *3d* (55.5 eV) electrons are −1 eV for $GeSe_2$, as reported by Ueno et al. [35]. Slight variation of binding energies is related to the material deposition process and conditions [36]. On the other hand, the O *1s* core-level electron signal is not observed in the Ge_xSe_{1-x} film. To search better thermal stability, other switching materials are discussed below.

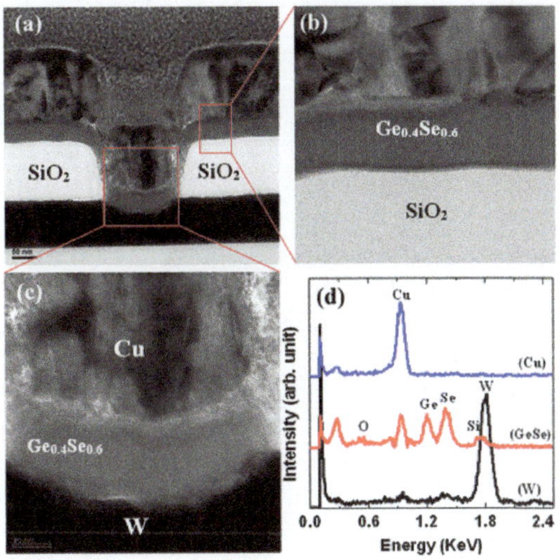

Figure 1: TEM images and EDX analysis of a Cu/$Ge_{0.4}Se_{0.6}$/W pristine memory device. **(a)** TEM image of a Cu/$Ge_{0.4}Se_{0.6}$/W pristine memory device with a scale bar of 50 nm. The $Ge_{0.4}Se_{0.6}$ film shows to be amorphous, as shown in the **(b)** outside and **(c)** inside of the via-hole region. **(c)** Different layers are clearly shown by the dark-field TEM image and **(d)** corresponding EDX analysis of the W, $Ge_{0.4}Se_{0.6}$, and Cu films from **(a)** [33].

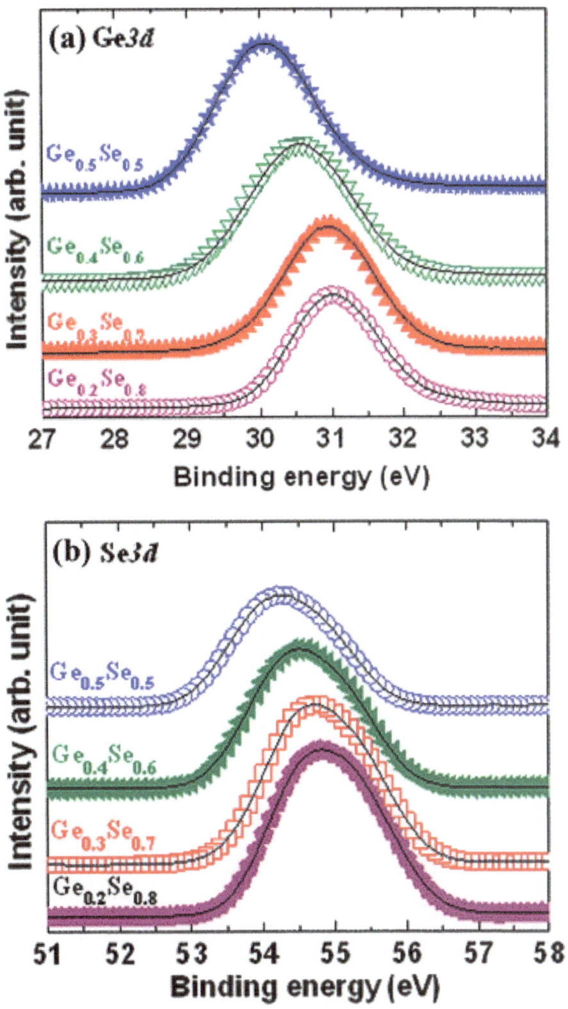

Figure 2: XPS spectra of Ge_xSe_{1-x} solid electrolytes for **(a)** Ge *3d* and **(b)** Se *3d* electrons. The binding energy increased with increasing Se contents.

Kozicki et al. [13] also reported the fabrication of a Ag (or Cu)/GeS/W structure. They deposited a W BE on SiO_2/Si substrates by chemical vapor deposition (CVD) method and then covered it with a 100-nm-thick SiO_2 layer. Via-holes with different diameters ranging from 180 nm to 5 μm were opened on the SiO_2 layer by optical projection lithography and RIE. Then, 40- and 50-nm-thick Ge_xS_{1-x} ($0.30 < x < 0.35$) films were deposited by PVD. Next, a 25-nm-thick Ag (or Cu) layer was deposited by PVD. The chamber pressure was maintained at 10^{-6} Torr. It is important to note that the deposition of the switching layer is a difficult process as the vapor pressure of the S component is very high. Vianello et al. [14] reported CBRAM devices using a Ag/Sb:GeS_2/W-plug structure, and 1T1R configuration in 8 × 8 arrays integrated in the complementary metal-oxide-semiconductor (CMOS) process was done. The plug size was 0. 2 μm. The thickness of the Ag-doped GeS_2 layer, which was deposited by co-sputtering, was 30 nm. The Sb material was doped 2% to 20% in GeS_2. A thin Ag layer was dissolved into GeS_2using a photo-diffusion process [13]. Then, the Ag layer as a TE was deposited. Choi et al. [15] reported a Cu/GeTe/TaN structure. A SiO_2 layer with a thickness of 100 nm was grown on a Si substrate. Corresponding materials such as TaN, GeTe, and SiO_2 were deposited (in the given order) onto the prepared substrate using a radio-frequency (RF) sputtering system. The thicknesses of TaN, GeTe, and SiO_2 layers were 30, 50, and 100 nm, respectively. Via-hole devices with different sizes of 200 nm, 400 nm, 10 μm, and 50 μm were patterned by e-beam lithography. Banno et al. [16] reported a Cu/Cu_{2-}S/Pt structure. Terabe et al. [8] reported a gap-type resistive switching memory device. At first, an Ag_2S-coated Ag wire was crossed by a Pt wire. They mentioned that the 'crossbar' structure was convenient for integrating switches to be used in actual devices. Some oxide-based materials for CBRAM devices are explained below.

Banno et al. [16] reported a Cu/Ta_2O_5/Pt structure for their CBRAM devices. A 15-nm-thick Ta_2O_5 film with an O/Ta ratio of 2.5 was deposited on a Pt electrode by RF sputtering. Next, a 50-nm-thick Cu electrode was deposited and covered by a 50-nm-thick Pt film to avoid oxidization of Cu. In our previous work [17], we have fabricated a Cu/Ta_2O_5/W structure. The via-holes were patterned as mentioned above in our work. Then, 15- to 20-nm-thick Ta_2O_5 films were deposited using an e-beam evaporator. The Ta_2O_5 granules were used. Next, a Cu TE with a thickness of 100 nm was deposited using a thermal evaporator. Finally, a lift-off process was done to obtain the memory devices. A post metal annealing treatment was performed at a temperature of 350°C for 1 min in N_2 ambient. Schindler et al. [18] reported a Cu/SiO_2/W structure. A 100-nm-thick W BE was deposited by CVD on SiO_2/Si substrates. Then, a SiO_2switching material with thicknesses varying from 12 to 50 nm was grown by electron beam evaporation. A Cu TE with a thickness of 35 to 45 nm was deposited. Then, Cu was diffused into SiO_2 by thermal annealing at 610°C to have a Cu/Cu:SiO_2/W structure. Li et al. [19] fabricated a Ag/ZrO_2/Au structure. First, a 50-nm-thick Ag layer was deposited using an electron beam evaporator. A ZrO_2 film was also deposited. The vacuum pressure and the deposition rate were 2×10^{-6} Torr and 1 Å/s, respectively. After that, a 50-nm-thick Au TE with different areas such as 100 × 100 μm² to 800 × 800 μm² was deposited. A GeO_x switching material as CMOS compatible was reported by us [20]. Via-holes with size ranging from 0.2 × 0.2 μm² to 8 × 8 μm² were reported. A GeO_x film with a thickness of approximately 10 nm was then deposited by RF sputtering. The GeO_x film showed to be polycrystalline. The chamber pressure was 2×10^{-5}Torr. The deposition power and pressure were 50 W and 2×10^{-2} Torr, respectively. Argon (Ar) gas with a flow rate of 25 sccm was used. A Ge target was used to deposit the GeO_x solid electrolyte. Then, a Cu TE with a thickness of approximately 200 nm was deposited using a thermal evaporator. A lift-off process was next performed to obtain the final via-hole devices in a Cu/GeO_x/W structure. Cross-point memory devices using an Al/Cu/GeO_x/W structure were also reported [21]. First, a SiO_2 layer with a thickness of approximately 200 nm was grown by a wet oxidation process. After the masking process, a GeO_x film with a thickness of 10 nm was deposited by the same RF sputtering. A Cu layer with a thickness of 40 nm was deposited using a thermal evaporator, and then, an Al layer with a thickness of approximately 160 nm was deposited *in situ* using a thermal evaporator. Finally, a lift-off process was performed to get the cross-point devices. A schematic illustration of the fabricated GeO_x-based cross-point memory device is shown in Figure 3a [21]. The GeO_x solid electrolyte was sandwiched in between Cu and W electrodes. An optical micrograph (OM) image of the 5 × 5 cross-points is shown in Figure 3b. The memory characteristics of this cross-point memory device are discussed later. Kim et al. [22] reported CBRAM characteristics using a Ag/a-Si/SiGe/W crossbar structure. Initially, a W BE with a thickness of approximately 20 nm was deposited by sputtering. In the next step, a boron-doped Si_xGe_{1-x} film of 20 nm was deposited by thermal CVD at 425°C. An amorphous Si (a-Si) as a resistive switching material was deposited by PECVD method. The thickness of the a-Si film was approximately 20 nm. Finally, a Ag TE was deposited. This complete stack was patterned by electron beam lithography and plasma etching process. The thickness of all the layers was 50 nm half-pitch arrays. Sun et al. [23] also reported bipolar CBRAM characteristics using a Ag/Si_3N_4/Pt memory device. The device size was 100 × 100 μm². Even though many groups reported a single-layer switching material, bilayer switching materials showed to be promising.

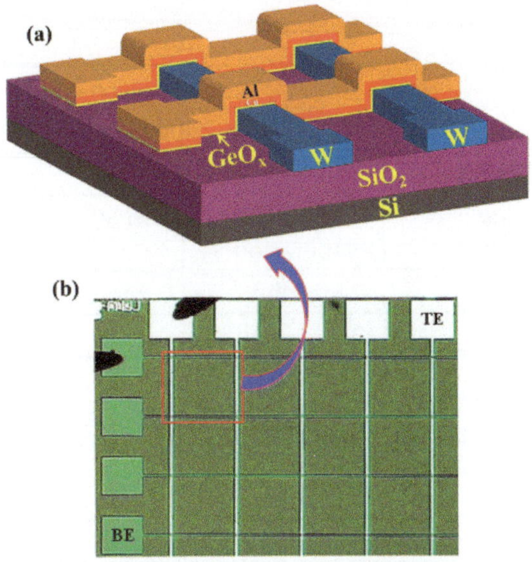

Figure 3: Schematic illustration **(a)** and optical image **(b)** of the cross-point memory devices for 3D architecture. The area of the active cross-point is approximately 1.2 × 1.2 μm². The thickness of the GeO_x solid electrolyte film is approximately 10 nm [21].

Aratani et al. [24] investigated CBRAM characteristics using a TE/Cu-Te/GdO_x/W structure. In this case, both TE and switching materials were deposited on the BE by a sputtering process. Device sizes of 20 and 50 nm were designed. The Cu-Te layer behaved as a buffer layer, and the GdO_x layer was used as a switching layer. The Cu filament formation/dissolution in the GdO_x layer was controlled by a buffer layer through external bias. Similarly, Cu-Te/SiO_x bilayers were also reported [25]. Yoon et al. [26] reported MoO_x/GdO_x bilayers in a Cu/MoO_x/GdO_x/Pt structure. A 10-nm-thick GdO_x layer was deposited on a Pt/Ti/SiO_2/Si substrate using an electron beam evaporator. A 100-nm-thick Pt was also deposited. After a lift-off process, a device size of 50 × 50 μm² was fabricated. A NanoBridge CBRAM device was reported by using a Cu/Ta/$TaSiO_y$/Ru structure [27]. The memory cell was integrated in the CMOS process. Via-holes with cell sizes ranging from 1 μm to 50 nm were completed. After deposition of a Ti or Ta barrier layer on Cu, a $TaSiO_y$ solid electrolyte was deposited by RF sputtering in Ar/O_2 gas. The CBRAM devices were connected through contact holes and controlled the switching through a CMOS transistor. Therefore, the 1T1R configuration was developed. In our previous work [28], we reported $GeSe_x$/TaO_x bilayer materials in an Al/Cu/$GeSe_x$/TaO_x/W structure. Although this CBRAM device using $GeSe_x$/TaO_x bilayers shows promising resistive switching characteristics, the thermal stability of the $Ge_{20}Se_{80}$ material is an issue. To have a CMOS-compatible structure, we reported a Al/Cu/TiO_2/TaO_x/W CBRAM device [29]. The via-hole devices were fabricated as follows [17,28]. A Ta_2O_5 film with a thickness of 18 nm was deposited from pure Ta_2O_5 granules using an electron beam evaporator. The resulting Ta_2O_5 film was mixed with Ta metal (i.e., TaO_x, where $x < 2.5$), as characterized by XPS. The Cu as mobile ions plays a major role in the Al/Cu/TaO_x/W structure resistive switching memory device. The Cu can be oxidized at the Cu/TaO_x interface during deposition and hinders the resistive switching memory performance. Therefore, Cu oxidation is expected to be avoided by inserting a Ti nanolayer at the Cu/TaO_x interface. A thin Ti layer with a thickness of approximately 3 nm was deposited *in situ* using an electron beam evaporator with Ti granules. Finally, a device with an Al/Cu/TiO_2/TaO_x/W structure was obtained. These memory devices were annealed by rapid thermal annealing (RTA) at 350°C in N_2 ambient for 1 min. Oxygen accumulated in the Ti nanolayer and formed TiO_2. Figure 4a shows a TEM image of the device with a Ti nanolayer at the Cu/TaO_x interface [29]. The device size was approximately 150 × 150 nm². All layers were covered well at the active and outer regions (Figure 4b,c). Oxygen accumulated in the Ti layer as expected, resulting in a thin TiO_2 layer of approximately 3 nm. This accumulated oxygen originated from the TaO_x layer and formed a more defective TaO_x film, which is in agreement with the Gibbs free energy and helps to have repeatable resistive switching characteristics. On the other hand, Al_2O_3-based bilayer materials were also reported. Goux et al. [30] reported the compositional effect of a Cu-Te buffer layer on the switching characteristics of a Pt/Cu_xTe_{1-x}/Al_2O_3/Si structure. The Al_2O_3 layer with a thickness of 3 nm was deposited by atomic layer deposition (ALD). Using a shadow mask, Cu_xTe_{1-x} dots were deposited by using a co-sputtering system. Cu and $Cu_{0.1}Te_{0.9}$ targets were used. The size of the dots was 3 mm wide and 50 nm thick. Belmonte et al. [31] fabricated a Cu/TiW/Al_2O_3/W structure in a 90 nm technology node. Firstly, a 3-nm-thick Al_2O_3 layer was deposited on a W BE by ALD. After that, a 10-nm-thick TiW alloy with 25 at % of Ti was deposited by using a sputtering system. Then, a Cu seed layer was deposited by a sputtering process. At last, a TiN TE was deposited to get the final device. Devulder et al. [32] reported a TiN/$Cu_{0.6}Te_{0.4}$-C/Al_2O_3/Si structure to enhance the thermal stability and switching behavior of memory devices. Table 1 presents a brief description about the material, structure, and fabrication procedure of CBRAM devices. Using chalcogenides, oxides, and bilayer materials in the above structures, the resistive switching memory characteristics are explained below.

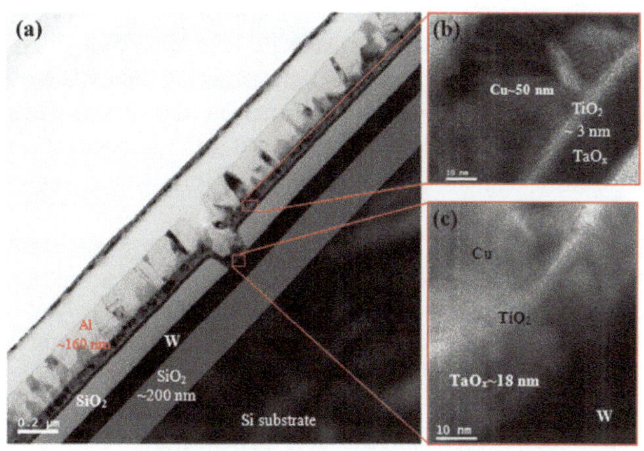

Figure 4: TEM image of an Al/Cu/Ti/TaO_x/W structure. **(a)** HRTEM images of the **(b)** outside and**(c)** inside of the via-hole device, as shown in **(a)**. After deposition of a Ti nanolayer, it became TiO_2 [29].

Table 1: CBRAM material and deposition methods

Device structure (TE/switching material/BE)	Deposition methods			Device size (μm)
	TE	Switching material	BE	
$Ag/Ag_{33}Ge_{20}Se_{47}/Ni$ [10]	-	Evaporation	-	40, 75
Ag/Ag-GeSe/Pt [11]	-	RF sputtering	Thermal evaporation	2 to 50
$Al/Cu/Ge_{0.5}Se_{0.5}/W$ [12]	Thermal	Electron beam	Sputtering	0.2 to 8.0
Ag or Cu/GeS/W [13]	PVD	PVD	CVD	0.18 to 5
$Ag/Sb:GeS_2/W$ [14]	-	RF PVD	-	0.2
$Cu/Ta_2O_5/Pt$ [16]	-	RF sputtering	-	-
$Cu/Ta_2O_5/W$ [17]	Thermal evaporator	E-beam evaporation	Sputtering	0.2 to 8.0
$Cu/Cu:SiO_2/W$ [18]	-	E-beam evaporation	CVD	0.35 to 5
Ag/a-Si/SiGe/W [22]	-	CVD	Sputtering	0.05
$Ag/Si_3N_4/Pt$ [23]	RF sputtering	PECVD	RF sputtering	10
$Cu/Cu\text{-}Te/GdO_x/W$ [24]	Sputtering	Sputtering	-	0.02, 0.05
$Cu/Ta/TaSiO_y/Ru$ [27]	-	PECVD	-	0.05 to 1
$Al/Cu/Ti/TaO_x/W$ [29]	Thermal	Electron beam	Sputtering	0.15
$Cu\text{-}Te/Al_2O_3/Si$ [30]	Co-sputtering	ALD	-	-
$Al/TiN/Cu/TiW/Al_2O_3/W$ [31]	-	ALD	Sputtering	0.09

MEMORY CHARACTERISTICS

Chalcogenide-Based Materials

Germanium (Ge)-based chalcogenide glasses have been widely studied because of their well-known mechanism and also they offer several advantages, whereas Cu^{z+} or Ag^+ ions exhibit high mobility. In our previous study [33], we have reported the bipolar current-voltage (*I-V*) characteristics of a $Cu/Ge_{0.4}Se_{0.6}/W$ structure, as shown in Figure 5a. The step voltage was 20 mV during voltage sweep. The current compliance (CC) was 200 μA. The size of the memory device was 200 × 200 nm^2. It is noted that the formation voltage is not required for Ge_xSe_{1-x} solid electrolyte-based CBRAM devices. The *I-V* hysteresis loop can be explained as follows by using arrows 1 to 9. Initially, a voltage applied to the TE is swept from 0 to +0.56 V (arrow 1); this is the HRS. Beyond the voltage of +0.56 V, there is an instantaneous switching from the HRS to the LRS (arrow 2). The resistive switching is observed beyond a voltage of 0.56 V, which is called SET voltage (V_{SET}). To form a Cu metallic filament into the $Ge_{0.4}Se_{0.6}$ solid electrolyte, the applied bias should be larger than the SET voltage (arrow 3). The memory device keeps the LRS up to a negative voltage of −0.18 V (arrows 3 to 6). The resistive memory device reaches a HRS if the voltage applied is more than the negative voltage or RESET voltage (V_{RESET}) < −0.16 V (arrow 7). A large RESET current (I_{RESET}) of −85 μA confirms that the strong Cu metallic filament is formed into the $Ge_{0.4}Se_{0.6}$ solid electrolyte. To reach a HRS, a negative voltage of −1.2 V should be large enough (arrow 8). The metallic filament is gradually dissolved by applying a higher negative voltage. To get a second switching, the sweeping voltage can be followed using arrows 9 and 1. The corresponding resistance vs sweeping voltage (*R-V*) hysteresis characteristics are plotted in Figure 5b. A high resistance ratio (HRS/LRS) of approximately 3 × 10^6 is observed, which is useful for MLC operation. Choi et al. [15] and Liaw et al. [37] also reported a high resistance ratio of >10^3 to 10^5. Kund et al. [9] have reported bipolar resistive switching phenomena using a Ag/GeSe/W structure at a CC of 2 μA. The V_{SET} was +0.2 V. Kozicki et al. [13] have reported resistive switching characteristics at a V_{SET} of +0.4 V and a CC of approximately 10 μA. One hundred consecutive *I-V* switching cycles at a CC of 100 μA are shown in Figure 6a. However, the step voltage was 0.1 V. The V_{SET} varied from 0.3 to 0.5 V. Variations of the resistance states and RESET current are also observed, due to the higher Se content (80%). Figure 6b shows the program/erase (P/E) cycles for every cycle. The P/E current and pulse width are 500 μA and 500 μs, respectively. The programming and erasing voltages are 1.25 and −1.0 V, respectively. The read voltage is 0.1 V. Even though large variations of resistance states are observed, the memory device will have 10^5 P/E cycles. All memory devices using Ge_xSe_{1-x} solid electrolytes are performed more than >10^5 P/E cycles, which is beneficial by using Ge_xSe_{1-x} solid electrolytes. Kozicki et al. [10] have reported >10^{10}read endurance using a $Ag/Ag_{33}Ge_{20}Se_{47}/Ag$ structure. Kund et al. [9] have investigated >10^6P/E endurance using a $GeSe_x$-based CBRAM device. Vianello et al. [14] have also reported the P/E endurance characteristics of a Sb-doped GeS_2-based chalcogenide-based CBRAM cell, as shown in Figure 7. The memory device performs >10^5 P/E cycles at 30-ns pulse width with a resistance ratio of >10. However, there is voltage-time dilemma. For a high-speed operation, the program/erase voltage needs to be higher. Hasegawa et al. [38] have studied the switching time of a Ag_2S- and Cu_2S-based gap-type atomic switch as a function of bias voltage measured at room temperature, as shown in Figure 8. The switching time

(t_s) is expressed as function of switching bias ($V_s > V_{SET}$), as shown in Equation 1:

$$t_s \infty \exp(-\beta V_s) \quad (1)$$

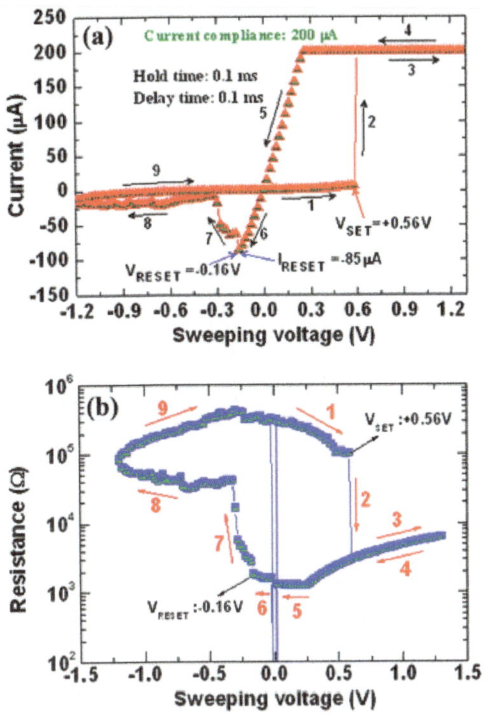

Figure 5: *I-V* and corresponding *R-V* hysteresis characteristics of an Al/Cu/$Ge_{0.4}Se_{0.6}$/W resistive memory device. **(a)***I-V* hysteresis characteristics of an Al/Cu/$Ge_{0.4}Se_{0.6}$/W resistive memory device. The device size was 0.2 µm. **(b)** Corresponding *R-V* hysteresis characteristics of the device [33].

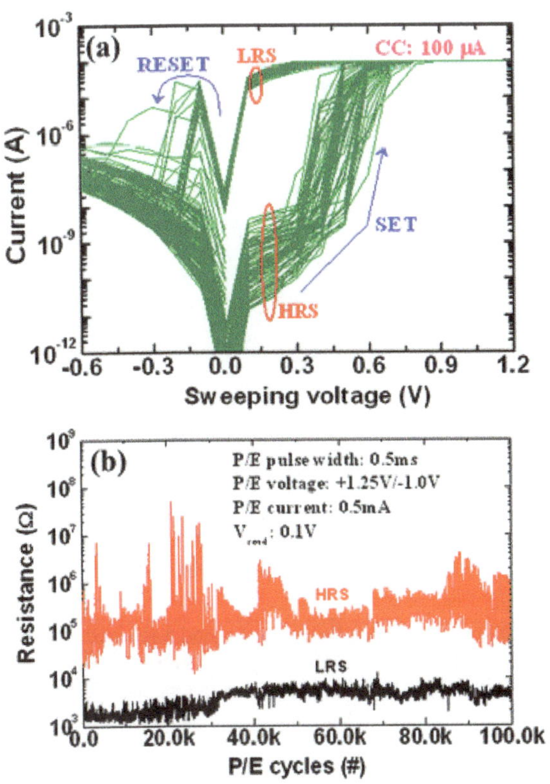

Figure 6: *I-V* hysteresis for a Cu/$Ge_{0.3}Se_{0.7}$/W memory device and typical P/E cycles of a Cu/$Ge_{0.2}Se_{0.8}$/W memory device. **(a)***I-V* hysteresis of more than 100 consecutive DC cycles for a Cu/$Ge_{0.3}Se_{0.7}$/W memory device. The size of the memory device is 4 × 4 µm².**(b)** Typical P/E cycles of a Cu/$Ge_{0.2}Se_{0.8}$/W memory device. Every P/E cycle was recorded. This suggests that the P/E cycles are more than 10^5.

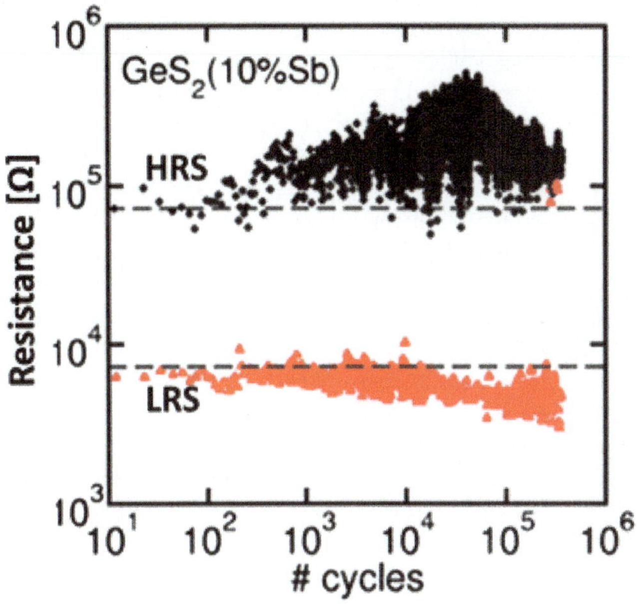

Figure 7: P/E endurance of >10^5 cycles of Ag/Sb-doped GeS_2/W-based 1S1R (one selector-one resistor configuration) CBRAM cell [14].

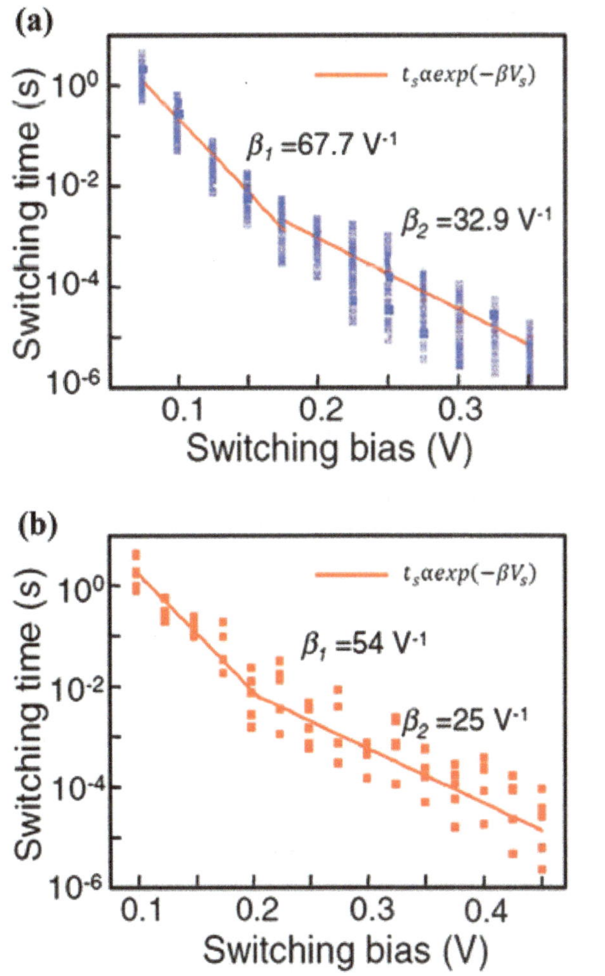

Figure 8: Switching time vs switching bias characteristics. For **(a)** Ag_2S-based gap-type atomic switch at room temperature and **(b)** Cu_2S-based gap-type atomic switch [38].

The exponential nature of the time vs voltage characteristics suggests that a higher SET voltage can drastically reduce the time to SET the memory cell. Even a 0.1-V increment can result in around one to two orders of reduction in SET time. The switching time can be distinguished into two decay components when the bias voltage is increased (Figure 8a). The values of decay factors $\beta 1$ and $\beta 2$ are 67.7 and 32.9 for low-voltage and high-voltage regions, respectively. Similar phenomena are also reported in a Cu_2S switching material, as shown in Figure 8b. In this case, the values of decay factors $\beta 1$ and $\beta 2$ are 54 and 25 for low-voltage and high-voltage regions, respectively. These two exponential decay factors exhibit different rate limiting processes for the low-voltage and high-voltage regions. It is well known that there are only few mechanisms such as metallic filamentary switching that require few nanoseconds (<10 ns) to switch a device from SET to RESET and vice versa. Thus, for the 10-year data retention criterion, the device needs to maintain the ratio of the retention time to the SET time of around 10^{16} and V_{SET}/V_{read} of around 10. Again, the read voltage needs to be high enough to maintain the ratio of $V_{program}/V_{read}$ at 10 for circuit constraint. It is also required that the cell withstand the read voltage for 10 years. Schindler et al. [39] have also reported this switching time-voltage dependence of SiO_2-based CBRAM cells. Therefore, CBRAM devices can be operated at low current and stable data retention is also an important part. We have reported data retention characteristics of >11 h at a CC of 200 μA for a Cu/$Ge_{0.4}Se_{0.6}$/W CBRAM device [40]. The memory device shows data retention at higher temperatures (85°C and 150°C) and maintains a large resistance ratio of >100; however, long-time data retention does not show to be stable. Vianello et al. [14] have shown a data retention of >10^5 s at 130°C using a Sb-doped GeS_2 solid electrolyte. Jameson et al. [41] have reported an excellent data retention of >10^6 s at 200°C in a Ag/GeS_2/W memory cell. In addition, MLC capability is also important for high-density memory application. Russo et al. [42] have reported multi-level data retention characteristics of >10^5 s with CC (or I_{compl}) varying from 5 to 500 μA, as shown in Figure 9. It can be observed that LRS or R_c is not so stable at 5 μA as compared to 50 and 500 μA. This suggests that the filament size is smaller for lower current compliance as compared to higher current compliance. However, HRS maintains the same level at different CCs for more than 10^4 s. Kund et al. [9] have studied the variation of LRS with varying CC ranging from 100 nA to 100 μA, leading to the fact that a large number of levels can be achieved. Figure 10a presents a resistive switching memory device using $Ge_{0.2}Se_{0.8}$ solid electrolytes in an Al/Cu/$Ge_{0.2}Se_{0.8}$/W structure which can be operated at a very low CC of 1 nA. The sweeping voltage direction is shown by arrows 1 to 2. Figure 10b shows the behavior of resistance states with CCs ranging from 1 nA to 1 mA for $Ge_{0.2}Se_{0.8}$ and $Ge_{0.5}Se_{0.5}$ devices. For both devices, LRS decreases with increasing CCs from 1 nA to 1 mA owing to filament diameter increase. At a CC of <10 μA, the LRS of the $Ge_{0.2}Se_{0.8}$ device is lower than that of the $Ge_{0.5}Se_{0.5}$ device. This suggests that Cu ions can be easily migrated through a higher Se content, which results in a large variation of HRS. However, the average HRS of $Ge_{0.2}Se_{0.8}$ devices decreases faster than that of $Ge_{0.5}Se_{0.5}$ devices. This indicates that the filament dissolution length of $Ge_{0.5}Se_{0.5}$ devices is larger than that of $Ge_{0.2}Se_{0.8}$ devices because of the thinner filament diameter or less overshoot effect. So resistive switching properties strongly depend on the compositional variations of Ge_xSe_{1-x} solid electrolytes. However, the SET voltage, endurance, data retention, and so on may be controlled using different compositions of $GeSe_x$ solid electrolytes as per the requirement of the applications.

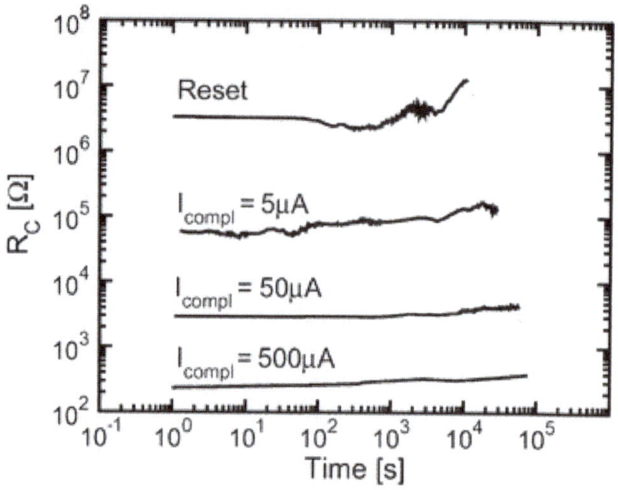

Figure 9: Excellent multi-level data retention of >10^5 s. With varying I_{compl} (CC) from 5 to 500 μA of memory cell which is very promising for high-storage multi-bit operation [42].

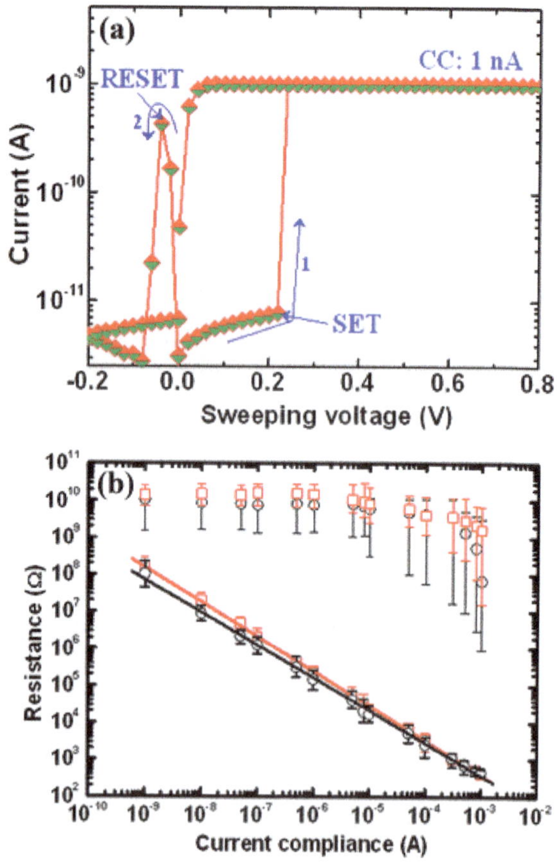

Figure 10: *I-V* hysteresis for $Ge_{0.2}Se_{0.8}$ and LRS/HRS vs CCs. **(a)** *I-V* hysteresis at a CC of 1 nA for a $Ge_{0.2}Se_{0.8}$ solid electrolyte. **(b)** LRS/HRS vs CCs for lower and higher Ge contents of the Ge_xSe_{1-x} solid electrolytes.

Oxide-based materials

It is observed that the chalcogenide-based CBRAM device shows great potential for next-generation non-volatile memory, but one of the discrepancies is compatibility with the CMOS process. To mitigate this issue, binary metal oxides like Ta_2O_5 [16,17], SiO_2 [18], ZrO_2 [19], GeO_x [20,21], and so on have been reported, and their memory characteristics are discussed here. Sakamoto et al. [43] have reported bipolar resistive switching behaviors using a Cu/Ta_2O_5/Pt structure at a CC of 100 μA. However, a large I_{RESET} of >1 mA was reported. In our previous study [17], we have reported bipolar resistive switching phenomena using an Al/Cu/TaO_x/W structure at a CC of 100 μA and a small operating voltage of ±1 V, as shown in Figure 11a. The voltage sweep is shown by arrows 1 to 8. The value of I_{RESET} is approximately 43 μA, and it is good for application. The values of V_{SET} and V_{RESET} are 0.75 and −0.2 V, respectively. This device is operated at a very low CC of 5 pA (Figure 11b). The I_{RESET} is approximately 3 pA. Although the device shows switching such a low current at 5 pA, it is difficult for real application because the operation speed will be also slower too. So far, it is operated at the lowest current in the literature. Schindler et al. [18,44] have investigated *I-V*characteristics using W/Cu:SiO_2/Cu and Ir/Cu:SiO_2/Cu cells and low operation currents of 1 nA and 5 μA, respectively. Tsunoda et al. [45] have reported CBRAM phenomena using a Ag/TiO_2/Pt structure at a CC of 100 μA. Statistical distribution of the HRS and LRS of Al/Cu/TaO_x/W memory devices shows also to be good in our previous study [17]. The memory device maintains an acceptable resistance ratio of cycle-to-cycle as well as device-to-device. Liu et al. [46] have reported the improvement of device-to-device uniformity introducing Cu nanocrystals (NCs) in Ag/ZrO_2/Cu-NC/Pt structures. The NCs guide conductive filament (CF) growth and reduce the randomness of the CF formation and rupture processes, leading to improved stability and uniformity of the memory devices. Li et al. [19] have reported good cumulative probability (device-to-device) of SET/RESET voltage and HRS/LRS in Au/ZrO_2/Ag devices. Even though *I-V* switching has shown to be good with low operation current, P/E endurance and data retention are also important, which are discussed below.

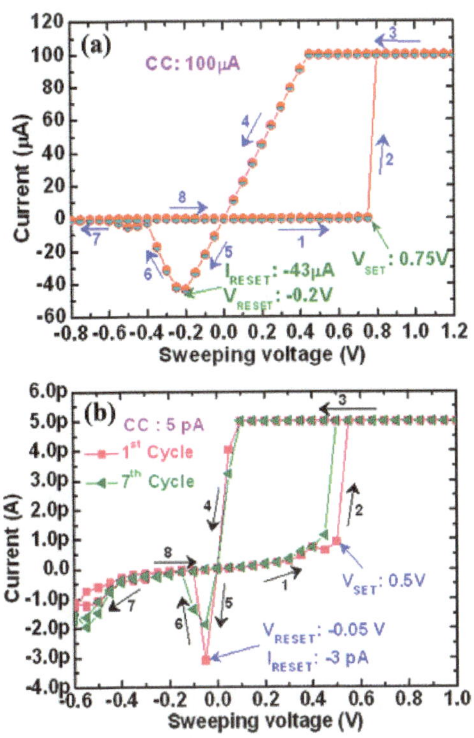

Figure 11

I-V hysteresis characteristics of a Cu/Ta_2O_5/W memory device. **(a)** Bipolar *I-V* hysteresis characteristics of a Cu/Ta_2O_5/W memory device. The memory device performs under a small operating voltage of ±1 V. Current compliance is 100 μA. **(b)***I-V* hysteresis characteristics with a low current compliance of 5 pA are also obtained [17].

Banno et al. [16] and Sakamoto et al. [47] have reported stable 10^4 P/E cycles with 100-μs pulse width, as shown in Figure 12. Balakrishnan et al. [48] have reported $>10^7$ P/E cycles in a Cu/SiO_2/W CBRAM cell. To analyze the data retention of an oxide-based CBRAM device, Schindler et al. [18] have reported $>10^5$ s data retention at a CC of 1 μA using a Cu/Cu:SiO_2/W structure. Both HRS and LRS exhibit stable data retention with a large resistance ratio of $>10^4$. Tsunoda et al. [45] have shown $>10^4$ s data retention at a CC of 100 μA in a Ag/TiO_2/Pt cell. We have reported good data retention at a CC of 50 μA of more than 200 h at 85°C with a high resistance ratio of $>10^8$ using a Cu/GeO_x/W structure [20]. Figure 13a shows the LRS vs current compliances. The LRS decreases with increasing CC from 5 pA to 700 μA, owing to the stronger Cu nanofilament formation in TaO_x films [17]. The value of LRS is independent of the thickness of TaO_x films. This suggests that our resistive switching memory device can be applicable for MLC. The RESET current increases with increasing CC from 5 pA to 700 μA, which indicates that a metallic filament can be formed by applying larger current. Figure 13b shows DC switching cycles, measured up to 1,000 cycles, with different CCs varying from 100 to 800 μA. The LRS decreases with increasing CCs because of the larger diameter of the filaments. However, the HRS decreases with increasing CCs, which is due to the remaining filament. This suggests that a single oxide-based switching layer has a cycle-to-cycle variation. Therefore, bilayer materials may have benefited to improve the switching cycles and can also maintain a high resistance ratio with demanded reliability too.

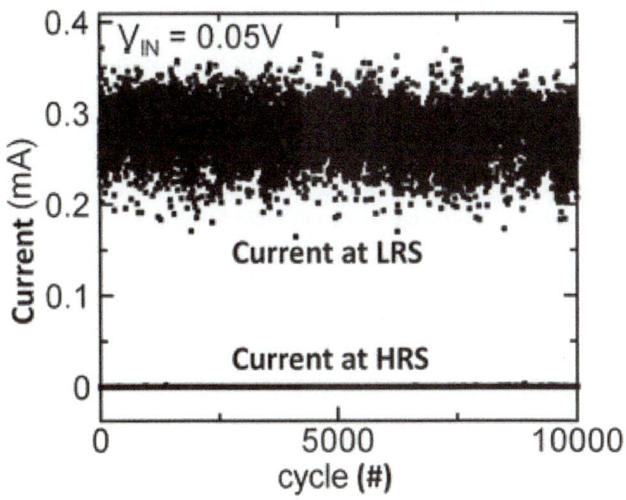

Figure 12: Program/erase endurance of $>10^4$ cycles of a Ta_2O_5-based CBRAM cell. This shows good endurance; however, the operation current is also high [47].

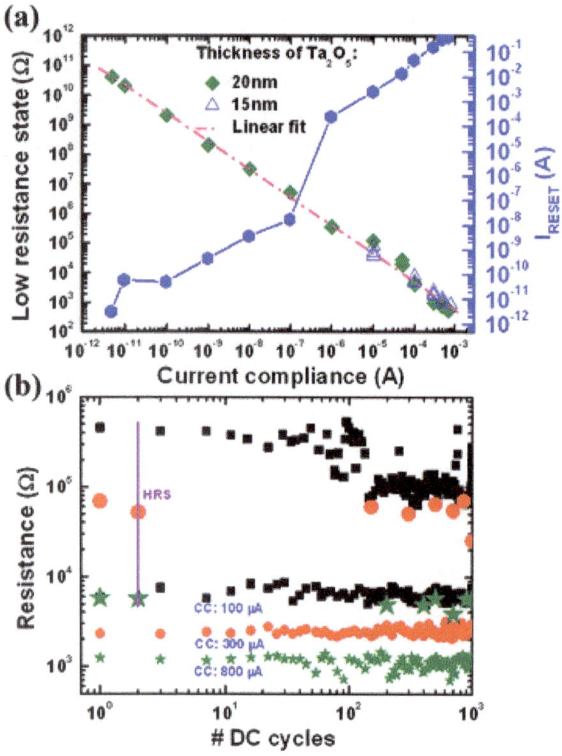

Figure 13: RESET current and DC switching cycles with current compliances. **(a)** RESET current (I_{RESET}) with current compliances. The LRS decreases with increasing CCs. **(b)** DC switching cycles (>10^3) with different current compliances. The thickness of the switching material is about 15 nm. A 100-Ω series resistance was used during endurance measurement. The device size is 1 × 1 μm² [17].

Bilayer Materials

Several researchers have studied CBRAM phenomena using bilayer materials to control Cu or Ag diffusion and improve switching uniformity and other electrical performances as well. Different bilayers such as Cu-Te/GdO_x [24], MoO_x/GdO_x [26], Ti/TaO_x [29], $GeSe_x$/TaO_x [28], Cu-Te/Al_2O_3 [30], TiW/Al_2O_3 [31], and so on are discussed below. Figure 14 shows a device-to-device cumulative probability. Improved uniformity by using Cu-Te/GdO_x bilayer materials in 1T1R configuration for 20-nm (Figure 14a) and 50-nm (Figure 14b) cell sizes is reported [24]. A cell size of 20 nm in diameter is also observed by using an atomic force microscopy image, as shown in the inset of Figure 14a. No such inherent differences of LRS are found in between two different cell sizes, which represents that switching characteristics are insensitive to the device area and indicates the filamentary switching mechanism. By using MoO_x/GdO_x bilayer materials, the I_{RESET} is reduced to 300 μA as compared to approximately 1 mA for a single MoO_x layer [26]. Improved switching characteristics by introducing a thin Ti nanolayer in Al/Cu/TiO_2/TaO_x/W structures have been observed by us [29]. Figure 15 shows the *I-V* switching characteristics of both resistive switching memory devices. One hundred consecutive switching cycles are indicated by arrows 1 → 5 under a CC of 500 μA. The memory device (Al/Cu/TaO_x/W structure) shows a leakage current of approximately 1.5 pA at a V_{read} of +0.1 V for a pristine device (Figure 15a), lower than that of the current (23 pA) for the device in an Al/Cu/TiO_2/TaO_x/W structure (Figure 15b). Because of a higher charge-trapping density by introducing a Ti nanolayer, the average leakage current is higher than that of the device without a Ti nanolayer. The I_{RESET} of the Al/Cu/TiO_2/TaO_x/W devices is lower than that of the Al/Cu/TaO_x/W devices (100 vs 1,000 μA). An improved LRS dispersion is also observed for Al/Cu/TiO_2/TaO_x/W devices owing to Cu migration controlled through a TiO_x nanolayer under external bias. A low operation current of 10 μA using a Cu/TiW/Al_2O_3/W structure is also reported by limiting the possible Cu diffusion and has shown resistive switching with limited variability [31]. The statistical distribution of HRS and LRS with different CCs of 25 and 10 μA is also reported. It can be observed that with reducing current compliance, the resistance ratio increases up to ten times, which may cause reduction of I_{RESET}. This is attributed to the higher value of HRS and broadened memory window. An Al/$GeSe_x$/TaO_x/W device shows a low operation current of 100 nA [28]. Compared to those of pure $GeSe_x$ in an Al/Cu/$GeSe_x$/W structure and TaO_x in an Al/Cu/TaO_x/W structure, improved uniformity of LRS/HRS with switching cycles is observed for $GeSe_x$/TaO_x bilayer materials, as shown in Figure 16. Considering 100 switching cycles at a CC of 100 nA, average SET voltages are found to be +0.2, 0.28, and >0.3 V for $GeSe_x$, TaO_x/$GeSe_x$, and TaO_x devices, respectively. A stable LRS/HRS of the $GeSe_x$/TaO_x device is observed, as compared to that of pure $GeSe_x$ and TaO_x devices. This is due to the Cu nanofilament formation in the TaO_x switching layer at the $GeSe_x$/W interface. The LRS values are different because of defect control in switching materials. The $GeSe_x$/TaO_x bilayer device can be operated with different CCs varying from 1 nA to 500 μA. The average (or standard deviation) values of the LRS for device-to-device are found to be 0.14 GΩ (4.7 × 10^7), 5.2 kΩ (1.7 × 10^3), and 601 Ω (117), and those of the HRS are 11.9 GΩ (9 × 10^9), 0.34 GΩ (5 × 10^8), and 9.6 MΩ (9.5 × 10^6) at CCs of 1 nA, 50 μA, and 500 μA, respectively. The average resistance ratios are 85, 6.5 × 10^4, and 1.6 × 10^4 at CCs of 1 nA, 50 μA, and 500 μA, respectively. Due to such a high resistance ratio, this resistive switching memory device can be useful for MLC applications in the future with a wide range of CCs from 1 nA to 500 μA.

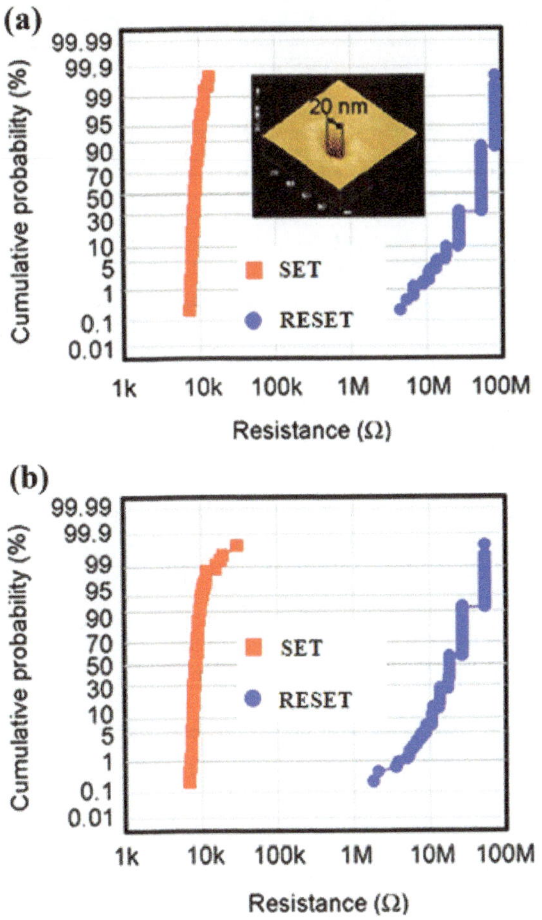

Figure 14: Statistical distribution of LRS under $V > V_{SET}$ (SET condition) and HRS under $V < V_{RESET}$(RESET condition). **(a)** 20-nm (shown in the inset) and **(b)** 50-nm cell in Cu-Te/GdO_xCBRAM devices [24].

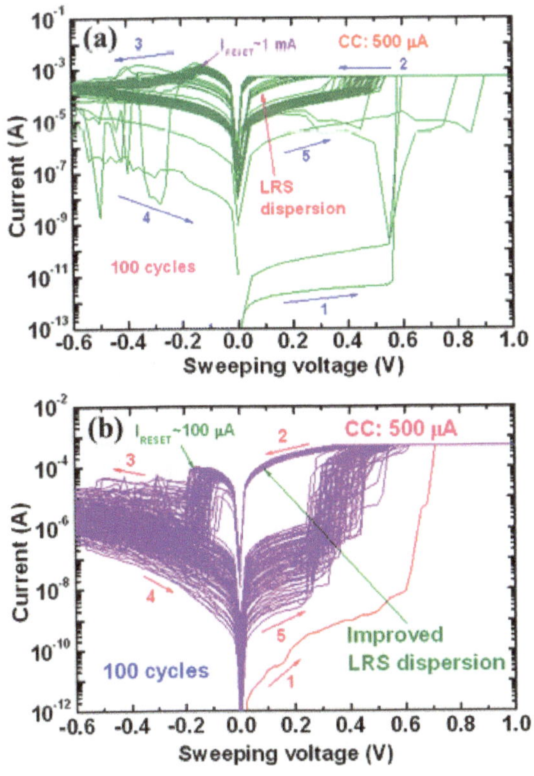

Figure 15: *I-V* characteristics of **(a)** Al/Cu/TaO_x/W and **(b)** Al/Cu/TiO_2/TaO_x/W structures. The memory device shows excellent uniformity at LRS after introducing a Ti nanolayer in the Cu/TaO_x interface [29].

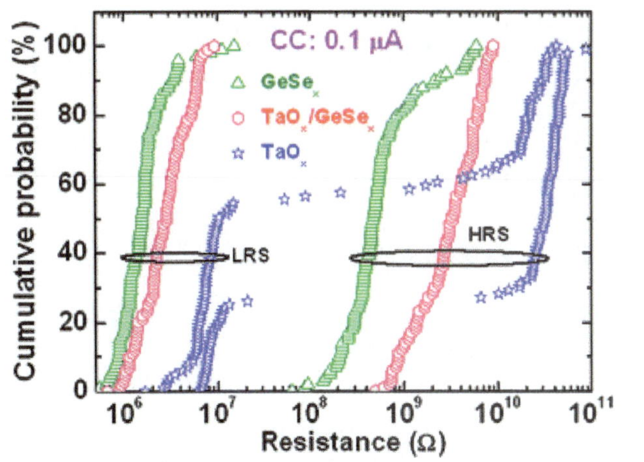

Figure 16: Cumulative probability of the LRS/HRS for the Al/Cu/$GeSe_x$/W, Al/Cu/$GeSe_x$/TaO_x/W, and Al/Cu/TaO_x/W CBRAM devices [28].

Introducing bilayer materials, long $>10^7$ P/E cycles are reported, as shown in Figure 17 [24]. The programming current and pulse width are 100 μA and 5 ns, respectively. A large resistance ratio of approximately 100 is also shown in the inset. The programming current is also reduced to 25 μA [31]. A small pulse width of 10 ns is used. To review the retention behavior of a bilayer CBRAM cell, long data retention of $>10^5$ s at 130°C in a Cu:Ge-Te/Al_2O_3/TiN CBRAM device with a small cell size of 20 nm is also reported [49]. Many devices are measured to observe the LRS values with elapsed time, as shown in Figure 18. A good resistance ratio of >10 is observed. Good data retention of $>10^5$ s at 85°C under a small CC of 50 μA is reported [17]. Figure 19represents the multi-level operation with variation of operation currents of Cu-Te/GdO_x bilayer materials [24]. Levels 1 to 4 are observed clearly. This depends on different programming currents, which can be used for MLC. In addition, the V_{SET} and V_{RESET} voltages are also important to control the switching characteristics as well as power. The device is operated with a high speed of 10 ns for the SET and RESET. Table 2 presents the V_{SET} and V_{RESET} of CBRAM devices with different switching materials. This indicates that bilayer materials show more reasonable V_{SET} and V_{RESET} voltages for application and the operation voltage of CBRAM devices is very good (±2 V).

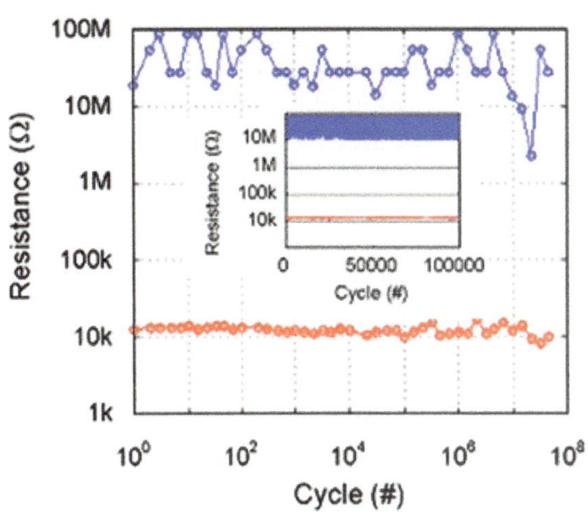

Figure 17: Pulse endurance of $>10^7$ cycles of a Cu-Te/GdO_x/W device. The pulse width was 10 ns. Every cycle up to 10^5 is shown in the inset [24].

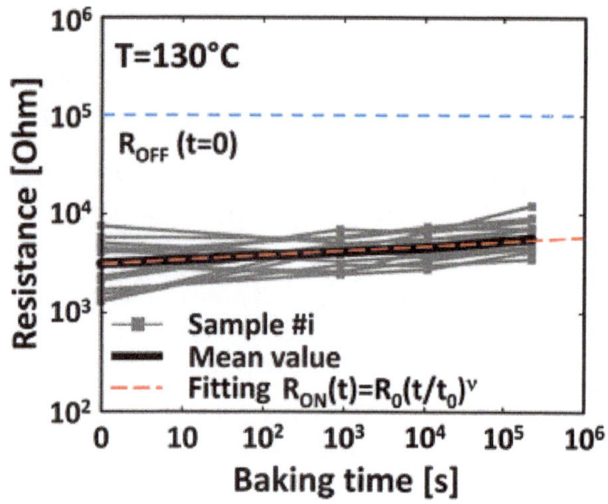

Figure 18: Good data retention characteristics of >10^5 s at 130°C of a Cu:Ge-Te/Al_2O_3/TiN CBRAM device [49].

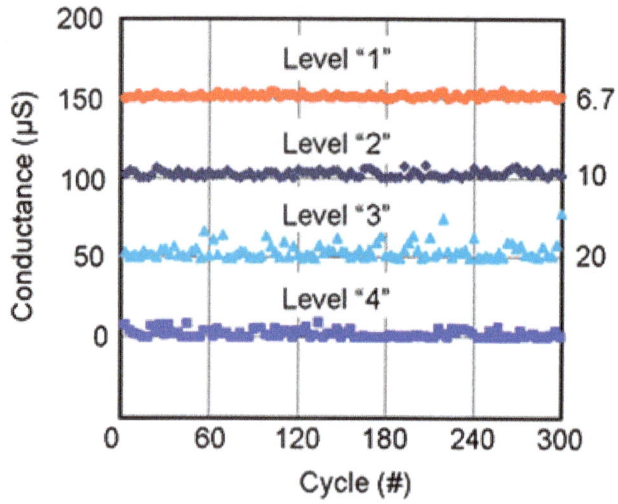

Figure 19: Four-level operations are shown by applying different operation currents. The first three levels are the SET with different programming currents, and the fourth level is the RESET condition. Pulse widths for SET/RESET are 10 ns/10 ns, respectively [24].

Table 2: SET and RESET voltages (V_{SET} and V_{RESET}) for various switching materials

Reference	$+V_{SET}$(V)	$-V_{RESET}$(V)
[9]	0.25	0.08
[10]	0.2	0.1
[11]	0.2	0.05
[12]	0.75	0.03
[13]	0.3	0.3
[14]	0.3	0.2
[16]	0.15	0.1
[17]	0.75	0.2
[18]	0.9	0.15
[20]	0.6	0.2
[23]	0.5	0.25
[24]	1.0	0.75
[26]	1.1	0.7
[29]	0.6	0.15
[30]	0.5	0.3
[31]	1.5	1.0
[33]	0.56	0.16
[42]	0.4	0.2
[46]	0.5	0.2
[49]	0.7	0.5

Switching Mechanism

Generally, it is known that the switching mechanism is based on the formation and removal of the metallic filament into the switching material, but there is one conflict about the growth kinetics of filaments. Therefore, filament formation/dissolution needs to be understood clearly for the mass production of CBRAM devices. When a positive bias is applied on the active electrode where the BE is grounded, then the Cu or Ag atoms are ionized and drift through the switching medium (solid electrolytes or oxide) under the electric field and finally reach the inert electrode. The filament formation/removal mechanism is reported first using a $Ag_{0.33}Ge_{0.20}Se_{0.47}$ solid electrolyte material, as shown in Figure 20 [50]. Symanczyk et al. used Ag and Au as active and inert electrodes, respectively. Under positive bias on the Ag electrode ($+V > V_{SET}$), Ag atoms become electrically oxidized and drift through the $Ag_{0.33}Ge_{0.20}Se_{0.47}$ solid electrolyte and finally reduce at the Au BE due to incoming electrons from the BE. As a result, the metallic filament forms due to the gradual process and the memory device switches from HRS to LRS, which is known as the SET process (Figure 20a). Ionization equations for the redox (reduction and oxidation) reaction are as follows:

$$\text{Oxidation}: \text{M} \rightarrow \text{M}^{z+} + ze^-,\ z = 1, 2 \quad (2)$$

$$\text{Reduction}: \text{M}^{z+} + ze^- \rightarrow \text{M} \quad (3)$$

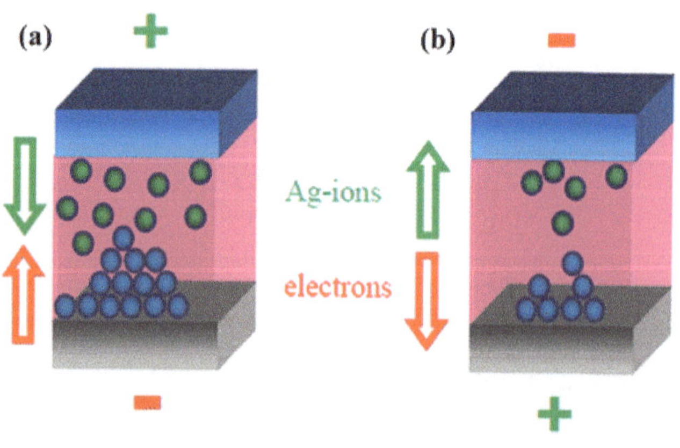

Figure 20: Simple schematic views of the devices at SET and RESET conditions. The device becomes **(a)** SET due to metallic path formation between the two electrodes, and **(b)** RESET is the result of the removal of the conductive path [50].

where 'M' stands for Ag or Cu. When negative bias is applied on TE ($-V < V_{RESET}$), Ag atoms in the filament become oxidized and migrate towards the TE, which results in the device to get back from LRS to HRS (Figure 20b). The resistive switching mechanism using a $Ge_{0.4}Se_{0.6}$ solid electrolyte in an Al/Cu/

$Ge_{0.4}Se_{0.6}$/W structure was also investigated by us [12,33]. For HRS to LRS ($+V > V_{SET}$), the Cu ions are generated at the Cu/$Ge_{0.4}Se_{0.6}$ interface from the Cu TE following the oxidation method as mentioned in Equation 2 and migrated towards the W BE due to the electric field, and then the Cu metal starts to grow from the W BE following the reduction method as mentioned in Equation 3 to form the Cu nanofilament in the Ge_xSe_{1-x} solid electrolyte through this redox reaction. It is reported that the base of the filament is at the cathode side and the thinner part of the filament at the anode side [33]. For LRS to HRS ($-V < V_{RESET}$), the electric field will be higher at the Cu/filament interface and the Cu filament starts to dissolve at the Cu/Ge_xSe_{1-x} interface by Joule heating first and then the electrochemical oxidation at the top of the filament or the Ge_xSe_{1-x}/filament interface. Basically, Joule heating at the weak part (i.e., the thinner part of a filament at the Cu/filament interface generally) of a metallic filament will play a role to dissolve the filament first. However, further study is necessary to understand the RESET mechanism. Then, the Cu ions are collected in the Cu electrode through the reduction method as mentioned in Equation 3. Therefore, the Cu filament length is gradually decreased (i.e., full removal of metals from the solid electrolyte), and the device is going back to pristine. It is interesting to note that the Cu ions can be migrated through porous regions in the solid electrolyte or defects. Therefore, defective solid electrolyte or oxide materials can be used to get such a resistive switching behavior. The Cu nanofilament is observed in Cu/$Ge_{0.4}Se_{0.6}$/W devices by TEM [33]. The device was prepared for TEM observation. Before going for TEM observation, the memory device was characterized with a program/erase of 4,000 cycles. Finally, the memory device was kept in a SET condition. The P/E current and pulse width applied are 500 μA and 500 μs, respectively. A huge Cu migration and filament formation are clearly observed. The $Ge_{0.4}Se_{0.6}$ film in the filament region shows to be crystalline owing to Cu ion migration and Cu filament formation after the SET condition. The diameter of the Cu nanofilament is approximately 11 nm. Some plausible explanations of Cu diffusion have also been discussed from other reported results in the literature. The diffusion occurs when copper or silver is in contact with chalcogenides, as reported by McHardy et al. [51]. The Cu or Ag reacts at room temperature with amorphous chalcogenides without any exposure to ultraviolet or electron radiation. After a few days, they have observed that copper diffuses throughout the $GeSe_2$ film and polycrystalline materials such as Cu_2Se (FCC) formed as small crystallites surrounding the amorphous film. In contrast, we observed that the $Ge_{0.4}Se_{0.6}$ film shows to be amorphous, which suggests that there is no diffusion of Cu initially unless external bias is applied. By applying external bias, structural changes (crystalline) occur in an amorphous $Ge_{0.4}Se_{0.6}$ solid electrolyte. A huge amount of Cu signal is obtained into the crystalline region, due to the formation of a Cu cluster or Cu nanofilament into the $Ge_{0.4}Se_{0.6}$ solid electrolyte. There are mainly two diffusion mechanisms involved: one is a fast interstitial mechanism and the other is a slow substitutional mechanism. However, Cu or Ag exhibits high ionic mobility in chalcogenides; therefore, Ag or Cu ions diffuse rapidly as interstitial species through these chalcogenides until a vacant site is encountered where substitution can occur [51,52]. Therefore, the defects play an important role for the formation/dissolution of a Cu filament through the solid electrolyte. Schindler et al. [53] have also reported the formation of Cu clusters on the surface of a $Ge_{0.3}Se_{0.7}$ film after deposition of a 160-nm $Ge_{0.3}Se_{0.7}$ film on a 50-nm-thick Cu layer. They also observed a Ag filament with a diameter of 20 nm in the $Ge_{0.3}Se_{0.7}$ solid electrolyte by conductive atomic force microscopy (CAFM) [53]. The Cu nanofilament was also observed by HRTEM after keeping the memory device at SET (Figure 21[20]). A typical device size was approximately 150×150 nm^2. The crystalline Cu nanofilament and GeO_x film (without Cu nanofilament) are shown. The clear crystallization in the GeO_x film shows a truncated conically shaped Cu nanofilament, which was also confirmed by EDX analysis (Figure 21f). The Cu signal calculated at the Cu electrode, GeO_x without filament, and Cu nanofilament in the GeO_x film are approximately 350, 200, and 400 at a typical energy of 0.92 keV, respectively. It is thus clear that the Cu signal is higher (400) in the Cu nanofilament than it would be without the Cu nanofilament in the GeO_x film. This nanofilament has Cu clusters (or nanocrystals) embedded in the GeO_x film or it is a Cu:GeO_x mixture. The Cu clusters are also confirmed by fast Fourier transform (FFT) analysis (Figure 21b,c,i), while the GeO_x film without Cu signals is shown in Figure 21d,h. Owing to the small crystals (nanograins) or amorphous GeO_x film, the FFT does not show a clear signal. Crystal *d*-spacing values of 2.0874 Å (Cu) and 2.4506 Å (Cu_2O) are observed inside the nanofilament (Figure 21b,i), which are very close to the *d*-spacing value of pure Cu (111) of 2.0880 Å [28] and similar to that of the Cu electrode (Figure 21e). The *d*-spacing value of 2.1953 Å was observed at the Cu filament/W interface, which is in between the *d*-spacing value of Cu (111) and W (110) and was approximately 2.2378 Å (Figure 21c). A truncated conically shaped crystalline Cu nanofilament, the widths of the base and top being 85 and 70 nm, respectively, was observed. After unipolar RESET, a small crystalline GeO_x film is also observed (Figure 21j). Further, small crystals (3- to 5-nm grains), an amorphous GeO_x film (Figure 21k,l), and the remaining Cu on the Cu electrode (Figure 21m) are also found. This suggests that the Cu nanofilament has been dissolved almost completely after the RESET of the unipolar resistive switching mode, and thus, the device starts to behave as a pristine device. However, understanding of the growth/dissolution kinetics of the metallic filament under bias is also important.

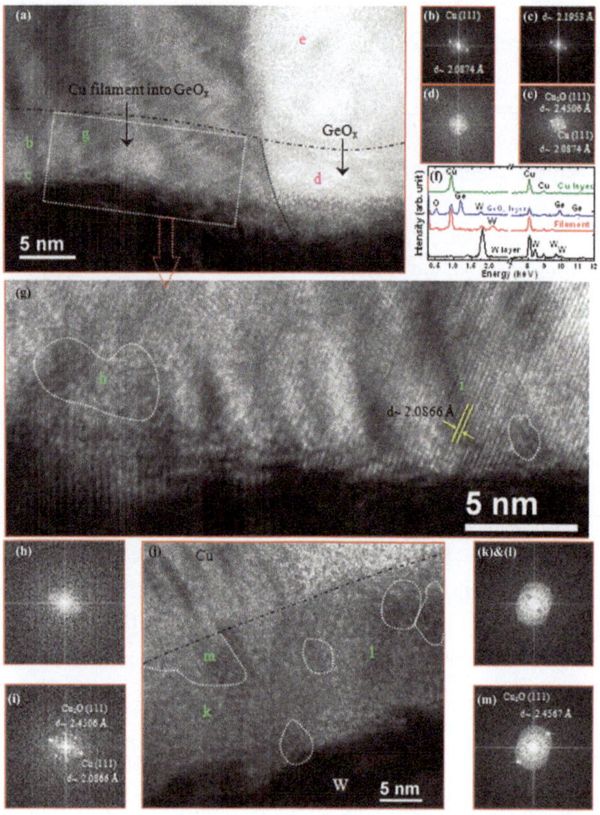

Figure 21: HRTEM images, FFT analysis, and EDX analysis of a Cu/GeO_x/W memory structure. **(a)**After SET operation (CC of >100 μA), the HRTEM image of a Cu/GeO_x/W memory structure. A crystalline Cu nanofilament is clearly observed in the GeO_x film. The FFTs confirm the **(b,g,i)** Cu nanofilaments, **(c)** Cu/W interface, **(d)** GeO_x layer without Cu nanofilament, and **(e)** Cu electrode. **(f)** All layers and filaments are also confirmed by EDX analysis. **(h)** Some amorphous regions are also observed. After the unipolar RESET operation, **(j)** an HRTEM image is shown. **(k, l)** An amorphous GeO_x layer with crystalline nanograins is observed because of the Ge-rich GeO_x film. **(m)** A small amount of Cu also remains on the Cu electrode because of higher Joule heating on the filament's weak points [20].

Yang et al. [54] have reported the growth kinetics of nanofilament formation using a Ag/SiO_2/Pt CBRAM device (Figure 22). When a positive voltage is applied to the Ag electrode, the Ag ion starts to inject into the SiO_2 dielectric layer and forms a well-defined conducting filament with a typical conical shape highlighted by the top arrow in Figure 22a. It is interesting to note that the thinnest part of the filament is apparently near the inert Pt electrode interface, with a much wider base that formed near the active Ag electrode which is contrary to solid electrolyte-based cells. When negative bias is applied on the TE, it has been observed that the filament at the narrowest region near the filament/inert electrode interface starts to dissolve and the rest of the filament is no longer electrically connected to the positively biased inert electrode and device switched to HRS, as shown in Figure 22b. Since the cation mobility inside the solid electrolyte-based CBRAM devices is higher, a large number of cations become reduced and the filament starts to grow from the inert electrode. Hence, a dendrite-like filament with the base at the active electrode and the thinnest part at the dielectric/inert electrode interface is formed, as presented in Figure 22c. To further investigate the role of cation transport in dielectric films, they have chosen a-Si as the switching medium due to lower cation mobility compared with a SiO_2-based device. When positive bias is applied on the TE, the filament starts to grow from the TE and a conducting path is formed in between two electrodes, as depicted in Figure 22d. This is due to very slow cation mobility through the a-Si by which before reaching to the inert electrode cations become reduced by the highly mobile incoming electrons from the inert electrode, as presented in Figure 22e. To unfold the switching mechanism of CBRAM devices, the exact filament dynamics is also very important, because the switching stability of the device depends on the filament geometry. So it is reported that the devices are initially at HRS or pristine, and after applying the external bias, the devices switch to LRS. This type of CBRAM device is gapless. On the other hand, there is another type of CBRAM devices where the devices are at LRS initially. This is known as gap-type atomic switch, whose switching mechanism is discussed below.

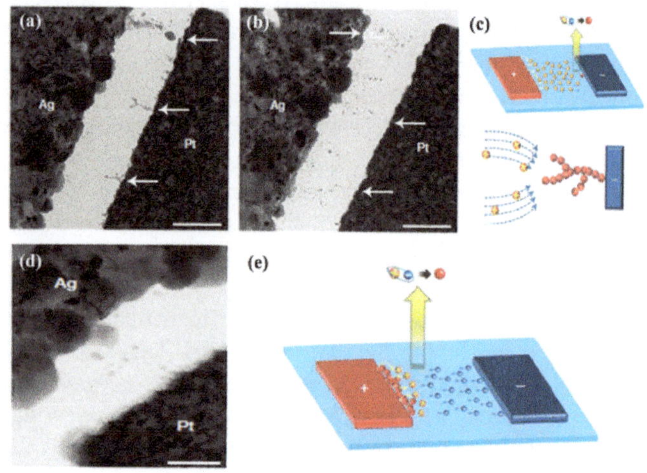

Figure 22: TEM images and schematic diagrams for the conducting filament kinetics. **(a)** TEM image of a SiO_2-based material after the formation of the conducting nanofilament. The arrows indicate complete and incomplete nanofilaments. The scale bar is 200 nm. **(b)** TEM image after RESET operation. **(c)** Schematic illustration of the filament growth process. **(d)** TEM image of a conducting filament in a-Si-based CBRAM devices. The scale bar is 50 nm. **(e)** Schematic diagram of the nanofilament growth process in an a-Si CBRAM device [54].

Hasegawa et al. [38] have investigated an implicit mechanism of the gap-type switching phenomenon, as shown in Figure 23. Initially, the device is at LRS. Applying positive bias on a Pt nanowire, Ag atoms become Ag^+ cations and diffuse through the Ag_2S layer. As a result, forming a 1-nm gap, the device goes to HRS (Figure 23a). After reversing the polarity of the external bias, a nanobridge is formed inside the gap and the device switches to LRS (Figure 23b). These oxidation and reduction phenomena have been explained clearly by the schematic illustration shown in Figure 24 [38]. In the equilibrium condition when no bias is applied, the electrochemical potential of the Ag^+ ions inside the Ag_2S surface and that of the Ag atoms on the surface are equal. Hence, the activation energy of the oxidation (E_O) and reduction (E_R) remain the same, as shown in Figure 24a. When a positive bias is applied on the Ag substrate, Ag^+ cations move towards the surface and hence the concentration of the Ag^+ ions increases at the Ag_2S surface. According to the formula of electrochemical potential, $\mu = \mu^0 + F + RT \ln \gamma C$, where μ^0 is the potential for $\gamma C = 1$, F is the Faraday constant, φ is the electrical potential, R is the molar gas constant, T is the absolute temperature, γ is the activity co-efficient, and C is the concentration of the metal cation, the potential of Ag^+ ions increases due to the increase of the concentration resulting in $E_R < E_O$, and the Ag nanowire starts to grow inside the gap (Figure 24b). Under negative bias, Ag^+ ions migrate towards the bottom side of the Ag_2S. As a result, the concentration of the Ag^+ ion decreases at the sub-surface, resulting in$E_o < E_R$, which enhances the oxidation rate of the Ag atoms and responsible for the HRS of the device (Figure 24c). Recently, Vianello et al. [14] have shown evidence of filamentary-based CBRAM operation using a Ag/Sb-doped GeS_2/W structure (Figure 25), which can be used for proto-typical production. In the energy-filtered TEM image, it is clearly shown that a metallic filament is formed under SET condition, which makes a conducting path inside the solid electrolyte. Two parameters, temperature and voltage, play an important role for this process. The positive voltage on the Ag active electrode is responsible for the nucleation of the activated Ag ions which migrate across the solid electrolyte. As a result, a filamentary path is formed from the W inert BE electrode towards the active TE. On the other hand, a reverse bias causes the dissolution of the existing filament and results in the RESET of the device. Therefore, it is observed that the operation mechanism of CBRAM devices depends on not only the oxidation-reduction of the active electrode but also the several ways of filament formation dynamics which induce an important role on overall device performances. The switching mechanisms in different references of similar structures are always various, and they need to be unique from understanding point of view as well as application. Various mechanisms are observed which are owing to different switching materials, the anode/switching material interface, and different deposition methods. Therefore, selecting the switching material and deposition methods also has a major role. Further study is needed to explore the growth/dissolution kinetics of the metallic filaments in the future.

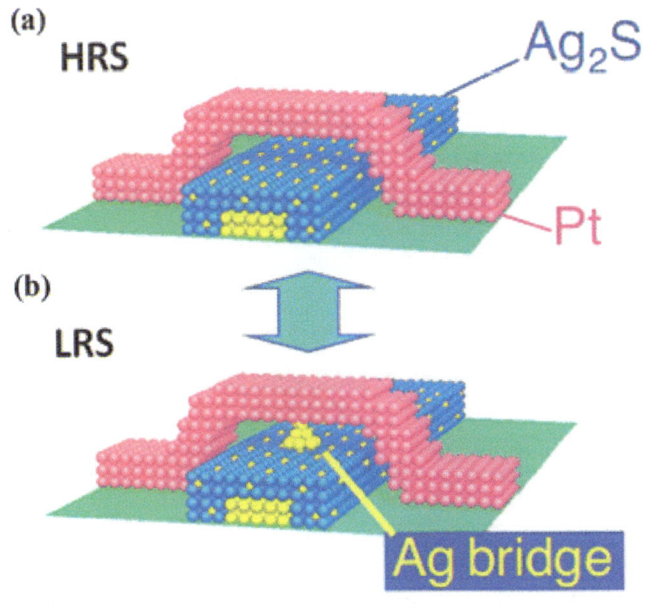

Figure 23: Schematic view of a crossbar structured for gap-type atomic switch. **(a)** A positive bias applied on the Pt nanowire makes the device in HRS by forming a 1-nm gap between the two electrodes. **(b)** A conducting Ag bridge is formed inside the gap by electrochemical reaction with the help of negative bias on the Pt nanowire [38].

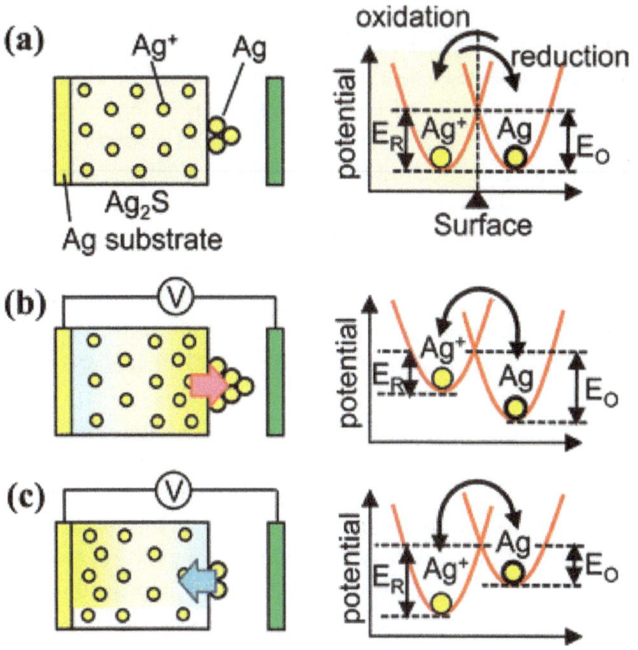

Figure 24: Simple schematic diagrams of the mechanism of Ag nanowire growth and dissolution process. **(a)** When no bias is applied, then the activation energies for reduction and oxidation are equal in the equilibrium condition. **(b)** Positive bias on the Ag_2S electrode causes the diffusion of Ag^+ cations towards the surface of the Ag_2S electrode, making $E_R$smaller than E_O. As a result, the reduction rate of Ag^+ cations enhances. **(c)** Due to the negative bias on the Ag_2S electrode, Ag^+ cations diffuse towards the bottom of the Ag_2S electrode, making E_O smaller than E_R. As a result, Ag atoms oxidize easily and diffuse inside the Ag_2S electrode [38].

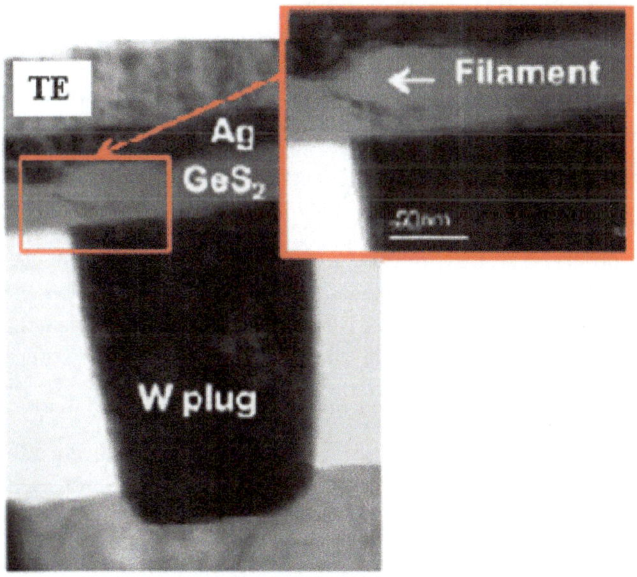

Figure 25: Cross-sectional TEM image of the memory device with a Ag/GeS_2/W structure. The energy-filtered TEM image shows the existence of the Ag metallic path during the SET condition of the device [14].

Crossbar Memory for 3D Architecture

From the above discussions, it is concluded that CBRAM technology is the most efficient one for future nanoscale non-volatile memory application; however, one of the key issues is the necessity of high-density storage. Among the different types of device structures, the cross-point CBRAM device in 3D architecture is the most possible way to compete with multi-bit storage NAND FLASH. In addition, this crossbar memory can be used as a logic operation also [55]. Kim et al. [22] have demonstrated CBRAM switching using a Ag/a-Si/SiGe/W stack in 40 top nanowire electrodes crossed with 40 bottom nanowire electrodes in crossbar architecture, as shown in Figure 26. The operation of each cross-point of the integrated crossbar array is designed by a binary bitmap image with 1,600 pixels (40 × 40) representing data 0, i.e., the 'HRS state', and pixels representing data 1, i.e., the 'LRS state'. For writing '1' into a cell inside the array, a 3.5-V, 100-μs pulse is applied across the selected cell through the CMOS decoder circuit while the other unselected electrodes in the 40 × 40 arrays are connected to a protective voltage with amplitude equaling half of the programming voltage to minimize disturbance of the unselected cells. A similar approach is used for writing '0' using a −1.75-V, 100-μs erase pulse. The programming/erasing is carried out based only on the input pattern and ignored the existing state of the memory cells, and a single programming/erase pulse is sufficient for each cell. Once all data are programmed in an array, the information in the array is then read out one cell at a time by applying a 1-V, 500-μs read pulse across the target cell, while grounding all unselected electrodes through the CMOS decoder. Stable switching with reasonable RESET current is needed to confirm the operations, which is not observed from their reports. Tada et al. [27] have also implemented a $TiO_x/TaSiO_y$ CBRAM stack in 4 × 4 crossbar architecture integrated with a CMOS circuit, and a higher operation current of >100 μA is also shown. According to our previous investigation [21], we have reported a cross-point CBRAM device using an $Al/Cu/GeO_x/W$ structure at a low current of 1 nA, as shown in Figure 27. Initially, all memory devices were in HRS. A formation voltage of >1 V is necessary to switch the memory device from HRS to LRS, which is shown in the first cycle and at a CC of 500 nA. After the formation process, the device shows normal bipolar resistive switching behavior. The memory device can be operated at a low CC of 1 nA, and a cylindrical-type filament can be expected because the HRS is the same after RESET operation. A change of HRS is observed at a CC of 50 μA. At a higher CC of 50 μA, the filament diameter is increased and the shape of the filament will be a conical type. After RESET operation, the Cu filament remains at the GeO_x/W interface. On the other hand, a high formation voltage of approximately 6 V is needed for Al TE [21]. In this case, the memory device can be operated at a low CC of 1 nA, but a large RESET current of >1 mA is needed to rupture the oxygen vacancy conducting filaments. Further, Al TE will react with GeO_x and form an AlO_x layer at the Al TE/GeO_x interface. Using Al TE, a high I_{RESET} of >20 mA was also reported by Kato et al. [56]. Lin et al. [57] have also reported high I_{RESET} for Al_2O_3-based resistive memory using a $Ti/Al_2O_3/Pt$ structure. According to several reported results, using an Al electrode is required for larger operation voltages as well as large RESET currents [56-58]. On the other hand, an excellent scaling of RESET current is observed for $Al/Cu/GeO_x/W$ cross-point memory devices with CCs from 1 nA to 50 μA. Furthermore, the RESET current is lower than the CC, which proves no current overshoot effect even 1R configuration or there is no parasitic effect. However, a GeO_x solid electrolyte-based cross-point memory in a $Cu/GeO_x/W$ structure shows a high resistance ratio of 10^8 for future MLC applications. This suggests that the Cu nanofilament diameter can be controlled by CCs using $Al/Cu/GeO_x/W$ cross-point memory devices. Recently, a cross-point memory using an AlO_x switching layer also shows good characteristics for 3D architecture [59]. However, one of the key issues for 3D CBRAM storage is sneak path leakage. This can be solved by using a 1TnR or 1S1R (one selector-one resistor) structure. Jo et al. [60] have reported the integration of a 3D-stackable 1S1R passive crossbar memory. The selector shows a sharp switching slope of <5 mV/dec, selectivity of 10^{10}, speed of <50 ns, and endurance of >10^8 cycles. A 4-Mb 1S1R crossbar array shows that the sneak path leakage current is suppressed below 0.1 nA. It is expected that a high-density memory can be utilized with a low power for using CBRAM devices.

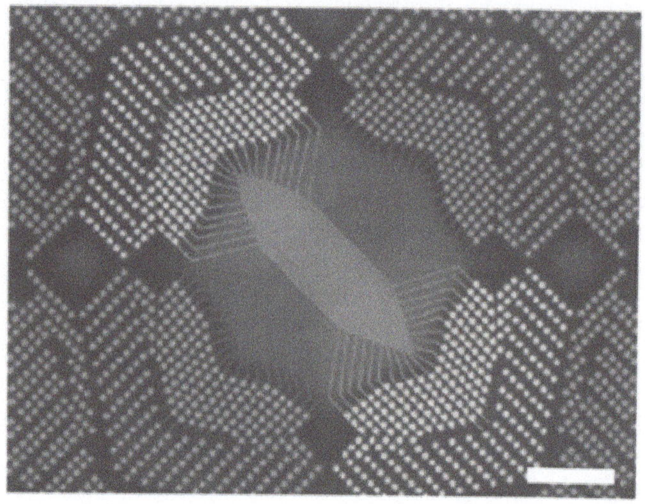

Figure 26: SEM image of a crossbar array fabricated on top of a CMOS chip [22]. This shows that high-density CBRAM devices could be produced.

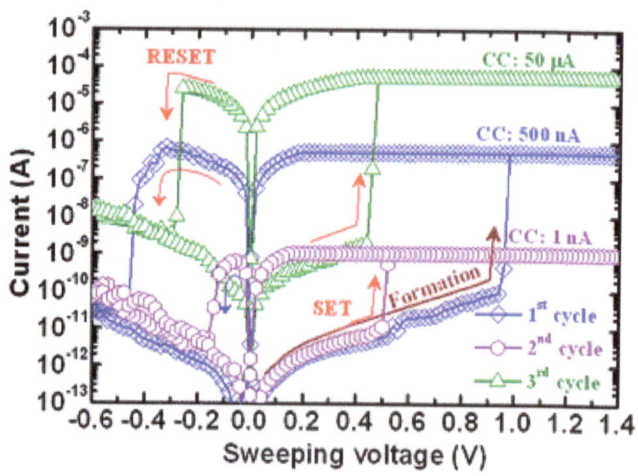

Figure 27: *I-V* switching characteristics of an $Al/Cu/GeO_x/W$ CBRAM device. The device can be operated at a low CC of 1 nA [21].

CBRAM as a Chip

A CBRAM with a 32-kb chip has been reported by using I2C (Inter-Integrated Circuit) and SPI (Serial Peripheral Interface) bus architecture [61]. The operation voltage is approximately 2.5 to 3.6 V. Chevallier et al. [62] have reported a 64-Mb CBRAM storage device with a 130 nm technology node. Otsuka et al. [63] have reported a 4-Mb storage device in a 180 nm technology node with promising 2.3 GB/s read and 216 MB/s write throughput. Figure 28 shows a schematic diagram of dual-cell CBRAM devices using Cu-Te/SiO_x (or a-Si) layers [64]. The operating parameters of the 168-mm^2 chips are the supply voltages of 1.2 and 5 V with an on-chip charge pump providing 6.6 V required for programming. It has two BEs which are driven by the two buried word line MOS transistors to write or erase the memory cells. The memory cell size is designed to be 6 F^2. This storage device can be combined high performance and high bit density whereas its other counterparts are failing to achieve the same goal. Those CBRAM chips can be used as an economical alternative to the conventional Electrically Erasable Programmable Read-Only Memory (EEPROM) technology and may be applicable to electronic toys, games, consumer electronics, wireless LAN, data storage, etc. The concept of this dual cell is similar to that of complementary resistive switching [65], where two cells are in a series. It is expected that the CBRAM device can be mass productive in the near future, if there is a solution of reliability under few microampere operation and a clear understanding of the switching mechanism.

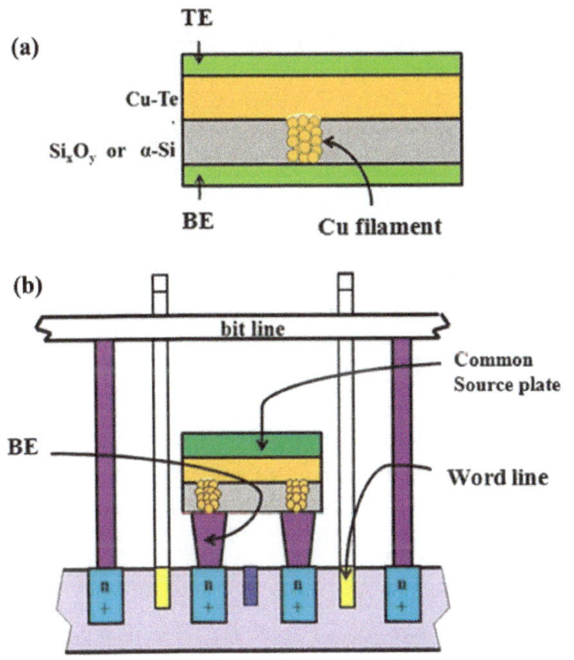

Figure 28: SET operation and dual cell of a CBRAM structure. **(a)** SET operation of a Si_xO_y-based CBRAM structure. **(b)** The dual cell of a CBRAM structure [64].

CONCLUSIONS

In this article, we have reviewed CBRAM switching characteristics and switching mechanism, introducing chalcogenide, oxide, and bilayer switching layers. In the first section, we review the switching performance of chalcogenide-based CBRAM. It is observed that a chalcogenide-based memory device shows good bipolar resistive switching, long P/E endurance of $>10^5$ cycles, and a good data retention of $>10^5$ s at >85°C. The memory device also shows multi-level operation with varying current compliance from CCs of 5 to 500 μA. It is observed that the oxide-based memory device performs at a low CC of 5 pA, good device-to-device uniformity, good data retention at low current of 1 μA, and good P/E cycles as well. It is found that bilayer switching materials show excellent device-to-device uniformity, robust data retention, and better P/E cycles of $>10^7$ with a high speed of few nanoseconds. The performances of CBRAM devices using different switching materials and structures are listed in Table 3. The switching mechanism is based on the formation and dissolution of the metallic filament depending upon electrical bias. The growth kinetics of the metallic path in different chalcogenide- and oxide-based switching materials are reported. However, it is still debated how to start growing/dissolving the metallic filament and which interface is responsible for different switching materials and structures also. It is suggested that further study is needed for the understanding of the switching mechanism. The cross-point memory in 3D architecture can have a promising solution for future high-density non-volatile memory with low power. CBRAM as a chip is showing to realize mass production in the future. However, the reliability issues in terms of stability of HRS, LRS, endurance, and data retention need to be solved and more studies are needed. It is expected that CBRAM devices have very good opportunity for scaled (<11 nm technology node) non-volatile memory devices or logic gate operations.

Table 3: Performances of CBRAM devices

Device structure (TE/switching material)/BE	Operation current (μA)	Retention time (s)	P/E endurance (cycles)
Ag/10% Sb-GeS_2/W [14]	100	10^5 at 150°C	$>10^5$
Cu/Cu-doped SiO_2/W [18]	5	10^5 at 1 μA	10^7
Cu/GeO_x/W [20]	50	$>10^6$ at 85°C	$>10^3$
Cu/Cu-Te/GdO_x/W [24]	110	3.6×10^5 at 130°C	~10^7
Al/Cu/Ti/TaO_x/W [29]	0.1 to 300	10^4	10^4
Al/TiN/Cu/TiW/Al_2O_3/W [31]	25	6×10^2 at 125°C	10^6

ACKNOWLEDGEMENTS

This work was supported by the National Science Council (NSC), Taiwan, under contract numbers NSC-97-2221-E-182-051-MY3, 98-2923-E-182-001-MY3, 98-2221-E-182-052-MY3, 101-2221-E-182-061, and 102-2221-E-182-057-MY2. The name of the NSC has been changed to Ministry of Science and Technology (MOST), Taiwan.

Competing interests

The authors declare that they have no competing interests.

REFERENCES

1. Hutchby J, Garner M. Assessment of the potential & maturity of selected emerging research memory technologies. Workshop & ERD/ERM Working Group Meeting (April 6–7, 2010). 2010.http://www.itrs.net/Links/2010ITRS/2010Update/ToPost/ERD_ERM_2010FINALReportMemoryAssessment_ITRS.pdf
2. Waser R, Aono M. Nanoionics-based resistive switching memories. Nat Mater. 2007;6:833.
3. Wong HSP, Lee HY, Yu S, Chen YS, Wu Y, Chen PS, et al. Metal-oxide RRAM. Proc IEEE. 2012;100:1951.
4. Chen A, Haddad S, Wu YC, Fang TN, Lan Z, Avanzino S, et al. Non-volatile resistive switching for advanced memory applications. In: Tech. Dig.-Int. Electron Devices Meet., Washington, DC, MD, 2005, p. 746.
5. Prakash A, Jana D, Maikap S. TaO_x-based resistive switching memories: prospective and challenges. Nanoscale Res Lett. 2013;8:418.
6. Hirose Y, Hirose H. Polarity-dependent memory switching and behavior of Ag dendrite in Ag-photodoped amorphous As_2S_3 films. J Appl Phys. 1976;47:2767.
7. Kozicki MN, Yun M, Hilt L, Singh A. Applications of programmable resistance changes in metal-doped chalcogenides. Pennington, NJ, USA: Electrochem. Soc; 1999. p. 298.
8. Terabe K, Hasegawa T, Nakayama T, Aono M. Quantized conductance atomic switch. Nature. 2005;433:47.
9. Kund M, Beitel G, Pinnow CU, Röhr T, Schumann J, Symanczyk R, et al. Conductive-bridging RAM (CBRAM): an emerging non-volatile memory technology scalable to sub 20 nm. In: Tech. Dig.-Int. Electron Devices Meet., Washington, DC, MD, 2005, p. 754.
10. Kozicki MN, Park M, Mitkova M. Nanoscale memory elements based on solid-electrolytes. IEEE Trans Nanotech. 2005;4:331.
11. Schindler C, Meier M, Waser R, Kozicki MN. Resistive switching in Ag-Ge-Se with extremely low write currents. Non-Volatile Memory Technology Symposium (NVMTS), 2007, p. 82.
12. Rahaman SZ, Maikap S, Das A, Prakash A, Wu Y, Lai CS, et al. Enhanced nanoscale resistive switching memory characteristics and switching mechanism using high-Ge-content Ge0.5Se0.5 solid electrolyte. Nanoscale Res Lett. 2012;7:614.
13. Kozicki MN, Balakrishnan M, Gopalan C, Ratnakumar C, Mitkova M. Programmable metallization cell memory based on Ag-Ge-S and Cu-Ge-S solid electrolytes. Non-Volatile Memory Technology Symposium (NVMTS), 2005, p. 83.
14. Vianello E, Molas G, Longnos F, Blaise P, Souchier E, Cagil C, et al. Sb-doped GeS_2 as performance and reliability booster in conductive bridge RAM. In: Tech. Dig.-Int. Electron Devices Meet., 2012, p. 31.5.1.
15. Choi SJ, Lee JH, Bae HJ, Yang WY, Kim TW, Kim KH. Improvement of CBRAM resistance window by scaling down electrode size in pure-GeTe film. IEEE Electron Device Lett. 2009;30:120.
16. Banno N, Sakamoto T, Iguchi N, Sunamura H, Terabe K, Hasegawa T, et al. Diffusivity of Cu ions in solid electrolyte and its effect on the performance of nanometer-scale switch. IEEE Trans Electron Devices. 2008;55:3283.
17. Maikap S, Rahaman SZ, Wu TY, Chen F, Kao MJ, Tsai MJ. Low current (5 pA) resistive switching memory using high-κ Ta_2O_5 solid-electrolyte. IEEE European Solid-State Device Conference, 2009, p. 217.
18. Schindler C, Thermadam SCP, Waser R, Kozicki MN. Bipolar and unipolar resistive switching in Cu-doped SiO_2. IEEE Trans Electron Devices. 2007;54:2762.
19. Li Y, Long S, Zhang M, Liu Q, Shao L, Zhang S, et al. Resistive switching properties of Au/ZrO_2/Ag structure for low-voltage nonvolatile memory applications. IEEE Electron Device Lett. 2010;31:117.
20. Rahaman SZ, Maikap S, Chen WS, Lee HY, Chen FT, Kao MJ, et al. Repeatable unipolar/bipolar resistive memory characteristics and switching mechanism using a Cu nanofilament in a GeO_x film. Appl Phys Lett. 2012;101:073106.
21. Rahaman SZ, Maikap S. Comparison of resistive switching characteristics using copper and aluminum electrodes on GeO_x/W cross-point memories. Nanoscale Res Lett. 2013;8:509.
22. Kim KH, Gaba S, Wheeler D, Cruz-Albrecht JM, Hussain T, Srinivasa N, et al. A functional hybrid memristor crossbar-array/CMOS system for data storage and neuromorphic applications. Nano Lett. 2012;12:389.
23. Sun B, Liu LF, Wang Y, Han DD, Liu X Y, Han RQ, et al. Bipolar resistive switching behaviors of Ag/Si_3N_4/Pt memory device. In: Tech Dig.-Solid-State and Integrated-Circuit Technology, 2008, p. 925.
24. Aratani K, Ohba K, Mizuguchi T, Yasuda S, Shiimoto T, Tsushima T, et al. A novel resistance memory with high scalability and nanosecond switching. In: Tech Dig.-International Electron Devices Meet., 2007, p. 783.
25. Zahurak J, Miyata K, Fischer M, Balakrishnan M, Chhajed S, Wells D, et al. Process integration of a 27nm, 16Gb Cu ReRAM. In: Tech. Dig.-Int. Electron Devices Meet., 2014, p. 140.
26. Yoon J, Choi H, Lee D, Park JB, Lee J, Seong DJ, et al. Excellent switching uniformity of Cu-doped MoO_x/GdO_x bilayer for nonvolatile memory applications. IEEE Electron Device Lett. 2009;30:457.

27. Tada M, Sakamoto T, Banno N, Aono M, Hada H, Kasai N. Nonvolatile crossbar switch using $TiO_x/TaSiO_y$ solid-electrolyte. IEEE Trans Electron Devices. 2010;57:1987.

28. Rahaman SZ, Maikap S, Chen WS, Lee HY, Chen F, Tien TC, et al. Impact of TaO_x nanolayer at the $GeSe_x/W$ interface on resistive switching memory performance and investigation of Cu nanofilament. J Appl Phys. 2012;111:063710.

29. Rahaman SZ, Maikap S, Tien TC, Lee HY, Chen WS, Chen FT, et al. Excellent resistive memory characteristics and switching mechanism using a Ti nanolayer at the Cu/TaO_x interface. Nanoscale Res Lett. 2012;7:345.

30. Goux L, Opsomer K, Degraeve R, Müller R, Detavernier C, Wouters DJ, et al. Influence of the Cu-Te composition and microstructure on the resistive switching of Cu-Te/Al_2O_3/Si cells. Appl Phys Lett. 2011;99:053502.

31. Belmonte A, Kim W, Chan BT, Heylen N, Fantini A, Houssa M, et al. 90 nm W/Al_2O_3/TiW/Cu 1T1R CBRAM cell showing low-power, fast and disturb-free operation. IEEE International Memory Workshop, 2013, p. 26.

32. Devulder W, Opsomer K, Seidel F, Belmonte A, Muller R, Schutter BD, et al. Influence of carbon alloying on the thermal stability and resistive switching behavior of copper-telluride based CBRAM cells. ACS Appl Mater Interfaces. 2013;5:6984.

33. Rahaman SZ, Maikap S, Chiu HC, Lin CH, Wu TY, Chen YS, et al. Bipolar resistive switching memory using Cu metallic filament in $Ge_{0.4}Se_{0.6}$ solid-electrolyte. Electrochem Solid-State Lett. 2010;13:H159.

34. Ueno T, Odajima A. Study of photo-induced effect in obliquely-deposited amorphous Ge-Se films by XPS. Jpn J Appl Phys. 1980;19:L519.

35. Ueno T, Odajima A. X-ray photoelectron spectroscopy of Ag-and Cu-doped amorphous As_2Se_3 and $GeSe_2$. Jpn J Appl Phys. 1982;21:230.

36. Moulder JF, Stickle WF, Sobol PE, Bomben KD. Handbook of x-ray photoelectron spectroscopy. Eden Prairie, MN: Perkin-Elmer Corp; 1992.

37. Liaw C, Kund M, Landsiedel DS, Ruge I. The conductive bridging random access memory (CBRAM): a non-volatile multi-level memory technology. IEEE European Solid-State Device Conference (ESSDERC), 2007, p. 226.

38. Hasegawa T, Terabe K, Tsuruoka T, Aono M. Atomic switch: atom/ion movement controlled devices for beyond Von-Neumann computers. Adv Mater. 2012;24:252.

39. Schindler C, Staikov G, Waser R. Electrode kinetics of Cu–SiO_2-based resistive switching cells: overcoming the voltage-time dilemma of electrochemical metallization memories. Appl Phys Lett. 2009;94:072109.

40. Rahaman SZ, Maikap S, Chiu HC, Lin CH, Wu TY, Chen YS, et al. Low power operation of resistive switching memory device using novel W/$Ge_{0.4}Se_{0.6}$/Cu/Al structure. IEEE International Memory Workshop (IMW), 2009.

41. Jameson JR, Blanchard P, Cheng C, Dinh J, Gallo A, Gopalakrishnan V, et al. Conductive-bridge memory (CBRAM) with excellent high-temperature retention. In: Tech. Dig.-Int. Electron Devices Meet., 2013, p. 30-1.

42. Russo U, Kamalanathan D, Ielmini D, Lacaita AL, Kozicki MN. Study of multilevel programming in programmable metallization cell (PMC) memory. IEEE Trans Electron Devices. 2009;56:1040.

43. Sakamoto T, Lister K, Banno N, Hasegawa T, Terabe K, Aono M. Electronic transport in Ta_2O_5 resistive switch. Appl Phys Lett. 2007;91:092110.

44. Schindler C, Weides M, Kozicki M, Waser R. Low current resistive switching in Cu–SiO_2cells. Appl Phys Lett. 2008;92:122910.

45. Tsunoda K, Fukuzumi Y, Jameson JR, Wang Z, Griffin PB, Nishi Y. Bipolar resistive switching in polycrystalline TiO_2 films. Appl Phys Lett. 2007;90:113501.

46. Liu Q, Long S, Lv H, Wang W, Niu J, Huo Z, et al. Controllable growth of nanoscale conductive filaments in solid-electrolyte-based ReRAM by using a metal nanocrystal covered bottom electrode. ACS Nano. 2010;4:6162.

47. Sakamoto T, Banno N, Iguchi N, Kawaura H, Sunmura H, Fujieda S, et al. A Ta_2O_5 solid-electrolyte switch with improved reliability. Symp. on VLSI Techno., 2007, p. 38.

48. Balakrishnan M, Thermadam SCP, Mitkova M, Kozicki. A low power non-volatile memory element based on copper in deposited silicon oxide. Non-Volatile Memory Technology Symposium (NVMTS), 2006, p. 104.

49. Guy J, Molas G, Vianello E, Longnos F, Blanc S, Carabasse C, et al. Investigation of the physical mechanisms governing data-retention in down to 10 nm nano-trench Al_2O_3/CuTeGe conductive bridge RAM (CBRAM). In: Tech. Dig.-Int. Electron Devices Meet., 2013, p. 30.2.2.

50. Symanczyk R, Balakrishnan M, Gopalan C, Happ T, Kozicki M, Kund M, et al. Electrical characterization of solid state ionic memory elements. Proceedings of the Non-Volatile Memory Technology Symposium, 2003, p. 17-1.

51. McHardy C, Fitzgerald A, Moir P, Flynn M. The dissolution of metals in amorphous chalcogenides and the effects of electron and ultraviolet radiation. J Phys C Solid State Phys. 1987;20:4055.

52. Phillips JC. Structural principles of alpha-AgI and related double salts. J Electrochem Soc. 1976;123:934.

53. Schindler C, Szot K, Karthäuser S, Waser R. Controlled local filament growth and dissolution in Ag-Ge-Se. Phys Status Solidi Rapid Res Lett. 2008;2:129.

54. Yang Y, Gao P, Gaba S, Chang T, Pan X, Lu W. Observation of conducting filament growth in nanoscale resistive memories. Nat Commun. 2012;3:732.

55. Borghetti J, Snider GS, Kuekes PJ, Yang JJ, Stewart DR, Williams RS. Memristive switches enable stateful logic operations via material implication. Nature. 2010;464:873.

56. Kato S, Nigo S, Lee JW, Mihalik M, Kitazawa H, Kido G. Transport properties of anodic porous alumina for ReRAM. J Phys Conf Ser. 2008;109:012017.

57. Lin CY, Wu CY, Wu CY, Lee TC, Yang FL, Hu C, et al. Effect of top electrode material on resistive switching properties of ZrO_2 film memory devices. IEEE Electron Device Lett. 2007;28:366.

58. Rahaman SZ, Maikap S, Ray SK, Lee HY, Chen WS, Chen F, et al. Record resistance ratio and bipolar/unipolar resistive switching characteristics of memory device using germanium oxide solid-electrolyte. Jpn J Appl Phys. 2012;51:04DD11.

59. Maikap S, Panja R, Jana D. Copper pillar and memory characteristics using Al_2O_3switching material for 3D architecture. Nanoscale Res Lett. 2014;9:366.

60. Jo SH, Kumar T, Narayanan S, Lu WD, Nazarian H. 3D-stackable crossbar resistive memory based on field assisted superlinear threshold (FAST) selector. In: Tech. Dig.-Int. Electron Devices Meet. (IEDM), 2014, p. 160.

61. Gopinath VP. http://sites.ieee.org/sfbanano/files/2013/11/PG_IEEE-Talk-Adesto_CBRAM_20131217_rev4_send.pdf (http://www.adestotech.com/cbram).

62. Chevallier CJ, Siau CH, Lim SF, Namala SR, Matsuoka M, Bateman BL, et al. A 0.13 μm 64Mb multi-layered conductive metal-oxide memory. IEEE International Solid-State Circuits Conference, 2010, p. 260.

63. Otsuka W, Miyata K, Kitagawa M, Tsutsui K, Tsushima T, Yoshihara H, et al. A 4 Mb conductive-bridge resistive memory with 2.3GB/s read-throughput and 216 MB/s program-throughput. IEEE International Solid-State Circuits Conference, 2011, p. 210.

64. Neale R. Showtime for the Micron-Sony 16Gb ReRAM. 2014.http://www.eetimes.com/author.asp?section_id=36&doc_id=1321276.

65. Linn E, Rosezin R, Kugeler C, Waser R. Complementary resistive switches for passive nanocrossbar memories. Nat Mater. 2010;9:403.

CHAPTER 16

CONDUCTANCE QUANTIZATION IN RESISTIVE RANDOM ACCESS MEMORY

Yang Li, Shibing Long, Yang Liu, Chen Hu, Jiao Teng, Qi Liu, Hangbing Lv, Jordi Suñé and Ming Liu

Keywords: Resistive random access memory (RRAM), Resistive switching (RS), Conductive filament (CF), Conductance quantization

INTRODUCTION

The persistent perusing of massive storage volume has been driving the scaling-down process of memory devices for decades. Memories characterized by low-power consumption and low fabrication cost are needed. Predominant flash memory has met a scaling-down limitation around 10-nm magnitude [1, 2]. Therefore, intensive studies have been carried out in seeking for the next-generation memories. Resistive random access memory (RRAM) has become one of the most promising candidates for the next-generation memory [3–14] because of the intrinsic excellent scalability, simple metal-insulator-metal (MIM) structure, low fabrication cost, 3D integration feasibility, and promising performances in speed, power, endurance, retention, etc. RRAM stores information based on the resistive switching effect. Under appropriate external electrical field, the resistance state of the RRAM device can be reversibly switched between a high resistance state (HRS) or OFF-state and a low resistance state (LRS) or ON-state. There are two resistive switching modes, i.e., unipolar and bipolar switching operations under the same or opposite bias polarities, respectively, which are closely related to the different material systems and the different switching mechanisms. The resistive switching can be a uniform or localized phenomenon. Uniform switching proportionally scales with the total area of the switching material, while localized switching is usually based on the formation and disruption of conductive filament (CF).

Abundant resistive switching materials, electrode materials, and their various interfaces are involved in RRAM switching mechanisms which are rather complex. Rich electrical, thermal, magnetic, and optical effects are therefore presented. Typical physical/chemical effects accompanied in resistive switching processes and in HRS/LRS states include electrochemical/thermochemical reactions [15–27], metal-insulator transition [28, 29], magnetic modulation [30–47], etc. In this regard, the RRAM device can serve as a rich platform for studying the multiple physical/chemical effects. In the CF-type RRAM device, when the CF in the resistive switching (RS) layer is formed, RRAM changes to LRS. If the CF is ruptured, the device switches back to HRS. The formation and rupture of the CF can be understood as cation or anion migration under applied voltage companied by electrochemical reaction of the metal or oxygen vacancies. Therefore, CF is believed to be consisted of metal or oxygen vacancies. The dimension of the CF can be electrically modulated to be in the order of several tens to a few nanometers, which has been evidenced by the observation of high-resolution transmission electron microscopy (HRTEM) [20, 48–60], scanning TEM (STEM) [59], and atomic force microscopy (AFM) [61–63]. In the localized filamentary switching, the scaling down of the RRAM device [64] would not influence its memory characteristics until the area is approaching the CF magnitude. As the CF size is in the range of nanoscale to atomic size, which is comparable to the mean free path (Fermi wavelength) of conduction electron, the scattering might be absent, resulting in ballistic electron transport [65] and the quantized conductance (QC) [66–68]. In recent studies, conductance quantization phenomena have been proved to exist in the atomic-sized CF in RRAM [69–72], and the interest for studying them continues. Revealing the QC effect is of great significance to deeply understand the

Citation: Yang Li, Shibing Long, Yang Liu, Chen Hu, Jiao Teng, Qi Liu, Hangbing Lv, Jordi Suñé, Ming Liu, Conductance Quantization in Resistive Random Access Memory, DOI: 10.1186/s11671-015-1118-6.

physics of RS mechanism in mesoscopic dimension, which is important to control the performance, reliability, and variability [73, 74] of RRAMs and to advance their practical application as non-volatile memories. At the same time, if the conductance quantization behaviors can be well modulated, it in turn can be utilized to realize the multi-level storage for ultra-high-density memory applications. Thus, summarizing and discussing the QC effect in RRAM is very necessary. In this review paper, we focus our attention on the recent development of the research on the QC effect in CF-based non-volatile RS devices including basic QC phenomenon in RRAM, RS mechanisms, device structures, materials, theory, and modeling of conductance quantization in RRAM.

Operating Principles and RS Mechanism of RRAM

In RRAM cell with MIM structure, non-volatile data storage is achieved through the reversible resistive switching between HRS and LRS, which are utilized to store the digits "0" and "1." RS is often based on the creation and partial destruction of CF. Dependent on the polarity of the external electrical field, RS is usually classified into two modes, unipolar and bipolar switching. The transition from HRS to LRS and that from LRS to HRS are called as the SET and RESET switching, respectively. In some cases, if the prepared RS layer in fresh cell is very insulating with low amount of defects, a forming process with high voltage is necessary to soft breakdown the RS layer to trigger the subsequent reproducible RESET and SET switching. Figure 1 shows the schematic *I–V* curves in RS and the corresponding states of CF-type RRAM device.

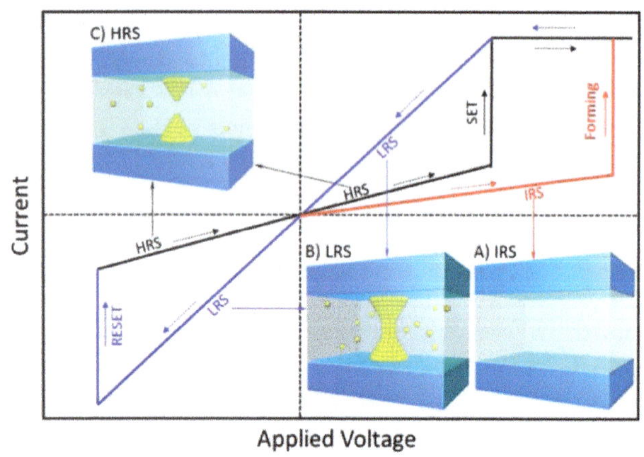

Figure 1: Schematic *I–V* curves of resistive switching process in a CF-type bipolar RRAM device. Insets *A–C* show the different resistance states of the device during the switching process. In most cases, the fresh RRAM device shows a very high initial resistance state (IRS) with few defects (inset *A*). In a positive bias sweep, when the voltage increases to a comparative high voltage ($V_{Forming}$), the device switches to the low resistance state (LRS) with a conducting filament formed in the RS layer (inset *B*). Then, in a negative voltage sweep, when the voltage reaches a critical value (V_{RESET}), the device switches from LRS to the high resistance state (HRS), corresponding to a RESET transition in which the CF is ruptured (inset *C*). At last, in another positive sweep, the device will switch to LRS again, with the filament reconnected (inset *B*). This process is called as SET, with the SET voltage (V_{SET}) much lower than ($V_{Forming}$). If the device has a good endurance, the above SET and RESET switching can be reproducibly and successively carried out for a large number of cycles.

Compared with the prototypical non-volatile memories (NVMs) such as magnetic random access memory (MRAM) based on the giant or tunneling magnetoresistance effect [75–79] and phase change random access memory (PRAM) based on the reversible phase transitions between amorphous and crystalline states of phase change materials [80–84], RRAM, an emerging NVM, has shown various complex resistive switching mechanisms, which is closely dependent on the different types of switching layer and electrode materials and also dependent on the different operation methods [85–87]. To date, the resistive switching mechanism in RRAM has been widely accepted to be mainly attributed to the reduction/oxidation (redox) electrochemistry mechanism, which can operate in the bulk RS layer, along CFs in the RS layer, and/or at the RS layer/metal contact interfaces in the MIM structure. Redox-based RRAM [3,15, 19, 88–95] can be further classified into two main types, "nanoionic" (including electrochemical metallization (ECM) [17–20, 90, 94, 96–101] and valence change mechanism (VCM) [5, 51, 61, 91, 102–112]) and "thermochemical mechanism (TCM)" (i.e., fuse/antifuse) [91, 113, 114]. In ECM and VCM devices, the diffusion or drift of charged species (Ag^+ or Cu^+cations), O anions or oxygen vacancies (Vos) in the RS layer are driven by the ion drift/redistribution and the redox electrochemistry mechanisms under an electrostatic drift field. While in TCM devices, it is driven by a thermal gradient diffusion mechanism. In fact, in a practical RRAM device, the RS process is very complicated, with multiple mechanisms simultaneously existing, but a certain one is predominant. It is worth pointing out that TCM is sometimes confused with unipolar VCM. In some cases, the ECM device is also called as conductive bridge random access memory (CBRAM) [115–119], programmable metallization cell (PMC) [120–122], and atomic switch [115–119, 123, 124]. If TCM dominates the RS, the resistance switching is unipolar. On the contrary, if ECM or VCM is dominant, the switching is usually bipolar. ECM devices are cation migration-based RRAMs, while VCM and TCM devices can be summed up into anion migration-based RRAMs. Plenty of dielectric materials have been found to show the redox-based resistance switching effect, including perovskites, solid-state electrolytes, chalcogenides, transition metal oxides, silicon dioxide, metal nitrides, organic complexes, polymers, etc., among which HfO_2 and TaO_x are most widely investigated and most competitive for the practical applications. Table 1 lists the three types of typical redox-based RS mechanisms and their corresponding material system, *I–V* curves, and operation polarity.

Table 1: The classification of redox-based resistive switching mechanisms and operation principles of redox-based RRAM

Switching mechanism	Electrochemical metallization (ECM)	Valence change mechanism (VCM)	Thermochemical mechanism (TCM)
Dominating charged species	Metal cations	O anions or oxygen vacancies (Vo)	O anions or oxygen vacancies (Vo)
Intrinsic nature of CF	Metal CF	Vo-CF (bipolar)	Vo-CF (unipolar)
Dominating driving force	External electric field	External electric field	Thermal gradient
Primary operation principle	SET process: (1) The active TE material (Ag, Cu, Ni) in the interface is oxidized under positive electric field; (2) the cations (Ag^+, Cu^+ or Cu^{2+}, Ni^+) drift into the RS layer; (3) the cations are reduced back from the BE/RS-layer interface or from the bulk RS layer or even from the TE/RS layer, depending on the difference between the drift velocity of cations and electrons; (4) metal CF is formed to connect BE and TE, with the reduction process continuing.	SET process: (1) Under positive electric field, TE material in the TE/RS-layer interface is oxidized and O^{2-}/Vo is generated; (2) O ions drift to TE or O vacancies drift to BE through the RS layer to form Vo-CF. The valence states of corresponding cations are changed.	SET (antifuse) process: O vacancies are generated, diffused, and redistributed to form Vo-CF in the bulk RS layer under the thermal gradient induced by electric field. The valence states of corresponding cations are changed.
	RESET process: Under the opposite electric field, the metal atoms in the CF are oxidized and drift away, thus CF is partially ruptured.	RESET process: Under the opposite electric field, O ions migrate back to the bulk RS layer to recombine with O vacancies in the CF, thus CF is partially ruptured.	RESET (fuse) process: CF is ruptured or fused as a result of joule heating along the CF through the thermal diffusion process of O vacancies.
Typical RS materials	Ion-conducting solid electrolyte (sulfides, selenides, or telluride of Ge, As, Sb, or Ga) such as Ag_2S [53], GeSe [224], Cu_2S [172,225], Ag_2Se [226], Ag-Ge-Se [227], $(AgI)_{0.5}(AgPO_3)_{0.5}$ [228], etc.; Binary or complex oxides such as HfO_2 [215, 222, 229,230], ZrO_2 [231, 232], SiO_2[233], WO_3 [234], TaO_x [235], GdO_x [236], etc.	Transition metal oxides (TMOs) such as TiO_2 [51, 87], HfO_2[106], ZrO_2 [112], $SrTiO_3$ [5,61], TaO_x [102], WO_3 [111], etc.; doped SiO_2 [104, 107–109]; amorphous C [103, 105, 110], etc.	Transition metal oxides (TMOs) such as HfO_2 [69, 70, 216, 237], NiO [22, 238–244], CoO [245], CuO [246], Fe_2O_3 [247], etc.
Typical electrode materials	(1) Top electrode (TE): an electrochemically active metal such as Ag, Cu, and Ni.	(1) Top electrode (TE): a low work function metal not easily reduced back after oxidation, such as Ti, Al, and Nb.	(1) Top electrode (TE): inert electrodes such as Pt, Pd, Ir, Ru, W, Au, etc.
	(2) Bottom electrode (BE): an electrochemically inert counter electrode such as Pt, Pd, Ir, Ru, W, Au, etc.	(2) Bottom electrode (BE): inert electrodes, such as Pt, Pd, Ir, Ru, W, Au, etc.	(2) Bottom electrode (BE): inert electrodes such as Pt, Pd, Ir, Ru, W, Au, etc.
Dominating material	Electrode	RS layer and electrode	RS layer

Typical *I–V*curve	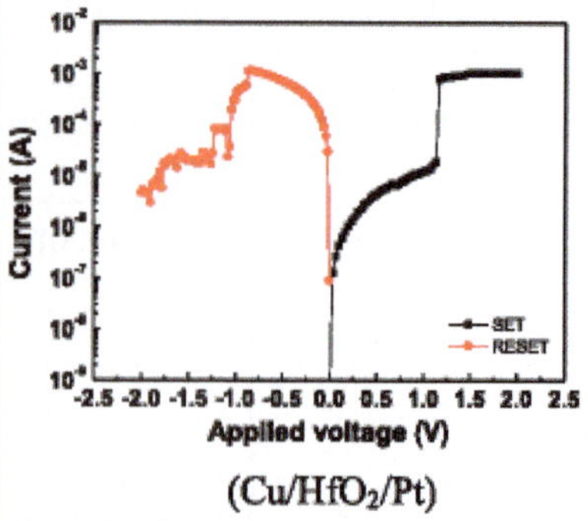 (Cu/HfO$_2$/Pt)	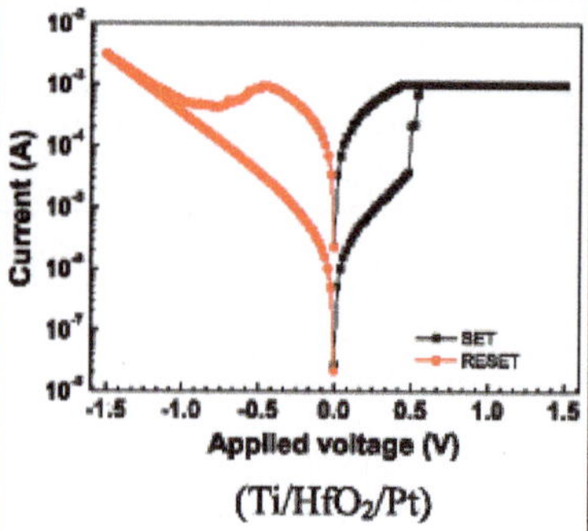 (Ti/HfO$_2$/Pt)	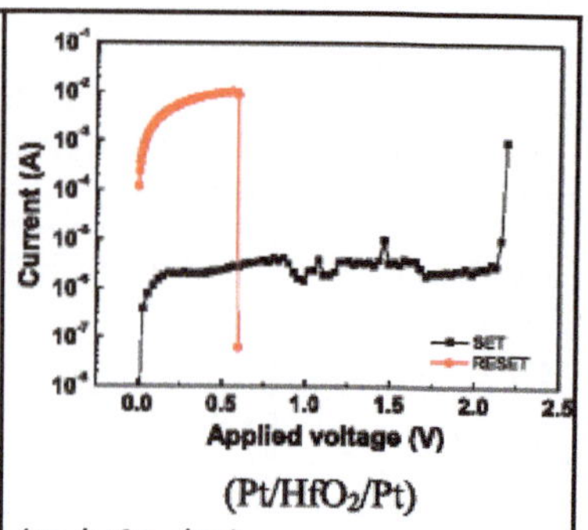 (Pt/HfO$_2$/Pt)
Operation polarity	Bipolar	Bipolar	Unipolar
RS type	Localized	Localized	Localized

Except the redox RS mechanism, the insulator-metal transition (IMT) or Mott transition in MIM structure can also contribute to RS effect. The corresponding resistive switching device is called as Mott memory and sometimes referred as correlated electron random access memory (CeRAM). In this type of memory device, the charge injection under the external electrical field induces the transition from weakly correlated electron state to strongly correlated electron state, which is activated by a critical electron population [125–127]. Electronic switches and memory elements based on the Mott transition have been explored using several typical material systems involved in Mott memory including VO_2 [128, 129], NiO [126, 127, 130], $SrTiO_3$ [131], $SmNiO_3$ [132], etc. The quasi two-dimensional electron gas (2DEG) formed at the interface between complex oxides has also been reported to show the metal-insulator transition effect [133–136]. However, in this paper, we will focus on the redox-based filamentary RS mechanisms that have been shown to support the conductance quantization effect.

Conductance Quantization in RRAM

The size of the CF can be modulated to the range of nanoscale to atomic size in both RESET and SET process. The modulation process can be achieved through specific electrical operations, especially in the RRAM devices showing progressive RESET/SET behavior [137]. This is similar to the performance of memristor [6, 138–142]. When CF is controlled to be thin enough to show atomic size, the quantum-sized effect [69] will appear in the CF-type RRAM devices. Conductance quantization phenomenon is an important representation of quantum-sized effect. Figure 2 gives an example of conductance quantization behavior observed in RRAM. The measured current–voltage (Fig. 2a) and corresponding conductance-voltage curves (Fig. 2b) show that the conductance quantization phenomenon appears in the RESET process of a Pt/HfO_2/Pt unipolar RRAM device. Obvious quantized conductance steps with multiples of G_0can be observed, in other words, abrupt conductance transitions of the order of G_0 between well-defined discrete states can be found in the final stages of the RESET transient. $G_0=2e^2/h$is the quantum of conductance, with the value of $12.9\,k\Omega^{-1}$ or $77.5\,\mu s$, where e is the electron charge and h is Planck's constant. Quantized conductance observed in the practical materials usually presents fluctuation with a certain degree, so statistical analysis on plenty of experimental data is often made use of to intrinsically reveal this effect. Figure 2c further shows the evolution of CF conductance of this device in the last stage of 100 successive RESET switching cycles. By collecting the conductance data at the step-like gradual RESET phase in the 100 successive RESET cycles in Fig. 2c, we can plot the histogram of normalized conductance, as shown in Fig. 2d. Conductance levels and peaks at 1 G_0, 2 G_0, 3 G_0, 4 G_0, etc. are clearly displayed in Fig. 2c, d, respectively.

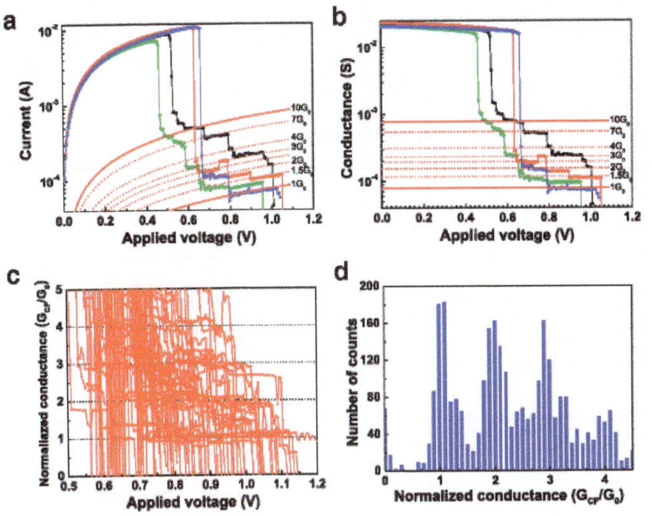

Figure 2: Conductance quantization phenomenon measured in the RESET process of a Pt/HfO_2/Pt unipolar RRAM device. **a** $I-V$ curves in four RESET cycles. **b** $G-V$ curves corresponding to (**a**). G is defined as I/V. The RESET transients of the device show discrete states with half-integer multiples of the conductance quantum. The *red lines* in (**a**) and (**b**) are "guides to the eye" and correspond to $I=nG_0V$ or $G=nG_0$ with $n=1$, 1.5, 2, 2.5, 3, 4, 7, and 10. Discrete conductance levels at about 1 G_0, 1.5 G_0, 2 G_0, 2.5 G_0, 3 G_0, 4 G_0, etc. are evident. **c** Evolution of CF conductance in the last stage of 100 successive RESET switching cycles of the Pt/HfO_2/Pt device. Discrete conductance levels at 1 G_0, 2 G_0, 3 G_0, 4 G_0, etc. are revealed. **d** Histogram of normalized conductance collected at the step-like gradual RESET phase in 100 successive RESET cycles. Conductance peaks at integer and semi-integer multiples of G_0 are clearly present.

Only the size of a conductor is small enough will the quantized conductance effect appear. In fact, as pointed out by Datta, if any of the three dimensions of a conductor is smaller than one of the three characteristic length scales [143]: (1) de Broglie wavelength of electrons; (2) the mean free path of electrons; and (3) the phase-relaxation length of electrons, the conductor will show conductance quantization behavior. Figure 3 shows the typical ranges of the three characteristic length in metal and semiconductor materials. In the devices with atomic-scale CF, the CF configuration is determined by the atomic granularity of the material. In this case, the transport through the CF is governed by the quantum nature of conductance, i.e., the current is carried along the discrete conductance channels. The reason for the occurrence of conductance quantization is that the electrons are not scattered when transporting through the atomic-scale conductor. The conductor behaves like a waveguide for electrons and does not follow the Ohm's law anymore. The waveguide could be understood as a ballistic transportation path made up of a bundle of discrete conductance channels, with each contributing a maximum amount of one G_0 to the total conductance of the conductor. The total conductance of the conductor is described by the Landauer formula, $G=G_0\sum_i T_i$ [144]. T_i is the transmission probability of each discrete conductance channel. If the channel is assumed to be fully transmitted, T_i equals to one. $T=\sum_i T_i$ is the transmission probability of the whole conductor, which is determined by the details of the conductor geometry and the electronic structure of the conductor material [145].

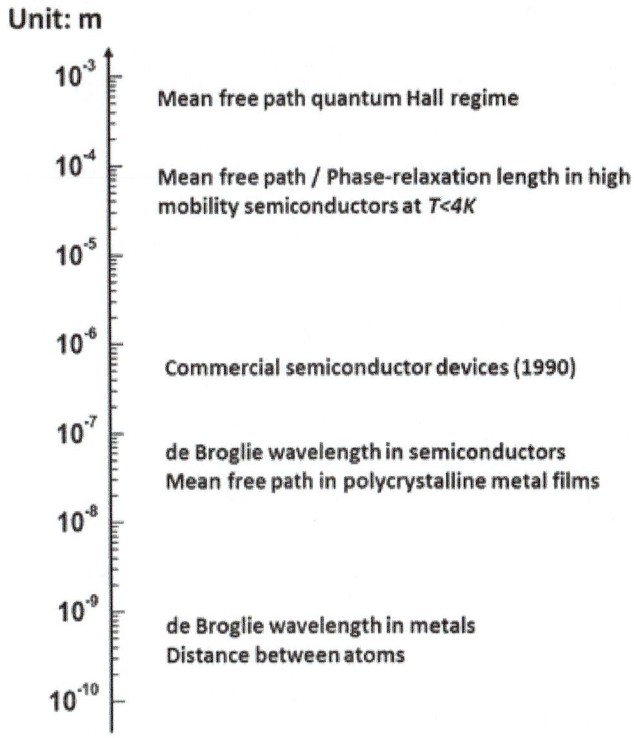

Figure 3: Three characteristic length scales related to quantum conductance phenomenon. The three characteristic length scales are the following: (1) the de Broglie wavelength, which is related to the kinetic energy of the electrons; (2) the mean free path, which is the distance that an electron travels before its initial momentum is destroyed; and (3) the phase-relaxation length, which is the distance that an electron travels widely from one material to another and is also strongly affected by temperature, magnetic field, etc. A conductor will show conductance quantization behavior if any of its three dimensions is smaller than the three characteristic length scales mentioned above. Reproduced with permission [143].

In fact, the phenomenon of quantum conductance was first observed by Van Wees et al. in a two-dimensional electron gas of a GaAs-AlGaAs heterostructure in 1988 [146]. Similar results were reported almost simultaneously by Wharam et al. [147], also using a 2D electron gas on a GaAs-AlGaAs heterojunction, at 0.1 K. Since then, the phenomenon of quantized conductance has been reported in various physical structures such as atomic quantum contact [68, 146,148–155], mechanically controllable break junctions [156–160], nanotubes [161–167], and the current-induced local oxidation of nanoscale constrictions [168].

As conductance quantization effect in RRAM has the potential applications of multi-level storage, it has attracted much attention in recent years. Conductance quantization phenomena were initially found in ECM devices [169] and then in VCM and TCM devices [69, 170, 171]. Conductance quantization effect in RRAM has been studied and reported in a series of literatures [69, 70, 126, 127, 129, 137, 140, 170–190], involving various materials and different RS mechanisms, CF types, SET, or RESET processes, as listed in Table 2. Some typical experimental observation results are given below as illustrations for quantum conductance phenomenon, as shown in Figs. 4, 5, 6, 7, and 8 [70, 171, 183, 186, 191].

Table 2: Different material systems showing conductance quantization effect

System	RS mechanism	CF	Quantization level	RS polarity	QC observed in SET or RESET
Ag_2S or Cu_2S (vacuum gap) [169]	ECM	Ag	G_0	Bipolar	SET
Ag/AgI/Pt [180]	ECM	Ag	G_0	Bipolar	SET
Ag/SiO_2/Pt [195,213]	ECM	Ag	0.5 G_0	Bipolar	SET
Ag/Ta_2O_5/Pt [181]	ECM	Ag	G_0	Bipolar	SET and RESET
Ag/Ag_2S/Pt (STM tip) [182]	ECM	Ag	G_0	Bipolar	RESET
Ag/P_3HT:PCBM/ITO [183]	ECM	Ag	G_0/0.5 G_0	Bipolar	SET and RESET
Ag/a-$La_{1-x}Sr_xMnO_3$/Pt [185]	ECM	Ag	G_0	Bipolar	SET
Ag/ionic conductor-layer/W tip [191]	ECM	Ag	G_0	Bipolar	SET
Ag/GeS_2/W [193]	ECM	Ag	G_0	Bipolar	SET
Cr/p^+-amorphous silicon/V [192]	ECM	Metal	0.5 G_0	Unipolar	SET
Nb/ZnO_x/Pt [170]	ECM	Nb or Vo	G_0/0.5 G_0	Bipolar	SET and RESET
Cu/HfO_x/Pt [190]	ECM	Cu	0.5 G_0	Bipolar	SET
Pt/HfO_2/Pt [69]	VCM	Vo	0.5 G_0	Unipolar	RESET
ITO/ZnO_x/ITO [170]	VCM	Vo	0.5 G_0	Unipolar	SET and RESET
Ti (Ta, W)/Ta_2O_5/Pt [171]	VCM	Vo	G_0	Bipolar	SET and RESET
V/V_2O_5/V [174]	-	-	0.5 G_0	-	SET
W/CeO_x/SiO_2/ $NiSi_2$[179]	VCM	Vo	0.5 G_0	Bipolar	RESET
n-Si/SiO_x/p-Si [186]	VCM	Vo	G_0/0.5 G_0	Bipolar and Unipolar	SET
Ti/HfO_2/TiN [188]	VCM	Vo	G_0	Bipolar	RESET
Ti/TiO_2/$SrTiO_3$/n-Si [194]	VCM	Vo	G_0	Bipolar	SET

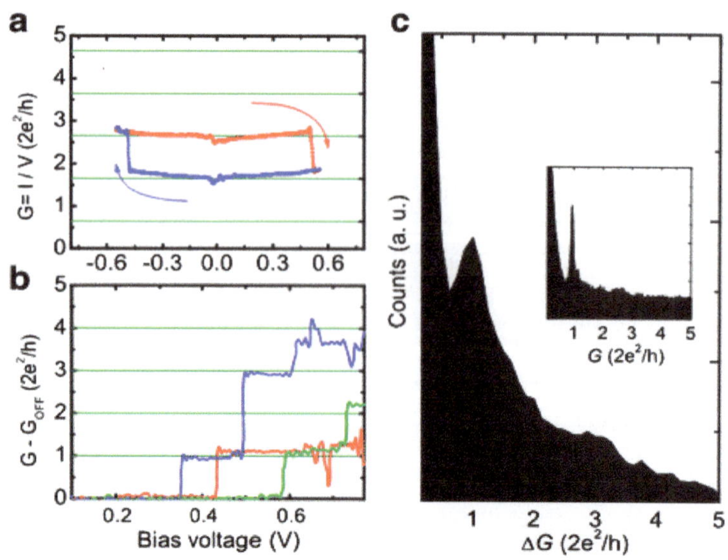

Figure 4: Switching characteristics and conductance quantization observed in nanoscale junctions with a structure of tungsten tip/ionic conductor layer/silver film [191]. **a** Conductance change during SET and RESET operation which shows atomic-scale conductance switching. *Green*

lines act as guides to the eye representing a series of conductance levels with equal interval of 1 G_0. **b** Conductance change from high resistance state to low resistance state following voltage sweep for three independent conductance states. **c**Histogram of the conductance difference between high resistance states and the low resistance states during the voltage sweep. The histogram consists of 130 independent $I-V$ curves with initial ON-state conductance smaller than 10 G_0. ΔG is the difference between the conductance and the zero-bias conductance. The inset histogram shows 5000 repeated and closing cycles at a constant bias voltage of 100 mV. Reproduced with permission.

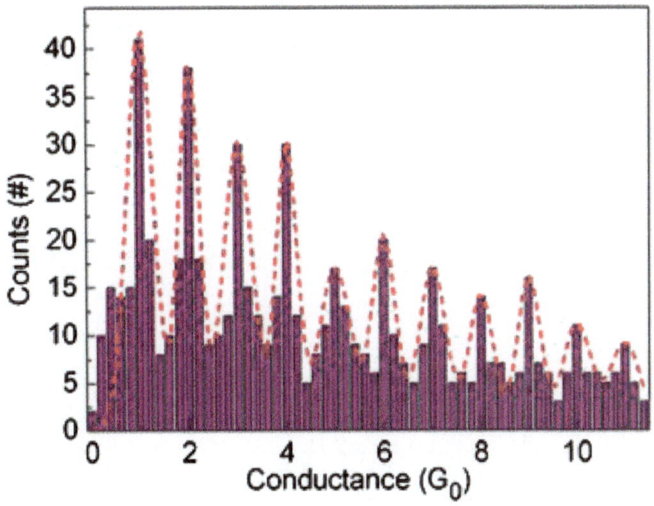

Figure 5: Histogram showing typical conductance quantization phenomenon [171]. The data are extracted from pulse stimuli results of Ti/Ta_2O_5/Pt-structured memory cells. The data are grouped in every 0.2 G_0. Histogram consists of 662 conductance values in 22 memory cells. The *dashed curve* which represents Gaussian fitting curve of the histogram acts as a guides to the eye. Reproduced with permission.

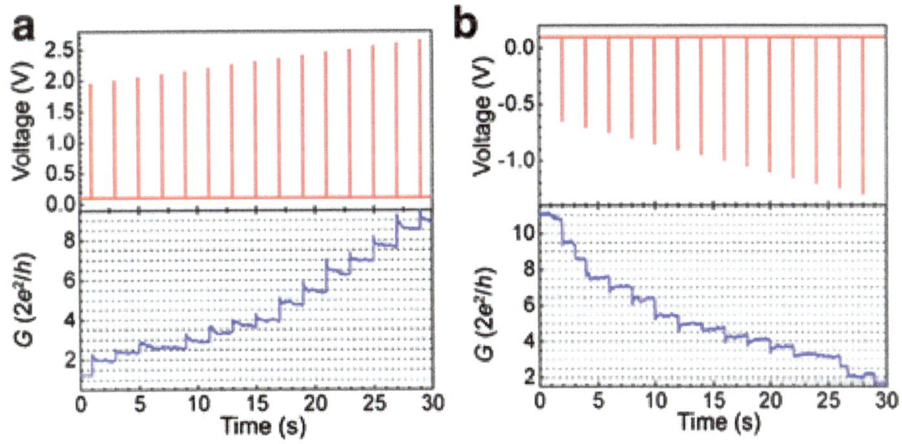

Figure 6: Quantized conductance phenomenon observed by pulse stimuli operation method in Ag/poly(3-hexylthiophene): [6, 6]-phenyl-C61-butyric acid methyl ester/indium–tin oxide sandwich structured devices [183]. Conductance quantization is observed under **a** successive positive pulses and **b** successive negative pulses. Positive voltage pulses are 1 μs wide and negative voltage pulses are 5 ms wide. Two adjacent positive or negative pulses are with an interval of 2 s and an increment of 0.05 V. The conductance is read under a basal voltage of 0.1 V. Reproduced with permission.

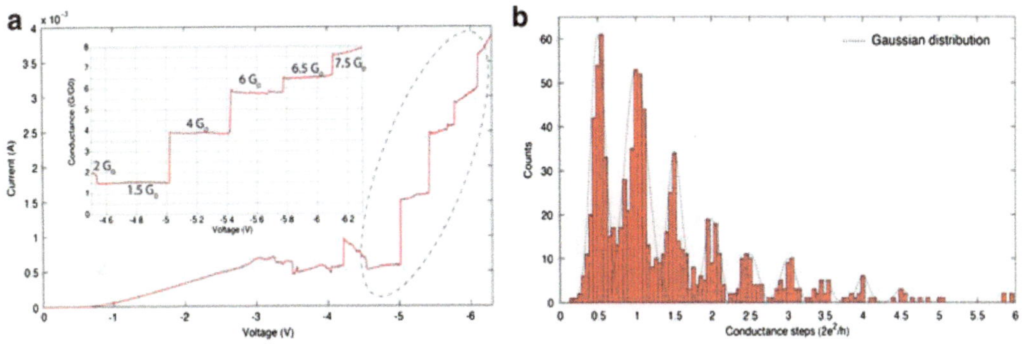

Figure 7: Quantized conductance steps observed in poly-Si/SiO_x/p-type Si-structured RRAM devices [186]. **a** Current–voltage curves showing nonlinear behavior with inset showing the relation between conductance and voltage. Several conductance quantization levels can be seen at both integer and half-integer multiples of G_0. **b** Histogram consisting of about 1000 conductance steps in which half-integer multiples of G_0 are clearly revealed. A series of Gaussian distributions act as guides to the eye (*dotted lines*). Reproduced with permission.

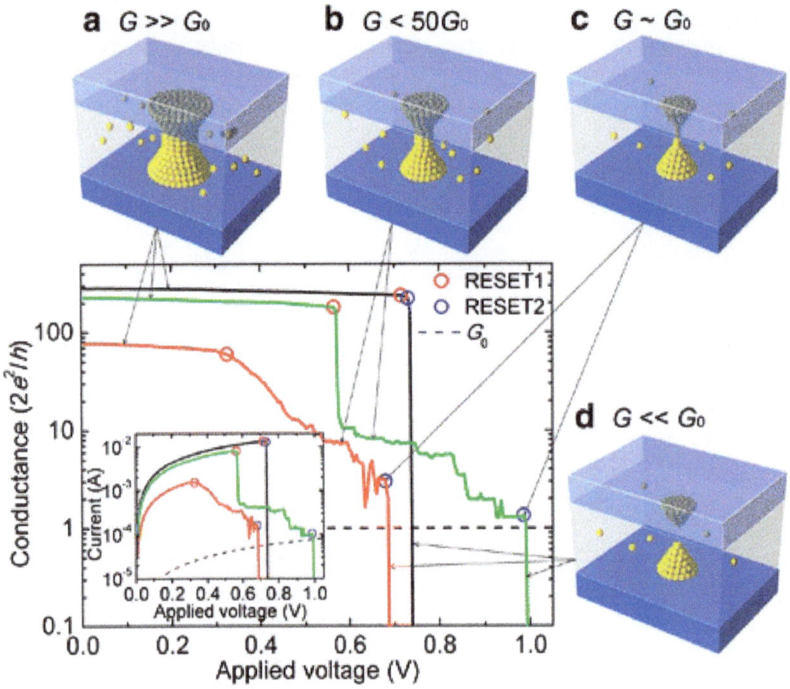

Figure 8: Typical conductance-voltage and current–voltage curves corresponding to RESET process of Pt/HfO_2/Pt devices [70]. *Black curves* show the usually observed abrupt RESET switching. *Green curves* display several successive jumps and *red curves* show progressive RESET process. Insets *A–D* show the different stages of the CF during the RESET process. The quantized conductance states in the step-like or progressive RESET processes are the intermediate states between low and high resistance states. A CF with conductance of the order of $G_0=2e^2/h$ is the natural boundary between the LRS and HRS states. The step-like or progressive RESET transition finalizes with an abrupt conductance drop of several orders of magnitude. This final drop corresponds to the opening of a spatial gap (potential barrier) in the CF. Discrete changes of conductance of the order of G_0 recorded during the step-like or progressive RESET transitions are interpreted as the signature of atomic-sized variations of the conducting filament (CF) nanostructure. Reproduced with permission.

The quantized conductance state of CF is practically an intermediate state or a specific LRS state with the conductance (G) in the order of G_0, i.e., integer multiples of G_0. One G_0 can be simply understood as being corresponding to a single atomic point contact or a nanowire. Conductance quantization effect indicates that the evolution of CF can be modulated to be in units of single atomic point contacts.

Structures, Materials, and Operation Methods of RRAM with QC Effect

There are many observations and reports of quantum conductance phenomenon in RRAM. Different device structures, switching and electrode materials, and operating methods are applied. They are summarized respectively as follows.

Device Structures

The basic structure of RRAM is a thin resistive switching layer, which is usually nanometers in thickness, sandwiched between two electrodes. Many derivative structures have been fabricated based on this basic structure. There are three kinds of structures of RRAM device exhibiting conductance quantization effect, as shown in Fig. 9. Figure 9a [171] shows a commonly used sandwich RRAM structure. It is a stack of thin films of bottom electrode, RS layer, and patterned top electrode. Figure 9b [180] shows a crossbar structure. Figure 9c [170] shows a structure similar to those in Fig. 9a, b, with the only difference being that a tip, such as a conductive atomic force microscopy (CAFM) tip, is used as the top electrode. More detailed fabrication information is not given here for it is not the main point of this paper.

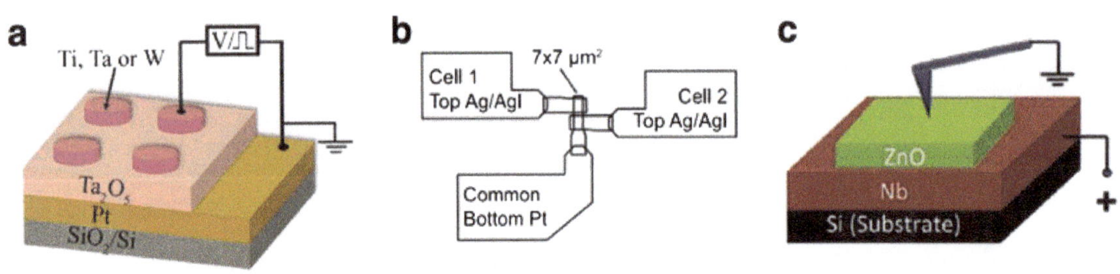

Figure 9: Typical RRAM device structures showing conductance quantization effect. **a** A commonly used sandwich RRAM structure [171]. **b** A crossbar structure [180]. **c** An ultra-small-sized RRAM device using CAFM tip as top electrode [170]. Reproduced with permission.

Material Systems

Since the quantized conductance phenomenon is the property of nanoscale conductors, the materials of RRAM devices especially those consisting of the CF play an important role to the quantized conductance behaviors. As has been discussed in "Introduction" section, the types of RS and electrode materials determine resistive switching mechanisms for the filament-based RRAM. For ECM devices, resistive layer is sandwiched between an inert electrode and an active electrode. During the forming and SET process, the metal ions of the active electrode are driven into the resistive layer, forming the CF. For VCM devices, both top and bottom electrode are inert and the CF is consisted of oxygen vacancies. So as for the ECM mechanism, the material of the active electrode is critical for the observation of the quantized conductance, while for the VCM mechanism, the material of the resistive layer plays a more important role.

In fact, the papers on conductance quantization in RRAM were initially published in 1991. Hajto et al. reported their observation of conductance quantization of RRAM for the first time. The studied device structure is Cr/α-Si:H/V [192]. Yun et al. showed similar result in the V/α-V_2O_5/V [174] device. Quantized conductance effect was observed in cation migration-based RRAMs [172, 173] and then extended to anion migration-based RRAMs [69, 70, 133, 169,170]. Since 2012, conductance quantization has attracted a lot of attention, mainly for its potential applications in the multi-level storage, and also for its interesting physics behind the phenomena. As a result, conductance quantization has been reported in more and more RRAM devices, as shown in Table 2. It should be noted that the reports of conductance quantization in sandwich structures with a vacuum gap [169] or a scanning tunneling microscope (STM) tip [170] are also included in this table.

In most cases of ECM RRAMs with QC effect, the reported material for active electrode is Ag [169, 180–183, 185, 191, 193]. Besides Ag, Nb [170] and V [174] have also been reported to be able to form quantum conductors in RRAM. The RS materials involving QC effect include traditional solid-state electrolyte materials such as AgI [180], Ag_2S [169, 182], Cu_2S [169], GeS_2[193], and transition metal oxides, such as Ta_2O_5 [171, 181], $LaSrMnO_3$ [185], α-Si [192], and polymer [183]. In bipolar and unipolar VCM devices with QC effect, the reported resistive switching material includes silicon oxide [186], transition metal oxides, such as HfO_2 [69, 188], Ta_2O_5 [171, 181], ZnO_x [170], etc., and bi-layered oxides, such as $TiO_2/SrTiO_3$ [194] and GeO_x/SiO_2 [179]. Among the reported RS materials which exhibit conductance quantization, Ta_2O_5 [171, 181] and HfO_2 [69, 188] show good performances irrespective of the type of the CFs, i.e., Vo-CF or metal CF. Both ECM and VCM devices made up of HfO_2 and Ta_2O_5 materials have shown obvious QC effect, as shown in Figs. 2, 5 [171], and 8 [70].

It can be seen from Table 2 that most reported devices show conductance step with integer multiples of G_0, while in some material systems, the conductance variation step may be half-integer multiples of G_0. The explanation of this difference will be discussed later in "Theory and Modeling of Quantum Transport in RRAM" section.

Operating Methods

To successfully observe the QC effect in RRAMs, it is of importance to make use of appropriate operating methods to the devices to accurately control the size of CFs to be close to the atomic scale. In this section, we will deal with all kinds of reported operating methods, including fresh device operation, voltage sweeping, voltage pulse operation, current sweeping, and constant voltage bias operation.

Fresh Device Operation

Before analyzing the detailed operation methods to achieve QC effect, we need to first discuss the different operating conditions for the RRAM devices with different initial resistance states (IRS). For the fresh RRAM devices, most of them show an initially very high resistance state and a forming operation is needed to form CF in the resistive layer. Since the fresh resistive layer is usually in high resistance state, a much higher voltage, compared with the SET process, is needed to form the CF in the resistive layer, as shown in Fig. 10a [170]. Compared to the forming process, the voltage amplitude applied in the SET process is lower, because the CF formed in the forming process will not be dissolved completely in the successive RESET operation, thus a lower voltage can program the device. But for some RRAMs, the forming process and the SET process show no obvious difference, i.e., the characteristics of the device in initial fresh state and high resistance state have no clear distinction, as shown in Fig. 10b [179]. In other words, this kind of RRAM device has the free-forming characteristics. Some RRAM devices also possibly show an initial low resistance state. For example, in the quantized conductance atomic switch (QCAS), as shown in Fig. 10c [169], it is initially in the ON-state. To start the switching cycles, the device needs to be RESET at first under a certain positive voltage bias. Due to the large amount of Ag atoms to be ionized to incorporate into the Ag_2S crystal in this first RESET process, the switching time of this process is quite long, lasting for a few seconds. But after this initial RESET operation, the device can work quickly with a high operation frequency of 1 MHz.

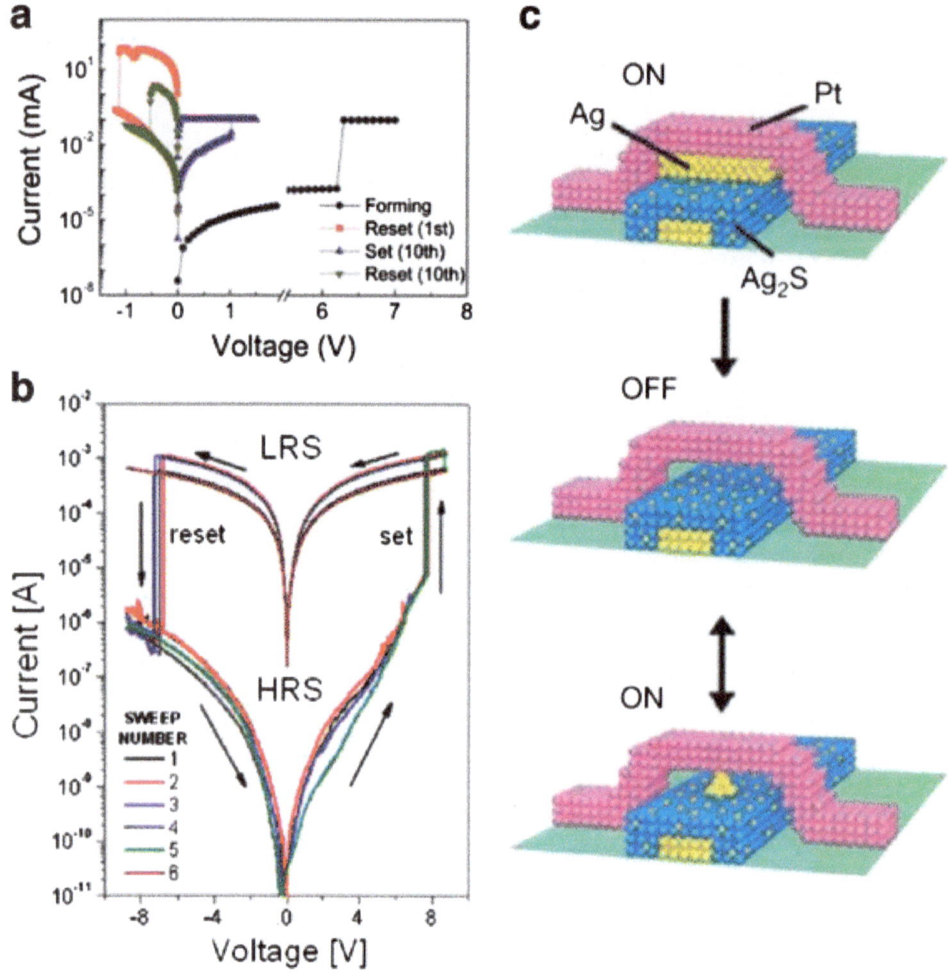

Figure 10: Typical forming, SET, and RESET characteristics of fresh RRAM devices with different initial resistance states. **a** Switching process illustration of Nb/ZnO/Pt device with an initial high resistance state [170]. Higher voltage in forming is needed compared to that in the SET process. **b** Switching process illustration of a free-forming $p^{+}Si/NiSi_2/SiO_2/CeO_x/W$ device [179]. The Forming process and the SET process show no obvious difference. **c** Schematic illustration of Ag_2S-based QCAS device and switching behavior between OFF- and ON-state [169]. The initial state of the device is ON-state and a RESET process is needed to start the switching cycles. Reproduced with permission.

Voltage Sweeping Mode

After the initial process which involves forming/SET, RESET or no particular operation method to start switching cycles, a certain operation method is needed to switch the device between ON-state and OFF-state. The most common operation method is voltage sweeping mode. The voltage sweeping in SET process induces electrochemical reactions resulting to the formation of CF. The voltage sweeping in RESET process contributes to joule-heating-assisted oxidation followed by the diffusion of metal ions or oxygen vacancies (Vo) under the concentration gradient and the applied electric field [181]. Many experimental results [69, 170, 171, 179, 180,188] showing conductance quantization under voltage sweeping mode have been reported. Figure 11 shows some of the reported experimental results in different device structures. Figure 11a [188] shows the progressive RESET process of the $Ti/HfO_2/TiN$-structured devices which exhibit bipolar RS behaviors. In the inset of Fig. 11a, zoomed current and voltage relation is shown and several current jumps which indicates the conductance quantization could be clearly seen. Figure 11b [170] shows the conductance change in bipolar characterized Nb/ZnO/Pt devices as a function of the bias voltage during the SET process. At least four conductance jumps are shown in the Fig. 11b and the conductance changed at a step of the integer multiples of quantum conductance G_0 from 1 to 8 G_0. The inset of Fig. 11b shows the current–voltage curve in a larger voltage range from 0 to 4 V. Figure 11c, d [69] shows the current–voltage curves in the RESET process of the unipolar $Pt/HfO_2/Pt$ devices. Figure 11dshows the detail of Fig. 11c during the last phase of the RESET transients. The dash line in Fig. 11c corresponds to the current–voltage curve of a conductance of 1 G_0.

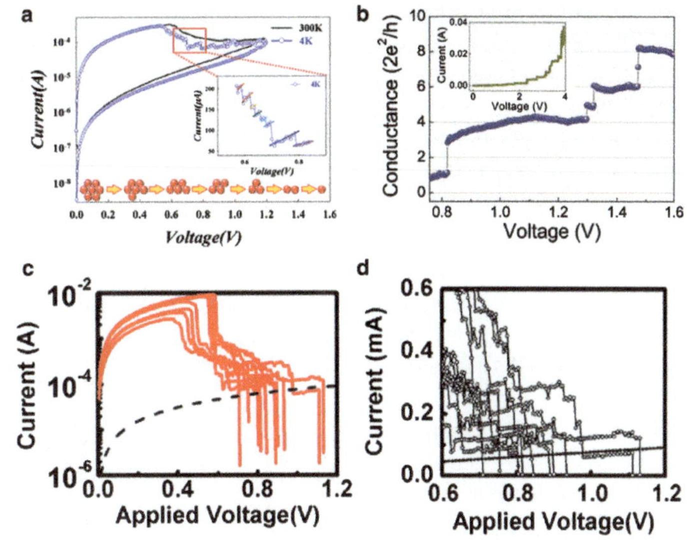

Figure 11: Typical conductance quantization phenomenon observed in different structured devices under voltage sweeping mode. **a** Current jump observed in Ti/HfO_2/TiN-structured memristor during RESET process. The inset diagram indicates the discrete resistance change due to quantum atomic reaction during RESET process [188]. **b** Conductance quantization observed in Nb/ZnO/Pt during SET process. [170] The inset shows the current–voltage curve in a larger voltage range from 0 to 4 V. **c** Progressive RESET process in Pt/HfO_2/Pt devices [69]. The *dashed line* corresponds to the current–voltage curve of 1 G_0. **d** Detail of the current–voltage evolution of (**c**). Reproduced with permission

Voltage Pulse Operation Mode.

By applying appropriate pulse voltage, the conductance state of the RRAM device could be changed at steps of quantum conductance. Both ECM and VCM RRAMs show quantized conductance step change under voltage pulse operation, as shown in Figs. 12 [181] and 13[171], respectively. For voltage pulse operation method, three parameters could be tuned, including the pulse amplitude, pulse width, and time interval between two adjacent voltage pulses.

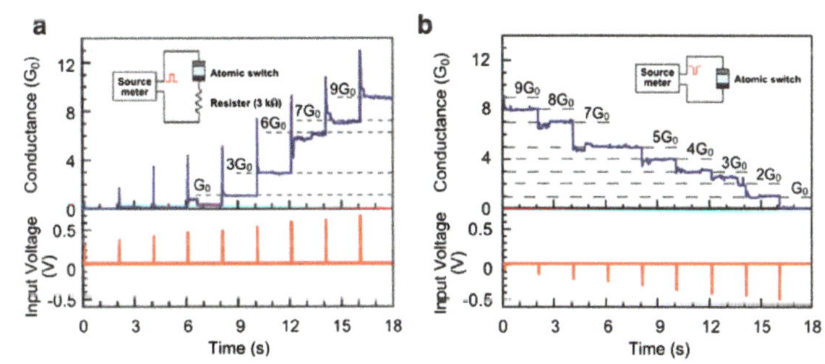

Figure 12: Quantized conductance observed in Ag/Ta_2O_5/Pt-structured ECM devices under voltage pulse operation mode [181]. **a** The value of conductance increases at steps of integer multiples of conductance quantum G_0 in the SET process under positive pulses with a width of 20 ms at an interval of 2 s. In order to prevent hard breakdown of RRAM device, a current-limiting resistor of 3 kΩ was connected in series with the device. **b** Quantized conductance decrease phenomenon observed in the RESET process under reversed voltage polarity. No current-limiting resistor is needed in the negative pulse stimuli mode. Reproduced with permission.

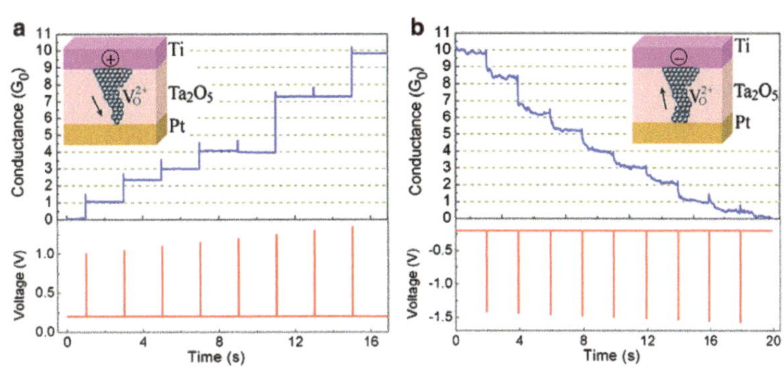

Figure 13: Quantized conductance observed in Ti/Ta_2O_5/Pt-structured VCM devices under voltage pulse operation mode [171]. **a** The value of conductance increases at steps of integer multiples of G_0 in the SET process under positive pulses with a width of 100 ns at interval of 2 s. **b** The value of conductance decreases at steps of integer multiples of G_0 in the RESET process under negative pulses with a width of 1 μs at an interval of 2 s. Reproduced with permission.

Take $Ag/Ta_2O_5/Pt$ RRAM as an example. Figure 12 shows the QC effect observed in this device. Under the pulse operation mode, the switching speed of the device is of the magnitude of μs to ns which is related to the values of the high resistance and the input pulse. The switching time is much shorter than the reaction time of the current compliance function, which results in the ineffectiveness of the current compliance function and leads to the further growth of the CF even after the current reaches the value of the compliance current. The solution for this problem is to insert a 3 kΩ resistor in series with the RRAM device, as shown in the inset of Fig. 12a. The current is limited by a 3 kΩ resistor during the stage that the RRAM switches from high resistance state to low resistance state. It is very important to limit the current value when the SET transition occurs. If there is no current limitation, the conductance of the device will abruptly jump to about 20 G_0, which indicates that the formed filament is rather thick and robust and quantized conductance step disappears. As can be seen from Fig. 12a, the quantized conductance increases from 0 to 9 G_0 at a step of conductance quantum under a series of increasing pulses from 0.3 to 0.7 V at a step of 0.05 V with time intervals of 2 s. Negative pulses with the same pulse width and interval time from −0.1 to −0.5 V were applied after the successive positive voltage pulses. In the negative pulse operation which corresponds to the RESET process, no series resistor is needed since the current in the circuit will decrease as the resistance value of RRAM increases in the RESET process. Similar quantized conductance change behavior has also been reported in the VCM RRAM with a structure of $Ti/Ta_2O_5/Pt$, as shown in Fig. 13.

The quantized conductance states change not only depending on the amplitude of the pulses but also depending on the time interval of the adjacent pulses. As shown in Fig. 14 [181], successive pulses with sufficiently long interval do not obviously change the conductance state. Whereas at short interval, pulses of the same amplitude and width make the conductance temporarily increase and gradually reach a constant value of G_0.

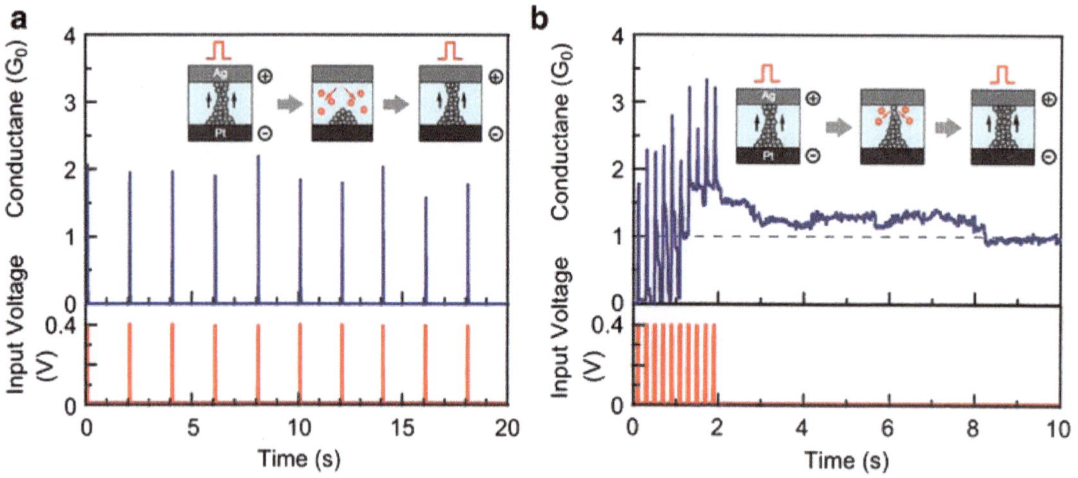

Figure 14: Quantized conductance change behavior under different time intervals [181]. **a** The conductance change under ten successive voltage pulses of 0.4 V with a pulse width of 20 ms at intervals of 2 s. The conductance state could increase to about 2 G_0 under the input pulses but immediately decays to zero after each input pulse is completed. **b** The conductance evolution under ten successive voltage pulses of 0.4 V with a width of 20 ms at intervals of 0.2 s. In this case, the conductance gradually increases and maintains at about 1 G_0 for more than 60 s after the tenth input pulse. Reproduced with permission.

Current Sweeping Mode

In some material systems, no more than one or two discrete conductance drops could be detected by conventional voltage sweep operation. Current sweep mode is utilized as an alternative operation mode. Through this method, more discrete conductance levels could be observed, as shown in Fig. 15 [180, 195]. The different quantization evolution behavior in SET process by taking voltage sweep mode and current sweep mode comes from the different formation process of the CF. In the voltage sweep mode, stepwise increased voltage is applied to the device. As the voltage increases to the SET point, the filament forms and the resistance of the device suddenly drops to a much lower value. This leads to an abrupt current jump and the switching time is faster than the response time of the current compliance current which causes a further growth of CF after the SET point. Both voltage and current increase at the SET point which results in a positive feedback to the CF formation, thus leading to a very fast CF formation process. Whereas in the current sweep mode, the current is programmed to increase stepwise. At the SET point, the voltage dropped across the device decreases abruptly, due to the sudden decrease of the device's resistance. This leads to a negative feedback to the CF formation, so the CF formation process is more gradual than that under voltage sweep. In this case, more discrete conductance levels could be observed.

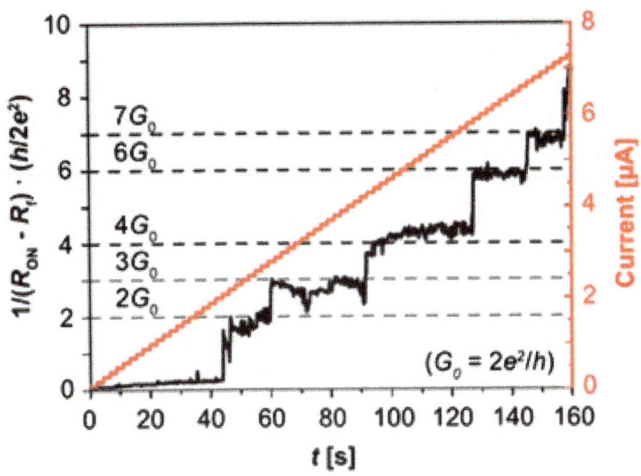

Figure 15: Quantized conductance phenomenon observed in Ag/AgI/Pt devices under current sweep mode [180, 195]. It can been seen that more than five resistance levels which are integer multiples of conductance quantum G_0 have been observed. Reproduced with permission.

Constant Voltage Bias Mode

Constant voltage bias is another way to observe quantized conductance. The bias value of voltage is critical to the observation of conductance quantization phenomenon. If the voltage amplitude is too large, the device may easily break down or the switching time may be too short to detect. On the other hand, if the amplitude of bias voltage is too small, the switching time of the device may be too long which is time consuming to observe conductance quantization or even no switching behavior could be observed, since the voltage is too small to drive the switching action to occur. Therefore, an appropriate bias voltage amplitude is needed for the observation of conductance quantization phenomenon, as shown in Figs. 16 [181] and Fig. 17[69].

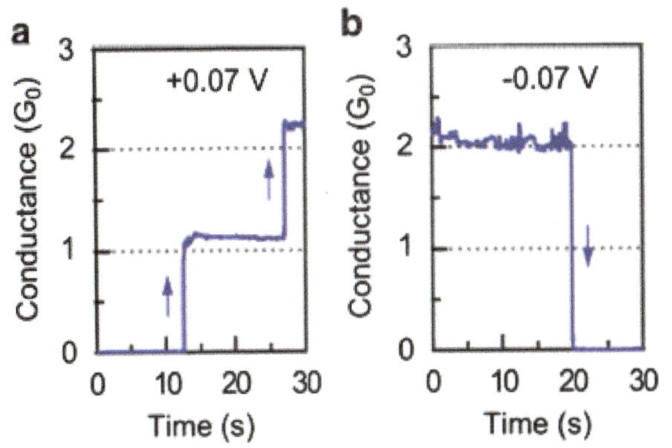

Figure 16: Conductance jump at the integer multiples of G_0 under voltage bias in $Ag/Ta_2O_5/Pt$ device [181]. **a** Conductance increase under positive voltage bias of +0.07 V. **b**Conductance decrease under negative voltage bias of −0.07 V. Reproduced with permission.

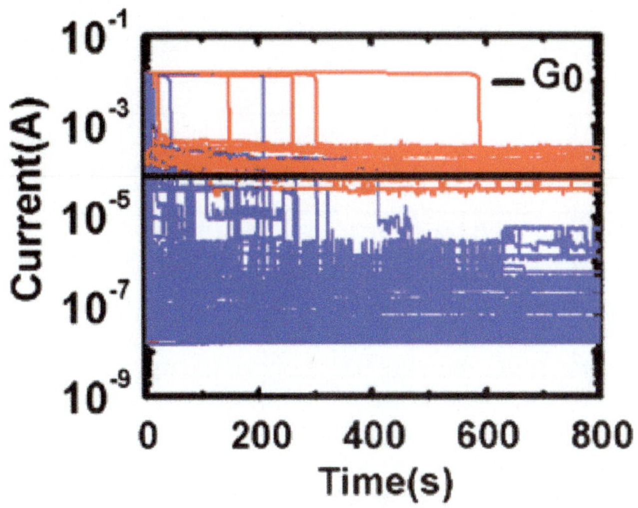

Figure 17: Current evolution of $Pt/HfO_2/Pt$ device under constant voltage bias for RESET process indicating quantum level change of the conductance [69]. Reproduced with permission.

Theory and Modeling of Quantum Transport in RRAM

Systems whose dimensions are much larger than microscopic objects like atoms but smaller than macroscopic objects are called mesoscopic systems [196]. When the dimension of the conductor is comparable to or smaller than the electron mean free path (mesoscopic scale), the classical Ohmic transport mechanism breaks down. When the conductor is under sufficiently high voltage, the Ohm's law does not apply for it either. When the above two conditions are satisfied, ballistic transportation begins to be dominant and the conductor behaves as an electron waveguide [197]. When the transmission probability T for the waveguide equals one, this waveguide or conduction channel contributes an amount of G_0 to the total conductance of the mesoscopic conductor.

The theory of electron transport mechanism in mesoscopic systems exhibiting quantized conductance has been gradually established through a time span of decades from the suggestion of the prototype by Rolf Landauer in 1957 [198] to the publication of the book "Electronic Transport in Mesoscopic Systems" by Datta in 1995 [143]. Many theoretical calculations about quantum conductance using different models have been reported [197, 199–209]. The quantized conductance steps were obtained in numerical and analytical calculations in a wide variety of materials [144, 205, 206, 208–211]. Review articles summarizing quantized conductance phenomenon in atomic-sized conductors [148] and nanowires [149] have also been published. At the same time, many experiments were also carried out to study quantized conductance [148, 156, 162, 163]. The conductance quantization effect was first observed in ballistic point contacts in the 2DEG of high-mobility GaAs-AlGaAs heterostructures in 1988 by Van Wees et al. [146], as shown in Fig. 18.

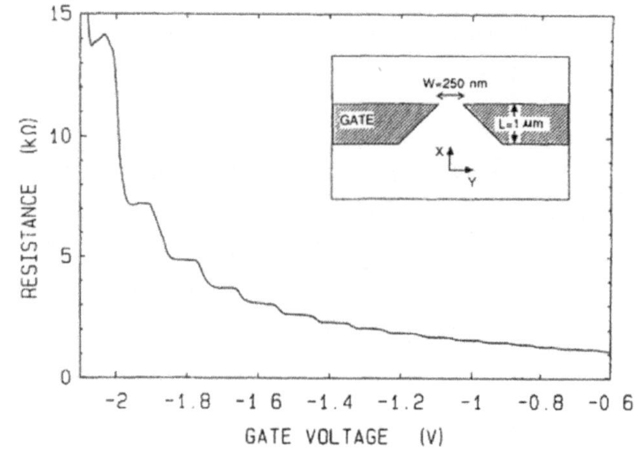

Figure 18: First observation of conductance quantization by Van Wees et al. [146]. The resistance of the point contact is a function of gate voltage at 0.6 K. The electron gas under the gate is depleted at −0.6 V when electrons only transport through the point contact and the contact is fully pinched off at −2.2 V. The inset shows the layout of the point contact. Reproduced with permission.

In RRAM, many theoretical works concerning calculation and simulation were reported. Several models have been proposed to describe and calculate the ballistic transportation and conductance quantization phenomenon in RRAM devices. As mentioned above, two conditions, i.e., small conductor dimension and high voltage bias, lead to ballistic transportation and conductance quantization, so in all proposed models, either the CF in resistive layer has been considered an atomic-sized conductor in insulator or the most constrictive part in CF is treated as being of atomic size. In this section, we will summarize four models. In the first one, the CF formed in the resistive layer is regarded as a one-dimensional linear atomic chain [175]. The second model treated the narrowest part of a CF as an atomic contact and calculated the electrical transport based on Landauer theory [179, 186, 212]. The third model focuses on CFs consisted of oxygen vacancies and calculated the quantized conductance of the filament with different space of oxygen vacancies through first principle calculation [69]. The fourth model is a circuit model, which ascribes the quantized conductance of the atom point contact in the ECM device to the discharge of the thermal emf voltage [195, 213].

One-Dimensional Linear Atomic Chain Model

In the work of Jameson et al. [175], a model based on one-dimensional metal atomic chains was proposed to relate to the quantized conductance phenomenon and to calculate the programming time of RRAM device whose resistance tends to be programmed to integer multiples of fundamental conductance G_0 under sufficiently high currents. This model is different from other previous models which considered the filament as a bulk-like (although small) object whose resistance is continuous and proportional to the bulk resistivity.

In this model, the filament is simply treated as a linear chain of atoms, with the conductance in the order of G_0. A schematic illustration of the model is shown in Fig. 19. The CF is assumed to form from cathode to anode since the metal cations are usually reduced at the cathode. The resistive layer where the CF is formed is treated as a one-dimensional series of N_w potential wells with a uniform height V_i and width d_i ("*i*" for internal). The last atom of the metal atom chain corresponds to the surface of the cathode. The resistive layer where CF is not formed yet is treated as a bulk electrolyte, and the resistive layer is separated by an "emission" barrier of height V_e and width d_e ("*e*" for emission) from the anode surface. When the positive forming or SET voltage V_{ac} is applied to the anode, the emission well is raised to an energy of eV_{ac}, while the well N_w is kept at zero. A metal ion with a charge of +*e* (e.g., Ag^+ for Ag/GeS_2/W cells) will be emitted from anode into the potential well by thermal excitation over the emission barrier. Then, it travels through the periodic potential barriers which represent the resistive layer with no CF inside and stacks up against the cathode to promote the stretch of the CF, i.e., an atomic chain that grows with time. When all N_w wells are filled with metal ions, the resistance state of RRAM will suddenly switch to low resistance state and the time needed is the programming time of the device.

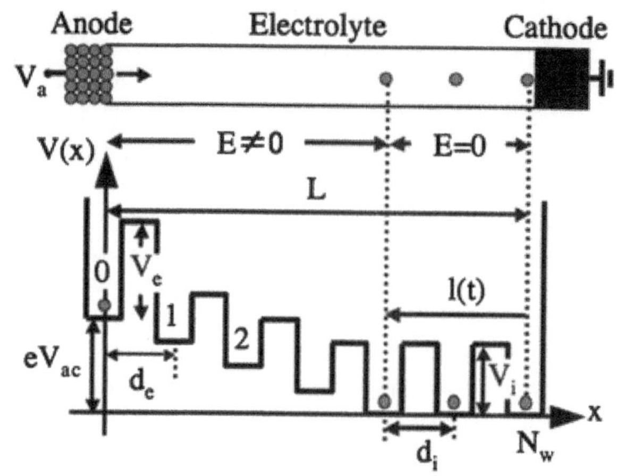

Figure 19: One-dimensional potential model for the forming process of a fresh RRAM cell [175]. The calculation was carried out under the assumption that the electric field *E* is zero within the filament and constant between the anode surface and the tip of the filament. The filament was treated as a one-dimensional atomic chain. Reproduced with permission.

The programming time (t_p) of a RRAM cell whose resistance tends to increase to the integer multiple of G_0 was calculated. It was found that several intrinsic material parameters, including V_e, d_i, $W_a - W_c$ (the difference between the work functions of the anode and cathode), influence the programming time of the virgin Ag/GeS_2/W cells. After selecting proper values for these parameters, the model can nicely describe the dependence of t_p on voltage, temperature, and GeS_2 thickness (for thick layers), which is in close relation to the quantization of the ON-state conductance. The model is effective for the cases of both constant voltage and ramped voltage programming. Further experimental study of the conductance quantization was reported by the same group in Ag/GeS_2/W RRAM device afterwards [193] and the result is listed in Table 2.

Quantum Point Contact Model

To deal with the post-breakdown (BD) conduction of gate dielectric of field effect transistor (FET), Suñé and Miranda have established the quantum point contact (QPC) model [214]. Recently, it has been found that the QPC model can be also made use of to describe the conduction of high and low resistance state in RRAM [71, 179, 182, 212, 215–219]. This model treats the thinnest part of the CF as a quantum point contact. It is able to explain the conductance quantization phenomenon in RRAM. The schematic illustration of the QPC model is shown in Fig. 20 [212]. The thinnest part of the CF is made up of a few atoms (Fig. 20a) and modeled as a potential barrier with several quantized subbands.

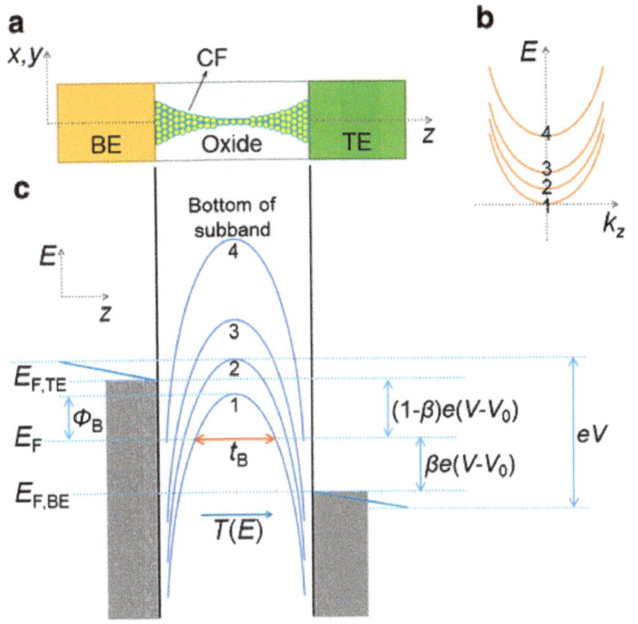

Figure 20: The schematic of the RRAM device with a narrow CF and the corresponding energy band diagram of the quantum point contact model. **a** Schematic structure of the RRAM device with a narrow CF. **b** The dispersion curves of the first four electronic subbands under the confinement of CF in certain z. **c** The dependence of the energy level of the bottom of the subbands on z. The transmission probability $T(E)$ of the bottom of ground quantized subband of a parabolic potential barrier is used for the calculation of electrical transport. The *shaded regions* are the states occupied by electrons. The number of the subbands isN_{ch}, with each one contributing to a conducting mode. In this figure, four subbands are shown. V is the applied voltage. V_0 is the voltage dropped on TE and BE, represented by the *two blue oblique lines*. Since V_0 is much lower than V, usually it can be neglected in the calculation. β is the fraction of voltage that drops at the BE interface, E_F is the Fermi level, $E_{F,TE}$ and $E_{F,BE}$ are the TE and BE quasi-Fermi levels, t_B is the width of the potential barrier at the equilibrium Fermi energy (E=0), and Φ_B is the height of the potential barrier, i.e., the bottom of the first subband. The barrier height is different between high resistance state and low resistance state, which leads to different current expressions. In the deep OFF-state, the barrier thickness t_B is equal to the gap length t_{gap}.

In the CF described by the QPC model, it was demonstrated that the Schrodinger equation can be decomposed into the transverse and longitudinal equations. Then, the quantum transport through a 3D tube-like constriction becomes a simple 1D tunneling behavior. The dispersion curve of the electronic subbands could be expressed as:

$$E(k_z, z) = \epsilon_n(z) + \frac{\hbar^2 k_z^2}{2m} \tag{1}$$

where z and k_z are the coordinates in the longitudinal direction in real-space and k-space, respectively, m is the mass of the electron, and $\hbar$ is the reduced Planck constant. If the confinement is in a rectangle shape, then [202]

$$\epsilon_n(z) = \frac{\pi^2\hbar^2}{2m}\left(\frac{n_x^2}{L_x(z)^2} + \frac{n_y^2}{L_y(z)^2}\right), \tag{2}$$

where $L_x(z)$ and $L_y(z)$ are the dimension of the constriction and n_x and n_y are integers. It means that the dispersion curve ($E-k_z$ relationship) consists of discrete parabolic subbands in each z, as shown in Fig. 20b. □$_n(z)$ strongly depends on the thickness of the filament. Thinner filament has stronger spacing out of the subbands.

The potential barrier for conducting transmission in the longitudinal direction (z) lies in the bottom of the subbands. When the filament is very thick, which means the confinement is weak, the dispersion curve will turn into the 3D bulk energy band. In that case, the bottoms of the subbands are in rather deep energy levels. Whereas, when the filament is very thin, the bottom of the subbands will be lifted. Consequently, the dependence of the energy level of the bottom of the subbands on z in a constricted tube is arch-shaped curves, as shown in Fig. 20c. The number of the subbands is just that of the conducting modes or conducting channels N_{ch}. Φ_B is the height of the first subband. The barrier width t_B is defined as the width of the first subband at E=0. If the CF is very thick, the barrier disappears. On the contrary, in the HRS state, the CF is ruptured and there is a gap in the CF region, so the barrier becomes very high. In the conduction, the injected electrons need to travel through the potential barrier, with a transmission probability $T(E)$.

In QPC model, based on the Landauer theory, the current flowing through the RRAM device can be calculated as [11]

$$I = \frac{2e}{h} N_{ch} \int_{-\infty}^{\infty} T(E)[f(E-\beta eV) - f(E + (1-\beta)eV)]dE, \tag{3}$$

where E is the energy, f is the Fermi-Dirac distribution function, and N_{ch} is the total number of 1D opened conducting channels connecting the electrodes. An inverted parabolic potential barrier is assumed to get an analytical expression for the transmission probability [71]:

$$T(E) = \{1 + \exp[-\alpha_B(E-\Phi_B)]\}^{-1}, \tag{4}$$

where α_B is related to the inverse of potential barrier curvature and is proportional to the thickness of the barrier, i.e., $\alpha_B = t_B\pi^2 h^{-1}\sqrt{2m^*/\Phi_B}$ [71, 214]. m^* is the effective electron mass. Inserting Eq. (4) into Eq. (3), we can get

$$I \approx \frac{2e}{h} N_{ch}\left\{eV + \frac{1}{\alpha}\mathrm{Ln}\left[\frac{1 + \exp\{\alpha_B[\Phi_B - \beta eV]\}}{1 + \exp\{\alpha_B[\Phi_B + (1-\beta eV)eV]\}}\right]\right\} \tag{5}$$

Equation (5) is applicable for both HRS and LRS, with the difference in the values of α_B and $\Phi_B$which represent the difference in potential barrier. In HRS, there is a gap in the CF region, so at low applied voltages (i.e., $V \to 0$), Eq. (5) can be simplified as

$$I \cong N_{ch} G_0 \exp(-\alpha_B \Phi_B)V. \tag{6}$$

So, the conduction in HRS is just determined by the barrier through the parameters α_B and Φ_B[215, 216, 219]. While in LRS, there is no spatial gap, so Eq. (5) converges to

$$I \cong N_{ch}\beta G_0 V, \tag{7}$$

which is a linear I–V, consistent with that usually observed in LRS. When the CF is very narrow, i.e., when N_{ch} is small, Eq.

(7) accounts well for the experimentally observed conductance quantization effects. The CF conductance is expressed as

$$G \cong N_{ch}\beta G_0, \tag{8}$$

showing that G is just the integer multiples of the quantum of conductance G_0, when the voltage drop at two interfaces is asymmetric, i.e., $\beta=1$. If N_{ch} is large, the model approaches the classical Ohmic regime, where quantization effect is less evident since CF conductance is high.

It is worth noting that there are some amounts of experimental points whose conductance is smaller than G_0, according to the reported results as shown in "Conductance Quantization in RRAM" and "Structures, Materials, and Operation Methods of RRAM with QC Effect" section. Values slightly different from G_0 are possible even when a continuous conducting channel connects the electrodes, since in Eq. (8), N_{ch} is an integer whereas $0<\beta<1$. In an atomic-scale conducting CF or quantum wire (QW), the voltage mainly drops at the interfaces with the external reservoirs and the value of β is the fraction of voltage that drops at the BE interface. The value of β might change with the actual geometry of the CF and with its coupling to the reservoirs. The presence of impurities in the QW or non-adiabatic coupling with the reservoirs might also explain a conductance smaller than G_0 for each conducting mode [153]. The adsorbed impurities on or in atom chains would change the CF constriction configuration and influence the electronic band structure.

In a subsequent work, Miranda et al. proposed a simple current–voltage model based on the quantized constriction of RRAM (Fig. 21) and explained the minimum unit of conductance of 0.5 G_0. The left-going current I^- and right-going current I^+ were respectively calculated as:

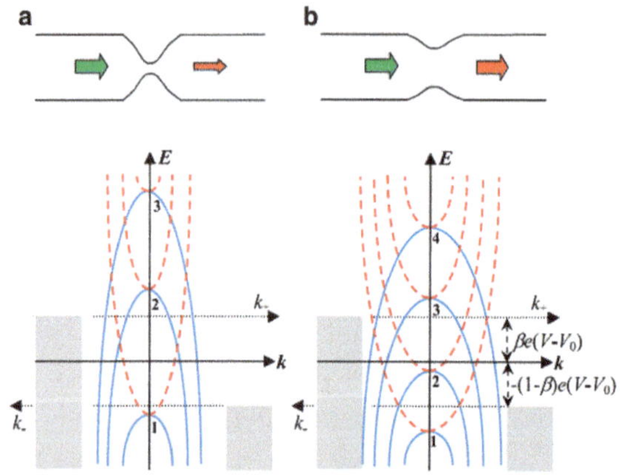

Figure 21: *E–k* relationship for **a** narrow and **b** wide constriction, respectively [179]. The *solid* and*dashed lines* correspond to the bottom of the longitudinal and transversal subbands, respectively. The *shaded regions* are the states occupied by electrons. The *indices*indicate the subband number. A tighter constriction leads to higher energy levels. Reproduced with permission.

$$I^+ = \frac{2e}{h}\int_{-\infty}^{\infty} T(E)M(E)f(E-\beta eV)dE \tag{9}$$

And

$$I^- = \frac{2e}{h}\int_{-\infty}^{+\infty} T(E)M(E)f[E+(1-\beta)eV]dE, \tag{10}$$

where M represents the number of conduction modes, and the voltage dropped in electrodes $V_0$is neglected. The total current $I=I^+-I^-$ is given as:

$$I = G_0[\beta N^+ + (1-\beta)N^-]V, \tag{11}$$

where N^+ and N^- are the number of right-going and left-going conduction modes, respectively, i.e., the number of subbands with $E\leq\beta e(V-V_0)$ and $E\leq-(1-\beta)e(V-V_0)$ in Fig. 21, respectively. For simplicity, considering the case of a symmetric potential drop at two ends of the constriction, β equals 0.5, thus Eq. (11) indicates that when the difference of $N^+ + N^-$ is an odd number, the conductance values of half-integer multiples of G_0 appear. Simulation results fit well with the experimental results measured in $W/CeO_x/SiO_2/NiSi_2$ devices [179].

A similar model was proposed by Mehonic et al. [186]. This model treats filaments as quantum constrictions within the framework of Landauer theory. In this model, the potential drop on the two ends of the constriction is not assumed to be symmetric.

The schematic illustration of this model is shown in Fig. 22. The lateral confined quantum constriction for carriers to flow through produces a set of discrete one-dimensional subbands in the conduction band. More conduction modes are allowed if the size of the constriction increases. Half-integer quantum conductance was studied under the quantum point conduction model. By assuming the transmission probability to be one for electrons with energy above the minimum energy value of the subband, to be zero with energy below this, and adopting the zero temperature limit, the total current is

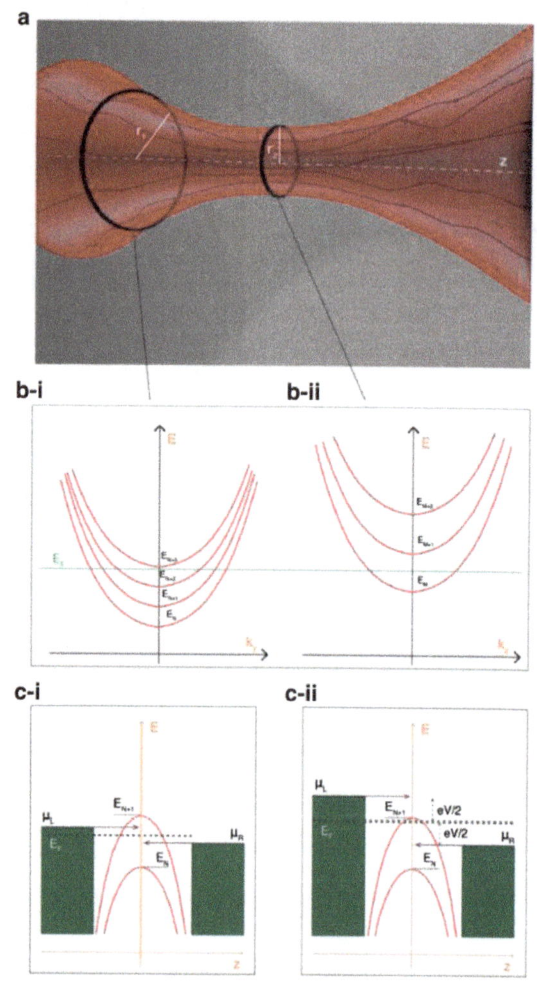

Figure 22: Quantized conductance effect based on the quantum point contact model [186]. **a**Schematic illustration of a conducting filament with a lateral constriction of one or several atoms at the narrowest

part of the filament. **b-I** Dispersion curves of the first four electronic subbands at the edge of the constriction. **b-II** Dispersion curves of the first three subbands at the center of the constriction where the confinement is stronger which leads to a spacing out of the sub-bands. **c-I** When the difference in chemical potential between the left and right reservoirs is small, both the left-going and right-going electron modes fall within the same subband. **c-II** When the difference in chemical potential between the two reservoirs is large, the left-going and right-going electron modes fall into different subbands. Reproduced with permission.

$$I = I_R - I_L = \frac{1}{2}(N_R + N_L)G_0 V, \qquad (12)$$

where N_R and N_L are the numbers of occupied subbands accessed from the right and left sides, respectively. Here, symmetric voltage drop in the TE and BE interfaces is assumed, i.e., $\beta=0.5$. Then, the half-integer quantum value appears when N_R+N_L is an odd number, which is in coherence with the result reported by Miranda et al. as described above. In all ECM devices, since the electron reservoirs are highly conductive, it is impossible to maintain a large difference in chemical potential, which is the reason for the difficult observation of half-integer quantization. While in another case, i.e., in VCM devices, most of them exhibit half-integer quantization. An important conclusion drawn in this paper is that the key quantity governing the type of quantization is the difference in chemical potentials between the two reservoirs.

The appearance of half-integer multiples of quantized conductance might also arise from the absence of the spin degeneracy. The quantum of conductance G_0, i.e., $2e^2/h$, is equally contributed by two spin-degenerate transport channels. Thus, in nonmagnetic materials where the spin degeneracy is reserved, the conductance is the integral multiple of G_0. While when the spin degeneracy is broken in magnetic systems, a single spin channel will contribute a conductance of 0.5 G_0, i.e., e^2/h, leading to the conductance of half-integral multiple of G_0. As easily inferred from Table 2, the half-integer G_0 appears mostly when the CF is composed of Vo. That is because the Vo can carry a weak magnetism in some cases, according to lots of previous studies [41, 220]. The magnetic CF may bring spin-splitting conductance channels, which give rise to the observed 0.5 G_0.

First Principle Calculation on the Quantized Conductance of Oxygen Vacancy Conductive Filament

As mentioned above, before switching to the high resistance state, the CF in RRAM behaves as a nanoscale conductive path with a few defects such as oxygen vacancies. It is necessary to explore whether the oxygen vacancy paths can explain the quantized conductance behavior. First principle calculations based on the density-functional theory (DFT) were carried out to get the quantized conductance of oxygen vacancy conductive path in crystalline HfO_2 matrix [69]. In this work, the zero-bias transmission probability $T(E)$ was calculated by using non-equilibrium Green's functions. The ballistic conductance was calculated through first principle method based on Landauer theory. The conductance of CF was then calculated via the Landauer formula, $G=T(E)G_0$. The generation of an oxygen vacancy is considered as the removal of an oxygen atom in a monoclinic-HfO_2 host. As a result, a filled impurity state is introduced in the band gap which is far from the band edges, as shown in Fig. 23a. The spatial spread of the impurity wavefunction determines that the states overlap between two neighboring oxygen vacancies, which further determines whether the carrier transport is hopping or band transport. As shown by the band structures of monoclinic-HfO_2 with a chain of oxygen vacancies in Fig. 23a–d, when oxygen vacancies are closer together, the overlap between the impurity wavefunctions increases, thus the impurity band width also increases. Therefore, the transition from hopping to band transport will occur, with a critical oxygen vacancy concentration of about 1.5×10^{21} cm^{-3} corresponding to a local composition HfO_{2-x} with $x=0.05$. Figure 23eshows the effect of atomic-sized changes in the CF diameter on the obvious change of the CF conductance. When one to three oxygen vacancy columns are formed, as shown by the inset of Fig. 23e, the conductance is observed to increase stepwise, with each transmitting channel contributing a quantum of conductance G_0. This result is in qualitative agreement with the interpretation that the filament with single- to few atom diameters behaves as a quantum wire and the observed conductance quantization originates from single or few atom changes in the atomic structure of the filament.

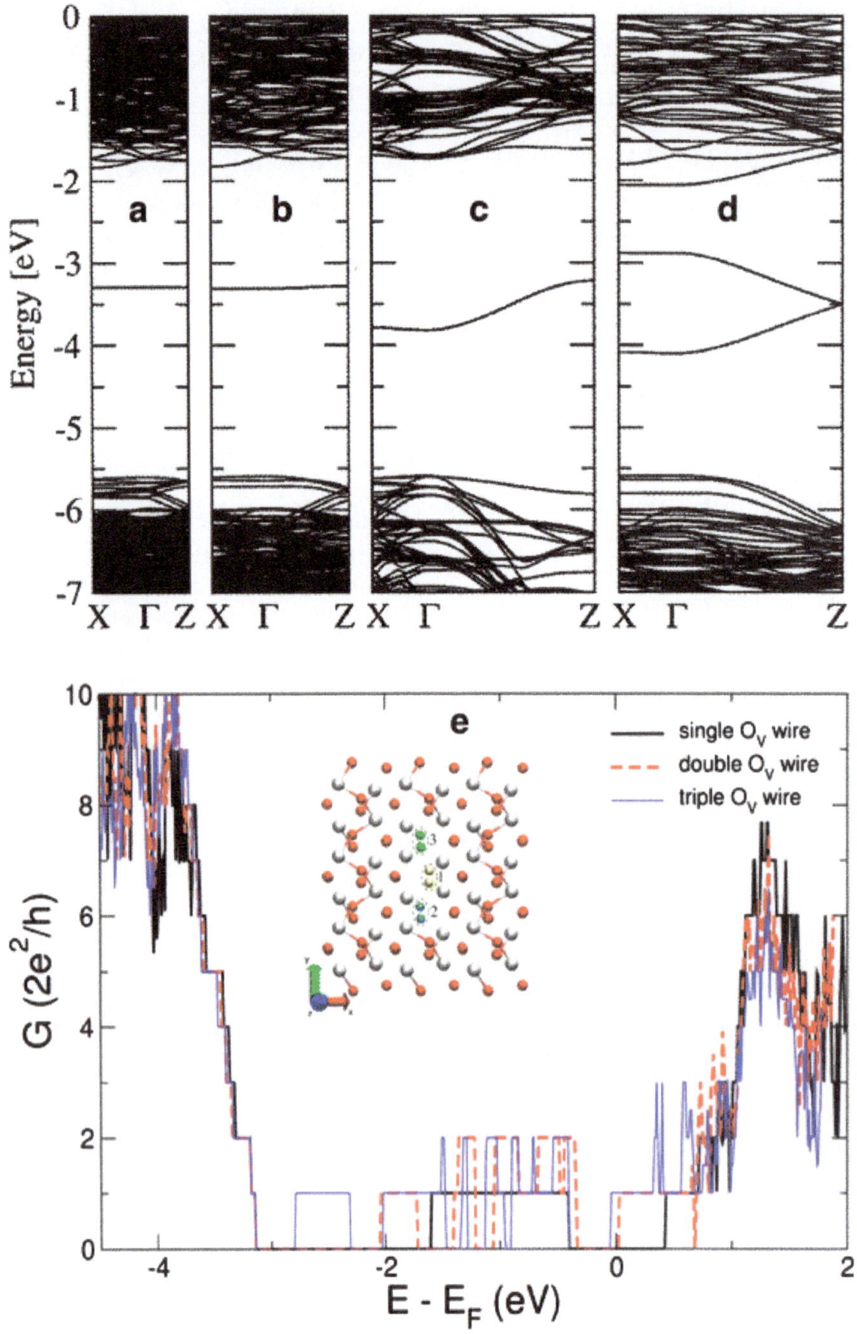

Figure 23: Calculated band structures for crystalline m-HfO_2 with O vacancies and the corresponding conductance [69]. **a–d** Crystalline m-HfO_2 band structure with different oxygen vacancies separated by 4 a_0, 2 a_0, a_0, and $a_0/2$. a_0 is the length of the c-axis vector for the m-HfO_2 primitive cell (0.5296 nm). **e** Conductance as a function of energy corresponding to a HfO_2 matrix where one, two, or three O atom rows are removed. The rows subsequently removed are shown in the instate (marked as "1," "2," and "3"), where*red* and *white spheres* correspond to O and Hf atoms, respectively. Reproduced with permission.

Equivalent Circuit Model for ECM Device Showing Conductance Quantization Effect

Valov and Waser et al. have found that the ECM device is inherently controlled by the non-equilibrium states, which are induced by several factors, including the chemical processes such as the dissolution of the active electrode materials into the electrolyte, the electrochemical processes, and the charge redistribution during operations [221]. The most distinct effect brought out by the non-equilibrium states is the generation of the electromotive force (emf) in the device, suggesting the presence of a nanobattery inside the device. On the basis of this work, recently they proposed a circuit model (Fig. 24a) for the ECM device accounting for quantized conductance [195, 213]. The discharging of the internal emf voltage (V_{emf}) can influence the device characteristics including ON-resistance (R_{ON}). In the model, they assumed that in the ON-state with atom point contact, the ON-resistance is restricted in the contact regime, being$R_{ON}=R_c=nG_0^{-1}$, and assumed that R_{ON} is directly dependent on the internal emf voltage ((V_{emf})). Figure 24b shows the simulation results about the impact of the external resistance R_{ext}on R_{ON}, exhibiting the staircase-like change of the cell conductance, which well accounts for the experimental results as shown in Fig. 24c. Moreover, the calculated time constants also fit the experimental data as shown in Fig. 24c quite well.

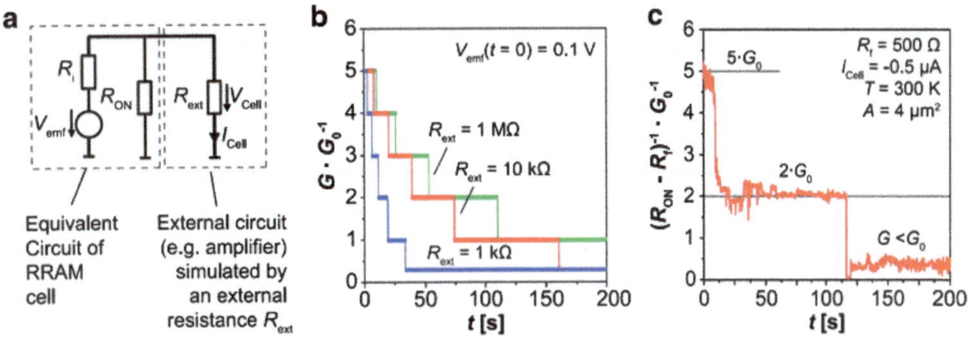

Figure 24: Equivalent circuit model for ECM device [195]. **a** Equivalent circuit model for an ECM cell including a nanobattery V_{emf} with an external circuit. R_i is the total resistance of the ionic current path. R_{ext} is the external resistance, e.g., from the neighboring cells in an array or a sense amplifier. **b** SPICE simulation results showing a staircase-like change of the cell conductance resulted from the discharging of V_{emf}. **c** Evolution of the conductance of a Ag/SiO_2/Pt cell in crossbar structure under a negative cell current I_{cell}. Reproduced with permission.

Prospects of Conductance Quantization in Applications

In these years, the conductance quantization phenomenon in RRAM has been widely investigated and developed, turning into an attractive and well-known effect. From the reported results on the conductance quantization in RRAM, the filament can be of atomic dimensions at the narrowest constriction, which shows that CF-type RRAM devices can still remain functional even if the diameter of CF scales down to the atomic size. Therefore, the ultimate scalability of RRAM is significantly higher than any current mainstream and emerging non-volatile memory.

As many reports have pointed out, one of the potential applications of conductance quantization effect is in multi-level ultra-high-density storage [8, 170]. According to the results in "Structures, Materials, and Operation Methods of RRAM with QC Effect" section, by carefully controlling the resistive switching process of RRAM device, for example, through accurately choosing appropriate compliance current, stop voltage, sweeping speed, pulse conditions, etc., the quantized conductance values can be well controlled in theory, thus multi-level storage can be realized. Since the controllability of quantized states of CF has been confirmed to gradually improve year after year, and at the same time the kinetics and physics of switching processes and conductance quantization have also been revealed to be much clearer [16, 222], the prospect of realizing ultra-high data storage by taking advantage of the phenomenon of conductance quantization has become more promising nowadays. Except the multi-level storage, basic logic circuits can also be realized. The best achievement is from the group of Prof. M. Aono [169]. Low-power logic gates such as AND, OR, NOT gate have been configured making use of quantized conductance atomic switches (QCASs), which were fabricated by crossing metal electrode wires with solid electrolyte wires. However, in order to advance the practical multi-level high-density storage or logic circuit application of conductance quantization in RRAM, in the future, deeper investigations should be focused on how to achieve the accurate control of the quantized conductance states, and great improvements should also be required in the multi-level storage performances including endurance, retention, etc., especially based on the pulse operations.

Another important aspect related to the conductance quantization effect in RRAM devices is that the quantized CF can be made use of to investigate any other novel physical effects, such as magnetic and thermoelectric properties. Some works on the magnetic modulation in RRAM have been reported [30–42, 44–46], most of which just studied the RRAM devices with usual oxygen vacancies of CF or metal CF. Our group has also investigated the intrinsic electron transport mechanism in the formed CF by measuring the thermoelectric Seebeck effect [223]. The small-polaron hopping model can well account for the electronic transport process for all resistance states of Ti/HfO_2/Pt device, although the corresponding resistance-temperature behaviors are contradictive. At the same time, from the point of view of device design, the controlled atomic-scale CF in simple two-terminal devices usually got at room temperature and in air can provide a media or platform to develop new one-dimensional nanodevices based on the quantum effects in CF. Here, we just show an interesting example. By replacing the electrode material with magnetic metals, we can configure the magnetic CF, and through electrical manipulation, we can try to control the CF size to get atom-sized magnetic CF. Thus, the magnetic properties such as the quantized anisotropic magnetoresistance (QAMR) effect can be studied in the atom-sized magnetic CF so as to deeply investigate the quantized transport of CF. This kind of works can provide a new characterization method for the research on the CF and the resistive switching mechanism. They might also provide a new idea of achieving stable QAMR effect in experiments and promote the deep understanding on the spin-dependent transport properties in atom-sized materials. In the long term, by simultaneously manipulating the resistance states and the ordered/disordered magnetic states, we might construct novel functional nanoscale electronic devices.

CONCLUSIONS

In this paper, we explained the resistive switching mechanism and operating principles of filamentary RRAM and analyzed their connection with the conductance quantization effect. Then, we introduced some typical researches on the conductance quantization effect of RRAM. The device structures, switching material system, and the operating methods of RRAM related to conductance quantization effect were summarized in detail. Next, the theory and modeling of quantum transport in the atomic CF of RRAM ascribing to the conductance quantization effect were discussed. Finally, we evaluated the opportunities and challenges of the quantized CF system in RRAM devices for the multi-level storage and any other applications in the future.

ACKNOWLEDGEMENTS

This work was supported by the National Natural Science Foundation of China under Grant Nos. 61322408, 61221004, 61574169, 61274091,61522048, 61422407, 61334007, and 61474135, Beijing Training Project for the Leading Talents in S & T under Grant No. ljrc201508, the National High Technology Research Development Program under Grant No. 2014AA032900, the Opening Project of Key Laboratory of Microelectronics Devices and Integrated Technology, Institute of Microelectronics, Chinese Academy of Sciences, the Spanish Ministry of Science and Technology under contract TEC2012-32305 (partially funded by the EU FEDER program), and the DURSI of the Generalitat de Catalunya under contract 2014SGR384.

Competing interests

The authors declare that they have no competing interests.

REFERENCES

1. Pirovano A, Schuegraf K (2010) Memory grows up. Nature Nanotech 5(3):177–8. doi:10.1038/nnano.2010.36
2. Fujisaki Y (2013) Review of emerging new solid-state non-volatile memories. Jpn J Appl Phys 52(4R):040001. doi:10.7567/jjap.52.040001
3. Waser R, Aono M (2007) Nanoionics-based resistive switching memories. Nature Mater 6(11):833–40
4. Sawa A (2008) Resistive switching in transition metal oxides. Mater Today 11(6):28–36. doi:10.1016/s1369-7021(08)70119-6
5. Waser R, Dittmann R, Staikov G, Szot K (2009) Redox-based resistive switching memories—nanoionic mechanisms, prospects, and challenges. Adv Mater 21(25–26):2632–63. doi:10.1002/adma.200900375
6. Yang JJ, Strukov DB, Stewart DR (2013) Memristive devices for computing. Nature Nanotech 8(1):13–24. doi:10.1038/nnano.2012.240
7. Jeong DS, Thomas R, Katiyar RS, Scott JF, Kohlstedt H, Petraru A et al (2012) Emerging memories: resistive switching mechanisms and current status. Rep Prog Phys 75(7):076502. doi:10.1088/0034-4885/75/7/076502
8. Pan F, Gao S, Chen C, Song C, Zeng F (2014) Recent progress in resistive random access memories: materials, switching mechanisms, and performance. Mater Sci Eng R-Rep 83:1–59. doi:10.1016/j.mser.2014.06.002
9. Pan F, Chen C, Wang Z-S, Yang Y-C, Yang J, Zeng F (2010) Nonvolatile resistive switching memories—characteristics, mechanisms and challenges. Progress in Natural Science: Materials International 20:1–15. doi:10.1016/s1002-0071(12)60001-x
10. Burr GW, Kurdi BN, Scott JC, Lam CH, Gopalakrishnan K, Shenoy RS (2008) Overview of candidate device technologies for storage-class memory. IBM J Res Dev 52(4/5):449–64
11. Akinaga H, Shima H (2010) Resistive random access memory (ReRAM) based on metal oxides. Proc IEEE 98(12):2237–51. doi:10.1109/JPROC.2010.2070830
12. Wong HSP, Heng-Yuan L, Shimeng Y, Yu-Sheng C, Yi W, Pang-Shiu C et al (2012) Metal-oxide RRAM. Proc IEEE 100(6):1951–70. doi:10.1109/JPROC.2012.2190369
13. Karg SF, Meijer GI, Bednorz JG, Rettner CT, Schrott AG, Joseph EA et al (2008) Transition-metal-oxide-based resistance-change memories. IBM J Res Dev 52(4.5):481–92
14. Prakash A, Jana D, Maikap S (2013) TaO_x-based resistive switching memories, prospective and challenges. Nanoscale Res Lett 8:418. doi:10.1186/1556-276X-8-418
15. Valov I, Waser R, Jameson JR, Kozicki MN (2011) Electrochemical metallization memories—fundamentals, applications, prospects. Nanotechnology 22(28):289502. doi:10.1088/0957-4484/22/28/289502
16. Valov I, Sapezanskaia I, Nayak A, Tsuruoka T, Bredow T, Hasegawa T et al (2012) Atomically controlled electrochemical nucleation at superionic solid electrolyte surfaces. Nature Mater 11(6):530–5. doi:10.1038/nmat3307
17. Ambrogio S, Balatti S, Choi S, Ielmini D (2014) Impact of the mechanical stress on switching characteristics of electrochemical resistive memory. Adv Mater 26(23):3885–92. doi:10.1002/adma.201306250
18. Tian X, Yang S, Zeng M, Wang L, Wei J, Xu Z et al (2014) Bipolar electrochemical mechanism for mass transfer in nanoionic resistive memories. Adv Mater 26(22):3649–54. doi:10.1002/adma.201400127
19. Valov I (2014) Redox-based resistive switching memories (ReRAMs): electrochemical systems at the atomic scale. Chem Electro Chem 1(1):26–36. doi:10.1002/celc.201300165
20. Yang Y, Gao P, Li L, Pan X, Tappertzhofen S, Choi S et al (2014) Electrochemical dynamics of nanoscale metallic inclusions in dielectrics. Nat Commun 5:4232. doi:10.1038/ncomms5232
21. Russo U, Ielmini D, Cagli C, Lacaita AL, Spiga S, Wiemer C et al (2007) Conductive-filament switching analysis and self-accelerated thermal dissolution model for reset in NiO-based RRAM. IEEE International Electron Devices Meeting (IEDM)
22. Russo U, Ielmini D, Cagli C, Lacaita AL (2009) Self-accelerated thermal dissolution model for reset programming in unipolar resistive-switching memory (RRAM) devices. IEEE Trans Electron Devices 56(2):193–200. doi:10.1109/TED.2008.2010584
23. Seungjae J, Siddik M, Wootae L, Jubong P, Xinjun L, Jiyong W et al (2011) Thermally-assisted Ti/$Pr_{0.7}Ca_{0.3}MnO_3$ ReRAM with excellent switching speed and retention characteristics. IEEE International Electron Devices Meeting (IEDM)
24. 2Kim S, Kim SJ, Kim KM, Lee SR, Chang M, Cho E et al (2013) Physical electro-thermal model of resistive switching in bi-layered resistance-change memory. Sci Rep 3:1680. doi:10.1038/srep01680

25. Peng S, Zhuge F, Chen X, Zhu X, Hu B, Pan L et al (2012) Mechanism for resistive switching in an oxide-based electrochemical metallization memory. Appl Phys Lett 100(7):072101. doi:10.1063/1.3683523
26. 2Shang J, Liu G, Yang H, Zhu X, Chen X, Tan H et al (2014) Thermally stable transparent resistive random access memory based on all-oxide heterostructures. Adv Funct Mater 24(15):2171–9. doi:10.1002/adfm.201303274
27. Pan L, Ji Z, Yi X, Zhu X, Chen X, Shang J et al (2015) Metal-organic framework nanofilm for mechanically flexible information storage applications. Adv Funct Mater 25(18):2677–85. doi:10.1002/adfm.201500449
28. Lee HS, Choi SG, Park H-H, Rozenberg MJ. A new route to the Mott-Hubbard metal-insulator transition: Strong correlations effects in $Pr_{0.7}Ca_{0.3}MnO_3$. Sci Rep. 2013; 3. doi:10.1038/srep01704.
29. Mottaghizadeh A, Yu Q, Lang PL, Zimmers A, Aubin H. Metal oxide resistive switching: evolution of the density of states across the metal-insulator transition. Phys Rev Lett. 2014; 112(6). doi:10.1103/PhysRevLett.112.066803.
30. Sokolov A, Sabirianov R, Sabirianov I, Doudin B (2009) Voltage-induced switching with magnetoresistance signature in magnetic nano-filaments. J Phys Condens Matter 21(48):485303. doi:10.1088/0953-8984/21/48/485303
31. Teixeira JM, Ventura J, Fermento R, Araujo JP, Sousa JB, Wisniowski P et al (2009) Electroforming, magnetic and resistive switching in MgO-based tunnel junctions. J Phys D Appl Phys 42(10):105407. doi:10.1088/0022-3727/42/10/105407
32. Pertsev NA, Kohlstedt H (2010) Resistive switching via the converse magnetoelectric effect in ferromagnetic multilayers on ferroelectric substrates. Nanotechnology 21(47):475202. doi:10.1088/0957-4484/21/47/475202
33. Son JY, Kim CH, Cho JH, Shin Y-H, Jang HM (2010) Self-formed exchange bias of switchable conducting filaments in NiO resistive random access memory capacitors. ACS Nano 4(6):3288–92. doi:10.1021/nn100323x
34. Chen G, Song C, Chen C, Gao S, Zeng F, Pan F (2012) Resistive switching and magnetic modulation in cobalt-doped ZnO. Adv Mater 24(26):3515–20. doi:10.1002/adma.201201595
35. Rubi D, Marlasca FG, Reinoso M, Bonville P, Levy P (2012) Magnetism and electrode dependant resistive switching in Ca-doped ceramic bismuth ferrite. Mater Sci Eng B 177(6):471–5. doi:10.1016/j.mseb.2012.02.022
36. Thakare V, Xing G, Peng H, Rana A, Game O, Anil Kumar P et al (2012) High sensitivity low field magnetically gated resistive switching in $CoFe_2O_4/La_{0.66}Sr_{0.34}MnO_3$heterostructure. Appl Phys Lett 100(17):172412. doi:10.1063/1.4707373
37. Chen G, Peng JJ, Song C, Zeng F, Pan F (2013) Interplay between chemical state, electric properties, and ferromagnetism in Fe-doped ZnO films. J Appl Phys 113(10):104503. doi:10.1063/1.4794882
38. Wang XL, Ku PS, Shao Q, Cheng WF, Leung CW, Ruotolo A (2013) Magnetism as a probe of the origin of memristive switching in p-type antiferromagnetic NiO. Appl Phys Lett 103(22):223508. doi:10.1063/1.4834795
39. Li Q, Shen TT, Cao YL, Zhang K, Yan SS, Tian YF et al (2014) Spin memristive magnetic tunnel junctions with CoO-ZnO nano composite barrier. Sci Rep 4:3835. doi:10.1038/srep03835
40. Otsuka S, Hamada Y, Shimizu T, Shingubara S (2014) Ferromagnetic nano-conductive filament formed in $Ni/TiO_2/Pt$ resistive-switching memory. Appl Phys A 118(2):613–9. doi:10.1007/s00339-014-8769-5
41. Ren SX, Sun GW, Zhao J, Dong JY, Wei Y, Ma ZC et al (2014) Electric field-induced magnetic switching in Mn:ZnO film. Appl Phys Lett 104(23):232406,http://dx.doi.org/10.1063/1.4883259
42. Wang J, Zhang X, Piao H-G, Luo Z, Xiong C, Wang X et al (2014) Magnetic field controllable nonvolatile resistive switching effect in silicon device. Appl Phys Lett 104(24):243511. doi:10.1063/1.4884771
43. Xiong YQ, Zhou WP, Li Q, He MC, Du J, Cao QQ et al (2014) Electric field manipulation of nonvolatile magnetization in Au/NiO/Pt heterostructure with resistive switching effect. Appl Phys Lett 105(3):032410. doi:10.1063/1.4891482
44. Xu Z, Gao M, Yu L, Lu L, Xu X, Jiang Y (2014) Co nanoparticles induced resistive switching and magnetism for the electrochemically deposited polypyrrole composite films. ACS Appl Mater Interfaces 6(20):17823–30. doi:10.1021/am5044399
45. Zhang Z, Jiang L (2014) Bias voltage induced resistance switching effect in single-molecule magnets' tunneling junction. Nanotechnology 25(36):365201. doi:10.1088/0957-4484/25/36/365201
46. Chen X, Zhu X, Xiao W, Liu G, Feng YP, Ding J et al (2015) Nanoscale magnetization reversal caused by electric field-induced ion migration and redistribution in cobalt ferrite thin films. ACS Nano 9(4):4210–8. doi:10.1021/acsnano.5b00456
47. Yang Z, Zhan Q, Zhu X, Liu Y, Yang H, Hu B et al (2014) Tunneling magnetoresistance induced by controllable formation of Co filaments in resistive switching Co/ZnO/Fe structures. EPL (Europhysics Letters) 108(5):58004. doi:10.1209/0295-5075/108/58004
48. Guo X, Schindler C, Menzel S, Waser R (2007) Understanding the switching-off mechanism in Ag^+ migration based resistively switching model systems. Appl Phys Lett 91(13):133513. doi:10.1063/1.2793686
49. Sakamoto T, Lister K, Banno N, Hasegawa T, Terabe K, Aono M (2007) Electronic transport in Ta_2O_5 resistive switch. Appl Phys Lett 91(9):092110,http://dx.doi.org/10.1063/1.2777170
50. Yang YC, Pan F, Liu Q, Liu M, Zeng F (2009) Fully room-temperature-fabricated nonvolatile resistive memory for ultrafast and high-density memory application. Nano Lett 9(4):1636–43. doi:10.1021/nl900006g
51. Kwon DH, Kim KM, Jang JH, Jeon JM, Lee MH, Kim GH et al (2010) Atomic structure of conducting

nanofilaments in TiO_2 resistive switching memory. Nature Nanotech 5(2):148–53. doi:10.1038/nnano.2009.456

52. Liu Q, Long S, Lv H, Wang W, Niu J, Huo Z et al (2010) Controllable growth of nanoscale conductive filaments in solid-electrolyte-based ReRAM by using a metal nanocrystal covered bottom electrode. ACS Nano 4(10):6162–8. doi:10.1021/nn1017582
53. Xu Z, Bando Y, Wang W, Bai X, Golberg D (2010) Real-time in situ HRTEM-resolved resistance switching of Ag_2S nanoscale ionic conductor. ACS Nano 4(5):2515–22. doi:10.1021/nn100483a
54. Miao F, Strachan JP, Yang JJ, Zhang MX, Goldfarb I, Torrezan AC et al (2011) Anatomy of a nanoscale conduction channel reveals the mechanism of a high-performance memristor. Adv Mater 23(47):5633–40. doi:10.1002/adma.201103379
55. Liu Q, Sun J, Lv H, Long S, Yin K, Wan N et al (2012) Real-time observation on dynamic growth/dissolution of conductive filaments in oxide-electrolyte-based ReRAM. Adv Mater 24(14):1844–9. doi:10.1002/adma.201104104
56. Yang Y, Gao P, Gaba S, Chang T, Pan X, Lu W (2012) Observation of conducting filament growth in nanoscale resistive memories. Nat Commun 3:732. doi:10.1038/ncomms1737
57. Park GS, Kim YB, Park SY, Li XS, Heo S, Lee MJ et al (2013) In situ observation of filamentary conducting channels in an asymmetric Ta_2O_{5-X}/TaO_{2-X} bilayer structure. Nat Commun 4:2382. doi:10.1038/ncomms3382
58. Celano U, Goux L, Belmonte A, Opsomer K, Franquet A, Schulze A et al (2014) Three-dimensional observation of the conductive filament in nanoscaled resistive memory devices. Nano Lett 14(5):2401–6. doi:10.1021/nl500049g
59. Privitera S, Bersuker G, Butcher B, Kalantarian A, Lombardo S, Bongiorno C et al (2013) Microscopy study of the conductive filament in HfO_2 resistive switching memory devices. Microelectron Eng 109:75–8, http://dx.doi.org/10.1016/j.mee.2013.03.145
60. Zhu X-J, Shang J, Liu G, Li R-W (2014) Ion transport-related resistive switching in film sandwich structures. Chin Sci Bull 59(20):2363–82. doi:10.1007/s11434-014-0284-8
61. Szot K, Speier W, Bihlmayer G, Waser R (2006) Switching the electrical resistance of individual dislocations in single-crystalline $SrTiO_3$. Nature Mater 5(4):312–20. doi:10.1038/nmat1614
62. Lee MH, Hwang CS (2011) Resistive switching memory: observations with scanning probe microscopy. Nanoscale 3(2):490–502. doi:10.1039/c0nr00580k
63. Lanza M (2014) A review on resistive switching in high-k dielectrics: a nanoscale point of view using conductive atomic force microscope. Materials 7(3):2155–82. doi:10.3390/ma7032155
64. Victor VZ, Roy M, Ralph KC, Gurtej S (2011) Scaling limits of resistive memories. Nanotechnology 22(25):254027
65. Sharvin YV (1965) A possible method for studying Fermi surfaces. Sov Phys JETP 21:655–656
66. Montie EA, Cosman EC, t Hooft GW, van der Mark MB, Beenakker CWJ (1991) Observation of the optical analogue of quantized conductance of a point contact. Nature 350(6319):594–5
67. Krans JM, van Ruitenbeek JM, Fisun VV, Yanson IK, de Jongh LJ (1995) The signature of conductance quantization in metallic point contacts. Nature 375(6534):767–9
68. Ohnishi H, Kondo Y, Takayanagi K (1998) Quantized conductance through individual rows of suspended gold atoms. Nature 395(6704):780–3,http://www.nature.com/nature/journal/v395/n6704/
69. Long S, Lian X, Cagli C, Cartoixà X, Rurali R, Miranda E et al (2013) Quantum-size effects in hafnium-oxide resistive switching. Appl Phys Lett 102(18):183505. doi:10.1063/1.4802265
70. Long S, Perniola L, Cagli C, Buckley J, Lian X, Miranda E et al (2013) Voltage and power-controlled regimes in the progressive unipolar RESET transition of HfO_2-based RRAM. Sci Rep 3:2929. doi:10.1038/srep02929
71. Lian X, Cartoixà X, Miranda E, Perniola L, Rurali R, Long S et al (2014) Multi-scale quantum point contact model for filamentary conduction in resistive random access memories devices. J Appl Phys 115(24):244507, http://dx.doi.org/10.1063/1.4885419
72. Zhao X, Xu H, Wang Z, Zhang L, Ma J, Liu Y (2015) Nonvolatile/volatile behaviors and quantized conductance observed in resistive switching memory based on amorphous carbon. Carbon 91:38–44, http://dx.doi.org/10.1016/j.carbon.2015.04.031
73. Long S, Cagli C, Ielmini D, Liu M, Suñé J (2012) Analysis and modeling of resistive switching statistics. J Appl Phys 111(7):074508. doi:10.1063/1.3699369
74. Zhang M, Long S, Wang G, Li Y, Xu X, Liu H et al (2014) An overview of the switching parameter variation of RRAM. Chin Sci Bull 59(36):5324–37. doi:10.1007/s11434-014-0673-z
75. Daughton JM (1992) Magnetoresistive memory technology. Thin Solid Films 216(1):162–8,http://dx.doi.org/10.1016/0040-6090(92)90888-I
76. Gallagher WJ, Parkin SS (2006) Development of the magnetic tunnel junction MRAM at IBM: from first junctions to a 16-Mb MRAM demonstrator chip. IBM J Res Dev 50(1):5–23
77. Kawahara T, Ito K, Takemura R, Ohno H (2012) Spin-transfer torque RAM technology: review and prospect. Microelectron Reliab 52(4):613–27,http://dx.doi.org/10.1016/j.microrel.2011.09.028
78. Khvalkovskiy A, Apalkov D, Watts S, Chepulskii R, Beach R, Ong A et al (2013) Basic principles of STT-MRAM cell operation in memory arrays. J Phys D Appl Phys 46(7):74001–20
79. Yuasa S, Fukushima A, Yakushiji K, Nozaki T, Konoto M, Maehara H et al. Future prospects of MRAM technologies. Proc IEEE-IEDM (Washington, DC, 9–11 December 2013). 2013:3.1.

80. Zalba B (2003) Marín JM, Cabeza LF, Mehling H. Review on thermal energy storage with phase change: materials, heat transfer analysis and applications. Appl Therm Eng 23(3):251–83

81. Farid MM, Khudhair AM, Razack SAK, Al-Hallaj S (2004) A review on phase change energy storage: materials and applications. Energy Convers Manage 45(9):1597–615

82. Sharma A, Tyagi V, Chen C, Buddhi D (2009) Review on thermal energy storage with phase change materials and applications. Renew Sust Energ Rev 13(2):318–45

83. Zilberberg O, Weiss S, Toledo S (2013) Phase-change memory: an architectural perspective. ACM Computing Surveys (CSUR) 45(3):29

84. Oike H, Kagawa F, Ogawa N, Ueda A, Mori H, Kawasaki M et al (2015) Phase-change memory function of correlated electrons in organic conductors. Phys Rev B Condens Matter 91(4):041101

85. Wentai L, Hangbing L, Liu Q, Long S, Wang W, Wang Y et al (2011) Improved resistive switching uniformity in Cu/HfO_2/Pt devices by using current sweeping mode. IEEE Electron Device Lett 32(8):1053–5. doi:10.1109/LED.2011.2157990

86. Hongtao L, Hangbin L, Baohe Y, Xiaoxin X, Ruoyu L, Qi L et al (2014) Uniformity improvement in 1T1R RRAM with gate voltage ramp programming. IEEE Electron Device Lett 35(12):1224–6. doi:10.1109/LED.2014.2364171

87. Wang G, Long S, Zhang M, Li Y, Xu X, Liu H et al (2014) Operation methods of resistive random access memory. Sci China Tech Sci 57(12):2295–304. doi:10.1007/s11431-014-5718-7

88. Waser R (2009) Resistive non-volatile memory devices (Invited Paper). Microelectron Eng 86(7–9):1925–8. doi:10.1016/j.mee.2009.03.132

89. Kim KM, Jeong DS, Hwang CS (2011) Nanofilamentary resistive switching in binary oxide system: a review on the present status and outlook. Nanotechnology 22(25):254002. doi:10.1088/0957-4484/22/25/254002

90. Lu W, Jeong DS, Kozicki M, Waser R (2012) Electrochemical metallization cells—blending nanoionics into nanoelectronics? MRS Bull 37(02):124–30. doi:10.1557/mrs.2012.5

91. Yang JJ, Inoue IH, Mikolajick T, Hwang CS (2012) Metal oxide memories based on thermochemical and valence change mechanisms. MRS Bull 37(02):131–7. doi:10.1557/mrs.2011.356

92. Valov I, Kozicki MN (2013) Cation-based resistance change memory. J Phys D Appl Phys 46(7):074005. doi:10.1088/0022-3727/46/7/074005

93. The international technology roadmap for semiconductors (ITRS), process integration, devices, and structures summary, 2013, http://www.itrs.net/

94. Xue WH, Xiao W, Shang J, Chen XX, Zhu XJ, Pan L et al (2014) Intrinsic and interfacial effect of electrode metals on the resistive switching behaviors of zinc oxide films. Nanotechnology 25(42):425204. doi:10.1088/0957-4484/25/42/425204

95. Tan H, Liu G, Zhu X, Yang H, Chen B, Chen X et al (2015) An optoelectronic resistive switching memory with integrated demodulating and arithmetic functions. Adv Mater 27(17):2797–803. doi:10.1002/adma.201500039

96. Feng P, Shong Y, Subramanian V (2011) A detailed study of the forming stage of an electrochemical resistive switching memory by KMC simulation. IEEE Electron Device Lett 32(7):949–51. doi:10.1109/LED.2011.2143691

97. 9Wu W, Wang ZL (2011) Piezotronic nanowire-based resistive switches as programmable electromechanical memories. Nano Lett 11(7):2779–85. doi:10.1021/nl201074a

98. Kim Y, Strelcov E, Hwang IR, Choi T, Park BH, Jesse S et al (2013) Correlative multimodal probing of ionically-mediated electromechanical phenomena in simple oxides. Sci Rep 3:2924. doi:10.1038/srep02924

99. Li L (2013) Electromechanically tuned resistive switching device. Appl Phys Lett 103(23):233512. doi:10.1063/1.4839415

100. Menzel S, Waser R (2013) Analytical analysis of the generic SET and RESET characteristics of electrochemical metallization memory cells. Nanoscale 5(22):11003–10. doi:10.1039/c3nr03387b

101. Onofrio N, Guzman D, Strachan A (2015) Atomic origin of ultrafast resistance switching in nanoscale electrometallization cells. Nature Mater 14(4):440–6. doi:10.1038/nmat4221

102. Wei Z, Kanzawa Y, Arita K, Katoh Y, Kawai K, Muraoka S et al (2008) Highly reliable TaO_XReRAM and direct evidence of redox reaction mechanism. IEEE International Electron Devices Meeting (IEDM)

103. Zhuge F, Dai W, He CL, Wang AY, Liu YW, Li M et al (2010) Nonvolatile resistive switching memory based on amorphous carbon. Appl Phys Lett 96(16):3. doi:10.1063/1.3406121

104. Chang K-C, Tsai T-M, Chang T-C, Syu Y-E, Wang C-C, Chuang S-L et al (2011) Reducing operation current of Ni-doped silicon oxide resistance random access memory by supercritical CO_2 fluid treatment. Appl Phys Lett 99(26):263501,http://dx.doi.org/10.1063/1.3671991

105. Di F, Dan X, Tingting F, Zhang C, Niu J, He Q et al (2011) Unipolar resistive switching properties of diamondlike carbon-based RRAM devices. IEEE Electron Device Lett 32(6):803–5. doi:10.1109/LED.2011.2132750

106. Yu S, Guan X, Wong HSP (2011) Conduction mechanism of TiN/HfO_X/Pt resistive switching memory: a trap-assisted-tunneling model. Appl Phys Lett 99(6):063507. doi:10.1063/1.3624472

107. Chang KC, Tsai TM, Chang TC, Wu HH, Chen KH, Chen JH et al (2013) Low temperature improvement method on Zn:SiO_x resistive random access memory devices. IEEE Electron Device Lett 34(4):511–3. doi:10.1109/LED.2013.2248075

108. Kuan-Chang C, Chih-Hung P, Ting-Chang C, Tsung-Ming T, Rui Z, Jen-Chung L et al (2013) Hopping effect of hydrogen-doped silicon oxide insert RRAM by

supercritical CO_2fluid treatment. IEEE Electron Device Lett 34(5):617–9. doi:10.1109/LED.2013.2251995

109. Zhang R, Chang K-C, Chang T-C, Tsai T-M, Chen K-H, Lou J-C et al (2013) High performance of graphene oxide-doped silicon oxide-based resistance random access memory. Nanoscale Res Lett 8(1):1–6. doi:10.1186/1556-276X-8-497

110. Chen Y-J, Chen H-L, Young T-F, Chang T-C, Tsai T-M, Chang K-C et al (2014) Hydrogen induced redox mechanism in amorphous carbon resistive random access memory. Nanoscale Res Lett 9(1):1–5. doi:10.1186/1556-276X-9-52

111. Hong DS, Chen YS, Li Y, Yang HW, Wei LL, Shen BG et al (2014) Evolution of conduction channel and its effect on resistance switching for Au-WO_{3-x}-Au devices. Sci Rep 4:4058. doi:10.1038/srep04058

112. Wang G, Long S, Yu Z, Zhang M, Ye T, Li Y et al (2015) Improving resistance uniformity and endurance of resistive switching memory by accurately controlling the stress time of pulse program operation. Appl Phys Lett 106(9):092103,http://dx.doi.org/10.1063/1.4907604

113. Raghavan N, Pey KL, Wu X, Wenhu L, Bosman M (2012) Percolative model and thermodynamic analysis of oxygen-ion-mediated resistive switching. IEEE Electron Device Lett 33(5):712–4. doi:10.1109/LED.2012.2187170

114. Wu PM, Paschoal W, Kumar S, Borschel C, Ronning C, Canali CM et al (2012) Thermoelectric characterization of electronic properties of GaMnAs nanowires. J Nanotechnol 2012:1–5. doi:10.1155/2012/480813

115. Otsuka W, Miyata K, Kitagawa M, Tsutsui K, Tsushima T, Yoshihara H et al. A 4 Mb conductive-bridge resistive memory with 2.3GB/s read-throughput and 216 MB/s program-throughput. IEEE International Solid-State Circuits Conference Digest of Technical Papers (ISSCC); 2011 Feb. 20–24.

116. Shimeng Y, Wong HSP (2011) Compact modeling of conducting-bridge random-access memory (CBRAM). IEEE Trans Electron Devices 58(5):1352–60. doi:10.1109/TED.2011.2116120

117. Seol C, Balatti S, Nardi F, Ielmini D (2012) Size-dependent drift of resistance due to surface defect relaxation in conductive-bridge memory. IEEE Electron Device Lett 33(8):1189–91. doi:10.1109/LED.2012.2199074

118. Zhang K-L, Liu K, Wang F, Yin F-H, Wei X-Y, Zhao J-S (2013) Modeling of conducting bridge evolution in bipolar vanadium oxide-based resistive switching memory. Chin Phys B 22(9):097101. doi:10.1088/1674-1056/22/9/097101

119. Belmonte A, Degraeve R, Fantini A, Kim W, Houssa M, Jurczak M et al (2014) Origin of the deep reset and low variability of pulse-programmed W\Al_2O_3\TiW\Cu CBRAM device. IEEE International Memory Workshop (IMW)

120. Russo U, Kamalanathan D, Ielmini D, Lacaita AL, Kozicki MN (2009) Study of multilevel programming in programmable metallization cell (PMC) memory. IEEE Trans Electron Devices 56(5):1040–7. doi:10.1109/TED.2009.2016019

121. Shimeng Y, Wong HSP (2010) Modeling the switching dynamics of programmable-metallization-cell (PMC) memory and its application as synapse device for a neuromorphic computation system. IEEE International Electron Devices Meeting (IEDM)

122. Puthentheradam SC, Schroder DK, Kozicki MN (2011) Inherent diode isolation in programmable metallization cell resistive memory elements. Appl Phys A 102(4):817–26. doi:10.1007/s00339-011-6292-5

123. Wang Z, Kadohira T, Tada T, Watanabe S (2007) Nonequilibrium quantum transport properties of a silver atomic switch. Nano Lett 7(9):2688–92. doi:10.1021/nl0711054

124. Wang Z, Gu T, Tada T, Watanabe S (2008) Excess-silver-induced bridge formation in a silver sulfide atomic switch. Appl Phys Lett 93(15):152106. doi:10.1063/1.2963197

125. Cario L, Vaju C, Corraze B, Guiot V, Janod E (2010) Electric-field-induced resistive switching in a family of mott insulators: Towards a new class of RRAM memories. Adv Mater 22(45):5193–7. doi:10.1002/adma.201002521

126. Celinska J, McWilliams C, Paz de Araujo C, Xue K-H (2011) Material and process optimization of correlated electron random access memories. J Appl Phys 109(9):091603,http://dx.doi.org/10.1063/1.3581197

127. McWilliams CR, Celinska J, de Araujo CA P, Xue K-H (2011) Device characterization of correlated electron random access memories. J Appl Phys 109(9):091608,http://dx.doi.org/10.1063/1.3581206

128. Kim H-T, Chae B-G, Youn D-H, Kim G, Kang K-Y, Lee S-J et al (2005) Raman study of electric-field-induced first-order metal-insulator transition in VO_2-based devices. Appl Phys Lett 86(24):242101–3

129. Ruzmetov D, Gopalakrishnan G, Deng J, Narayanamurti V, Ramanathan S (2009) Electrical triggering of metal-insulator transition in nanoscale vanadium oxide junctions. J Appl Phys 106(8):083702, http://dx.doi.org/10.1063/1.3245338

130. Xue K-H, de Araujo CAP, Celinska J, McWilliams C (2011) A non-filamentary model for unipolar switching transition metal oxide resistance random access memories. J Appl Phys 109(9):091602

131. Meijer G, Staub U, Janousch M, Johnson S, Delley B, Neisius T (2005) Valence states of Cr and the insulator-to-metal transition in Cr-doped $SrTiO_3$. Phys Rev B Condens Matter 72(15):155102

132. Ha SD, Aydogdu GH, Ramanathan S (2011) Metal-insulator transition and electrically driven memristive characteristics of $SmNiO_3$ thin films. Appl Phys Lett 98(1):012105

133. Ohtomo A, Hwang H (2004) A high-mobility electron gas at the $LaAlO_3$/$SrTiO_3$heterointerface. Nature 427(6973):423–6

134. Thiel S, Hammerl G, Schmehl A, Schneider C, Mannhart J (2006) Tunable quasi-two-dimensional electron gases in oxide heterostructures. Science 313(5795):1942–5

135. Niranjan MK, Wang Y, Jaswal SS, Tsymbal EY (2009) Prediction of a switchable two-dimensional electron

gas at ferroelectric oxide interfaces. Phys Rev Lett 103(1):016804

136. Park J, Bogorin D, Cen C, Felker D, Zhang Y, Nelson C et al (2010) Creation of a two-dimensional electron gas at an oxide interface on silicon. Nat Commun 1:94

137. Villena MA, Roldán JB, Jimenez-Molinos F, Suñé J, Long S, Miranda E et al (2014) A comprehensive analysis on progressive reset transitions in RRAMs. J Phys D Appl Phys 47(20):205102. doi:10.1088/0022-3727/47/20/205102

138. Szot K, Rogala M, Speier W, Klusek Z, Besmehn A, Waser R (2011) TiO_2—a prototypical memristive material. Nanotechnology 22(25):254001. doi:10.1088/0957-4484/22/25/254001

139. Sacchetto D, De Micheli G, Leblebici Y (2012) Multiterminal memristive nanowire devices for logic and memory applications: a review. Proc IEEE 100(6):2008–20. doi:10.1109/JPROC.2011.2172569

140. Indiveri G, Linares-Barranco B, Legenstein R, Deligeorgis G, Prodromakis T (2013) Integration of nanoscale memristor synapses in neuromorphic computing architectures. Nanotechnology 24(38):384010. doi:10.1088/0957-4484/24/38/384010

141. Sacchetto D, Gaillardon P, Zervas M, Carrara S, De Micheli G, Leblebici Y (2013) Applications of multi-terminal memristive devices: a review. IEEE Circuits Syst Mag 13(2):23–41. doi:10.1109/MCAS.2013.2256258

142. Hu SG, Wu SY, Jia WW, Yu Q, Deng LJ, Fu YQ et al (2014) Review of nanostructured resistive switching memristor and its applications. Nanosci Nanotechnol Lett 6(9):729–57. doi:10.1166/nnl.2014.1888

143. Datta S (1995) Electronic transport in mesoscopic systems. Cambridge University Press, New York

144. Landauer R (1970) Electrical resistance of disordered one-dimensional lattices. Philos Mag 21(172):863–7

145. Scheer E, Agraït N, Cuevas JC, Yeyati AL, Ludoph B, Martín-Rodero A et al (1998) The signature of chemical valence in the electrical conduction through a single-atom contact. Nature 394(6689):154–7

146. Van Wees B, Van Houten H, Beenakker C, Williamson JG, Kouwenhoven L, Van der Marel D et al (1988) Quantized conductance of point contacts in a two-dimensional electron gas. Phys Rev Lett 60(9):848

147. Wharam D, Thornton T, Newbury R, Pepper M, Ahmed H, Frost J et al (1988) One-dimensional transport and the quantisation of the ballistic resistance. J Phys C Solid State Phys 21(8):L209

148. Agrait N, Rodrigo J, Vieira S (1993) Conductance steps and quantization in atomic-size contacts. Phys Rev B Condens Matter 47(18):12345

149. Pascual JI, Méndez J, Gómez-Herrero J, Baró AM, García N, Binh VT (1993) Quantum contact in gold nanostructures by scanning tunneling microscopy. Phys Rev Lett 71(12):1852–5. doi:10.1103/PhysRevLett.71.1852

150. Fendley P, Ludwig AWW, Saleur H (1995) Exact conductance through point contacts in the $\nu=1/3$ fractional quantum hall effect. Phys Rev Lett 74(15):3005–8. doi:10.1103/PhysRevLett.74.3005

151. Ott F, Barberan S, Lunney J, Coey J, Berthet P, de Leon-Guevara A et al (1998) Quantized conductance in a contact between metallic oxide crystals. Phys Rev B Condens Matter 58(8):4656

152. Yang C-S, Thiltges J, Doudin B, Johnson M (2002) In situ monitoring of quantum conductance in electrodeposited magnetic point contacts. J Phys Condens Matter 14(50):L765

153. Havu P, Puska MJ, Nieminen RM, Havu V. Electron transport through quantum wires and point contacts. Phys Rev B: Condens Matter. 2004; 70(23). doi:10.1103/PhysRevB.70.233308.

154. Vandersypen LMK, Elzerman JM, Schouten RN, Willems van Beveren LH, Hanson R, Kouwenhoven LP (2004) Real-time detection of single-electron tunneling using a quantum point contact. Appl Phys Lett 85(19):4394. doi:10.1063/1.1815041

155. Xie FQ, Nittler L, Obermair C, Schimmel T. Gate-controlled atomic quantum switch. Phys Rev Lett. 2004; 93(12). doi:10.1103/PhysRevLett.93.128303.

156. He J, Sankey O, Lee M, Tao N, Li X, Lindsay S (2006) Measuring single molecule conductance with break junctions. Faraday Discuss 131:145–54. doi:10.1039/b508434m

157. Reed MA, Zhou C, Muller C, Burgin T, Tour J (1997) Conductance of a molecular junction. Science 278(5336):252–4

158. He H, Li C, Tao N (2001) Conductance of polymer nanowires fabricated by a combined electrodeposition and mechanical break junction method. Appl Phys Lett 78(6):811–3

159. Gonzalez JI, Lee T-H, Barnes MD, Antoku Y, Dickson RM (2004) Quantum mechanical single-gold-nanocluster electroluminescent light source at room temperature. Phys Rev Lett 93(14):147402

160. Tian J-H, Liu B, Li X, Yang Z-L, Ren B, Wu S-T et al (2006) Study of molecular junctions with a combined surface-enhanced Raman and mechanically controllable break junction method. J Am Chem Soc 128(46):14748–9

161. Chico L, Benedict LX, Louie SG, Cohen ML (1996) Quantum conductance of carbon nanotubes with defects. Phys Rev B Condens Matter 54(4):2600

162. Frank S (1998) Carbon nanotube quantum resistors. Science 280(5370):1744–6. doi:10.1126/science.280.5370.1744

163. Poncharal P, Wang Z, de Heer W (1999) Conductance quantization in multiwalled carbon nanotubes. Eur Phys J D 9(1):77–9

164. Choi HJ, Ihm J, Louie SG, Cohen ML (2000) Defects, quasibound states, and quantum conductance in metallic carbon nanotubes. Phys Rev Lett 84(13):2917

165. Sanvito S, Kwon Y-K, Tománek D, Lambert CJ (2000) Fractional quantum conductance in carbon nanotubes. Phys Rev Lett 84(9):1974

166. Kong J, Yenilmez E, Tombler TW, Kim W, Dai H, Laughlin RB et al. Quantum interference and ballistic transmission in nanotube electron waveguides. Phys Rev Lett. 2001; 87(10). doi:10.1103/PhysRevLett.87.106801.

167. Urbina A, Echeverría I, Pérez-Garrido A, Díaz-Sánchez A, Abellán J. Quantum conductance steps in solutions of multiwalled carbon nanotubes. Phys Rev Lett. 2003; 90(10). doi:10.1103/PhysRevLett.90.106603.

168. Agraıt N, Yeyati AL, Van Ruitenbeek JM (2003) Quantum properties of atomic-sized conductors. Phys Rep 377(2):81–279

169. Terabe K, Hasegawa T, Nakayama T, Aono M (2005) Quantized conductance atomic switch. Nature 433(7021):47–50

170. Zhu X, Su W, Liu Y, Hu B, Pan L, Lu W et al (2012) Observation of conductance quantization in oxide-based resistive switching memory. Adv Mater 24(29):3941–6. doi:10.1002/adma.201201506

171. Chen C, Gao S, Zeng F, Wang GY, Li SZ, Song C et al (2013) Conductance quantization in oxygen-anion-migration-based resistive switching memory devices. Appl Phys Lett 103(4):043510. doi:10.1063/1.4816747

172. Sakamoto T, Sunamura H, Kawaura H, Hasegawa T, Nakayama T, Aono M (2003) Nanometer-scale switches using copper sulfide. Appl Phys Lett 82(18):3032. doi:10.1063/1.1572964

173. Liao ZM, Hou C, Zhao Q, Wang DS, Li YD, Yu DP (2009) Resistive switching and metallic-filament formation in Ag_2S nanowire transistors. Small 5(21):2377–81. doi:10.1002/smll.200900642

174. 1Yun E-J, Becker MF, Walser RM (1993) Room temperature conductance quantization in V||amorphous-V2O5||V thin film structures. Appl Phys Lett 63(18):2493. doi:10.1063/1.110459

175. Jameson JR, Gilbert N, Koushan F, Saenz J, Wang J, Hollmer S et al (2011) One-dimensional model of the programming kinetics of conductive-bridge memory cells. Appl Phys Lett 99(6):063506. doi:10.1063/1.3623485

176. Xue K-H, Paz de Araujo CA, Celinska J, McWilliams C (2011) A non-filamentary model for unipolar switching transition metal oxide resistance random access memories. J Appl Phys 109(9):091602. doi:10.1063/1.3581193

177. Cartoixà X, Rurali R, Suñé J. Transport properties of oxygen vacancy filaments in metal/crystalline or amorphous HfO_2/metal structures. Phys Rev B: Condens Matter. 2012; 86(16). doi:10.1103/PhysRevB.86.165445.

178. Liu X, Liu Y, Chen W, Li J, Liao L (2012) Ferroelectric memory based on nanostructures. Nanoscale Res Lett 7(1):285. doi:10.1186/1556-276X-7-285

179. Miranda E, Kano S, Dou C, Kakushima K, Suñé J, Iwai H (2012) Nonlinear conductance quantization effects in CeO_x/SiO_2-based resistive switching devices. Appl Phys Lett 101(1):012910. doi:10.1063/1.4733356

180. Tappertzhofen S, Valov I, Waser R (2012) Quantum conductance and switching kinetics of AgI-based microcrossbar cells. Nanotechnology 23(14):145703. doi:10.1088/0957-4484/23/14/145703

181. Tsuruoka T, Hasegawa T, Terabe K, Aono M (2012) Conductance quantization and synaptic behavior in a Ta_2O_5-based atomic switch. Nanotechnology 23(43):435705. doi:10.1088/0957-4484/23/43/435705

182. Wagenaar JJT, Morales-Masis M, van Ruitenbeek JM (2012) Observing "quantized" conductance steps in silver sulfide: Two parallel resistive switching mechanisms. J Appl Phys 111(1):014302. doi:10.1063/1.3672824

183. Gao S, Zeng F, Chen C, Tang G, Lin Y, Zheng Z et al (2013) Conductance quantization in a Ag filament-based polymer resistive memory. Nanotechnology 24(33):335201. doi:10.1088/0957-4484/24/33/335201

184. Ielmini D, Cagli C, Nardi F, Zhang Y (2013) Nanowire-based resistive switching memories: devices, operation and scaling. J Phys D Appl Phys 46(7):074006. doi:10.1088/0022-3727/46/7/074006

185. Liu D, Cheng H, Zhu X, Wang G, Wang N (2013) Analog memristors based on thickening/thinning of Ag nanofilaments in amorphous manganite thin films. ACS Appl Mater Interfaces 5(21):11258–64. doi:10.1021/am403497y

186. Mehonic A, Vrajitoarea A, Cueff S, Hudziak S, Howe H, Labbe C et al (2013) Quantum conductance in silicon oxide resistive memory devices. Sci Rep 3:2708. doi:10.1038/srep02708

187. Saura X, Miranda E, Jiménez D, Long S, Liu M, Rafí JM et al (2013) Threshold switching and conductance quantization in Al/HfO_2/Si(p) structures. Jpn J Appl Phys 52(4S):04CD6. doi:10.7567/jjap.52.04cd06

188. Syu Y-E, Chang T-C, Lou J-H, Tsai T-M, Chang K-C, Tsai M-J et al (2013) Atomic-level quantized reaction of HfO_X memristor. Appl Phys Lett 102(17):172903. doi:10.1063/1.4802821

189. Hu C, McDaniel MD, Posadas A, Demkov AA, Ekerdt JG, Yu ET (2014) Highly controllable and stable quantized conductance and resistive switching mechanism in single-crystal TiO_2resistive memory on silicon. Nano Lett 14(8):4360–7. doi:10.1021/nl501249q

190. Lv H, Xu X, Sun P, Liu H, Luo Q, Liu Q et al (2015) Atomic view of filament growth in electrochemical memristive elements. Sci Rep 5:13311. doi:10.1038/srep13311

191. Geresdi A, Halbritter A, Gyenis A, Makk P, Mihály G (2011) From stochastic single atomic switch to nanoscale resistive memory device. Nanoscale 3(4):1504–7

192. Hajto J, Rose MJ, Snell AJ, Osborne IS, Owen AE, Lecomber PG. Quantised electron effects in metal/a-Si:H/metal thin film structures. J Non-Cryst Solids. 1991;137–138, Part 1(0):499–502. http://dx.doi.org/10.1016/S0022-3093(05)80164-4.

193. Jameson JR, Gilbert N, Koushan F, Saenz J, Wang J, Hollmer S et al (2012) Quantized conductance in Ag/GeS_2/W conductive-bridge memory cells. IEEE Electron Device Lett 33(2):257–9

194. Hu C, McDaniel MD, Ekerdt JG, Yu ET (2013) High ON/OFF ratio and quantized conductance in resistive switching of on silicon. IEEE Electron Device Lett 34(11):1385–7

195. Tappertzhofen S, Linn E, Menzel S, Kenyon AJ, Waser R, Valov I (2015) Modeling of quantized conductance effects in electrochemical metallization cells. IEEE Trans Nanotechnol 14(3):505–12. doi:10.1109/TNANO.2015.2411774

196. Van KAMPENNG (1976) The expansion of the master equation. Adv Chem Phys 34:245–309. doi:10.1002/9780470142530.ch5

197. Xia J-B (1992) Quantum waveguide theory for mesoscopic structures. Phys Rev B Condens Matter 45(7):3593–9. doi:10.1103/PhysRevB.45.3593

198. Landauer R (1957) Spatial variation of currents and fields due to localized scatterers in metallic conduction. IBM J Res Dev 1(3):223–31

199. Büttiker M, Imry Y, Landauer R, Pinhas S (1985) Generalized many-channel conductance formula with application to small rings. Phys Rev B Condens Matter 31(10):6207–15. doi:10.1103/PhysRevB.31.6207

200. Landauer R (1989) Conductance determined by transmission: probes and quantised constriction resistance. J Phys Condens Matter 1(43):8099

201. Ciraci S, Tekman E (1989) Theory of transition from the tunneling regime to point contact in scanning tunneling microscopy. Phys Rev B Condens Matter 40(17):11969–72. doi:10.1103/PhysRevB.40.11969

202. Brandbyge M, Schiøtz J, Sørensen MR, Stoltze P, Jacobsen KW, Nørskov JK et al (1995) Quantized conductance in atom-sized wires between two metals. Phys Rev B Condens Matter 52(11):8499–514. doi:10.1103/PhysRevB.52.8499

203. Milliken F, Umbach C, Webb R (1996) Indications of a Luttinger liquid in the fractional quantum Hall regime. Solid State Commun 97(4):309–13

204. Brandbyge M, Jacobsen KW, Nørskov JK (1997) Scattering and conductance quantization in three-dimensional metal nanocontacts. Phys Rev B Condens Matter 55(4):2637

205. Mozos J-L, Wan C, Taraschi G, Wang J, Guo H (1997) Quantized conductance of Si atomic wires. Phys Rev B Condens Matter 56(8):R4351

206. Lang N, Avouris P (1998) Oscillatory conductance of carbon-atom wires. Phys Rev Lett 81(16):3515

207. Brandbyge M, Sørensen MR, Jacobsen KW (1997) Conductance eigenchannels in nanocontacts. Phys Rev B Condens Matter 56(23):14956

208. Todorov T, Sutton A (1993) Jumps in electronic conductance due to mechanical instabilities. Phys Rev Lett 70(14):2138

209. García-Martín A, Torres J, Sáenz J (1996) Finite size corrections to the conductance of ballistic wires. Phys Rev B Condens Matter 54(19):13448

210. Lang N (1995) Resistance of atomic wires. Phys Rev B Condens Matter 52(7):5335

211. Maslov DL, Stone M (1995) Landauer conductance of Luttinger liquids with leads. Phys Rev B Condens Matter 52(8):R5539–R42. doi:10.1103/PhysRevB.52.R5539

212. Miranda EA, Walczyk C, Wenger C, Schroeder T (2010) Model for the resistive switching effect in MIM structures based on the transmission properties of narrow constrictions. IEEE Electron Device Lett 31(6):609–11

213. Tappertzhoven S, Linn E, Menzel S, Waser R, Valov I. Quantum size effects and non-equilibrium states in nanoscale silicon dioxide based resistive switches. IEEE Silicon Nanoelectronic Workshop; Jun 8–10 Honolulu, Hawaii, USA 2014.

214. Miranda E, Sune J (2001) Analytic modeling of leakage current through multiple breakdown paths in SiO_2 films. IEEE International Reliability Physics Symposium

215. Zhang M, Long S, Wang G, Xu X, Li Y, Liu Q et al (2014) Set statistics in conductive bridge random access memory device with Cu/HfO_2/Pt structure. Appl Phys Lett 105(19):193501. doi:10.1063/1.4901530

216. Shibing L, Xiaojuan L, Cagli C, Perniola L, Miranda E, Ming L et al (2013) A model for the set statistics of RRAM inspired in the percolation model of oxide breakdown. IEEE Electron Device Lett 34(8):999–1001. doi:10.1109/LED.2013.2266332

217. Degraeve R, Goux L, Clima S, Govoreanu B, Chen YY, Kar GS et al (2012) Modeling and tuning the filament properties in RRAM metal oxide stacks for optimized stable cycling. IEEE International Symposium on VLSI Technology, Systems, and Applications (VLSI-TSA)

218. Degraeve R, Roussel P, Goux L, Wouters D, Kittl J, Altimime L et al (2010) Generic learning of TDDB applied to RRAM for improved understanding of conduction and switching mechanism through multiple filaments. IEEE International Electron Devices Meeting (IEDM)

219. Zhang M, Wang G, Long S, Yu Z, Li Y, Xu D et al. Unpublished: a physical model for the statistics of set switching time of resistive RAM measured by width-adjusting pulse operation method (accepted, manuscript ID: EDL-2015-07-1267.R3.). IEEE Electron Device Lett. 2015.

220. Bauer U, Yao L, Tan AJ, Agrawal P, Emori S, Tuller HL et al (2015) Magneto-ionic control of interfacial magnetism. Nature Mater 14(2):174–81. doi:10.1038/nmat4134

221. Valov I, Linn E, Tappertzhofen S, Schmelzer S, van den Hurk J, Lentz F et al (2013) Nanobatteries in redox-based resistive switches require extension of memristor theory. Nat Commun 4:1771. doi:10.1038/ncomms2784

222. Lv H, Xu X, Liu H, Liu R, Liu Q, Banerjee W et al. Evolution of conductive filament and its impact on reliability issues in oxide-electrolyte based resistive random access memory. Sci Rep. 2015; 5. doi:10.1038/srep07764

223. Wang M, Bi C, Li L, Long S, Liu Q, Lv H et al (2014) Thermoelectric Seebeck effect in oxide-based resistive switching memory. Nat Commun 5:4598

224. Rahaman SZ, Maikap S, Chiu H-C, Lin C-H, Wu T-Y, Chen Y-S et al (2010) Bipolar resistive switching memory using Cu metallic filament in $Ge_{0.4}Se_{0.6}$ solid electrolyte. Electrochem Solid-State Lett 13(5):H159–H62

225. Kaeriyama S, Sakamoto T, Sunamura H, Mizuno M, Kawaura H, Hasegawa T et al (2005) A nonvolatile programmable solid-electrolyte nanometer switch. IEEE J Solid State Circuits 40(1):168–76. doi:10.1109/JSSC.2004.837244

226. Schoen DT, Xie C, Cui Y (2007) Electrical switching and phase transformation in silver selenide nanowires. J Am Chem Soc 129(14):4116. doi:10.1021/ja068365s

227. Schindler C, Szot K, Karthäuser S, Waser R (2008) Controlled local filament growth and dissolution in Ag–Ge–Se. physica status solidi (RRL). Rapid Res Lett 2(3):129–31. doi:10.1002/pssr.200802054

228. Guo HX, Yang B, Chen L, Xia YD, Yin KB, Liu ZG et al. Resistive switching devices based on nanocrystalline solid electrolyte $(AgI)_{0.5}(AgPO3)_{0.5}$. Appl Phys Lett. 2007; 91(24). doi:10.1063/1.2825273.

229. Wang Y, Liu Q, Long S, Wang W, Wang Q, Zhang M et al. Investigation of resistive switching in Cu-doped HfO_2 thin film for multilevel non-volatile memory applications. Nanotechnology. 2010; 21(4). doi:10.1088/0957-4484/21/4/045202.

230. Zhang M, Long S, Wang G, Liu R, Xu X, Li Y et al (2014) Statistical characteristics of reset switching in Cu/HfO_2/Pt resistive switching memory. Nanoscale Res Lett 9:694

231. Guan W, Long S, Liu Q, Liu M, Wang W (2008) Nonpolar nonvolatile resistive switching in Cu doped ZrO_2. IEEE Electron Device Lett 29(5):434–7. doi:10.1109/LED.2008.919602

232. Sun J, Liu Q, Xie H, Wu X, Xu F, Xu T et al (2013) In situ observation of nickel as an oxidizable electrode material for the solid-electrolyte-based resistive random access memory. Appl Phys Lett 102(5):053502, http://dx.doi.org/10.1063/1.4790837

233. Schindler C, Thermadam SCP, Waser R, Kozicki MN (2007) Bipolar and unipolar resistive switching in Cu-doped SiO_2. IEEE Trans Electron Devices 54(10):2762–8. doi:10.1109/TED.2007.904402

234. Kozicki MN, Gopalan C, Balakrishnan M, Mitkova M (2006) A low-power nonvolatile switching element based on copper-tungsten oxide solid electrolyte. IEEE Trans Nanotechnol 5(5):535–44. doi:10.1109/TNANO.2006.880407

235. Sakamoto T, Banno N, Iguchi N, Kawaura H, Sunamura H, Fujieda S et al (2007) A Ta_2O_5solid-electrolyte switch with improved reliability. IEEE International Symposium on VLSI Technology, Systems, and Applications (VLSI-TSA)

236. Aratani K, Ohba K, Mizuguchi T, Yasuda S, Shiimoto T, Tsushima T et al (2007) A novel resistance memory with high scalability and nanosecond switching. IEEE International Electron Devices Meeting (IEDM)

237. Shibing L, Xiaojuan L, Tianchun Y, Cagli C, Perniola L, Miranda E et al (2013) Cycle-to-cycle intrinsic RESET statistics in HfO_2 based unipolar RRAM devices. IEEE Electron Device Lett 34(5):623–5. doi:10.1109/LED.2013.2251314

238. Jung K, Seo H, Kim Y, Im H, Hong J, Park J-W et al (2007) Temperature dependence of high- and low-resistance bistable states in polycrystalline NiO films. Appl Phys Lett 90(5):052104, http://dx.doi.org/10.1063/1.2437668

239. Jung R, Lee M-J, Seo S, Kim DC, Park G-S, Kim K et al (2007) Decrease in switching voltage fluctuation of Pt/NiO_X/Pt structure by process control. Appl Phys Lett 91(2):022112. doi:10.1063/1.2755712

240. Yun J-B, Kim S, Seo S, Lee M-J, Kim D-C, Ahn S-E et al (2007) Random and localized resistive switching observation in Pt/NiO/Pt. physica status solidi (RRL). Rapid Res Lett 1(6):280–2. doi:10.1002/pssr.200701205

241. Long S, Cagli C, Ielmini D, Liu M, Sune J (2011) Reset statistics of NiO-based resistive switching memories. IEEE Electron Device Lett 32(11):1570–2. doi:10.1109/LED.2011.2163613

242. Russo U, Ielmini D, Cagli C, Lacaita AL (2009) Filament conduction and reset mechanism in NiO-based resistive-switching memory (RRAM) devices. IEEE Trans Electron Devices 56(2):186–92. doi:10.1109/TED.2008.2010583

243. Cagli C, Nardi F, Ielmini D (2009) Modeling of set/reset operations in NiO-based resistive-switching memory devices. IEEE Trans Electron Devices 56(8):1712–20. doi:10.1109/TED.2009.2024046

244. Ielmini D, Nardi F, Cagli C (2011) Physical models of size-dependent nanofilament formation and rupture in NiO resistive switching memories. Nanotechnology 22(25):254022. doi:10.1088/0957-4484/22/25/254022

245. Shima H, Takano F, Akinaga H, Tamai Y, Inoue IH, Takagi H (2007) Resistance switching in the metal deficient-type oxides: NiO and CoO. Appl Phys Lett 91(1):012901–3

246. Yasuhara R, Fujiwara K, Horiba K, Kumigashira H, Kotsugi M, Oshima M et al (2009) Inhomogeneous chemical states in resistance-switching devices with a planar-type Pt/CuO/Pt structure. Appl Phys Lett 95(1):012110

247. Inoue IH, Yasuda S, Akinaga H, Takagi H (2008) Nonpolar resistance switching of metal/binary-transition-metal oxides/metal sandwiches: homogeneous/inhomogeneous transition of current distribution. Phys Rev B Condens Matter 77(3):035105

CHAPTER 17

ELECTRICAL INSTABILITY IN PENTACENE TRANSISTORS WITH MYLAR AND PMMA/MYLAR GATE DIELECTRICS TRANSFERRED BY LAMINATION PROCESS

Abdou Karim Diallo, Abdoul Kadri Diallo, Diouma Kobor, Marcel Pasquinelli

Keywords: Organic Transistor, Mylar, Lamination, Pentacene, Bias Stress

INTRODUCTION

Organic semiconductors have enabled the emergence of new exciting research field called organic electronics. This enthusiasm for devices based on organic materials is partly driven by the increased interest in both academic and industry thanks to a low manufacturing cost, mass production, compatibility with flexible substrates compared to their equivalent inorganic materials. Transistor remains unambiguously a key element of electronics devices and is especially present in most new technologies. The considerable development of organic field effect transistors (OFETs)—in which the active layer is organic semiconductor, enables to foresee applications in fields such as flat display driving, radio frequency identification (RFID) tag and sensors [1] - [4] .

Many interesting techniques had been used for fabrication of organic transistor devices, and each of them showed advantages and disadvantages. Although organic semiconductor can be processed on flexible substrate they generally do not withstand conventional lithography techniques. This gives rise the development of alternative deposition and patterning methods leading to the concept of soft lithography which is introduced in the end of the 1990s [5] and since highly used. Among these different techniques we could mention soft contact lamination which was applied to organic light emitting diodes [6] and OFETs [7] . Another method such as microcontact printing has been used for patterning organic FETs [8] . Other transfer methods have been reported in the literature as electrode peeling transfer [9] [10] or metal transfer printing [11] . The lamination technique was proposed to transfer an organic semiconductor active layer previously deposited on a donor layer onto a receiver layer acting as the gate insulator [12] . Previous works have shown that Mylar foil can be successfully laminated serving as dielectric [13] and some studies use it both as substrate and dielectric to fabricate transistors [14] - [16] .

However, a major drawback of OFETs is the lack of stability making them a critical issue that must be addressed before their commercialization. Hysteresis is an instability that can be frequently observed during device operation. It appears as a difference in the source-drain current values observed during forward and backward scans of the gate or drain voltages. Therefore, minimizing the hysteresis effect is a priority in electronic circuits.

It was also reported that a prolonged polarization of gate electrode named bias stress tends to shift the threshold voltage, degrades the subthreshold parameters [17] reducing the reliability. This complex bias stress phenomenon has been a subject of intense research during several years and remains poorly understood. Under gate bias-stress, the drain current decreases with time due to charge trappings. Several authors considered these traps located at the semiconductor/dielectric interface, or in the gate dielectric, or within the semiconductor active layer [18] - [20] . The bias stress is highly dependent on the

Citation: Karim Diallo, A. , Kadri Diallo, A. , Kobor, D. and Pasquinelli, M. (2016) Electrical Instability in Pentacene Transistors with Mylar and PMMA/Mylar Gate Dielectrics Transferred by Lamination Process. Journal of Applied Mathematics and Physics, 4, 1202-1209. doi: 10.4236/jamp.2016.47125.

involved materials, the device geometry and configuration, the polarization conditions, the gate dielectric and semiconductor deposition temperatures, and the atmosphere in which the device is operating.

In this paper, we report on electrical instabilities in pentacene-based transistors with Mylar and PMMA/Mylar gate dielectrics transferred by a lamination process in ambient environment. Special emphasis is given on comparing the two types of devices in ON state and subthreshold regions under gate bias stress. Device with PMMA/Mylar dielectric exhibited good performances and stability in ON state, while in depletion regime, the subthreshold parameters were degraded. In contrast, the device using only Mylar as dielectric has an opposite behavior. In addition, both devices showed onset voltage shifts in opposite direction.

EXPERIMENTAL DETAILS

The substrate was a 175 nm thick PET foil (DuPont Melinex ST 504) thoroughly sonically cleaned. The different transistor fabrication steps are described in Figure 1(b). Figure 1(a) is a photograph of transistors realized on PET substrate. Two different types of devices in bottom gate configurations were processed; Device 1 and Device 2. A 70 nm thick Al gate was first deposited by vacuum evaporation and patterned. Then a 900 nm thick Mylar foil (DuPont) was stacked on the patterned substrate and pressed between two hot plates. The quality of the transfer was observed to be strongly relying on both the temperature (in the range of 70°C - 110°C) and the pressure. This Mylar is a thin polyethylene terephthalate sheet with dielectric constant close to silicon oxide.

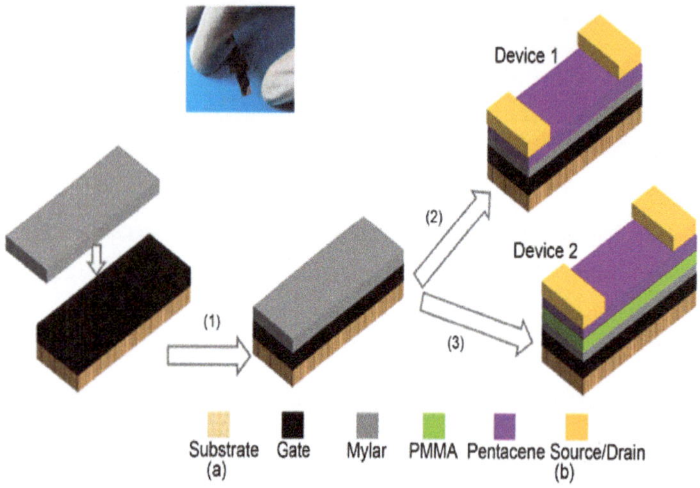

Figure 1. Schematic illustration of Mylar lamination for transistor fabrication: (a) photograph of electrical devices, (b) Pro- cess flow of transistor devices fabricated on a PET substrate, on which Al gate was deposited. (1) Mylar was then pressed; (2) Device 1 obtained after pentacene and source/drain evaporation; (3) PMMA was spun on the top of the laminated Mylar dielectric, Device 2 is completed by evaporating pentacene and source and drain contacts.

After pressure release, a 70 nm pentacene film is evaporated onto the Mylar foil at a substrate temperature and a deposition rate of 70°C and 0.02 nm/respectively. The device is completed by the deposition of Au source and drain electrodes through a shadow mask. The channel width and length were, respectively, 2 mm and 50 μm. Device 2 is different to Device 1 by spin coating 80 nm thick PMMA layer on the top of the transferred Mylar foil. PMMA was annealed in air for 20 min at 120°C. Pentacene was then deposited on PMMA followed by the evaporation of top Au source and drain contacts. PMMA is also expected to slightly planarize the Mylar surface and thus lead to a smoother surface. The process flow Mylar lamination for transistors fabrication is represented inFigure 1. Current—voltage characteristics were obtained at room temperature in air and in the dark with a probe station and two Keithley 2400 sourcemeter under LABVIEW® environment. The mobility was μ_{sat} extracted from the saturation region of the transfer curves with the equation:

$$I_{D,sat} = \frac{W}{2L} \mu_{sat} C \left(V_G - V_T\right)^2$$

where $I_{D,sat}$ is drain current in the saturation regime; W/L is the width to length ratio; C is the capacitance per unit area; V_G the gate voltage and V_T the threshold voltage.

The morphology of the pentacene thin films was studied in an Amplitude Modulation mode Atomic Force Microscopy (AM-AFM) using the AFM Solver P-7, “stand alone” Smena-B (NT-MDT, Russia), with typical spring constant k of 22 N/m and tip radius of 10 nm. This mode was shown to be more appropriate than the contact mode to image soft materials such as polymers, functionalized surface and biological objects, in air.

RESULTS AND DISCUSSION

Electrical Characterization and Pentacene Morphology

Bottom-Gate/Top-Contact (BG/TC) thin-film transistors were fabricated as described in the experimental part. All measurements were performed in air at room temperature and in the dark. For all two devices, the organic transistors operate in the accumulation mode since the gate electrode is biased negatively with respect to the grounded source electrode. Drain current (I_D) is almost linear with drain voltage at low V_D, whereas it tends to saturate at higher drain voltage due to the pinch off of the accumulation layer. Figure 2(a) and Figure 2(b) show typical output characteristics curves of Device 1 and Device 2 with well-defined linear and saturation regimes.

To study the electrical instability related to hysteresis, we performed electrical measurements in forward and back scans. As we can see, Device 1 exhibits a hysteresis in forward and reverse characteristics (Figure 2(a)) while for Device 2 no hysteresis is recorded. The Mylar film is a low-k (dielectric constant) polymer dielectric—with k = 3.25 at 1 kHz and the molecular structure shows that it is non-polar—the observed hysteresis could be attributed to: 1) impurities present in the semiconductor or at the dielectric interface, such as water molecules diffusing through grain boundaries [21] (small grains of pentacene were observed during the growth on Mylar) or 2) structural defects initially present on the Mylar surface. Indeed, when Mylar foil was pressed on the aluminum gate, the mechanical stress made the surface rougher. These structural defects and a rough surface constitute a prosperous environment for charge trapping. Drain current in forward is higher than in reverse for the same gate voltage, a sign of hole trapping. The lack of hysteresis observed for Device 2 confirms a good interface between PMMA and pentacene as observed by several authors [22] [23] . The electrical performance parameters are summarized in Table 1. A key parameter that assesses the interface quality is the subthreshold slope (S); it is extracted from the transfer characteristic in logarithmic scale in saturation regime. The values are 19 V/decade and 8 V/decade respectively for Device 1 and Device 2. The highest S value is attributed to the device with Mylar dielectric, which is a confirmation of poor interface. Threshold voltage, I_{on}/I_{off} ratio and mobility are −25 V, 1.3 × 10^2 and 4.1 × 10^{-3} cm^2·V^{-1}·s^{-1} respectively for Device 1 while Device 2 exhibited −20 V, 5 × 10^3 and 5.1 × 10^{-2} cm^2·V^{-1}·s^{-1} respectively for the threshold voltage, I_{on}/I_{off} ratio and mobility.

To better understand this difference, AFM images were performed in order to identify the morphology of pentacene layer that can explain the relation between they nanostructures and the charges carriers mobility.

Figure 3(a) gave AFM image of pentacene on Mylar with a surface roughness about 17 nm and peak-to-val- ley value up to 200 nm, we also observe small grain sizes (200 - 350 nm) with low connectivity, as consequence,

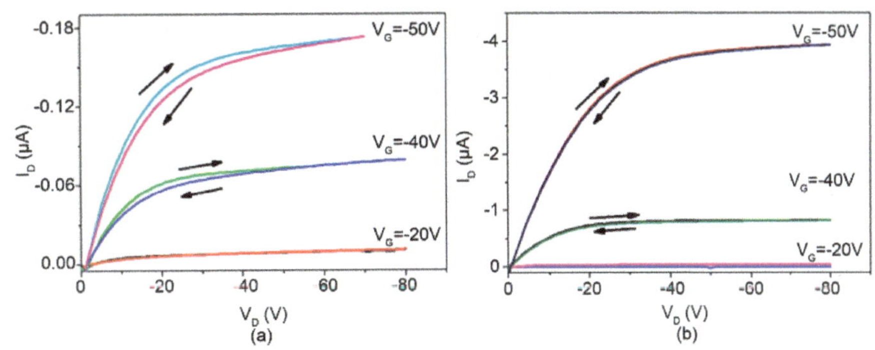

Figure 2. Forward and reverse output characteristics of transistor devices using: (a) Mylar and (b) PMMA/ Mylar as dielectrics at different gate voltages.

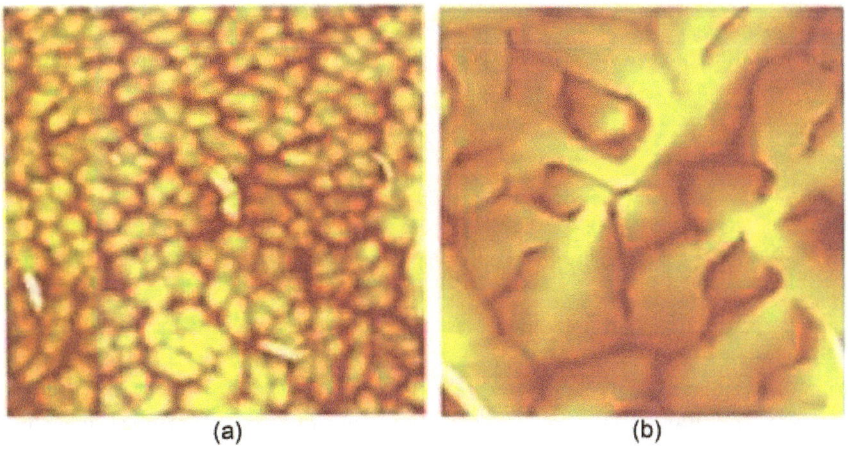

Figure 3. AFM images (4 × 4 µm^2) of 70 nm of pentacene film evaporated on (a) Mylar and (b) PMMA/Mylar.

Table 1. Summary of the electrical parameters of Device 1 and 2 before bias stress.

Devices	V_T (V)	µ (cm^2·V^{-1}·s^{-1})	I_{on}/I_{off}	S (V/decade)
Device 1	−25	4.1 × 10^{-3}	1.3 × 10^2	19
Device 2	−20	5.3 × 10^{-2}	5 × 10^3	8

the combination of these two effects leads to electrical parameters degradation [24] [25] such as mobility, threshold voltage and I_{on}/I_{off} ratio. It has been reported that, the diffusion of impurities such as polar water molecules into grain boundaries changes the intermolecular interactions in those regions, increasing energy barrier for charge carrier intergrain transport or ions associated with water screen the electric field at the channel and lower the concentration of gate-induced mobile carriers [26] . All of the mechanisms would lead to reduce the performances of Device 1.

In contrast, thanks to the PMMA layer, Device 2 exhibits improved performances. Figure 3(b) showed large dendritic grain size (2 - 5 μm) with surface roughness of PMMA/Mylar around 12 nm. The largest grain sizes allow reducing the number of grain boundaries acting as charge traps during the transport. Since the carrier mobility is very sensitive to the molecular ordering of the pentacene and the carrier trapping at grain boundaries, the large grain sizes of pentacene on PMMA give rise to high field-effect mobility [25]. These results clearly point out the highly beneficial PMMA layer on the transport properties in the pentacene channel throughout the ON state regime and as long as bias stress was not applied.

Bias Stress Effect

To study the bias stress-induced degradation in transistor devices, a constant gate voltage was applied to the organic transistors. To determine the changes in the transfer curves, the applied gate voltage is interrupted at short time intervals by a sweep of the gate. This technique allowed us to obtain the electrical parameters. Figure 4(a) shows the square root of transfer characteristics before and after different stress times for Device 1. We observe that the applied stress shifts the transfer characteristics in the negative gate voltage direction (with $\Delta V_T = -10$ V) and the shape does not appreciably change with stress time. From the plot, it is clear that the slope of the linear part of the curve, which is proportional to the field effect mobility in the saturation regime, is similar for all curves. The extracted mobility remains minimally affected after the application of gate bias stress (−50 V/3000 s).

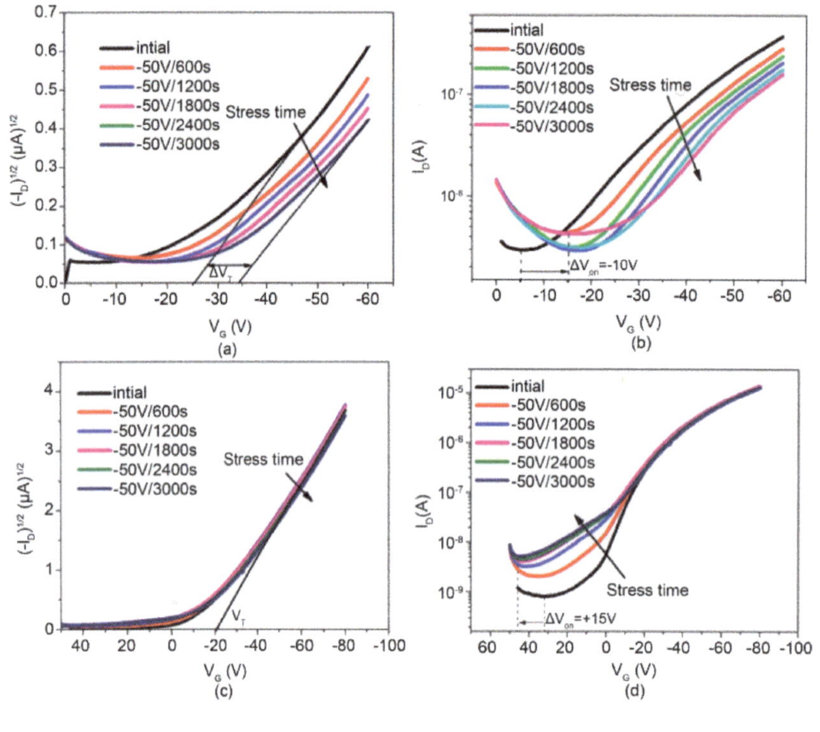

Figure 4. Transfer characteristics of Device 1 and Device 2 before and after bias stress for 3000 s. −50 V was applied on the gate voltage (a) square root for Device 1, (b) logarithmic scale for Device 1, (c) square root for Device 2, (d) logarithmic scale for Device 2. The curves are obtained in saturation regime at $V_D = -70$ V.

The negative threshold voltage shift associated to negative gate bias is a consequence of holes being trapped. Figure 4(b) is the transfer characteristics in logarithmic scale of Device 1 before and after the bias stress, all curves have a similar shape. It is worth noting that the subthreshold slope of the device does not change even after the device has undergone bias stressing. This indicates that no additional interface states are created at the channel/dielectric interface after the device was stressed; however, a slight increase of I_{off} current was recorded with noticeable onset voltage shift.

Regarding the transistor with PMMA/Mylar as dielectric, both mobility and threshold voltage remain unchanged after negative bias stress of −50 V for 3000 s as it was operating in ON state (Figure 4(c)). A clear difference appears between the device using Mylar as dielectric and that with PMMA/Mylar under bias stress in ON state.

The grain boundaries play a crucial role in the bias stress effect [26] [27] , and it was shown that in pentacene- based OTFTs the charge trapping was found to occur preferentially in intergrain regions of organic thin film [28] . One plausible explanation results from the difference of morphologies as observed by AFM images. Pentacene morphology on Mylar being less ordered with small grain size, leads to increase the amount of grain boundaries.

These latter are synonyms of traps, a prolonged polarization of the gate electrode permanently creates a constant electric field, which promotes a continuous charge trapping. However, the large grain sizes (less grain boundaries) on PMMA/Mylar reduce the number of traps, accordingly, the charge trappings decrease.

Figure 4(d) is the transfer characteristics of Device 2 in logarithmic scale after several stresses. As long as bias stress was applied, the electrical parameters in subthreshold region were degraded. As can be seen, I_{off} current increases nearly by one order of magnitude and the S is increased from 8 V/decade to 20 V/decade (while for Device 1, S remains unchanged with a slight increase of I_{off} current). The degradation of subthreshold slope as well as the subthreshold current increase can be explained by interface state creation. The I_{off} current increase cannot be attributed to the gate leakage as it was enough thick in both Device 1 and Device 2; moreover all devices were characterized in air with constant humidity rate. These different behaviors observed in subthreshold region for both devices might be the bias-assisted moisture/oxygen diffusion [17] . The combination of moisture and bias stress leads to increase the depletion current and the subthreshold slope; this is exactly what we observe in Figure 4(d).

Another key parameter that must be addressed is the onset voltage V_{on} which is critically influenced by the semiconductor/dielectric interface. The onset voltage is determined as the minimum of the drain courant versus gate voltage in logarithmic scale and it is highly desirable to get a near-zero V_{on} in circuitry. After bias stress the onset voltage of Device 2 is shifted towards positive gate values with $\Delta V_{on} = +15$ V, while for Device 1 it is negatively shifted and $\Delta V_{on} = -10$ V after 3000 s under bias stress, showing an opposite behavior. The onset voltage shift of Device 1 is exactly similar to the threshold voltage shift $\Delta V_{on} = \Delta V_T = -10$ V. According to Knipp et al. [29] , the threshold voltage shift could be attributed to donor states. Water molecules were shown to be localized between pentacene molecules and create trap states in the band gap [30] , and the orientation of water dipoles changes the local polarization of the pentacene monolayers, creating trap states in the band gap [31] . In contrast, Oxygen is known to create deep acceptor levels and to increase the residual doping level of the organic semiconductor, both contributing to a positive shift of the onset voltage [32] [33] , this may explain what we observed in Device 2.

CONCLUSION

Electrical performances and instabilities of transistors based on pentacene semiconductor with laminated Mylar and PMMA/Mylar dielectrics have been studied. Hysteresis has been observed with the device using only Mylar as dielectric. In contrast, no hysteresis has been recorded for the device using PMMA/Mylar double layer. Hysteresis apparition was related to the pentacene grain sizes; the large gain of pentacene obtained with PMMA prevent any moisture diffusion compared to pentacene growth on Mylar showing small grain sizes and more grain boundaries. These latter are considered as an ideal environment for charge trapping. Negative bias stress performed in ambient air leads to electrical parameters degradation depending on operating regimes. In ON state the transistor with Mylar exhibited a high threshold voltage shift with a minimal change of mobility and I_{on}/I_{off} ratio after bias stress. However, the transistor using PMMA/Mylar showed good electrical performances in ON state but in subthreshold regime, the depletion current increased as well as the subthreshold swing due to the doping and bias-assisted moisture/oxygen diffusion leading to defect creation. Both devices showed onset voltage shift in opposite direction. Mylar lamination is a powerful technique for transistor fabrication whether in bottom or top gate configurations for mass fabrication in ambient environment. Our research is in progress to optimize the electrical parameters especially by improving the dielectric lamination and to understand the complex bias stress phenomenon that remains so far poorly understood in order to minimize the transistor instability.

ACKNOWLEDGEMENTS

A. K. D acknowledges the FIRST (Fonds d'Impulsion pour la Recherche Scientifique et Technique) program for financial support.

REFERENCES

1. Sekitani, T., Zschieschang, U., Klauk, H. and Someya, T. (2010) Flexible Organic Transistors and Circuits with Extreme Bending Stability. Nature Materials, 9, 1015-1022. http://dx.doi.org/10.1038/nmat2896
2. Cho, J.H., Lee, J., Xia, Y., Kim, B., He, Y. and Renn, M.J. (2008) Printable Ion-Gel Gate Dielectrics for Low-Voltage Polymer Thin-Film Transistors on Plastic. Nature Materials, 7, 900-906. http://dx.doi.org/10.1038/nmat2291
3. Sokolov, A.N., Roberts, M.E. and Bao, Z. (2009) Fabrication of Low-Cost Electronic Biosensors. Materials Today, 12, 12-20. http://dx.doi.org/10.1016/S1369-7021(09)70247-0
4. Huang, W.G., Diallo, A.K., Dailey, J.L., Besar, K. and Katz, H.E. (2015) Electrochemical Processes and Mechanistic Aspects of Field-Effect Sensors for Biomolecules. Journal of Materials Chemistry C, 3, 6445-6470. http://dx.doi.org/10.1039/C5TC00755K
5. Xian, Y. and Whitesides, G. (1998) Soft Lithography. Angewandte Chemie International Edition, 37, 550-575. http://dx.doi.org/10.1002/(SICI)1521-3773(19980316)37:53.0.CO;2-G
6. Lee, T.-W., Zaumseil, J., Bao, Z., Hsu, J.W.P. and Rogers, J.A. (2004) Organic Light-Emitting Diodes Formed by Soft Contact Lamination. Proceedings of the National Academy of Sciences of the United States of America, 101, 429-433. http://dx.doi.org/10.1073/pnas.0304179101
7. Ling, M.M. and Bao, Z. (2004) Thin Film Deposition, Patterning, and Printing in Organic Thin Film Transistors. Chemistry of Materials, 16, 4824-4840. http://dx.doi.org/10.1021/cm0496117
8. Parashkov, R., Becker, E., Riedl, T., Johannes, H.-H. and Kowalsky, W. (2005) Microcontact Printing as a Versatile Tool for Patterning Organic Field-Effect Transistors. Advanced Materials (Weinheim), 17, 1523-1527.
9. Fujita, K., Yasuda, T. and Tsutsui, T. (2003) Flexible Organic Field-Effect Transistors Fabricated by the ElectrodePeeling Transfer with an Assist of Self-

Assembled Monolayer. Applied Physics Letters, 82, 4373. http://dx.doi.org/10.1063/1.1583867

10. Parashkov, R., Becker, E., Ginev, G., Riedl, T., Johannes, H.-H. and Kowalsky, W. (2004) Flexible All-Organic FieldEffect Transistors Fabricated by Electrode-Peeling Transfer. Japanese Journal of Applied Physics, 43, L130. http://dx.doi.org/10.1143/JJAP.43.L130

11. Wang, Z., Yuan, J., Zhang, J., Xing, R., Yan, D. and Han, Y. (2003) Metal Transfer Printing and Its Application in Organic Field-Effect Transistor Fabrication. Advanced Materials (Weinheim), 15, 1009-1012. http://dx.doi.org/10.1002/adma.200304846

12. Ofuji, M., Lovinger, A.J., Kloc, C., Siegrist, T., Maliakal, A.J. and Katz, H.E. (2005) Organic Semiconductor Designed for Lamination Transfer between Polymer Films. Chemistry of Materials, 17, 5748-5753. http://dx.doi.org/10.1021/cm051616w

13. Diallo, K., Erouel, M. and Tardy, J. (2006) Pentacene Field-Effect Transistors with a Laminated Mylar™ Foil as Gate Dielectric. Applied Physics Letters, 89, Article ID: 233512. http://dx.doi.org/10.1063/1.2402901

14. Bonfiglio, A., Mameli, F. and Sanna, O. (2003) A Completely Flexible Organic Transistor Obtained by a One-Mask Photolithographic Process. Applied Physics Letters, 82, 3550. http://dx.doi.org/10.1063/1.1577216

15. Loi, A., Manunza, I. and Bonfiglio, A. (2005) Flexible, Organic, Ion-Sensitive Field-Effect Transistor. Applied Physics Letters, 86, Article ID: 103512. http://dx.doi.org/10.1063/1.1873051

16. Cosseddu, P. and Bonfiglio, A. (2006) Soft Lithography Fabrication of All-Organic Bottom-Contact and Top-Contact Field Effect Transistors. Applied Physics Letters, 88, Article ID: 023506. http://dx.doi.org/10.1063/1.2166487

17. Diallo, K., Erouel, M., Tardy, J., Andre, E. and Garden, J.-L. (2007) Stability of Pentacene Top Gated Thin Film Transistors. Applied Physics Letters, 91, Article ID: 183508. http://dx.doi.org/10.1063/1.2802039

18. Zschieschang, U., Weitz, R.T., Kern, K. and Klauk, H. (2009) Bias Stress Effect in Low-Voltage Organic Thin-Film Transistors. Applied Physics A, 95, 139-145. http://dx.doi.org/10.1007/s00339-008-5019-8

19. Powell, M.J., van Berkel, C., French, I.D. and Nichols, D.H. (1987) Bias Dependence of Instability Mechanisms in Amorphous Silicon Thin-Film Transistors. Applied Physics Letters, 51, 1242-1244. http://dx.doi.org/10.1063/1.98692

20. Conley Jr., J.F. (2010) Instabilities in Amorphous Oxide Semiconductor Thin-Film Transistors. IEEE Transactions on Device and Materials Reliability, 10, 460-475. http://dx.doi.org/10.1109/TDMR.2010.2069561

21. Simeone, D., Rapisarda, M., Fortunato, G., Valletta, A. and Mariucci, L. (2011) Influence of Structural Properties on Environmental Stability of Pentacene Thin Film Transistors. Organic Electronics, 12, 447-452. http://dx.doi.org/10.1016/j.orgel.2010.12.013

22. Deman, A.L. and Tardy, J. (2005) PMMA-Ta2O5 Bilayer Gate Dielectric for Low Operating Voltage Organic FETs. Organic Electronics, 6, 78-84. http://dx.doi.org/10.1016/j.orgel.2005.03.002

23. Ukah, N.B., Granstrom, J., Sanganna Gari, R.R., King, G.M. and Guha, S. (2011) Low-Operating Voltage and Stable Organic Field-Effect Transistors with Poly (Methyl Methacrylate) Gate Dielectric Solution Deposited from a High Dipole Moment Solvent. Applied Physics Letters, 99, Article ID: 243302. http://dx.doi.org/10.1063/1.3669696

24. Steudel, S., De Vusser, S., De Jongue, S., Janssen, D., Verlaak, S., Gence, J. and Herremans, P. (2004) Influence of the Dielectric Roughness on the Performance of Pentacene Transistors. Applied Physics Letters, 85, 4400-4402. http://dx.doi.org/10.1063/1.1815042

25. Gundlach, D.J., Lin, Y.Y., Jackson, T.N., Nelson, S.F. and Schlom, D.G. (1997) Pentacene Organic Thin-Film Transistors-Molecular Ordering and Mobility. IEEE Electron Device Letters, 18, 87-89. http://dx.doi.org/10.1109/55.556089

26. Li, D., Borkent, E.-J., Nortrup, R., Moon, H., Katz, H. and Bao, Z. (2005) Humidity Effect on Electrical Performance of Organic Thin-Film Transistors. Applied Physics Letters, 86, Article ID: 042105. http://dx.doi.org/10.1063/1.1852708

27. Häusermann, R. and Batlogg, B. (2011) Gate Bias Stress in Pentacene Field-Effect-Transistors: Charge Trapping in the Dielectric or Semiconductor. Applied Physics Letters, 99, Article ID: 083303. http://dx.doi.org/10.1063/1.3628297

28. Tello, M., Chiesa, M., Duffy, C.M. and Sirringhaus, H. (2008) Charge Trapping in Intergrain Regions of Pentacene Thin Film Transistors. Advanced Functional Materials, 18, 3907-3913. http://dx.doi.org/10.1002/adfm.200800009

29. Knipp, D., Street, R.A., Völkel, A. and Ho, J. (2003) Pentacene Thin Film Transistors on Inorganic Dielectrics: Morphology, Structural Properties, and Electronic Transport. Journal of Applied Physics, 93, 347-355. http://dx.doi.org/10.1063/1.1525068

30. Tsetseris, L. and Pantelides, S.T. (2007) Intercalation of Oxygen and Water Molecules in Pentacene Crystals: FirstPrinciples Calculations. Physical Review B, 75, Article ID: 153202. http://dx.doi.org/10.1103/PhysRevB.75.153202

31. Pernstich, K.P., Oberhorff, D., Goldmann, C. and Batlogg, B. (2005) Modeling the Water Related Trap State Created in Pentacene Transistors. Applied Physics Letters, 89, 213509. http://dx.doi.org/10.1063/1.2396924

32. Knipp, D., Muck, T., Benor, A. and Wagner, V. (2006) Environmental Stability and Electronic Transport of Pentacene Thin Film Transistors. Journal of Non-Crystalline Solids, 352, 1774-1777. http://dx.doi.org/10.1016/j.jnoncrysol.2005.11.145

33. Meijer, E.J., Detcheverry, C., Baesjou, P.J., van Veenendaal, E., de Leeuw, D.M. and Klapwijk, T.M. (2003) Dopant Density Determination in Disordered Organic Field-Effect Transistors. Journal of Applied Physics, 93, 4831-4835. http://dx.doi.org/10.1063/1.1559933

CHAPTER 18 DESIGN AND ANALYSIS OF LOW POWER HYBRID MEMRISTOR-CMOS BASED DISTINCT BINARY LOGIC NONVOLATILE SRAM CELL

Veerappan Saminathan and Kuppusamy Paramasivam

Keywords: Memristor-CMOS, Nonvolatile Memory, Power Consumption, DBL, PWL Input

INTRODUCTION

SRAM is used as a memory element because of faster operation and better performance. However, data retention leads a main problem in SRAM since data is lost when power is switched off. The volatility characteristic of conventional SRAM rises the booting time for systems utilizing SRAM either as main memory or cache. Introduction of nonvolatile memory into the main memory or cache architectures can be an effective means to decrease booting time and energy consumption. As the technology developed, new storage capabilities are required. Memristor-CMOS based SRAM cell can be a capable circuit component that would permit conventional SRAM cells to retain data when power is off without need of extra circuitry. A memristor is a fourth fundamental circuit element and can be used as a memory because of its nonvolatile characteristics, better scalability, higher packing density and low power consumption. It can be characterized by Loen Chua in 1971 as a fourth fundamental missing circuit component [1] . The other three basic components are resistor, capacitor and inductor. The main difference between the memristor and three fundamental circuit elements (R, L, and C) is its nonlinear input-output characteristics. In 2008, Hewlett Packard lab declared that the memristor was fabricated physically using two terminal titanium-di-oxides (TiO_2) [2] . HP lab clarified the first physical demonstration of a memristor, conforming Chua's theory [3] and flashing in the electronic industry.

Basically the memristor is a memory resistor whose resistance changes when a voltage is applied to this element and maintained on that particular value when the source is removed. It is suitable to use the name memristor to explain a "memristive device". It links magnetic flux to electric charge in similar to the resistor links voltage to current, capacitor links voltage to charge and inductor links flux to current. By exploiting memristor as storage element in a memory is a clear choice. The most emerging memory technologies, which are measured as potential replacement for Flash, DRAM and SRAM, are based on memristors. These technologies are somewhat unformed and are not yet fully commercial [4] . The ability to control and alter their current-voltage characteristics can be used for performing different computational operations. Current CMOS based memory technology looks many challenges to meet the increasing demand for refining the processing speed and higher data size. For instance, static random access memory (SRAM), the most widely used on-chip memory, is facing its physical limits in attaining higher densities and lowering power consumption. In addition, progress in using other emerging memory technologies, such as eDRAM, MRAM, PCRAM, is prohibited by their lack of compatibility with CMOS, their slow access time, and their limited scalability [5] . Memristor is a two terminal device that works in one of two nonvolatile resistive states R_{ON} or R_{OFF}. It consume less power than transistors as they do not require a minimum voltage to sense their state or need of power to retain [6] .

Citation: Saminathan, V. and Paramasivam, K. (2016) Design and Analysis of Low Power Hybrid Memristor-CMOS Based Distinct Binary Logic Nonvolatile SRAM Cell. Circuits and Systems, 7, 119-127. doi: 10.4236/cs.2016.73012.

The paper planned with introduction of memristor and modelling of memristor with nonlinear dopant drift with new parameter [7] . Then it goes directly into the structure of the proposed circuit, its working principle and its functionality. Then it confers the nonvolatile property and power dissipation of memristor-CMOS based memory cell and finally it concludes with the future views of the circuit.

HP Labs Model of the Memristor

The physical model of the memristor involves two-layer of thin film (size D ≈ 10 nm) made up of TiO_2, sandwiched between platinum (Pt) contacts. The structure of model is shown in Figure 1. One of the layers is doped with oxygen vacancies and thus it acts as a semiconductor. The second layer is an undoped region has an insulating property. As a significance of complex material processes, the width w of the doped region is modulated depending on the amount of electric charge passing through the memristor. With electric current passing in a given direction, the boundary between the two regions is moving in the same direction.

The total resistance of the memristor R_{MEM} is a sum of the resistances of the doped and undoped regions. The relation is given in Equation (1) as follows.

$$R_{mem}(x) = R_{ON}(x) + R_{OFF}(1-x) \quad (1)$$

where, $x = w/D$ is the width of the doped region, referenced to the total length D of the TiO_2layer, and R_{OFF} is the bound values of the memristor resistance for w = 0 and R_{ON} are the bound values of the memristor resistance for w = D. The Ohm's law is valid between the memristor voltages and current [7] as shown in Equation (2)

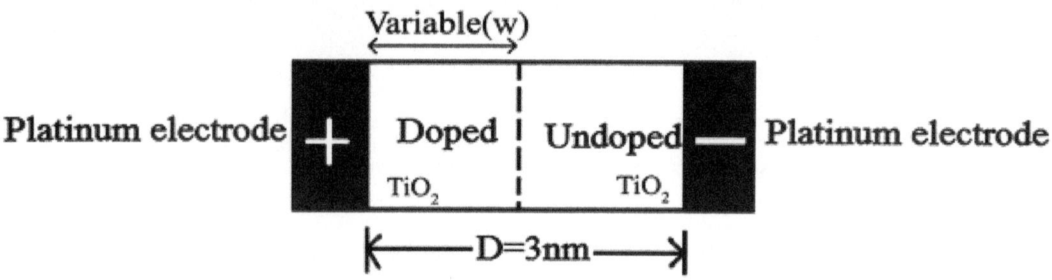

Figure 1. Memristor model based on [2] .

$$v(t) = R_{mem}(w)\cdot i(t) \quad (2)$$

The speed of the movement of the boundary between the doped and undoped regions depends on the resistance of doped area, on the passing current, and on other factors according to the state Equation (3).

$$\frac{dx}{dt} = k\cdot i(t) \quad (3)$$

$$k = \frac{\mu_v R_{ON}}{D^2}, \quad (4)$$

where $\mu_v = 10^{-14} m^2 \cdot s^{-1} \cdot V^{-1}$ is the so-called dopant mobility.

Small voltages can yield massive electric fields, which can secondarily produce significant nonlinearities in ionic transport. These nonlinearities clear themselves particularly at the thin film edges, where the speed of the boundary between the doped and undoped regions slowly decreases to zero. This phenomenon, called nonlinear dopant drift, can be modeled by the so called window function f(x). The paper [8] proposes the window function in the form of following Equation (5).

$$f(x) = 1-(2x-1)^{2p}, \quad (5)$$

where p is a positive integer. This window function confirms zero drift at the boundaries but it grieves with terminal state problem. Biolek [7] proposed an alternative window function that contains memristive current i as an additional parameter. The following Biolek window function in Equation (6) is used for simulation.

$$f(x) = 1-(x-stp(i))^{2p}, \quad (6)$$

where $stp(i) = 1 \text{ for } i \geq 0$ and

$$stp(i) = 0 \text{ for } i \leq 0$$

The current is considered to be positive if it increases the width of the doped layer. Function (6) is zero at either edge.

RELATED WORKS

SRAM is a semiconductor memory commonly used in electronics industry and general computing applications. It is a volatile because when the power is removed from the memory device, the data will disappear. The most commonly used SRAM type is the 6T SRAM which offers better speed of operation, noise immunity and standby current. The main disadvantage of the 6T SRAM structure is its large size and high power consumption. To overcome these restrictions, hybrid memristor-CMOS SRAM memories are being developed recently. There are many researches on memristor-CMOS based memories. Barker Mohammad et al. [6] examined the write and read performance for a memristor based memory, based on surveying three different types of existing mathematical: linear, nonlinear, and exponential drift models. In [9] Syed Shakib Sarwar et al. proposed a new idea of NVRAM cell using memristor. The read time is much faster compared to a conventional SRAM and the power consumption is also much smaller. In [10] E. Linn et al. presented a complimentary resistive switch and it consists of two anti-serial memristive elements which validates the construction of large passive crossbar arrays with a drastic reduction in power consumption. Eshraghian et al. in [11] provided the nonvolatile characteristics and nanoscale geometry of the memristor together with its compatibility with CMOS process technology increases the memory cell packing density, reduces power dissipation and provides for new approaches towards power reduction and management through disabling blocks of MCAM cells without loss of stored data.

PROPOSED MEMRISTOR-CMOS HYBRID MEMORY CELL

The main feature of proposed circuits is as follows.

1) Create a spice model of memristor that closely matches the mathematical model by using Biolek window function [7] that includes memristor current as an additional parameter.
2) Design of new nonvolatile memory cell that consists of four transistors and it is compared with existing memristor based memory cell [9] and conventional 6T SRAM cell [12] .
3) No need of generating external "comb" signal for enabling data input and data out of the memory cell [9] . The same read and write signal within the memory cell is used for enabling data in and out.
4) No need of safety margin is allotted for each logic output [13] . By using DBL method it can be removed and provides better read/write data integrity.
5) Analyses the nonvolatile property, read time, write time and power dissipation of proposed memory cell.

DBL Method

In this method, a ground potential is raised to slightly negative rather than zero. During read operation, the memory cell output goes to positive volts when data input is logic 1 and negative volts when data input is logic zero for the corresponding data input Din = vdd (logic 1) and Din = 0 volt (logic 0). It will eliminate a safety margin is left for each logic output [13] and enables a better Read/write data integrity. This voltage can be easily interpreted as a logic1 and logic 0 by any one of the sensing amplifier or comparator techniques.

The circuit schematic of proposed SRAM cell is shown in Figure 2. The NMOS transistor M1 and M2 are used to Read the data from memristor and write the data into the memristor respectively. The transistor M3 and M4 are used to isolate a memory cell from other memory cell during read and write operation and enables the data input into the memory cell and data out from the memory cell. No extra signal is needed such as comb in [9] . Two memristors U1 and U2 are used as a storage element.

During write operation, two memristors U1 and U2 are comes under parallel connection as shown in Figure 3(a) and during read operation, they are comes under serial connection as shown in Figure 3(b). When a bit is to be written, RD is connected to low state and WR is connected to high state. The data input Din = 2 volts (logic 1) and Din = 0 volt (logic 0). If a bit is to be read, RD is connected to high state and WR is connected to low state, then the voltage at data out as [9] following in Equation (7).

$$\text{Data out} = \{(Vdd/2) - (-Vdd/10)\} \times \{R2/(R2 + R1)\} + (-Vdd/10) \quad (7)$$

where, R1 and R2 are the resistances of memristor U1 and U2 respectively. As a polarities of the memristors are opposite, flux produced at U1 and U2 are also opposite in direction. If Din = logic 1, flux at U2 is higher than U1 and Din = logic 0, flux at U1 is higher than U2.

SIMULATION RESULTS AND ANALYSIS

Simulation arrangement is formed with additional transistor for the verification of nonvolatile property of a

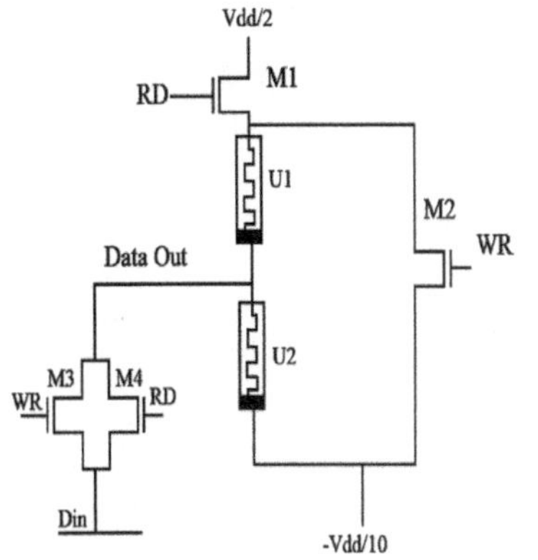

Figure 2. Proposed SRAM Cell.

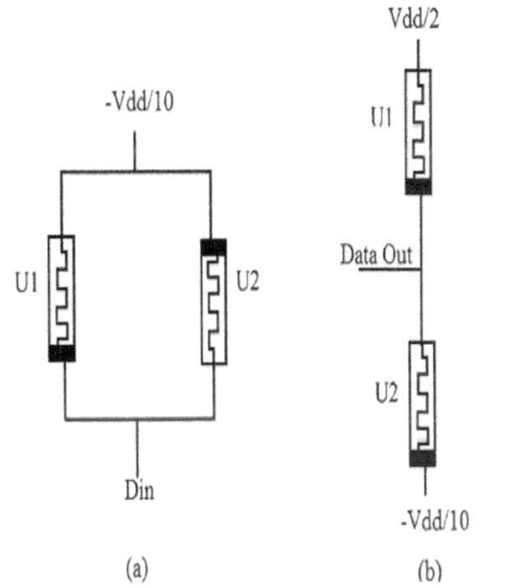

Figure 3. Circuit schematic while (a) Write operation, (b) Read operation.

memristor. MOSFETs of the tsmc-180 nm technology are used for simulation. The proposed circuit is simulated by using LTspice IV 4.1 simulation tool. It is a fourth generation switching regulator design program from Linear Technology. The program consists of a high performance SPICE simulator protracted with a mixed mode simulation capability. The Program's integrated classified schematic capture and SPICE simulator are available for general use. The improved performance of the SPICE Simulation engine is an advantage for simulating general analog circuits and should be of attention to all electronic engineers [14] . In this paper, during write operation −0.2 volt (−vdd/10) is applied at one end of memory cell and other end having 2 volts for logic 1 and 0 volt for logic 0. During read operation, voltage applied between the memory cell is vdd/2 and (−vdd/10). A voltage −0.2 volt is selected based on the best output pattern and simulation parameter of the memristor. The memory cell output bounces distinct binary logic i.e. 1.043 volts for logic 1 and −0.571 volt for logic 0. This voltage can be easily understood by using comparator with zero reference voltage. The following parameters are used for circuit simulation: R_{ON} = 1 kΩ, R_{OFF} = 100 KΩ, R_{INIT} = 80 KΩ, D = 3 nm, u_v = 350e−9 and p = 25. The above parameters are chosen based on the biolek window function and best output pattern after the number of iterations are completed.

Write and Read Operation (When Din = 1)

During write1operation, logic 1 is written in to the memory cell. For that, data input Din = 1, write signal WR = 1 is activated in the interval 2 ns to 4.5 ns by applying PWL input voltage as shown in Figure 4(a). The input data for logic 1 is appeared at memristor as shown in Figure 4(b). The write time interval is found based on the best flux pattern created when writing logic 1. At that time flux at memristor U2 is higher than U1, i.e. in opposite direction as shown Figure 4(c). After write operation, all the power sources are switched off for the interval 20 ns to 40 ns, the power supply Vdd/2 and (−Vdd/10) goes to low state as shown in Figure 5. Next read signal RD = 1 is applied continuously five pulses and each pulse having 0.1ns duration to examine the saved data as shown in Figure 6(a). After that the data is present i.e. data out = 1.043 volts as shown in Figure 6(b). The results reveal that the reading operation does not interrupt the internal state of the memory cell.

Write and Read Operation (When Din = 0)

During write 0 operation, logic 0 is written in to the memory cell. For that, Din = 0, WR = 1 is activated in the interval 2 ns to 4.5 ns by applying PWL input voltage as shown in Figure 7(a). The input data for a logic 0 is appeared at memristor as shown in Figure 7(b). When writing logic 0, flux at memristor U1 is higher than U2 i.e. in opposite direction as shown Figure 7(c). After write 0 operations, all the power sources are switched off for the interval 20 ns to 40 ns. Next data is to be read by activating RD = 1 for a continuously five pulses having 0.1 ns duration as shown in Figure 8(a). Data is present i.e. data out = −0.571 volt as shown in Figure 8(b).

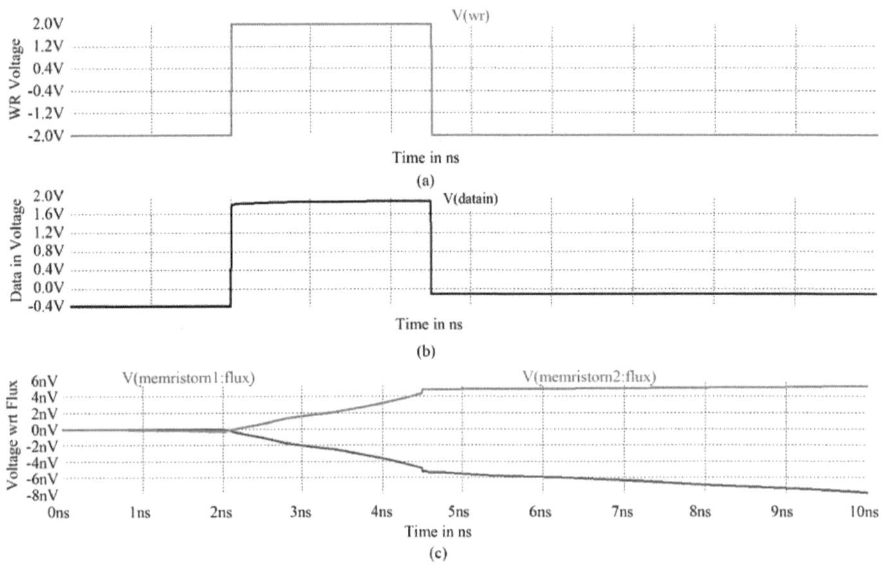

Figure 4. Timing diagram of write operation when Din = 1. (a). Write signal, (b). Data input at memristor, (c). Flux of memristor U2 is higher than U1.

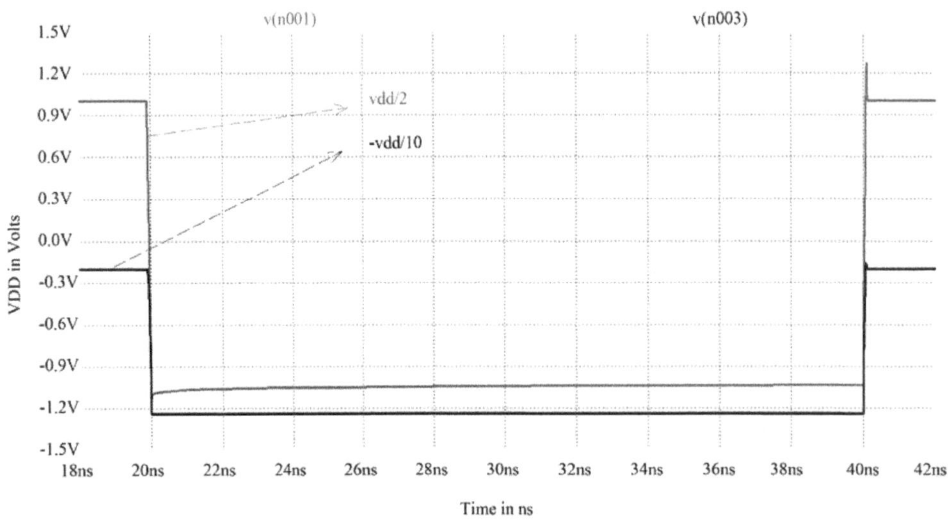

Figure 5. All the power sources switched off during the time interval 20 - 40 ns for investigate nonvolatility of memristor.

This voltage is interpreted as logic 0 and the reading operation does not disturb the internal state of the memory cell.

Observations

This proposed memristor-CMOS memory cell is nonvolatile in nature. A read operation is done after switching

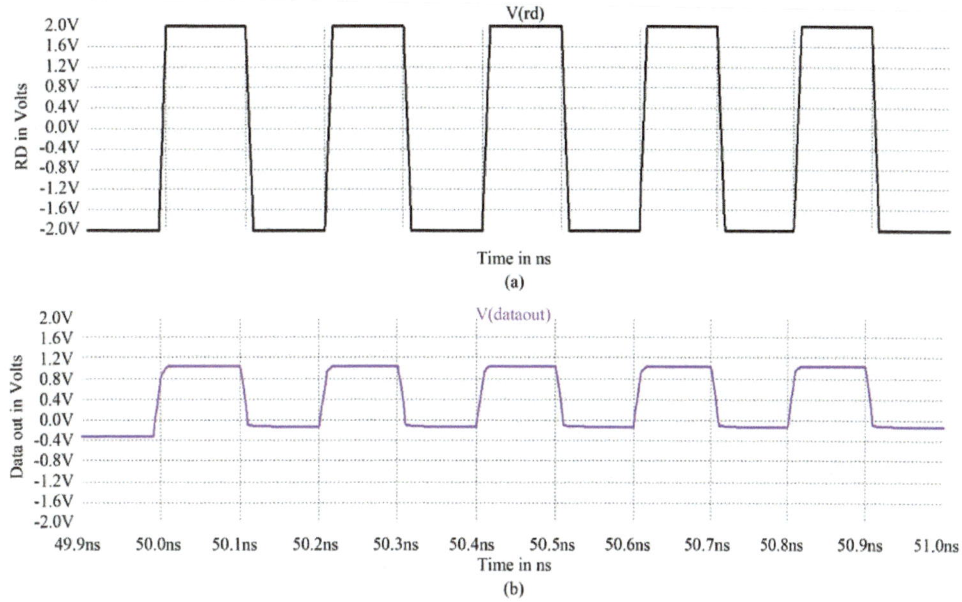

Figure 6. Timing diagram of read operation when Din = 1. (a). Read signal, (b). Data output.

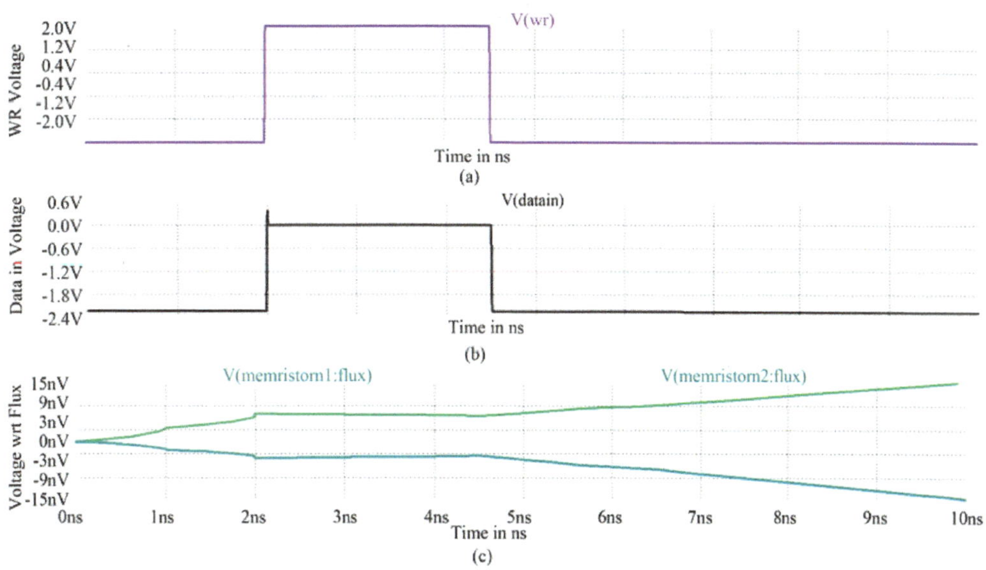

Figure 7. Timing diagram of write operation when Din = 0. (a). Write signal, (b). Data input at memristor, (c). Flux of memristor U1 is higher than U2.

ON the power sources and found "1(0)" in the memory cell. The Write and read time of the proposed memory cell is compared with memristor based memory cell [9] as shown Table 1.

The proposed memory cells entails a less read and write time compared with [9] and more write time than the 6T SRAM cell [12] . By further growing the mobility of the memristor [9] and new type of window function incorporated in memristor modelling, the write time can be considerably reduced. From simulation the power dissipation curve was found and integration was done to get the energy dissipated for write 0, write 1, read 0 and read 1 operation. The corresponding power dissipations are obtained by dividing energy values in particular operation cycle as shown in Table 2. The total power dissipation is calculated by averaging the obtained values as given in Table 3.

Power consumption is less than memristor based memory SRAM cell and conventional 6T cell. Our proposed memory cell consists of four transistors and two memristors. So the predicted area is less than 6T SRAM cell and higher than the existing memristor based memory cell. The width of the memristor is 3nm, so the area can be further reduced by adopting recent CMOS technologies.

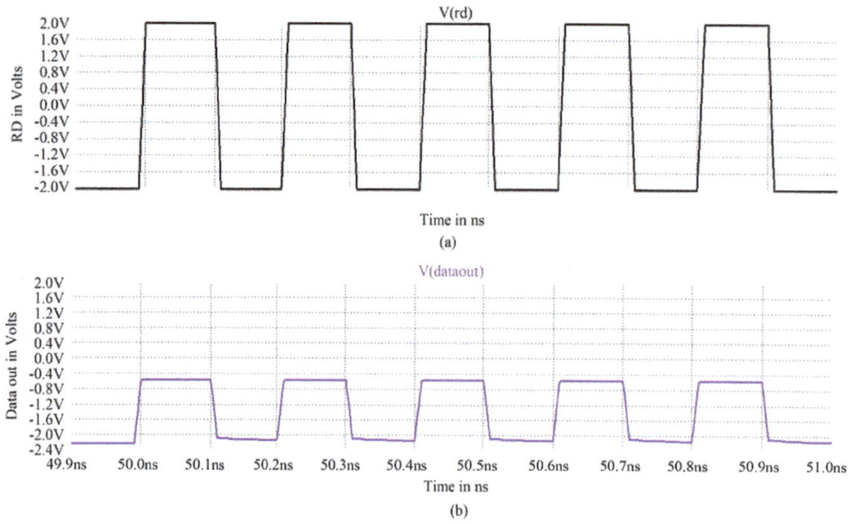

Figure 8. Timing diagram of write operation when Din = 0 (a). Read signal, (b). Data output.

Table 1. Write and read time comparison.

Operation	Proposed SRAM Cell (ns)	Memristor Based SRAM Cell [9]	6T-Cell [10] (ns)
Write time	2.5	5.9	0.85
Read time	0.1	0.2	1.23

Table 2. Power dissipation during write and read operation.

Operation	Write 0	Write 1	Read 0	Read 1
Energy (fj/cycle)	45.271	17.218	10.330	12.545
Average power (μW)	18.108	6.887	103.388	125.448

Table 3. Power comparison.

Operation	Proposed SRAM Cell (mW)	Memristor Based SRAM Cell (mW) [9]	6T-Cell [10] (mW)
Power	0.127	0.407	10.373

CONCLUSION

In this paper, we proposed a memristor-CMOS based nonvolatile SRAM cell and investigated a nonvolatility of memristor. Safety margin left for each output logic will be avoided by DBL method. Power consumption is significantly reduced. Read and write time is much smaller compared with memristor based memory cell but write time is much higher than the conventional SRAM cell. It can be further reduced by incorporating recent memristor modelling and CMOS fabrication technology. There are further scopes to work on design of SRAM array, reduction of write time and power consumption as well.

REFERENCES

1. Chua, L.O. (1971) Memristor—The Missing Circuit Element. IEEE Transactions on Circuit Theory, 9, 507-519. http://dx.doi.org/10.1109/TCT.1971.1083337
2. Strukov, D.B., Snider, G.S., Stewart, D.R. and Williams, R.S. (2008) The Missing Memristor Found. Nature, 453, 80-83. http://dx.doi.org/10.1038/nature06932
3. Chen, G. (2008) Leon Chua's Memristor. IEEE Circuits and Systems Magazine, 8, 55-56. http://dx.doi.org/10.1109/MCAS.2008.925116
4. Kvatinsky, S., Friedman, E.G., Kolodny, A. and Weiser, U.C. (2013) The Desired Memristor for Circuit Designers. IEEE Circuits and System Magazine, 13, 17-22. http://dx.doi.org/10.1109/MCAS.2013.2256257
5. (2010) International Technology Roadmap for Semiconductor. http://www.itrs.net/
6. Mohammed, B., Homouz, D. and Elgabra, H. (2013) Robust Hybrid Memristor-CMOS Memory: Modeling and Design. IEEE Transactions on VLSI Systems, 21, 2069-2079. http://dx.doi.org/10.1109/TVLSI.2012.2227519

7. Biolek, Z., Biolek, D. and Biolkova, V. (2009) SPICE Model of Memristor with Nonlinear Dopant Drift. Radio Engineering, 18, 210-214.

8. Joglekar, Y.N. and Wolf, S.J. (2009) The Elusive Memristor: Properties of Basic Electrical Circuits. European Journal of Physics, 30, 661-675. http://dx.doi.org/10.1088/0143-0807/30/4/001

9. Sarwar, S.S., Saqueb., S.A.N., Quaiyum, F. and Harun-Ur Rashid, A.B.M. (2013) Memristor-Based Nonvolatile Random Access Memory: Hybrid Architecture for Low Power Compact Memory Design. IEEE Access, 1, 29-34. http://dx.doi.org/10.1109/ACCESS.2013.2259891

10. Linn, E., Rosezin, R., Kugeler, C. and Waser, R. (2010) Complementary Resistive Switches for Passive Nanocrossbar Memories. Nature Materials, 9, 403-406. http://dx.doi.org/10.1038/nmat2748

11. Eshraghian, K., Cho, K.R., Kavehei, O., Kang, S.K., Abbot, D. and Kang, S.M.S. (2010) Memristor MOS Content Addressable Memory (MCAM): Hybrid Architecture for Future High Performance Search Engines. IEEE Transactions on VLSI Systems, 19, 1407-1417.

12. Sreerama Reddy, G.M. and Reddy, P.C. (2009) Design and Implementation of 8K-Bits Low Power SRAM in 180 nm Technology. Proceedings of the International Multiconference of Engineers and Computer Scientists, Vol. 2, 1-8.

13. Ho, Y., Huang, G.M. and Li, P. (2009) Nonvolatile Memristor Memory: Device Characteristics and Design Implications. IEEE/ACM International Conference on Computer-Aided Design Digest of Technical Papers, San Jose, 2-5 November 2009, 485-490.

14. Linear Technology Corporations. http://www.linear.com

CHAPTER 19

SCANNING HEAD FOR THE APERTURELESS NEAR FIELD OPTICAL MICROSCOPE

D. V. Kazantsev and H. Ryssel

Keywords: Near-Field Optics, Apertureless SNOM, Scanning Head

INTRODUCTION

A powerful tool for the local investigation of surface optical properties is apertureless scanning near field optical microscopy [1] ASNOM (scattering SNOM, s-SNOM). It is a modification of SNOM [2] in which the light scattered by an AFM [3] tip is collected. The tip in this instrument can be considered to be a macroscopic dipole antenna which re-emits the radiation to the environmental space due to dipole oscillations, excited in the tip by the external electromagnetic field. The "grounding" conditions of this antenna are determined by the "capacitor" of the slit between the tip and the surface being investigated. This is the reason why the ASNOM is able to provide a map representing the dielectric properties of the sample and the local field at the tip location. The lateral resolution of the instrument is determined by the tip radius. The tip which scatters the electromagnetic radiation is typically covered with a metal [4] due to its higher polarizability. The spatial resolution of the AS-NOM technique was proven to be as high as 1 nm [5], so that it seems to be sufficient to select nanoparticles [6], viruses [7] and even single molecules [8] on the sample surface for the investigations.

With exception of the tip-enhanced Raman spectroscopy [9-11], the light scattered by the tip has the same wavelength as the excitation light. In such a case the interferometer used to detect the signal [12], which was already mentioned in the very first paper on ASNOM, is a vital part [1]. Without it, the dynamic range of the photodetector may be not sufficient [13] to recognize the signal component caused by the really near-field (non-radiative) part of the tip-sample electromagnetic interaction. The method allows, due to implementation of the interferometer detecting scheme, to store both amplitude and phase [14,15] of the light scattered by the tip. Most ASNOMs are experimental setups [13,16,17], but there are also commercially available versions [18]. There are several artifacts which distort the data collected with the help of ASNOM. First, the optical signal recovered at the oscillation frequency of the tip [19] results mainly from scattered light caused by the interference between the tip and its image in the surface. These variations of the photocurrent are useless in the terms of near-field (nonradiative) electromagnetic interaction of the tip and the surface. To get rid of this background, the signal must be recovered using higher harmonics of the tip oscillation frequency [20], because the modulation of the tip polarizability caused by the non-radiative electromagnetic tip-sample interaction is the only effect in the system which depends in a non-linear way on the tip position.

Second, the ASNOM is rather sensitive [21] to the polarization of the light driving the tip (the E-vector must be along the tip, i.e. normal to the surface). In some home-built [16,17] and commercial [22,23] systems the light comes along the tip i.e. normally to the surface. In such a geometry just a small fraction of the focused radiation is able to excite the dipole oscillations along the tip, namely that focal cone beam fraction which illuminates the tip from aside due to the high numerical aperture of the objective used. The optical signal acquired in

Citation: D. Kazantsev and H. Ryssel, "Scanning Head for the Apertureless near Field Optical Microscope," Modern Instrumentation, Vol. 2 No. 2, 2013, pp. 33-40. doi: 10.4236/mi.2013.22006.

such a geometry is rather weak. Operation of the ASNOM is based on demodulation of the tip scattering variations caused by the tip tapping motion, which means that the conditions of the tip mechanical oscillation affect strongly the amplitude and phase of the recovered optical signal. In principle, the cantilever bending can be used for the lock-in reference [13], but not all lock-in models are able to track the fast variations of the reference signal phase. Also, the variations in the tip-sample bounce conditions (which are strongly nonlinear in the terms of the tip oscillation, see e.g. [24, 25]) may affect significantly the amplitude of the nearfield component in the scattered light variations. Unavoidable mechanical vibrations, caused by the water cooling of the main laser and acoustical-optical modulator in the heterodyne interferometer detection scheme, are present in the setup. Therefore, the tip oscillation conditions have to be rather well stabilized by the feedback, which is not always the case.

Another requirement to the scanning head being designed is obviously its ability to allow the optical access to the tip. The coherent light beam is focused onto the tip, and the light scattered by the tip is collected back to the interferometer. The probing beam must come along the surface, but not completely parallel to it (the efficiency of the dipole antenna located near the ground plane is zero in the exactly "horizontal" direction, see [21]). One also has to keep in mind to provide optical access to the sample from the top or from the substrate, so that the mechanical parts of the scanning head must not shadow the beams.

Possibly, the problems mentioned above explain why ASNOM still did not get out of the stage "nice trick providing sharp lateral resolution", to the state of widely used instrument, providing the quantitative data. Just a few papers are devoted to the design and utilization of ASNOMs.

The aim if this work was to make a user-friendly, flexible ASNOM system for everyday's use in the research lab. In particular, it means that 1) sample loading, optical alignment and operation must be simple; 2) the fabrication of the hand-made tips (see e.g. [17,19]) should be excluded from the operation; the commercial tips should be used instead.

MECHANICAL DESIGN

General Approach

A schematic view of the scanning head which fulfills the requirements listed above is presented in **Figure 1**.

The tip is always fixed in the focus of the objective, so that the amplitude-phase conditions of the tip scattering in the optical path of the interferometer are constant. The focus of the work beam is aimed onto the tip by the objective XYZ translation stage during the initial setup alignment after the cantilever chip is reloaded. The optical nature of the ASNOM experiments brings some special requirements to the mechanical design, compared to the other SPM techniques. Namely, an optical access has to be provided from many different directions. The requirement to make the instrument to be wavelength- independent leads to the implementation of a mirror [26] instead of a lens to focus the external probing beam onto the cantilever. The option to choose the area on the sample surface under visual control, as well as the option to implement a high-NA objective to collect the radiation scattered in the direction normal to the sample surface has to be implemented. Also, the experimental geometry, in which the tip vicinity area is illuminated from beneath the sample, was included as a possible mode of operation. To make operation convenient, tip replacement, sample replacement, and the optical realignment procedures were designed to reduce their interaction as much as possible.

Sample Raster Scanning and "Height" Positioning

The raster scanning of the surface is provided by a com-

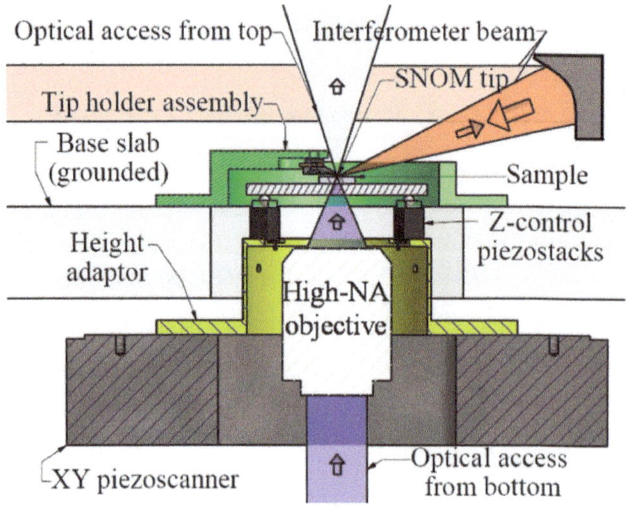

Figure 1. Cross-sectional view of the tip scattering microscope scanning head. The cantilever control unit lays on the base slab, and the sample is fixed on the scanner flange, which comes from beneath of that slab through appropriate hole in it. The sample positioning parts (Z-approach and lateral transfer) are not shown.

mercial scanner P-541 (Physik Instrumente, GmbH, Germany). The control electronics includes additional feedback loop, based on the XY displacement capacitive sensors integrated in the scanner table. At the same time, the sample positioning normal to the surface is provided by three piezos, which drive up/down thin and lightweight coin-like sample holder under the control of the dynamic mode AFM feedback system. This allows to achieve a rather large feedback loop bandwidth which is limited in this case just by the mass of the parts involved in the motion normal to the surface. At the expense of reduced metrological quality, such a solution allows to set faster scanning speed values, without damage to the cantilever and without noticeable variations in the tipsample interaction conditions.

To provide a coarse tip-sample approach, the scanner is driven upwards and downwards together with the sample fixed on the top of it. The approach is provided by three screws with contact spheres on their ends, driven by three step motors. No plates or levers were necessary to be installed between the motor-driven screw shafts and the point balls fixing the scanner. As it is shown below, the topography noise and long-term stability of the system is very good. These three approach screws are fixed on the base plate by the traditional way "hole-slit-plane". To choose the sample area being scanned, the piezo scanner is mounted on top of the flat XY micrometer stage (OWIS GmbH).

Cantilever Control Assembly

The cantilever bending oscillation is excited by a small piezo, driven at the cantilever resonance frequency. Commercial AFM cantilevers (Budget Sensors and Micro Masch) covered by platinum were used. The cantilever deflection sensor is based on the optical lever principle [27,28]. The beam of a small semiconductor laser with feedback-controlled output power (SPMC01, Power Technologies Inc., 1 mW at 670 nm) is focused onto the back side of the cantilever. To bring the focal spot onto the cantilever, the laser is fixed in a tilting-type kinematic mount. This control beam, being reflected from the cantilever, is redirected by the mirror onto a quad photodiode pad, so that the spot displacement on the photodiode is determined by the cantilever deformation. The mirror which redirects the beam onto the quad photodiode is also fixed on a tilting-type mount. These two custommade alignments make it easy to compensate the cantilever spatial and angular position tolerance after a cantilever replacement. The photodiode is mounted in a holder, which allows its rotation around the beam direction. Thus, it is possible to optimize the setup, by alignment of the photodiode segments along the lateral trajectory of the beam spot to separate "lateral force" and "normal force" signals.

The conventional AFM optical lever electronics was splitted into two PCBs. To reduce the heat dissipation on the AFM head as well as its weight, most of the electronics ("normal" and "lateral" difference stages, sum stage, analog divider(s), buffer stages for the tip vibration piezoplate excitation, tip bias control, voltage regulators for the OpAmps, diode laser, and quad photodiode bias) arc mounted on the PCB located 10 - 15 cm apart. The transimpedance stages are mounted on the same small PCB (11 mm diam) on which the quad photodiode is located. Therefore, the bandwidth of the optical lever sensor is at least 3 MHz, and allows to monitor quantitatively the non-linearities in the tip oscillation. It means also that the phase of the observed tip oscillation contains no parasite shift for the frequencies below approx. 1 MHz, due to the Kramers-Kronig relations.

To leave open the optical path from the top to the sample surface, the laser beam of the cantilever control is declined at 45° to the vertical axis of the setup. This geometry reduces the sensitivity of the control by 30%, but this plays just a minor role, because the noise level of the setup is low enough. The angular shape of the probe beam in the optical lever scheme is far from being perfectly circular. It is close to elliptical with an excentricity of about 1:6. Simultaneously, the focus of the beam is also elongated in its shape. By focusing the spot onto the cantilever along its axis [29], we get an additional possibility to improve the sensitivity of the cantilever control scheme. The narrower the angular shape of the reflected beam, the higher the sensitivity of the quad photodiode scheme, regardless to the distance from the cantilever [30].

The parts related to the AFM operation were assembled in a stiff unit (See Figures 1 and 2). Once aligned after a new cantilever is glued onto the piezo-driven cantilever holder, the assembly does not require any optical realignment and sensor sensitivity recalibration even if it is removed from the system for sample reloading. The position of the cantilever control assembly is determined by three radial slits in its bottom surface, corresponding

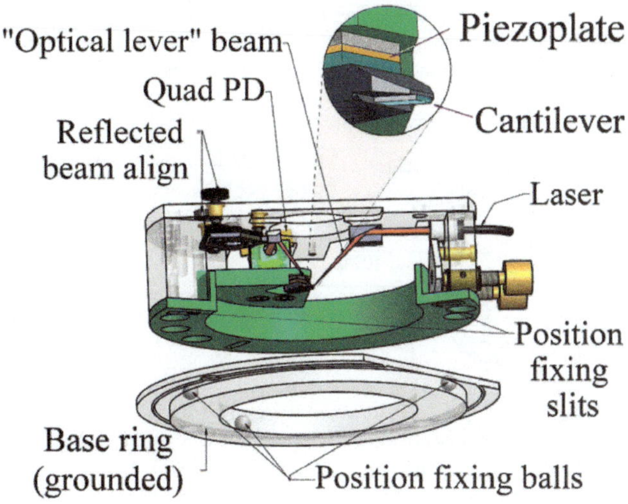

Figure 2. Cantilever control assembly of the ASNOM head.

to three steel bearing balls, pressed into the adapter ring. The ring is fixed, in turn, on the base slab. The precision of the tip position in the focus after reinstallation of the cantilever control assembly is better than 1 μm.

Lightweight but Stiff Tip Holder

To suppress the sensitivity of the instrument to vibrations of the optical table, the cantilever assembly was designed to be very lightweight and stiff. The optical table is much heavier than the scanning instrument, so we assumed the "seismic" noise motion of the optical table to be completely independent on the presence of the scanner. Therefore, the seismic noise can be described just by the following vibration law

$$x_j(t), \quad (1)$$

and consequently

$$v_j(t) \equiv \dot{x}_j(t), a_j(t) \equiv \ddot{x}_j(t) \quad (2)$$

where $v_j(t)$, and $a_j(t)$, denote velocity and acceleration of the optical table, respectively (here $j \in \{x, y, z\}$. In such a model (See **Figure 3**), the vibration of the main optical table can be considered as just the noise in the weight of the parts (in particular, of the tip holder). The external inertial force applied to the part is expressed as $F_j(t) = ma_j(t)$. Below the tip holder resonant frequency, the less the mass of the parts, the less the system sensitivity to external vibrations. On the other hand, the higher the stiffness of the tip holder assembly, the less the sensitivity of the system to external vibrations. In our model, the largest contribution to the tip-sample separation distance is provided by the tip carrier unit of the mass m, with the rigidity characterized by the value k (the softest mechanical link in the system, including also the elasticity of the tip holder parts themselves, see **Figure 3**). There-

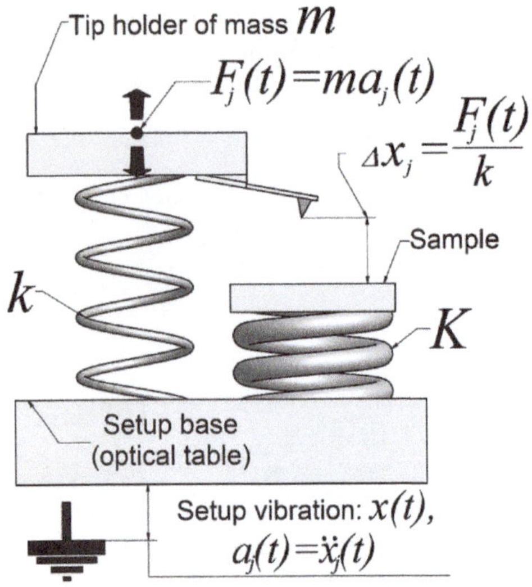

Figure 3. Model of the instrument to consider its sensitivity to the vibration of the main optical table. The stiffness K of the sample holder is considered to be much higher than the stiffness k corresponding to the tip holder parts.

fore, the sensitivity of the tip control assembly to the external vibrations can be expressed by the ratio m/k. In contrast to other systems [22,31-34], we designed this assembly to be lightweight, wide, and rigid. The low mass of the tip control assembly reduces sensitivity of the tip-sample position to the vibrations of the base, wide feet span prevents it from the wobbling mode of displacement, walls and frames prevent it from the membrane-like modes of oscillation.

Cantilever or Tuning Fork?

There are clear reasons why the AFM-like cantilever should be strongly preferred to the tuning fork in the case of ASNOM design. The demodulation of the interferometer output signal at higher harmonics of the tip tapping frequency is used in ASNOM. The recovery of the signal part related solely to the near-field optical interacttion of the tip and surface in the scattered radiation requires to provide a clear difference between the states "close" and "far" of the tip. Therefore, the larger the tip oscillation amplitude, the more pronounced the collected near-field-optical signal. Practically, an amplitude of 40 - 60 nm is used with solid state samples. In such a case, the forces arising at the moment of the tipsurface impact are rather high, so that it may lead to a tip destruction and cause fast wearing of the probe. The typical mass of the AFM cantilever is much less than the mass of the tuning fork prong which can be estimated by the dimensions of the objects: the dimensions of the cantilever are about $3 \times 30 \times 200$ m^3 while the prong of the tuning fork has dimensions of approx. $500 \times 500 \times 3000$ m^3. Usually, these two mechanical oscillators have a similar resonant frequency (30 - 100 kHz) which means that the forces to drive the mass of the oscillating part are much larger in the case of the tuning fork. Thus, to modify noticeably the amplitude/phase of the probe oscillation much more energy is needed every cycle in the case of the tuning fork oscillator.

Since AFM cantilever is chosen as the concept, the optical lever (beam deflection) is the optimum solution to monitor the cantilever oscillation. It is sensitive, simple, reliable, and it does not require any special measures as piezoresistive cantilevers do [35]. In addition, it is sensitive to the DC-mode deflections of the cantilever (and that can be utilized for the absolute calibration of the tip displacement by the scanning of the grating sample with a known step height). These various reasons make the AFM cantilever monitored by the optical lever scheme an optimum solution for the flexible application.

CHARACTERIZATION

Topography Noise Level

The mechanical quality of the system is demonstrated in **Figure 4** where the AFM image of a DNA molecule on a mica substrate is shown. The characteristic deviation in the topography height histogram (unfiltered) is about 0.34 nm (See Figures 4(a) and (b)). The RMS deviation of the height error signal is 0.26 nm (See Figures 4(c) and (d)). It should be pointed out, that the image was obtained in the presence of vibrations caused by a water pump.

These results prove that the concept "light but stiff tip holder on a heavy base" which we kept, in order to suppress the influence of external vibrations. The map of the tip vibration amplitude and its histogram (observed simultaneously with topography) is also presented in Figure 4. One can see that the level of the tip-sample contact is kept rather stable in the experiment. The noise properties of the cantilever control unit alone (without tipsample contact) are even better. The RMS deviation of the vibration amplitude measured without contact to the surface is no more than 0.07 nm.

One could expect, as a side effect of the general concept "heavy basement - lightweight roof", a sensitivity of the instrument to acoustic noise (microphone effect). This matter was taken into account during the characterization and implementation of the instrument. It was observed that sometimes the human speech near the setup can be found in the topography map, as additional noise of no more than 0.2 nm RMS. It was found that the cantilever control assembly itself has no noticeable sen-

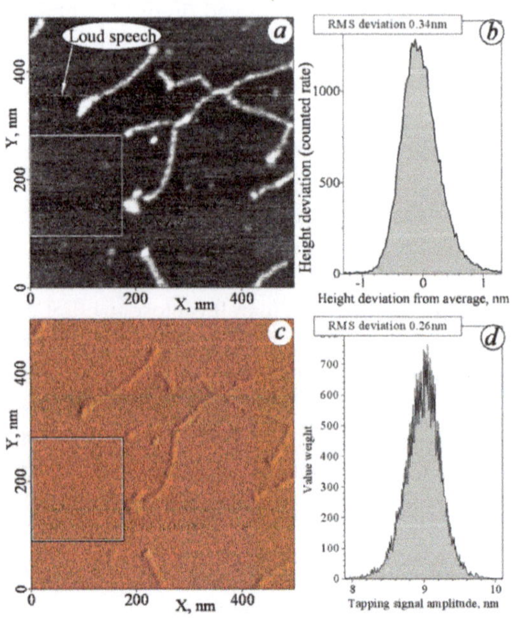

Figure 4. DNA image on a mica substrate. (a) Topography image; (b) Height histogram of the free mica area; (c) Tip vibration sensor signal amplitude; (d) Amplitude histogram of the same area. Feedback setpoint corresponds to 10nm of the vibration amplitude for this cantilever.

sitivity against the acoustic noise. There is also almost no microphone effect in the DC mode of the feedback, in which the tip displacement is directly used as the feedback input. The microphone effect appears mainly if the tip-sample distance is stabilized by the feedback in dynamic mode. One can explain this by the strong nonlinearity of the tip-sample interaction potential, and by the high value of the cantilever quality factor Q. To quench the oscillation, just a single period of the cantilever frequency might be enough (nonlinear effect), while stable amplitude and phase of the driven oscillation of the cantilever gets restored only after approximately Q oscillation cycles (linear effect). It was found that the sensitivity mainly depends on the way how the sample ($7 \times 7 \times 0.5$ mm^3, or even $10 \times 10 \times 0.15$ mm^3 in the case of mica substrate) is fixed on the holder. In any case the acoustic sensitivity of the instrument is sufficiently low, and acoustic noise is by far not the main source of noise in the system. An acoustic isolation cage during operation is, therefore, not necessary.

Long-Time Mechanical Stability

The ASNOM images of metal dots (Au) on the dielectric surface (quartz) taken with an interval of 18 hours with no adjustment of the setup are presented in **Figure 5**.

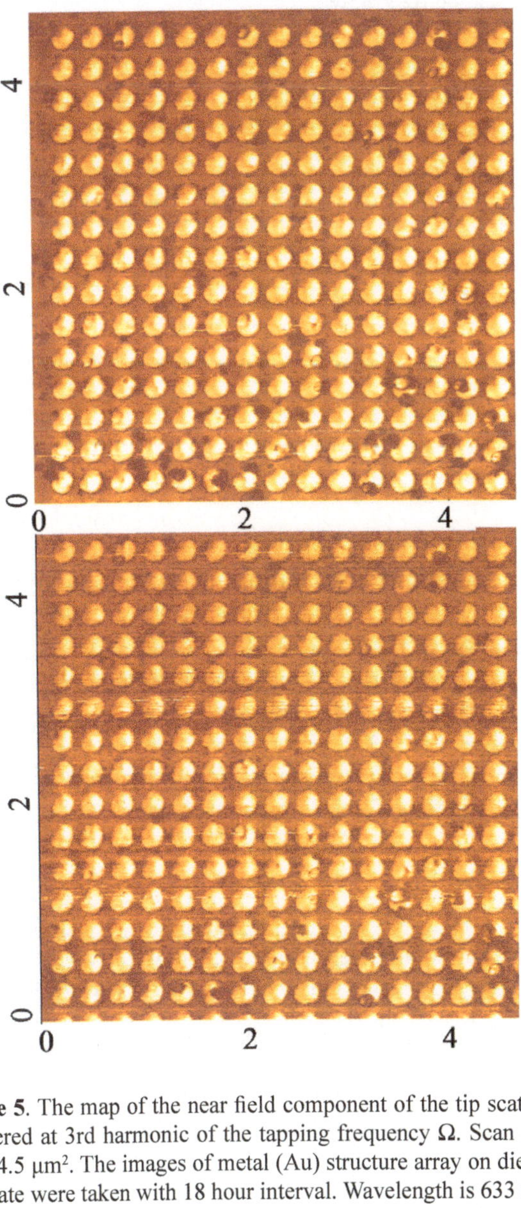

Figure 5. The map of the near field component of the tip scattering, recovered at 3rd harmonic of the tapping frequency Ω. Scan area is 4.5×4.5 μm^2. The images of metal (Au) structure array on dielectric substrate were taken with 18 hour interval. Wavelength is 633 nm.

The light scattered by the tip was detected by the interferometer at the wavelength of 633 nm. A Mach-Zehnder-like interferometer heterodyne scheme was used [14] with demodulation of the interferometer output signal at a frequency corresponding to the third harmonic of the tip oscillation. One can see that there is no significant drift of the surface with respect to the tip during that period; the presence of the clear light scattering signal in the second image means also that the tip does not drift away from the caustic of the main beam focus (which has a characteristic size of about the wavelength used, namely 600 - 800 nm).

The Light Field Distribution around the Tip

In many practical applications it is very useful to know the real configuration of the electromagnetic field at the very tip vicinity. **Figure 6** represents the map of the scattered light amplitude and phase in the case when the tip stays fixed in respect to the sample while the focus of the main beam objective scans around the position of the best scattering signal. The mapping was performed using a crystalline SiC sample which provides a strong enhancement of the tip scattering at the CO_2 laser band due to the presence of lattice resonance. The smooth variation of the amplitude shows an independence of the amplitude output on the signal phase. The period in the phase variations corresponds to the geometrical change of the optical path from the interferometer to the tip, due to the objective positioning. In particular, the phase period $p_0 = 8.3$ μm in the section corresponding to the purely vertical motion of the objective (perpendicular to the beam of the interferometer signal arm) corresponds to an angle (≈40°) of the deflection in the vertical plane: $p_0 \cos(\theta) = \lambda/2$. The ramp-like phase behavior is simply caused by the change of the optical path from the interferometer to the tip, due to the objective motion. The scattering amplitude and phase images demonstrate a high quality of the wave fronts in the tip irradiating beam. It means that the tip illumination conditions can be well described by a plane wave. It is important in such experiments, in which the tip is just a probe to measure the collective response of the surface, e.g. in the investigations of the surface polariton waves.

KNOWN PROBLEMS

Focus Position Drift

The instrument is very sensitive to the focus position of the light beam on the tip (i.e. the focus of the light being used for the tip-scattering in the experiment, not the additional beam to monitor the AFM cantilever deflection). Due to the interference origin of this sensitivity (the light scattered by the tip is detected afterwards with the help of the interferometer), the mechanical stability of the tip

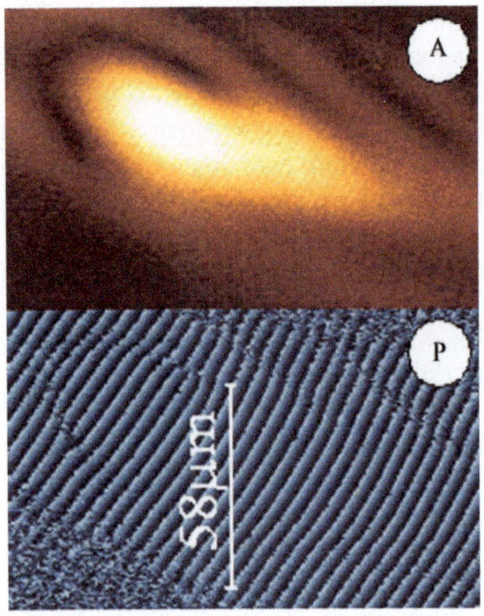

Figure 6. The variations in the detected second harmonic component amplitude (A) and phase (P), caused by the objective mirror scanning. The Pt-coated tip is brought to a contact with the SiC surface under the feedback control. No lateral scanning signal is applied to the sample/tip piezo table. Vertical axis corresponds to the vertical motion of the objective (perpendicular to the interferometer arm beam) horizontal direction is along the cantilever. Objective scan span is 95 × 95 μm². Wavelength is 10.85 μm.

position in the main beam focus is more critical at the visible wavelength band and less important at the infrared band.

Light Polarization

The polarization of the electromagnetic radiation which excites the response of the scattering tip (modulated with the tip tapping frequency due to the tip-sample near-field interaction being investigated) has to be always normal to the surface. The tip acts just as an antenna rod in conventional radio physics, so its ability to scatter the radiation (as well as its sensitivity to the driving electromagnetic field) is maximum if the E vector of the driving field is parallel to the tip's long axis.

Signal Recovery

The subject of the acquisition in the ASNOM setup are the higher harmonics of the tip tapping oscillation frequency in the detector photocurrent. DC component and fundamental do not carry any relevant information. The amplitude of the useful component in the detector signal is, typically, more than 100 times less than the amplitude of the parasitic signals in the interferometer's detector photocurrent caused by the tip oscillation and by other displacements of the parts. Even in the best case of strong resonant dielectric response of the sample the useful component in the detector output is hardly visible on the screen of the oscilloscope. Thus, one should have a very good idea about the expected signal. The subjects of acquisition are relatively short variations of the photocurrent which occur mainly at the very moment when the tip-sample distance is minimum.

One must keep in mind that the photocurrent of the detector in the interferometer is proportional to the intensity of the field. The variations in the photocurrent, although they are being caused by the same modification of the tip scattering amplitude/phase, depend strongly on the reference beam phase. The length of the reference arm is initially arbitrary in the interferometer. To avoid that, the average over the reference beam phase is necessary for each ASNOM map point [1,14]. Phase-sensitive lock-in recovery at the difference frequency must be arranged using the "heterodyne" scheme [13,14] (in case if Mach-Zehnder-like interferometer with acoustical-optical scattering is used), or phase modulation [15] (in case a Michelson interferometer scheme is used) has to be implemented to average the recovered signal over all possible phases of the reference beam. In general the Michelson interferometer is preferable for the ASNOM operation, because its alignment is wavelength-independent.

Data Interpretation

The amplitude of the light scattered by the tip (mainly scattered by the tip and the cantilever, with a tiny additional fraction caused by the tip-sample near-field interaction at the moment of the very contact) is a product of the local electromagnetic field at the tip location and of rather complicated expressions containing the tip/sample dielectric value, tip radius and the distance from the tip to the surface. In case all these values vary due to the tip oscillation, the final image data, consisting of recovered higher harmonic components, may become rather difficult to be interpreted.

CONCLUSION

A cantilever control unit for the operation of an ASNOM was designed and characterized. The requirements specific for ASNOM operation (wide optical access, good optical stability) were fulfilled. The design concept (framelike light-weight detachable assembly), contrary to the common style (compact and relatively heavy assembly), was proven to be successful. The sensitivity of the instrument against vibrations was successfully suppressed. Fast (seismic-like vibrations) and long-term (thermal drift) noises were proven to be low enough. The instrument is wavelength-independent. Its operation was demonstrated in visible (λ = 633 nm) and mid-IR ($\lambda \approx$ 10.6 μm) light bands.

ACKNOWLEDGEMENTS

The author thanks Dr. Keilmann and Dr. Hillenbrand for their advising during the work presented, Dr. Ocelic who fabricated the non-linear signal recovery electronics, and Mr. Gatz for his perfect mechanical works. The author gratefully acknowledges funding of the Erlangen Graduate School in Advanced Optical Technologies (SAOT) in the framework of the excellence initiative. The work was supported by the German National Science Foundation (DFG grant KA3105/1-1).

REFERENCES

1. F. Zenhausern, M. P. O'Boyle and H. K. Wickramasinghe, "Apertureless Near-Field Optical Microscope," Applied Physics Letters, Vol. 65, 1994, No. 13, pp. 1623-1625. doi:10.1063/1.112931
2. E. Betzig, M. Isaacson and A. Lewis, "Collection Mode Near-Field Scanning Optical Microscopy," Applied Physics Letters, Vol. 51, No. 25, 1987, pp. 2088-2090. doi:10.1063/1.98956

3. G. Binnig, C. F. Quate and C. Gerber, “Atomic Force Microscope,” Physical Review Letters, Vol. 56, No. 9, 1986, pp. 930-933. doi:10.1103/PhysRevLett.56.930

4. U. C. Fischer and D. W. Pohl, “Observation of SingleParticle Plasmons by Near-Field Optical Microscopy,” Physical Review Letters, Vol. 62, No. 4, 1989, pp. 458- 461.

5. F. Zenhausern, Y. Martin and H. K. Wickramasinghe, “Scanning Interferometric Apertureless Microscopy: Optical Imaging at 10 Angstrom Resolution,” Science, Vol. 269, No. 5227, 1995, pp. 1083-1085. doi:10.1126/ science.269.5227.1083

6. A. Cvitkovic, N. Ocelic, J. Aizpurua, R. Guckenberger and R. Hillenbrand, “Infrared Imaging of Single Nanoparticles via Strong Field Enhancement in a Scanning Nanogap,” Physical Review Letters, Vol. 97, No. 6, 2006, Article ID: 060801.doi:10.1103/ PhysRevLett.97.060801

7. M. Brehm, T. Taubner, R. Hillenbrand and F. Keilmann, “Infrared Spectroscopic Mapping of Single Nanoparticles and Viruses at Nanoscale Resolution,” Nano Letters, Vol. 6, No. 7, 2006, pp. 1307-1310. doi:10.1021/nl0610836

8. Y. Martin, F. Zenhausern and H. K. Wickramasinghe, “Scattering Spectroscopy of Molecules at Nanometer Resolution,” Applied Physics Letters, Vol. 68, No. 18, 1996, pp. 2475-2477. doi:10.1063/1.115825

9. R. Stöckle, Y. Suh, V. Deckert and R. Zenobi, “Nanoscale Chemical Analysis by Tip-Enhanced Raman Spectroscopy,” Chemical Physics Letters, Vol. 318, No. 1-3, 2000, pp. 131-136. doi:10.1016/S0009-2614(99)01451-7

10. L. T. Nieman, G. M. Krampert and R. E. Martinez, “An Apertureless Near-Field Scanning Optical Microscope and Its Application to Surface-Enhanced Raman Spectroscopy and Multiphoton Fluorescence Imaging,” Review of Scientific Instruments, Vol. 72 No. 3, 2001, pp. 1691- 1699. doi:10.1063/1.1347975

11. B. Pettinger, B. Ren, G. Picardi, R. Schuster and G. Ertl, “Nanoscale Probing of Adsorbed Species by Tip-Enhanced Raman Spectroscopy,” Physical Review Letters, Vol. 92, No. 9, 2004, Article ID: 096101. doi:10.1103/PhysRevLett.92.096101

12. J. S. Batchelder and M. A. Taubenblatt, “Interferometric Detection of Forward Scattered Light from Small Particles,” Applied Physics Letters, Vol. 55, No. 3, 1989, pp. 215-217.doi:10.1063/1.102268

13. A. Bek, R. Vogelgesang and K. Kern, “Apertureless Scanning near Field Optical Microscope with Sub-10 nm Resolution,” Review of Scientific Instruments, Vol. 77, No. 4, 2006, Article ID: 43703. doi:10.1063/1.2190211

14. R. Hillenbrand and F. Keilmann, “Complex Optical Constants on a Subwavelength Scale,” Physical Review Letters, Vol. 85, No. 14, 2000, pp. 3029-3032. doi:10.1103/PhysRevLett.85.3029 [Paper reference:1]

15. N. Ocelic, A. Huber and R. Hillenbrand, “Pseudoheterodyne Detection for Background-Free Near-Field Spectroscopy,” Applied Physics Letters, Vol. 89, No. 10, 2006, Article ID: 101124. doi:10.1063/1.2348781

16. G. Wurtz, R. Bachelot and P. Royer, “A Reflection-Mode Apertureless Scanning Near-Field Optical Microscope Developed from a Commercial Scanning Probe Microscope,” Review of Scientific Instruments, Vol. 69, No. 4, 1998, pp. 1735-1743. doi:10.1063/1.1148834

17. R. Bachelot, P. Gleyzes and A. C. Boccara, “ReflectionMode Scanning Near-Field Optical Microscopy Using an Apertureless Metallic Tip,” Applied Optics, Vol. 36, No. 10, 1997, pp. 2160-2170. doi:10.1364/AO.36.002160

18. Neaspec GmbH. http://www.neaspec.com

19. A. Lahrech, R. Bachelot, P. Gleyzes and A. C. Boccara, “Infrared Reflection-Mode Near-Field Microscopy Using an Apertureless Probe with a Resolution of λ/600,” Optics Letters, Vol. 21, No. 17, 1996, pp. 1315-1317. doi:10.1364/OL.21.001315

20. M. Labardi, S. Patanè and M. Allegrini, “Artifact-Free Near-Field Optical Imaging by Apertureless Microscopy,” Applied Physics Letters, Vol. 77, No. 5, 2000, pp. 621- 623.doi:10.1063/1.127064

21. O. J. F. Martin and C. Girard, “Controlling and Tuning Strong Optical Field Gradients at a Local Probe Microscope Tip Apex,” Applied Physics Letters, Vol. 70, No. 6, 1997, pp. 705-707. doi:10.1063/1.118245

22. NT-MDT (Moscow) NTegra SNOM System. http:// www.ntmdt.com/device/ntegra-spectra

23. JPK GmbH, TAO SNOM System. http://www.jpk.com/ tao.download.61f23e409a2a8af9654595a415025092. pdf

24. U. D. Schwarz, H. Hölscher and R. Wiesendanger, “Atomic Resolution in Scanning Force Microscopy: Concepts, Requirements, Contrast Mechanisms, and Image Interpretation,” Physical Review B, Vol. 62, No. 19, 2000, pp. 13089-13097.doi:10.1103/ PhysRevB.62.13089

25. F. J. Giessibl, “Advances in Atomic Force Microscopy,” Reviews of Modern Physics, Vol. 75, No. 3, 2003, pp. 949-983. doi:10.1103/RevModPhys.75.949

26. F. Keilmann and R. Hillenbrand, German Patent DE10-2006002461A1.

27. G. Meyer and N. M. Amer, “Novel Optical Approach to Atomic Force Microscopy,” Applied Physics Letters, Vol. 53, No. 12, 1988, pp. 1045-1047. doi:10.1063/1.100061

28. S. Kitamura and M. Iwatsuki, “Observation of 7 × 7 Reconstructed Structure on the Silicon (111) Surface Using Ultrahigh Vacuum Noncontact Atomic Force Microscopy,” Japanese Journal of Applied Physics, Vol. 34, Part 2, No. 1B, 1995, pp. L145-L148.doi:10.1143/ JJAP.34.L145

29. US Patent RE37, 299E.

30. C. Putman, B. Grooth, N. Hulst and J. Greve, “A Detailed Analysis of the Optical Beam Deflection Technique for Use in Atomic Force Microscopy,” Journal of Applied Physics, Vol. 72, No. 1, 1992, pp. 6-12. doi:10.1063/1.352149

31. G. Merritt, E. Monson, E. Betzig and R. Kopelman, “A Compact Fluorescence and Polarization Near-Field Scanning Optical Microscope,” Review of

Scientific Instruments, Vol. 69, No. 7, 1998, pp. 2685-2690. doi:10.1063/1.1148999

32. P. G. Gucciardi, M. Labardi, S. Gennai, F. Lazzeri and M. Allegrini, "Versatile Scanning Near-Field Optical Microscope for Material Science Applications," Review of Scientific Instruments, Vol. 68, No. 8, 1997, pp. 3088-3092. doi:10.1063/1.1148246

33. A. Naber, H.-J. Maas, K. Razavi and U. C. Fischer, "Dynamic Force Distance Control Suited to Various Probes for Scanning Near-Field Optical Microscopy," Review of Scientific Instruments, Vol. 70, No. 10, 1999, pp. 3955- 3961. doi:10.1063/1.1150019

34. Nanosurfr Mobile S Atomic Force Microscope.

35. M. Tortonese, R. C. Barrett and C. F. Quate, "Atomic Resolution with an Atomic Force Microscope Using Piezoresistive Detection," Applied Physics Letters, Vol. 62, No. 8, 1993, pp. 834-836. doi:10.1063/1.108593

CHAPTER 20

MEASUREMENT OF GAMMA-RAYS USING SMARTPHONES

Sandy Tith and Nares Chankow

Keywords: Gamma-Ray, Dose Rate, Smartphones, Radiation Safety, Industrial Radiography, CMOS Image Sensor, Cesium-137, Iridium-192, Android

INTRODUCTION

Gamma rays have been extensively used in medicine, industry, agriculture, education and research. Gamma-rays are electromagnetic radiation like visible light but they have energy approximately $10^4$to 10^6 times higher than that of the visible light. They can penetrate matter making them very useful in various ways. In medicine, gamma rays play important roles in diagnosis and therapy. Various gamma-ray applications in industry have been proven and widely used, for example radiography, thickness measurement, level measurement, sterilization of products, etc. Radiation workers are trained to protect themselves and the public from receiving radiation dose beyond the allowable limits. The whole body annual dose limit for radiation workers is 50 millisievert (mSv) or 5 rem but must not exceed the average of 20 mSv per year during any 5 consecutive years. For better radiation safety control, the dose rate of 10 microsievert per hour (mSV/hr) is limited so that working totally 2000 hours per year (8 hours a day, 5 days a week for 50 weeks) will not exceed the annual dose limit of 20 mSv. Radiation workers must wear personal dosimeter like film badges which are completely replaced by thermoluminescent dosimeters (TLD's) and more recently optically stimulated luminescent dosimeters (OSL's) for measurement of cumulative dose from 1 month up to 3 months. In normal practice, however, dosimeters for short term measurement are also required such as digital personal dosimeter and survey meters which are not available for all workers.

It has been known and tested that the CMOS (Complementary Metal Oxide Semiconductor) image sensor chip in digital camera is sensitive to gamma and X-rays [1] -[4] . All modern smartphones come with high quality digital cameras which may be used for measurement of gamma-rays. In this research, four Android smartphones are selected according to their CMOS sensitive areas, number of image pixels, cost and availability, to test their responses to gamma-rays at different energies and dose rates. Application software is also developed so that the smartphones can display the counting rate and the dose rate simultaneously.

Materials

Four smartphones including Samsung Galaxy GT-S5570 mini, Samsung Galaxy SII, Samsung Galaxy SIII and Huawei Ascend P7, are first selected. Their specifications are shown in Table 1below. They are tested in measurement of gamma-rays from the commonly used gamma-ray sources including cobalt-60 (^{60}Co), Cesium-137 (^{137}Cs) and Iridium-192 (^{192}Ir) with known activities to compare their sensitivities and to investigate relationship between the dose rate and the counting rate. The characteristics of the gamma-ray sources are showed in Table 2.

Citation: Tith, S. and Chankow, N. (2016) Measurement of Gamma-Rays Using Smartphones. Open Journal of Applied Sciences, 6, 31-37. doi: 10.4236/ojapps.2016.61004.

Methodology

The selected smartphones were first exposed to gamma-rays from an ^{192}Ir industrial radiography source at vari-

Table 1. Specifications of the selected smartphones.

Smartphones (back camera)	Max. image pixels (photo mode)	Frame rate	Max. image pixels (video mode)	Aperture
Samsung Galaxy GT-S5570 mini	3.15 MP	14 fps	320 × 240	No data
Samsung Galaxy SII	8 MP	24 fps	1920 × 1080	f/2.7
Samsung Galaxy SIII	8 MP	30 fps	1920 × 1080	f/2.6
Huawei Ascend P7	13 MP	30 fps	1920 × 1080	f/2.0

Table 2. Gamma-ray sources and their prominent gamma-rays [5] [6].

Gamma-ray sources	Half-life	Gamma-ray energy (keV)	Emission probability	Γ^* (mSv/hr-Ci at 1 m)
^{137}Cs	30 years	662	85%	33
^{192}Ir	74.2 days	296 308 317 468 589 604 612	29% 30% 81% 49% 4% 9% 6%	5
^{60}Co	5.26 years	1173 1332	100% 100%	133

*Specific gamma-ray constant is the dose rate at 1 m distance from 1 curie (Ci) source.

ous distances to obtain the dose rates up to about 240 mSv/hr. The measurements were carried out in video mode with black adhesive tape on the camera lenses making the image background dark. Interactions of gamma- rays with the CMOS chips appear as flashing bright spots on dark background which could be clearly observed as showed in Figure 1. After measurement, the video file was transferred to a personal computer then converted to pictures by using the freely downloadable AOAO Video to Picture Converter software [7] . The numbers of bright spots in all pictures were finally counted using the Image J software [8] which could also be freely available. The total number of count was divided by the counting time to express in count rate. Importantly, from our investigation one interaction of gamma photon could cause the bright spots size up to 10 pixels depending on type of CMOS chip. Thus a cluster of bright spots from 1 to 10 pixels were counted as 1 count. The calibration procedure is summarized in Figure 2. It was found that the Samsung Galaxy GT-S5570 mini had lowest sensitivity among the four smartphones while Huawei Ascend P7 had the highest sensitivity as illustrated graphically in Figure 3. After that smartphones were tested with gamma-rays from ^{137}Cs and ^{60}Co sources at various dose rates in order to test their sensitivities to high energy gamma-rays. The results are illustrated graphically in Figures 4-6. The sensitivity to high energy gamma-rays was found to be less than to ^{192}Ir due to lower interaction probability as predicted. Application software was finally developed for the Android smartphones so that they could be used to measure gamma-rays and display the results in both counting rate and the dose rate on the

Figure 1. Showing bright spots on dark background image caused by interaction of gamma- rays with digital camera's CMOS chip at low (left) and high (right) gamma dose rates.

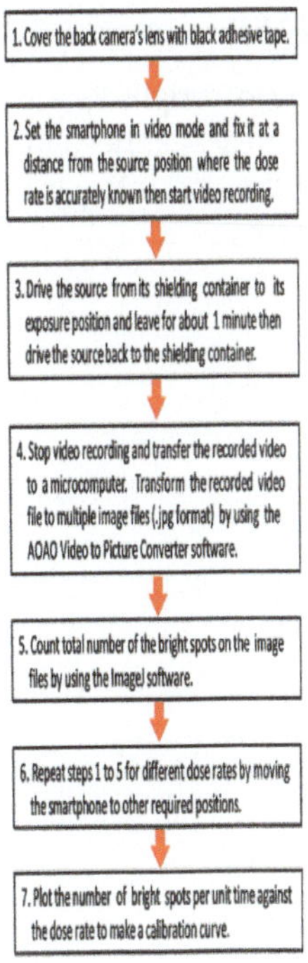

Figure 2. Calibration procedure.

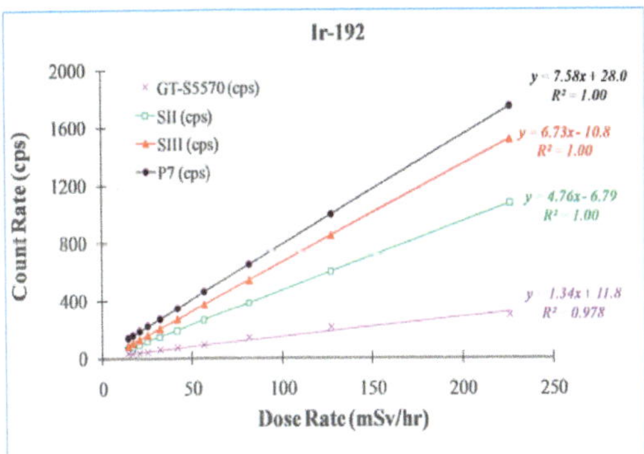

Figure 3. Gamma count rates vs the dose rates from Ir-192 industrial radiography source.

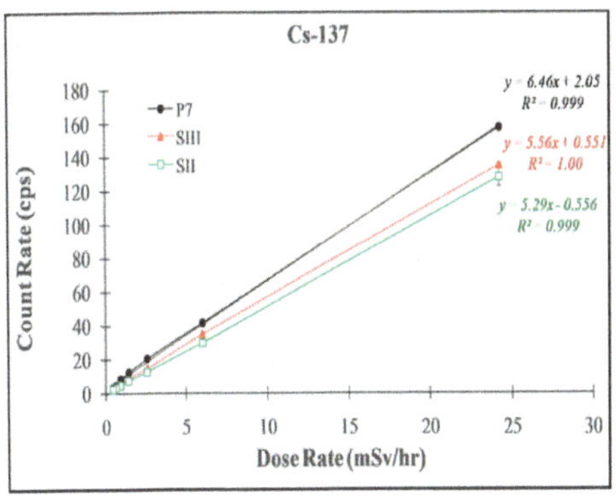

Figure 4. Gamma count rates vs the dose rates from Cs-137 source.

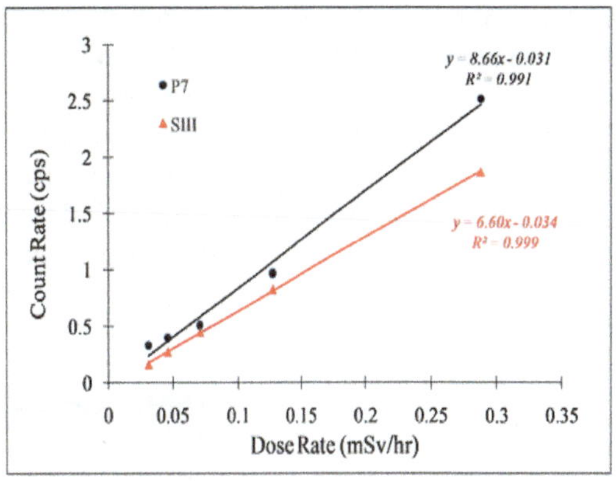

Figure 5. Gamma count rates vs the dose rates from low activity Co-60 source.

screens. The application software allowed user to perform counting from the saved video file or near real-time during measurement. User could input calibration equation to convert the counting rate to dose rate. Parts of the Application software menus are illustrated in Figure 7 and Figure 8. The application software has setting

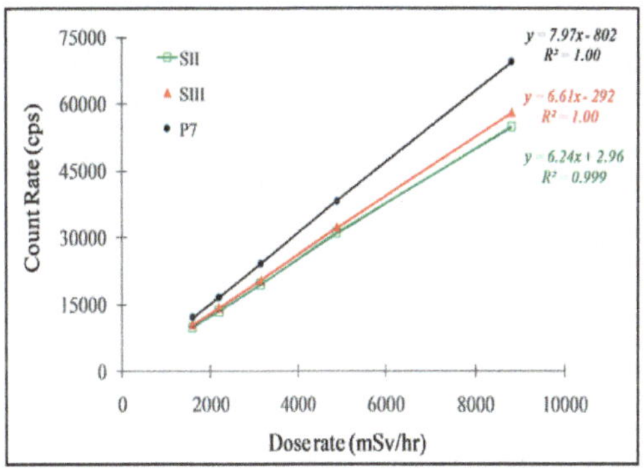

Figure 6. Gamma count rates vs the dose rates from high activity Co-60 source.

Figure 7. Set-up menus of the application software run on an Android smartphone.

parameters for threshold level to cut-off noisy signals, minimum and maximum pixel size to be counted, counting time and counting cycle. The number of counts is displayed near real-time but the dose rate will be displayed after the preset counting cycle is completed. Dose rate measurements by using the smartphones with the application software were conducted at the calibration facility of the Office of Atoms for Peace (OAP) in comparison with the standard ionization chamber. The results were very satisfactory as illustrated in Table 3.

RESULTS AND DISCUSSION

The video images showed flashing of bright spots randomly appeared on the screen. It could be clearly noticed that number of the bright spots increased with the dose rate. It was found that the count rates increased linearly

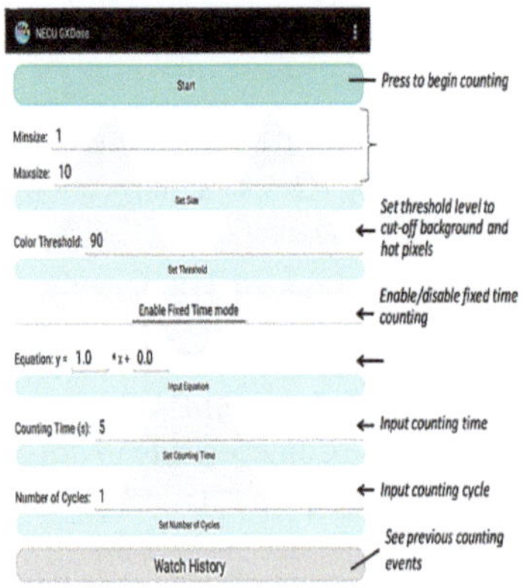

Figure 8. Brief description of the application software run on an Android smartphone.

Table 3. Comparison the dose rates from ^{137}Cs source read by Huawei Ascend P7 in comparison with the known values and the values obtained from a standard ionization chamber.

Actual dose rate (mSv/hr)	Read by survey meter		Read by Huawei Ascend P7	
	Dose rate (mSv/hr)	Discrepancy	Dose rate (mSv/hr)	Discrepancy
0.545	0.547	+0.002	0.525	−0.020
0.346	0.350	+0.004	0.352	+0.006
0.239	0.241	+0.002	0.235	−0.004
0.176	0.176	0.000	0.177	+0.001
0.133	0.132	−0.001	0.132	−0.001

with increase of the dose rates in wide dose rate ranges for all tested smartphones as illustrated in Figures 3-6. Even at very high dose rate from ^{60}Co source. Sensitivities of the smartphones were found to be quite different. Among the four tested smartphones, Huawei Ascend P7 had highest sensitivity, the second was Samsung Galaxy S III, the third was Samsung Galaxy S II and the least sensitive was Samsung Galaxy GT-S5570 mini. Therefore, each smartphone needs to be calibrated to obtain its own calibration equation as normally do with radiation survey meters and dosimeter. It should be kept in mind that, like all survey meters and dosimeters, if they are calibrated with ^{137}Cs they can only give correct dose rate for ^{137}Cs gamma-rays. Correction factors are requires for other gamma-ray energies. This is because all detectors are energy dependent.

CONCLUSION

The tested results clearly confirmed that CMOS chip of smartphone's digital camera was sensitive to gamma- rays. When the camera lens was covered with black adhesive tape, interaction of gamma-rays with CMOS chip caused flashing bright spots on dark background which could be counted with appropriate software to obtain count rate as in radiation counters. Furthermore, the linear relationship between the count rate and the count rate in wide range has made conversion from count rate to dose rate simpler. However, other smartphones may be more sensitive than the ones tested in this research but they were not bought to test due to limited budget at this preliminary investigation. From our experience, the sensitivity depends very much on size of the CMOS chip not the image resolution or total number of image pixels. Our brief test of some of newly released smartphones with 13 MP (megapixels) camera showed that they were much less sensitive to gamma-rays than Huawei Ascend P7(13 MP) and Samsung Galaxy S III (8 MP). However, with the in-house developed Application software and appropriate calibration, any Android smartphones can be used to measure gamma-rays but having different sensitivities. The smartphones are therefore useful for radiation workers particularly for those who work with high activity radiation sources like industrial radiographers, radiologists, radiation technicians, gamma irradiation plant workers and others. For example, industrial radiography normally uses ^{192}Ir with activity of about 100 curies (Ci). From Table 2, 1 Ci ^{192}Ir gives 5 mSv/hr at 1 meter. Thus 100 Ci will give the dose rate at 1 meter as high as 500 mSv/hr or 20 mSv/hr at 5 meters without shielding materials. If the worker stays at 5 meters for an hour, he will receive a total dose of 20 mSv which is equivalent to annual dose limit for radiation workers. Radiation workers with smartphones and the Application software can, therefore, have their own radiation measuring devices for better radiation safety and control. In-house developed Application software is now being improved and the one for iPhone is also under development.

ACKNOWLEDGEMENTS

We are grateful to the Department of Medical Sciences and the Office of Atoms for Peace (OAP) for allowing us to use their dose calibration facilities and materials. We would like to express our deepest thanks to Assistant Professor Attaporn Pattarasumunt for valuable discussions and suggestion as well as to Mr. Chalaempong Polee and Miss Nuatawan Thamrongsiripak for their suggestions and assistance during experiments. Thanks are also extended to Miss Tayida Tapjinda for her valuable assistance in developing the in-house Android application software.

REFERENCES

1. Akaranate, A., Chankow, N. and Pattarasumunt, A. (2013) Fast and Low Cost X-Ray Fluoroscopy Displayed on a 3D Monitor. Open Journal of Applied Sciences, 3, 308-311.http://dx.doi.org/10.4236/ojapps.2013.34039
2. Iamsamang, K. and Chankow, N. (2013) Low Cost Gamma-Ray Fluoroscopic System for Se-75 and Ir-192 Industrial Radiography Sources. NDT Spectra (Official Journal of Malaysian Society for Non-Destructive Testing), No. 7, 209- 216.
3. Gumiela, M. and Kozik, R. (2012) Studies of the Applicability of CMOS and CCD Sensors for Detection, Dosimetry and Imaging of Alpha, Beta, Gamma, X-Ray and Proton Beam Spots. Studies of the Applicability of CMOS and CCD Sensors for Detection, Dosimetry and Imaging of Alpha, Beta, Gamma, X-Ray and Proton Beam Spots, 2012.
4. Cogliati, J.J., Derr, K.W. and Wharton, J. (2014) Using CMOS Sensors in a Cellphone for Gamma Detection and Classification. Using CMOS Sensors in a Cellphone for Gamma Detection and Classification, 26.
5. Lederer, C.M., Hollander, J.M. and Perlman, I. (1967) Table of Isotopes. 6 Edition, John Wiley & Sons, New York.
6. (1992) International Atomic Energy Agency, Industrial radiography. IAEA, Vienna.
7. Studio, A.D. (2008) Aoaophoto Video to Picture Converter 4.0.
8. Ferreira, T. and Rasband, W. (2012) ImageJ User Guide: IJ1.46r.http://imagej.nih.gov/ij/docs/guide

CHAPTER 21

DESIGN OF LOW POWER LEVEL SHIFTER CIRCUIT WITH SLEEP TRANSISTOR USING MULTISUPPLY VOLTAGE SCHEME

Raveendran Arun Prasath and Parasuraman Ganesh Kumar

Keywords: Level shifter (LS), MultiSupply Voltage Design (MSVD), Subthreshold operation, Ultralow Power

INTRODUCTION

MultiSupply Voltage Domain (MSVD) technique is an effective method to reduce both dynamic and leakage powers in modern system-on-chips. It consists of partitioning the design into separate voltage domains. Based on timing requirements, each module can be operated at a proper power supply voltage level [1] - [3] . Time critical domains run at higher power supply voltage (VDDH) to maximize performance, while noncritical domains run at lower power supply voltage (VDDL), to reduce static and dynamic power without impacting on the overall circuit performance. In MSVD system, level shifters (LSs) are required on the boundaries between the circuit subsections operating at different power supply voltages to convert signals from the VDDL to the VDDH voltage level.

Multithreshold CMOS technique is mainly used to optimize the delay and power of the circuit. It consists of two levels of threshold voltage such as low threshold voltage and high threshold voltage. Low threshold voltage is used to increase the speed and reduce the clock period. High threshold voltage is used to reduce the static power leakage. Sleep transistors are series to the pull-up and/or the pull-down of the logic gates, and turn them off when the circuit is idle, thereby decreasing the leakage component due to sub-threshold currents. When an nMOS sleep transistor is used on the pull-down path, a SLEEP signal controls its active/standby mode (i.e., SLEEP during standby and SLEEP during active mode) [4] - [6] . In the standby mode, the sleep transistor is off, thus disconnecting its insertion point, called virtual ground, from the physical ground. In active mode, the gated circuit operates normally, but it incurs delay degradation due to the series resistance of the sleep transistor.

The common LS is the differential cascade voltage switch (DCVS) circuit that is typically used for changing signals between the two different above-threshold voltage domains. Drawback of the DCVS-LS is the conflict between the pull-up and pull-down networks which becomes severe when input signals are in the subthreshold range, thus making the conventional sizing methods impractical to get a properly functioning circuit. To address this problem, in order to improve the performance several conventional DCVS circuits have been proposed. The four-stage cascaded DCVS circuit was designed for robust level up-conversion from the subthreshold region. But it introduced large power penalties to the intermediate power supplies and also limited speed performance [7] - [9] .

A two-stage LS was proposed, the first stage exploits a DCVS circuit with an always-on diode-connected nMOS transistor on the top; whereas, the second one is a conventional DCVS stage that achieves rail to rail swing. It avoids intermediate power lines, but again it is not enough to reach high-speed performances. The LS was designed by two pMOS current limiters to reduce the half-latch pull-up strength within the conventional DCVS structure. This circuit is able to convert subthreshold input signals and it requires reasonably

Citation: Prasath, R. and Kumar, P. (2016) Design of Low Power Level Shifter Circuit with Sleep Transistor Using MultiSupply Voltage Scheme. Circuits and Systems, 7, 1132-1139. doi: 10.4236/cs.2016.77097.

sized pull- down transistors. Also this LS is relatively slow and energy-hungry. A circuit based on current mirrors has been designed for better speed performance, but the current mirror output floats when the input voltage signal is high [10] - [12] . This causes a detrimental effect on subthreshold leakage of the output buffer.

Recently, LS was designed, it is suitable for fast and wide range voltage conversion from the subthreshold domain to above-threshold domain, but it has high power consumption. A new low-power LS is designed to improve the speed and reduce the power by using multithreshold CMOS technique and sleep transistor.

This paper is organized as follows. Firstly, the related work is briefly explained in Section 2; next, the proposed methodology is described in Section 3. Then, the results and discussion are described in Section 4. Finally, the content is concluded in Section 5.

RELATED WORK

The Level shifter in the system is mainly used for fast and wide range voltage conversion in Multi Supply Voltage Domain [7] . Multi Threshold CMOS (MTCMOS) technique is used in the architecture of level shifter circuit. These circuit which gives robust voltage shifting from the deep sub-threshold to the above-threshold domain, while demonstrating fast response and low energy consumption [13] - [15] .

The LS circuit is shown in Figure 1 is designed by using the multi-threshold CMOS design technique along with novel topological strategies. The circuit consists of an input inverter, a main voltage conversion stage and an output inverting buffer. The input inverter (MP_1/MN_1) is designed using low threshold voltage (lvt) transistors. To have higher strength of the pull-down network, also MN2 and MN3 are lvt transistors. Then, two lvt pMOS devices (MP_2 and MP_3) are added to both the branches of the circuit.

These devices limit the cross-bar current that is the current flowing in the pull-up network and opposing to the discharge of NH (or NL) node at the beginning of their high to low transition. To further facilitate the high to low transition at the nodes NH and NL, MP_4 and MP_5 are high threshold voltage (hvt) transistors. This choice also reduces leakage current flowing through the pull-up networks when MP_4 or MP_5 are turned off. However, using hvt, pMOS transistors has the counter effect of slowing the low to high transition of the nodes NH and NL. Therefore, to reduce the switching delay, a pull-up network able to self-adapt its strength to the actually occurring desirable transition [16] - [18] .

This behavior was obtained by introducing the parallel connected hvt devices MP_6-MP_{10} and MP_7-MP_{11},

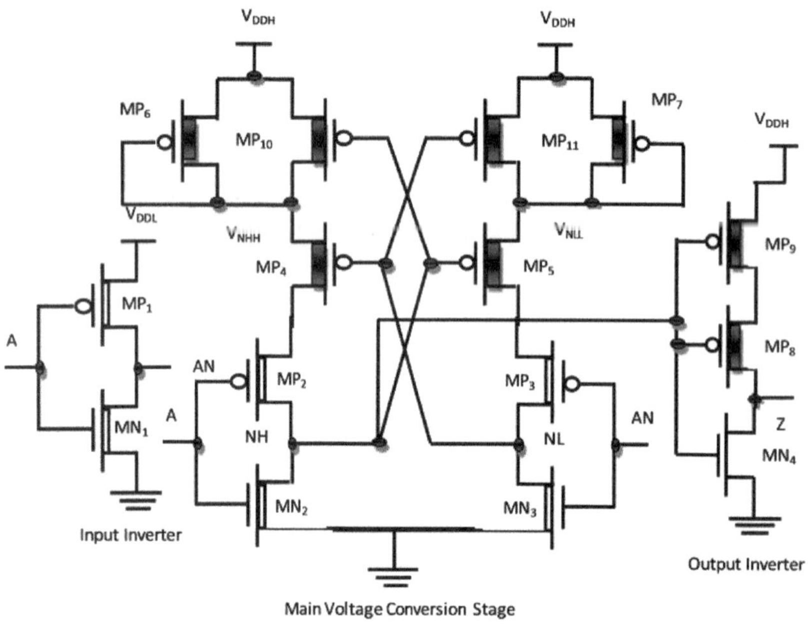

Figure 1. Existing level shifter.

with MP_6 and MP_7 being diode connected transistors. The two variable virtual power supplies on the two branches of the circuits are V_{NHH} and V_{NLL}. Therefore, the strength of pull-up networks is adapted to the next output switching transition. Assuming that the output Z is initially low (high), the pull-up network of the left branch is weakened (strengthened) and that of the right branch is strengthened (weakened), thus speeding-up a low-to-high (high-to-low) output transition.

The introduction of hvt MP_{10} and MP_{11} devices controlled by NH and NL voltages. To assure a rail-to-rail conversion, an output inverting buffer is connected to the node NH. This inverter uses a standard threshold voltage (svt) nMOS (MN_4) device and two stacked hvt pMOS transistors ($MP_8$and MP_9). The used pull-up configuration allows static current when NH is high to be significantly reduced. In such a condition, the MP_9 device has the effect of maintaining the source node of MP_8below V_{DDH}. As the source terminal of MP_8 results to be at lower voltage than its bulk node, the MP_8 threshold voltage increases. The source-gate voltage of MP_8 is also reduced, thus further reducing static current in the above referred condition. These LS is suitable for robust logic voltage shifting from near/ sub-threshold to above threshold domain has been presented.

This circuit exploits proper design strategies to increase the operating speed while maintaining low energy consumption and large voltage conversion range [19] [20] .

Table 1 represents the sizing of transistor in existing system. Proper transistor sizing can be achieved by changing the aspect ratio of the transistors, i.e. width and length of the transistors. MN_1 to MN_2 denotes the nMOS transistors and MP_1 to MP_{11} denotes the pMOS transistors.

PROPOSED WORK

The proposed low power level shifter is mainly used for power reduction and wide range voltage conversion in MultiSupply Voltage Domain. Novel level shifter architecture is combined with Multi Threshold CMOS (MTCMOS) technique can be achieved by proper sizing of transistors and sleep transistor. This circuit provides robust voltage shifting from the deep sub-threshold to the above-threshold domain with low power consumption.

Sleep Transistor

Sleep transistor is an effective approach to minimize standby leakage current. Sleep transistor is added to the circuit serially. Sleep transistor is used as a switch to shut off power supplies to parts of a design in standby mode. A sleep transistor is referred to either a pMOS or nMOS high threshold transistor that connects the permanent power supply to circuit power supply which is commonly called "virtual power supply". The sleep transistor is controlled by a power management unit to switch ON and OFF power supply to the circuit. The nMOS sleep transistor is used to switch VSS supply and it is called footer switch. The header switch is pMOS has lower drive current than footer switch is NMOS of a same size. Due to this, a header switch implementation usually consumes more area than a footer switch implementation. Because of this problem, footer switch is used in the proposed level shifter. When the state of footer switch is "1" then it is in turn on condition to block the leakage current to the ground.

Table 1. Transistor sizing of existing system.

Device	W/L (μm)	Device	W/L (μm)	Device	W/L (μm)
MN_1	0.3/0.1	MP_2	1.2/0.2	MP_7	0.18/0.2
MN_2	1.2/0.2	MP_3	0.18/0.2	MP_8	1.2/0.2
MN_3	0.5/0.2	MP_4	0.3/0.2	MP_9	1.2/0.2
MN_4	1.1/0.2	MP_5	0.18/0.2	MP_{10}	0.12/0.2
MP_1	0.45/0.1	MP_6	0.3/0.2	MP_{11}	0.12/0.2

Circuit Diagram

The proposed LS circuit is shown in Figure 2 is designed by using the multi-threshold CMOS design technique along with sleep mode strategies. The circuit diagram consists of an input inverter, a main voltage conversion stage, sleep transistor and an output inverting buffer.

Table 2 represents that the sizing of transistors in proposed level shifter by changing the width and length of the transistors. The circuit has been sized to achieve the minimum energy delay product in the following operating condition: V_{DDH} = 1 V, V_{DDL} = 0.1 V, 25 fF load capacitance on the output node, T = 25°C, TT process corner.

Input Inverter

The input inverter MP_1 and MN_1 are designed for low threshold voltages (lvt) transistors. It provides fast differential low voltage input signal to main voltage conversion stage.

Main Voltage Conversion Stage

The MN_2 and MN_3 are lvt transistors used to achieve high strength in pull down network. Then two lvt pMOS devices MP_2 and MP_3 are added to both the branches of the circuit. MP_4 and MP_5 are high threshold voltage (hvt) transistors. It is used to reduce leakage current from the pull up transistors when MP_4 and MP_5 are turned off. And also this hvt pMOS transistor are used to reduce the switching delay and a pull-up network are able to self-adapt its strength for occurring actual transition.

This behavior was obtained by using a parallel connected hvt devices MP_6-MP_{10} and MP_7-MP_{11}, with MP_6 and MP_7 being diode connected transistors. It is used to adapt the strength of the pull up networks into next output switching transition. The hvt is applied to MP_{10} and MP_{11} devices and it is controlled by High Node (NH) and Low Node (NL) voltages. The use of diode-connected pMOS transistors MP_6 and MP_7 limits the output range of the main conversion stage. MN_5 (nMOS) transistor act as a sleep transistor which is used to block the leakage current from the main voltage conversion stage to the ground.

Output Inverting Buffer

To secure a rail-to-rail conversion, an output inverting buffer is connected to the node NH. The output inverter consists of standard threshold voltage (svt) in MN_4 (nMOS) and to stacked hvt $MP_8$and MP_9 (pMOS).

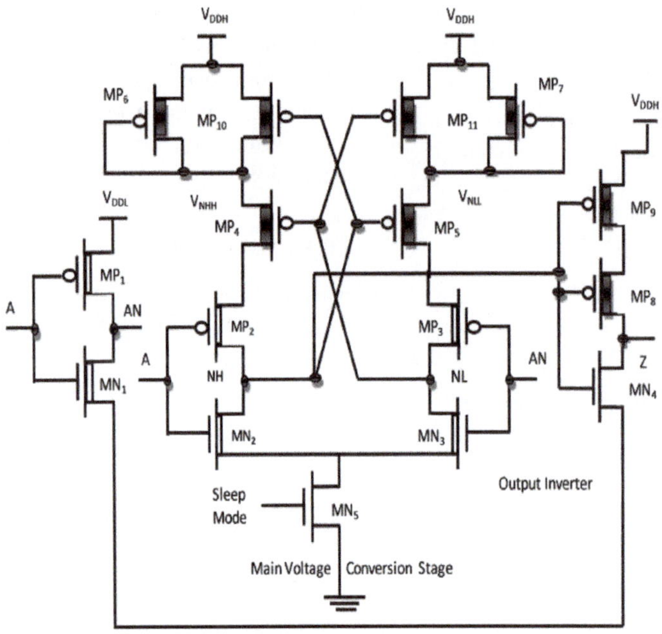

Figure 2. Proposed level shifter.

Table 2. Transistor sizing of proposed system.

Device	W/L (μm)	Device	W/L (μm)	Device	W/L (μm)
MN_1	0.3/0.1	MP_2	0.3/0.2	MP_7	1.2/0.2
MN_2	1.5/0.2	MP_3	1.2/0.2	MP_8	1.2/0.2
MN_3	1.1/0.2	MP_4	1.2/0.2	MP_9	1.2/0.2
MN_4	1.1/0.2	MP_5	1.2/0.2	MP_{10}	1.2/0.2
MP_1	1.2/0.2	MP_6	1.2/0.2	MP_{11}	1.2/0.2

Operation

As the input signal switches, the transistor MN_2 is turned on and the node NH starts to be discharged. Due to this, the source to gate voltage of MP_4 is reduced and increases its threshold voltage. Owing to this, MP_4 is weakened, thus increasing the discharge of the node NH. In the meantime, the stronger pull-up of the right branch charges the node NL and causing MP_4 to be turned off. This allows the discharging of NH to be further accelerated. As NH approaches the ground voltage, the positive feedback becomes to explode, causing MP_5 to be fully activated. Therefore, NL is fully charged at the low voltage level. Once transitions in the nodes NH and NL are completed, the virtual power supplies $V_{NHH} = V_{DDH}$ and $V_{NLL} = V_{DDH}$? V_{dsat}, MP_7 are established to provide fast switching in the subsequent input transition.

The used pull-up configuration allows static current when NH is high to be significantly reduced. In such a condition, the MP_9 device has maintaining the source node of MP_8 below V_{DDH}. As the source terminal of MP_8 results to be at lower voltage than its bulk node, the MP_8 threshold voltage increases. The source-gate voltage of MP_8 is also reduced, thus further reducing static current in the above condition. As 3 V external power supply is supplied to MN_5 (sleep transistor) and it is always in turn on condition to block the leakage power from the main voltage conversion stage.

IMPLEMENTATION

Simulation Results

Figure 3 represents that simulation diagram of existing level shifter. In this diagram, appropriate nMOS and pMOS transistors are selected and then fix the width and length of the transistor and connected by using wire. High voltage nMOS transistor is used as a sleep transistor. Voltage pins are used to calculate the voltage conversion and wattmeter is used to calculate the power.

Figure 4 shows the waveform of voltage conversion and power analysis of proposed level shifter. PM1 denotes the power of 2.56 nW. VF1 denotes voltage converted to 1 V. VF2 denotes the voltage at NH. VF3 denotes the voltage at NL. VG1 denotes the voltage in sleep transistor.

Performance Analysis

Table 3 compares the existing system and proposed system. In both the systems, voltage conversion can be achieved from 100 mV to 1 V but low power can be achieved in proposed system than the existing system.

The Figure 5 shows that comparison chart of power analysis for existing and proposed system. The proposed system consumes 2.56 nW power. The existing system consumes 5.66 nW power. When compared to existing system, the proposed system consumes low power.

CONCLUSIONS

The proposed level shifter is suitable for robust logic voltage shifting from near/subthreshold to above-threshold domain. The proposed circuit exploits proper design strategies to maintain very low energy consumption and large voltage conversion range. Sleep transistor technique is used which is able to further reduce the total leakage power from the circuit to the ground by breaking the physical connection to the ground. And also low

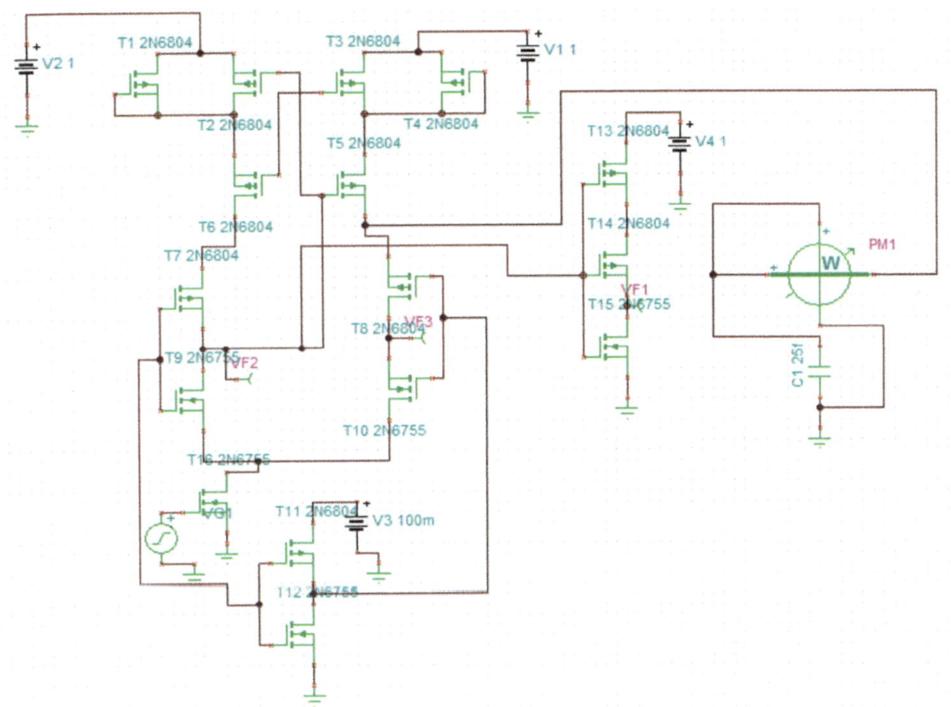

Figure 3. Simulation diagram of proposed level shifter.

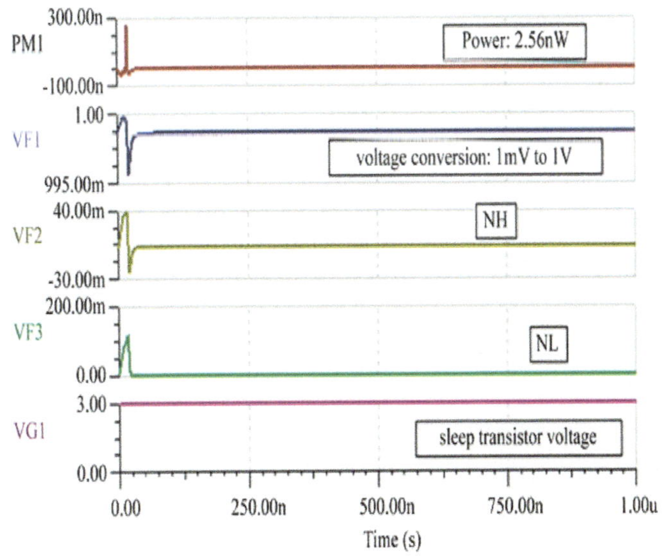

Figure 4. Output waveform of proposed level shifter.

Table 3. Comparison table of power analysis.

Sl. No.	Methodology	V_{DDH} (V)	V_{DDL} (V)	Power (nW)
1	Existing Method	1	0.1	5.66
2	Proposed Method	1	0.1	2.56

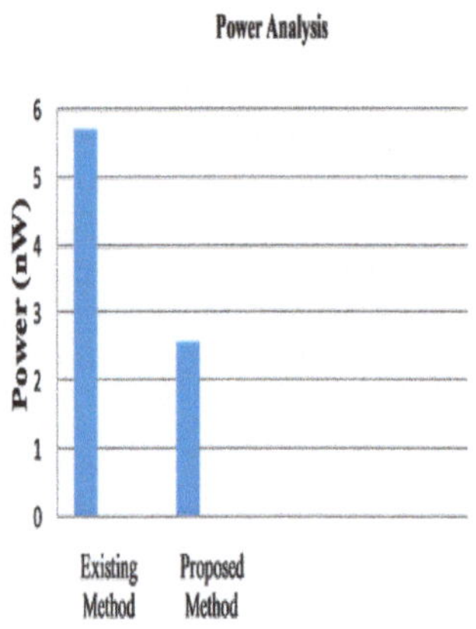

Figure 5. Comparison chart of power analysis.

power consumption can be achieved by proper sizing of transistor. The proposed design reliably converts 100 mV input voltage into 1 V output voltage. The simulation results demonstrate that the new LS shows power of 2.56 nW compared to existing system power of 5.66 nW.

The proposed level shifter is used in many applications. In future, the proposed level shifter is implemented in flip-flop circuit for further reducing the power and achieves the voltage conversion from subthreshold to above- threshold.

ACKNOWLEDGEMENTS

The authors are grateful to the Dean of Anna University, Regional Office-Madurai and management and principal of K.L.N. College of Engineering for providing all facilities for their research work.

REFERENCES

1. Chang, I.J., Kim, J.J. and Roy, K. (2006) Robust Level Converter Design for Sub-Threshold Logic. ACM in Proceedings of the 2006 International Symposium on Low Power Electronics and Design, Tegernsee, 14-19 October 2006, 14-19. http://dx.doi.org/10.1109/lpe.2006.4271800
2. Chavan, A. and MacDonald, E. (2008) Ultra Low Voltage Level Shifters to Interface Sub and Super Threshold Reconfigurable Logic Cells. 2008 IEEE Aerospace Conference, 1-8 March 2008, 1-6. http://dx.doi.org/10.1109/AERO.2008.4526473
3. Chi, J.C., Lee, H.H., Tsai, S.H. and Chi, M.C. (2007) Gate Level Multiple Supply Voltage Assignment Algorithm for Power Optimization under Timing Constraint. IEEE Transactions on Very Large Scale Integration (VLSI) Systems, 15, 637-648. http://dx.doi.org/10.1109/TVLSI.2007.898650
4. Clark, L.T., Deutscher, N., Demmons, S. and Ricci, F. (2002) Standby Power Management for a 0.18 μm Microprocessor. ACM in Proceedings of the 2002 International Symposium on Low Power Electronics and Design, 12-14 August 2002, 7-12. http://dl.acm.org/citation.cfm?doid=566408.566413
5. Corsonello, P., Lanuzza, M. and Perri, S. (2014) Gate-Level Body Biasing Technique for High-Speed Sub-Threshold CMOS Logic Gates. International Journal of Circuit Theory and Applications, 42, 65-70. http://dx.doi.org/10.1002/cta.1838
6. Ishihara, F., Sheikh, F. and Nikolic, B. (2004) Level Conversion for Dual-Supply Systems. IEEE Transactions on Very Large Scale Integration (VLSI) Systems, 12, 185-195. http://dx.doi.org/10.1109/TVLSI.2003.821548
7. Hasanbegovic, A. and Aunet, S. (2009) Low-Power Subthreshold to above Threshold Level Shifter in 90 nm Process. IEEE in NORCHIP, Trondheim, 16-17 November 2009, 1-4. http://dx.doi.org/10.1109/norchp.2009.5397793
8. Koo, K.H., Seo, J.H., Ko, M.L. and Kim, J.W. (2005) A New Level-Up Shifter for High Speed and Wide Range Interface in Ultra Deep Sub-Micron. IEEE International Symposium on Circuits and Systems, 23-26 May 2005, 1063-1065.
9. Lanuzza, M., Corsonello, P. and Perri, S. (2015) Fast and Wide Range Voltage Conversion in MultiSupply Voltage Designs. IEEE Transactions on Very Large Scale Integration (VLSI) Systems, 23, 388-391. http://dx.doi.org/10.1109/TVLSI.2014.2308400
10. Lanuzza, M., Corsonello, P. and Perri, S. (2012) Low-Power Level Shifter for Multi-Supply Voltage Designs. IEEE Transactions on Circuits and Systems II, 59, 922-926. http://dx.doi.org/10.1109/TCSII.2012.2231037
11. Lin, Y.S. and Sylvester, D.M. (2008) Single Stage Static Level Shifter Designs for Subthreshold to I/O Voltage Conversion. 2008 ACM/IEEE International Symposium on Low Power Electronics and Design (ISLPED), Banglore, 11-13 August 2008, 197-200.
12. Lütkemeier, S. and Rückert, U. (2010) A Subthreshold to Above-Threshold Level Shifter Comprising a Wilson Current Mirror. IEEE Transactions on Circuits and Systems II, 57, 721-724. http://dx.doi.org/10.1109/TCSII.2010.2056110
13. Osaki, Y., Hirose, T., Kuroki, N. and Numa, M. (2012) A Low-Power Level Shifter with Logic Error Correction for Extremely Low-Voltage Digital CMOS LSIs. IEEE Journal of Solid-State Circuits, 47, 1776-1783. http://dx.doi.org/10.1109/JSSC.2012.2191320
14. Shao, H. and Tsui, C.Y. (2007) A Robust, Input Voltage Adaptive and Low Energy Consumption Level Converter for Sub-Threshold Logic. 33rd European Solid State Circuits Conference, Munich, 11-13 September 2007, 312-315.
15. Soeleman, H., Roy, K. and Paul, B.C. (2001) Robust Subthreshold Logic for Ultra-Low Power Operation. IEEE Transactions on Very Large Scale Integration (VLSI) Systems, 9, 90-99. http://dx.doi.org/10.1109/92.920822
16. Wang, W.T., Ker, M.D., Chiang, M.C. and Chen, C.H. (2001) Level Shifters for High-Speed 1 V to 3.3 V Interfaces in a 0.13 μm Cu-Interconnection/Low-k CMOS Technology. 2001 International Symposium on VLSI Technology, Systems, and Applications, Honolulu, 18-20 April 2001, 307-310.
17. Wooters, S.N., Calhoun, B.H. and Blalock, T.N. (2010) An Energy-Efficient Subthreshold Level Converter in 130-nm CMOS. IEEE Transactions on Circuits and

Systems II, 57, 290-294. http://dx.doi.org/10.1109/TCSII.2010.2043471

18. Yu, C.C., Wang, W.P. and Liu, B.D. (2001) A New Level Converter for Low-Power Applications. The 2001 IEEE International Symposium on Circuits and Systems, San Diego, 6-9 May 2001, 113-116.

19. Zhai, B., Pant, S., Nazhandali, L., Hanson, S., Olson, J., Reeves, A. and Blaauw, D. (2009) Energy-Efficient Subthreshold Processor Design. IEEE Transactions on Very Large Scale Integration (VLSI) Systems, 17, 1127-1137. http://dx.doi.org/10.1109/TVLSI.2008.2007564

20. Prasath, R.A. and Kumar, P.G. (2015) A 8b 20MS/s 1.2 V Digital Ramp Asynchronous SAR ADC Using MTS Algorithm IN 0.13 nm CMOS Process. Journal of Revista Cientifica-Facultad De Ciencias Veterinarias, 25, 21-27. http://www.cientificaonline.com/index.php/cientifica/article/view/726

INDEX

N

O

P

Q

R

S

T

U

V

W